AF616821

TEXTBOOK OF BIOTECHNOLOGY

TEXTBOOK OF BIOTECHNOLOGY

G.R. CHHATWAL
M. Sc., Ph.D., FIAHPS
Member, Royal Society of Chemists, London (U.K.)
Hon. Professor in Environmental Science,
I.I.E.E., New Delhi
Reader in Chemistry, D.S. College, New Delhi
(University of Delhi)

ANMOL PUBLICATIONS PVT LTD
New Delhi - 110 002

ANMOL PUBLICATIONS PVT LTD
4374/4B, Ansari Road, Daryaganj
New Delhi-110 002

Textbook of Biotechnology
First Edition 1995
Reprint - 1998 , 2003

ISBN 81-7488-000-3

PRINTED IN INDIA

Published by J.L. Kumar for Anmol Publications Pvt. Ltd
New Delhi - 110 002 and Printed at Mehra Offset Press, Delhi.

Preface

The growth of biotechnology as an organised discipline has been remarkable in the past one decade. This has been the direct result of the revolutionary changes in various biological and scientific techniques. This book has been carefully designed, structured, compiled and edited to provide the students of biotechnology, at different levels, with a handy and reliable source for definitions, statements, laws, theories, principles and concepts, and various experimental techniques. The presentation throughout has been aimed at sustaining the interest of the readers while enriching their vocabulary and comprehension of technical terms and expressions.

Each chapter included in this book is self-sufficient in itself. Wherever necessary, suitable diagrams have been added to make the concept more clear. The concepts included in each chapter of this book have been presented in brief analytical phrases thereby avoiding the more comprehensive type of treatment appropriate to large reference works.

This book will provide a reference source as well as a textbook for students of Biotechnology, Medical, Plant Breeding, Botany, Zoology, Pharmacy, Genetics, Cytology, Cell Biology, Molecular Biology, Cytogenetics, Agriculture and Environmental Science. The man with very little background in biotechnology will find this book invaluable and it will be exceptionally useful to the students as well as to the expert.

G.R. Chhatwal

Contents

Contents

1

Introduction

Biotechnology is a very vast and developing branch of science. It can be defined in a variety of way. However, the most common definition of it as given by a leading biotechnologist is "The applications of biological organisms, systems or processes to manufacturing and service industries." From such a definition it appears that biotechnology is, "biology applied for profit, either financial or less often, humanitarian."

The emergence of biotechnology in its present form can be traced back to 1970's when it emerged as a separate branch of study because of the increased potential for applications of the emerging techniques of molecular biology. Historically speaking the word biotechnology was used in 1920 by Leeds city council (U.K.) in 1920 and an institute in biotechnology was set up by them. Going back still further we can say that a biotechnological process (viz. production of alcoholic beverages by fermentation) was in actual use some 5000 years ago.

Thus, we can say that Biotechnology is an old science—we find applications in the Bible and applications in the antiquities. Biotechnology is a new science, being applied to numerous sectors of industry, science, medicine. Biotechnology is an effervescent science, a verdant field of research; a promise of a promise and its ferment for a bright future is everywhere.

Biotechnology today is undergoing a resurgence in a wide range of applications and the tremendous increase all over the world. It has resulted in a proliferation of new ideas and concepts as also a large increase in available information and data in all of the biotechnologies.

The recent dramatic developments in biotechnology have stirred stockbrokers and company directors, beguiled politicians with the prospect of new sources of wealth, and inspired journalists to write of 'cures for cancer.' But the main story of biotechnology does not revolve around these people, nor even the scientists whose genius has brought them Nobel prizes. The real stars of biotechnology can only be seen with the help of a microscope—tiny microbes, and cells taken from plants and animals.

One aspect of biotechnology, genetic engineering, is also becoming a topic of considerable public interest and debate because of it long-term possibilities in altering and improving the genetic heritage of agricultural plants, farm animals and even human beings. This potential raises a number of important moral, ethical and ecological questions apart from posing a formidable scientific challenge.

In laboratories around the world genetic engineers have already designed microbes to manufacture dozens of potentially invaluable substances. Insulin produced by microbes has now been approved for use in diabetics in the UK and US. Interferon, which certainly fights viral diseases and may combat cancer, is now being tested on volunteers, while growth hormone, which reverses one of the major causes of dwarfism, will soon be available in large quantities from genetically engineered microbes. These are only part of the first wave of products from the fledgling genetic engineering industry. Other medical products set to follow include drugs to treat strokes, burns, nerve damage and perhaps, even obesity, plus a range of vaccines. With all these medical possibilities on the horizon, as well as the production of fuels and chemicals for industry, it is not surprising that genetic engineering has received the lion' share of headlines in recent years. However, other dramatic advances in biology will prove equally influential in the bioindustrial revolution. Cell culture—the cultivation of fragments of plants or animals in the laboratory—has opened up countless exciting possibilities. The impact of research in plant cell culture may enable plant breeders to create new crops which grow more rapidly, require less fertilizer and thrive in poor soils. Monoclonal antibodies, the marker molecules manufactured by white blood cells, have already begun to revolutionize medical diagnosis.

Historical Background of Biotechnology

Most types of fermented fords that are available these days were widely used and consumed during the Roman times. Processes such as baking, brewing and cheese making have been in use for many years and were developed long before man knew of the existence of micro-organisms involved. Louis Pasteur was interested in discovering why the local wines were souring and in a series of classic experiments he demonstrated that alcohol could be produced from grape-yeast mixture but not from albumin—yeast mixtures. He could also show that fermentation occurs in presence of yeast. He also discovered that each type of fermentation is mediated by a specific micro-organism which gains energy during the process.

The innate skills of minute living cells are truly astounding. Millions of years of evolution have endowed them with a staggering versatility and resilience. Microbes can be found almost everywhere— in boiling water, locked in ice, immersed in oil. Some can feed on the apparently most unnutritious materials—petrol, wood, plastic, even solid rock. When, in addition, the substances microbes can manufacture are examined the immense potential of biotechnology starts to be revealed. Antibiotics, insecticides, fuels, dyes, industrial chemicals and vitamins are just a few of the multitude of valuable materials that can be obtained from microbes.

These facts would amply justify an intense interest in developing new industries employing workforces of millions of microbes busily manufacturing substances we need. However, the biotechnology boom has been truly sparked by the advent of genetic engineering. It is only a decade since scientists first discovered that they could graft totally foreign pieces of genetic formation into microbes. The exploits of genetic engineers have given a new twist to Haldane's words: if you can't find a bug that makes what you want, then create one that will!

Biotechnology in Medicines

New and improved treatments for the three major killers in developed countries: diseases of the heart and blood vessels, cancer and diabetes. Better and cheaper antibiotics to counter the spread of infectious organisms that have developed a resistance to conventional antibiotics. Vaccines to protect against viral diseases, such as hepatitis, influenza and rabies, and parasitic diseases, including malaria and sleeping sickness, which strike down millions of people each year.

Rapid tests that will aid doctors to make accurate diagnoses of many diseases.

Improved methods for matching organs for transplantation.

Techniques for correcting the body's chemistry to cure hereditary diseases, such as haemophilia.

Biotechnology in Agriculture

The creation of crops which make their own fertilizers, with immense savings in costs to the farmer.

Plants which can thrive on land that presently lies barren because the soil lacks water or is too salty.

Substances that can speed the growth of farm animals.

Vaccines to protect cattle from foot-and-mouth disease.

Cheaper forms of animal feed, consisting of microbes grown on waste materials.

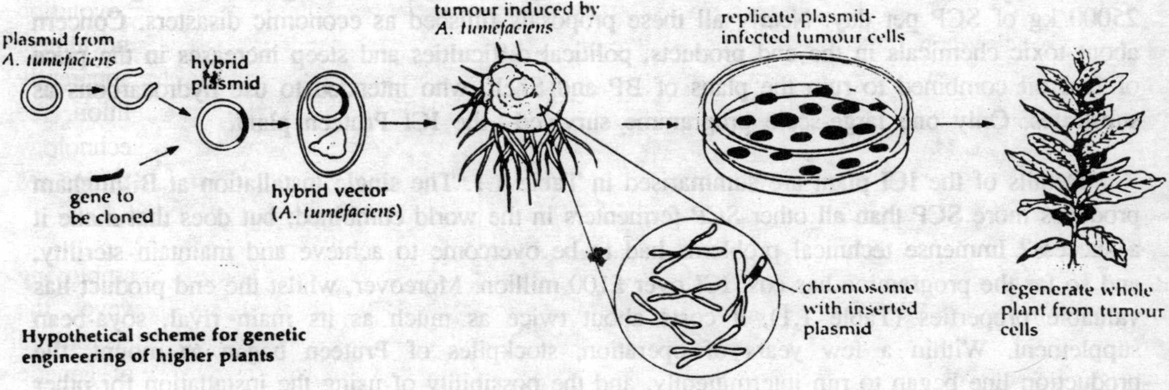

Fig. 1.1. Hypothetical scheme for genetic engineering of higher plants.

Nitrogen Fixation

Cereals such as wheat, corn, rice and barley form a major part of the staple diet. However, modern high-yielding cultivars have high nitrate requirements. Fertiliser is expensive, and its production requires substantial amounts of energy. About 10% of the world's oil consumption goes into fertiliser production. Attempts are being made to reduce dependence on artificial fertiliser by various means such as introducing nitrogen-fixing genes into cereals and by encouraging mutualistic associations between cereal roots and free-living nitrogen fixers.

Nitrogen fixing *(Nif)* genes can be transferred from one bacterium to another, but they have not yet been successfully incorporated into higher plants. One possible mechanism for gene transfer is to exploit the plasmid in *Agrobacterium tumefaciens* (Fig. 1.1).

Turning the Tables on Pathogens

The hypothetical scheme for the genetic engineering of higher plants described above exploits the plant pathogenic bacterium *A. tumefaciens*. This organism causes *crown gall disease* on many important plants, such as peach. The bacterium inserts its plasmid into a host cell chromosome, which induces cell division and tumour formation. Infected cells produce opines (nitrogenous compounds) which are required by the bacterium. About 140 genera of plants are potential hosts. Transmission is by cuts and biting insects. Control is achieved by using resistant varieties, by sanitation, or by dipping roots into a solution containing *A. radiobacter* which is harmless and antagonistic to *A. tumefaciens*.

Biotechnology in Food Production

Background. During the 1970s, several industrial giants investigated the possibility of converting cheap organic materials into protein using micro-organisms. It was envisaged that Single-Cell Protein (SCP) could replace imported protein-rich soya-bean and fish-meal supplements for animal feed. The concept was attractive; 0.25 kg of (some) micro-organisms can grow into 25000 kg of SCP per day. Nearly all these proposals finished as economic disasters. Concern about toxic chemicals in the end products, political difficulties and steep increases in the price of fuel oil combined to ruin the plans of BP and Shell, who intended to use hydrocarbons as substrates. Only one large-scale programme survived—the ICI Pruteen plant.

Details of the ICI plant are summarised in Table 1.1. The single installation at Billingham produces more SCP than all other SCP fermenters in the world combined; but does that make it a success? Immense technical problems had to be overcome to achieve and maintain sterility, and so far the programme has cost ICI over £100 million. Moreover, whilst the end product has valuable properties (Table 1.1), it costs about twice as much as its main rival, soya-bean supplement. Within a few years of operation, stockpiles of Pruteen began to mount, the production line began to run intermittently, and the possibility of using the installation for other purposes had to be considered.

TABLE 1.1

Technical data Relating to the ICI Pruteen Process

Inoculum	*Methylphilus methylotrophus*
Substrate	Methanol, plus NH_4^+ (nitrogen source), trace elements, oxygen
Fermenter	Steam-sterilised 1500 m^3 air-lift continuous-culture reaction vessel. Cell concentration maintained at about 3%
Temperature	35-42°C
pH	6.5-7.0
Doubling time	5 h
Production capacity	7 x 10^4 tonnes $year^{-1}$
Products	CO_2 (bottled and sold) Single-cell protein, marketed as animal feed supplement Pruteen (70% protein and rich in the essential amino acids lysine and methionine; energy content, 1500 kJ per 100 g—about the same as rice; shelf life, about 7 years)

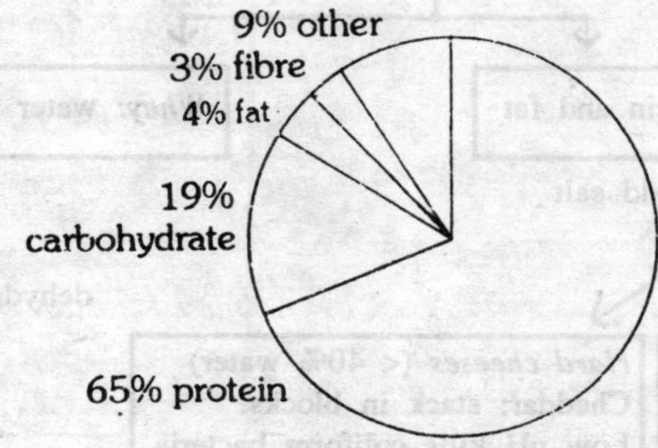

Fig. 1.2. Food *from blue-greens. Spirulina*, a cyanobacterium, grows in warm mineral-rich ponds in the tropics. In Chad and Mexico it is harvested and dried to form a biscuit-like food. Its protein content is 35 times that of maize, and its growth rate is 7 times greater. Trials are under way to investigate the possibility of developing its potential as SCP.

Other Bugs. While most European consumers will only accept a few genera of fungi in their meals, in other parts of the world very different microbes are eaten. The possibility of scaling up production of these alternative sources is being investigated. In place where they would not be accepted for human consumption, there is still the potential to divert the material into animal feed.

Food Industry (e.g. cheese)

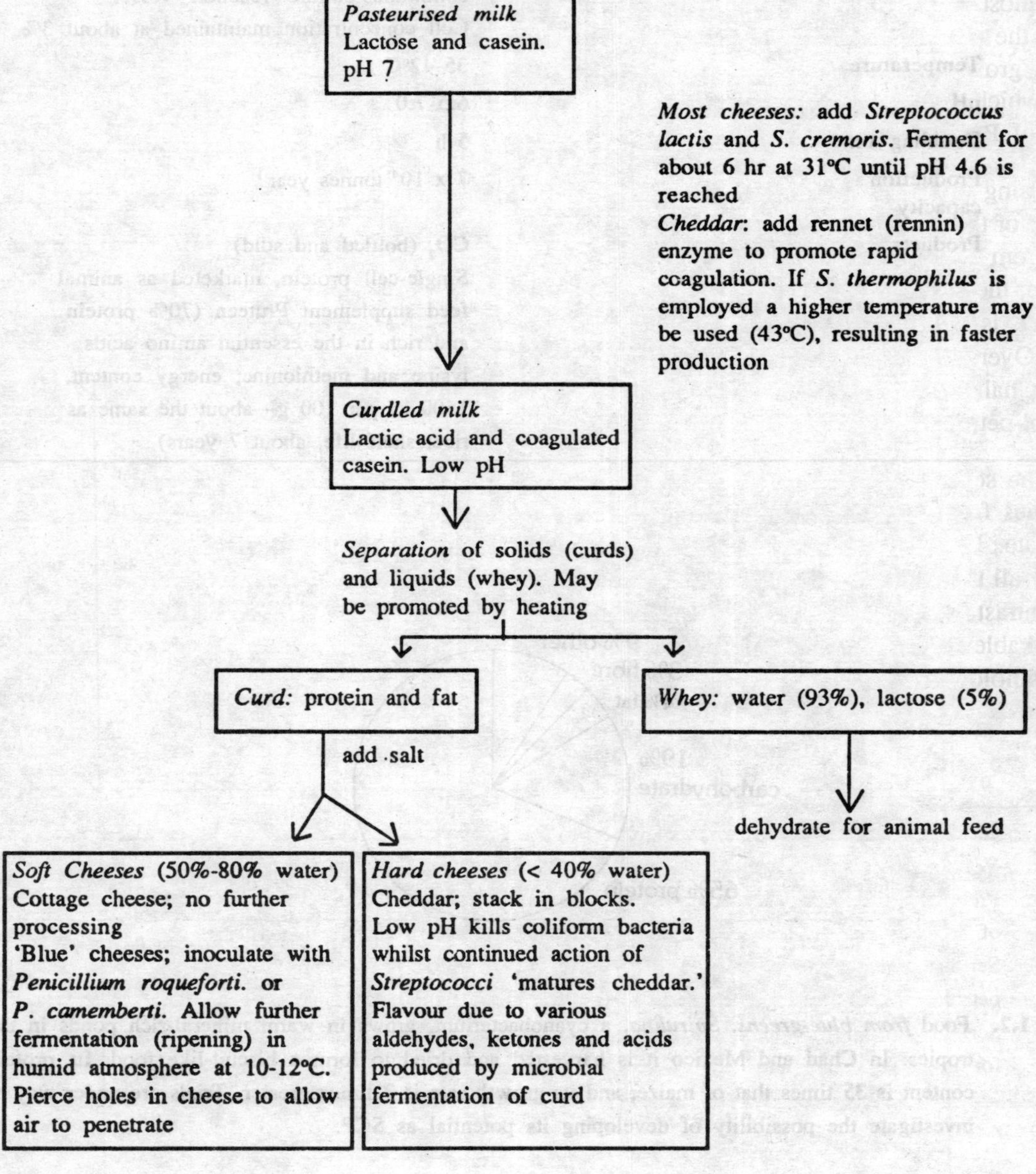

Biotechnology in Energy Production

Renewable fuels, including methane and hydrogen gases, and fuel alcohol for domestic and industrial use.

Substances manufactured by microbes that will help to extract oil locked underground.

Fuel production: the gasohol programme

Background. The economic consequences of the steep rise in oil prices during the 1970s were most severe in the oil-importing countries of the Third World. Brazil's response was to begin the world's largest single biotechnology programme. The concept was extremely simple. Brazil grows sugarcane—lots of it. Yeast converts sugar to alcohol. Alcohol is a combustible fuel which, unlike petrol, is both renewable and home produced. The Brazilian National Alcohol Programme (the 'gasohol programme') started in 1975.

Using orthodox plant-breeding techniques the yield in sugar-cane rose by 30% in the first decade of the programme. Simultaneously, improvements in fermenter design reduced fermentation time from 24 hours to 6 hours, and the efficiency of conversion rose by 10%. Now in its second decade, the scale of the operation is impressive. It accounts for about 5% of the gross domestic product, is worth some £400 billion and generates about 2500,000,000 gallons of alcohol per year. Over 400 distillaries are in operation and more are planned. The scheme has created almost half a million jobs. All Brazilian cars now run either on pure alcohol (95%) or on an alcohol-petrol mixture. About half a million alcohol-only new cars are sold each year.

The scheme has not been without its problems (Table 1.2), but this has not deterred other countries from looking at the possibility of generating fuel from biomass. Brazil is, of course, fortunate. Even by the year AD 2000, when it is hoped that the programme will be able to supply all the country's energy needs, less than 0.5% of its land will be required for the scheme. In contrast, abundant fossil fuels and shortage of land would make a similar programme unworkable in, say, the UK. Here, success is more likely to be achieved at the hi-tech end of biotechnology.

TABLE 1.2

Problems Associated with the Gasohol Programme

1. Gasohol costs about twice as much petrol.
2. Volume for volume, gasohol releases less energy, so that about 20% more fuel is needed.
3. Producing and using 1 l alcohol creates about 13 l of high BOD effluent which must be disposed of at some cost to the environment.
4. Serious corrosion problems were associated with early vehicles (which were essentially modified petrol engines). Engines specifically designed for alcohol have proved more successful.
5. Serious social and ecological problems are being caused by commandeering large tracts of land for growing sugarcane.

Biotechnology in Industry

In Industry. New sources of raw materials for the manufacture of plastics, paints, artificial fibres and adhesives.

Microbes which can extract metals from solid rock.

New systems for controlling pollution.

This list gives just a few of the benefits biotechnology can bring. It is not unreasonable for the public to treat the prophets of new technologies with a healthy scepticism and, indeed, the era of 'bio-hype' has arrived, with many extravagant claims being made for the future of biotechnology. Such inflation is both foolish and unnecessary the realistic expectations for the bioindustrial revolution are quite impressive enough to command the attention of everyone who wants to know how our world will change in the next few years.

Other Industrial Uses

Antibiotic production

Industrial chemicals

Animal feed.

Biodegradable Plastics

Most plastics degrade exceedingly slowly, and disposing of plastic waste can be a serious problem. One answer may be to use biodegradable plastic. The storage compound of most bacteria, polyhydroxybutyrate, is sometimes extruded . This can be processed to form plastic products which degrade more quickly, so making disposal easier.

Copper Mining

Deposits of many important high grade ores are diminishing at an alarming rate, and traditional methods of mining low grade ores are often prohibitively expensive. Microbial mining may provide a viable alternative in some cases. The sulphur bacterium, *Thiobacillus ferro-oxidans* oxidises insoluble ore (chalcopyrites, $CuFeS_2$) and converts it into soluble copper sulphate ($CuSO_4$). Sulphuric acid is a byproduct of the reaction, and this helps to maintain the extremely acidic conditions in which the organism thrives. Needless to say, *Thiobacillus* does not perform the oxidation as a whim. The reaction yields energy which it then uses to fix CO_2; it is a chemoautotroph. Using *Thiobacillus*, 'tailings' (copper waste tips) containing as little as 0.25% copper can be economically mined.

Uranium Mining

Low grade uranium ore (0.02% uranium) has been mined in India, Russia and Canada using microbes. At Stanrock (Canada) the ore was originally mined by traditional mechanical methods. However, the accumulating underground pools were, as a result of microbial action, an even richer source of the mineral. Indeed, mining costs were cut by 75% by simply hosing down the rock face (to encourage microbial growth) and pumping out the resulting solution. Feasibility studies are now being undertaken to explore the possibility of using microbes to extract cobalt, nickel and other metals. Organisms as diverse as *Pseudomonas* and baker's yeast actively absorb or adsorb heavy metals.

Oil Recovery

Xanthan gum is a polysaccharide produced by the bacterium *Xanthamonas campestris*. The gum is an inert compound which thickens water and improves its ability to drive out oil trapped underground. When mixed with drilling muds, it also serves as a lubricant for the giant drills as they penetrate the rock.

Fig. 1.3. Mining with microbes.

Sewage Disposal

A modern sewage plant is essentially a highly efficient form of traditional cess pit, scaled up to cope with the enormous and diverse demands of an industrial urban society.

Oil Pollution

Various species of micro-organisms such as *Pseudomonas* can consume the hydrocarbons from which oil is composed. Since each species only consumes a very limited range of hydrocarbons two strategies have been adopted:

(i) Use a mixture of strains. The method has been successfully used to clear up oil-contaminated water in derelict ships and in cleaning up water supplies.

(ii) Genetically engineer a 'superbug' so that all the oil-consuming genes are in one strain. This has been accomplished, but its usefulness under field conditions has yet to be tested. Most biologists agree that it is unlikely that bacteria could ever cope with the very large oil spills produced by tankers. In this case, prevention is certainly better than cure.

Food Industry Waste

Sugar-rich waste from the food industry can create enormous environmental problems if carelessly discharged and is costly to dispose of by traditional methods. Fermenter technology

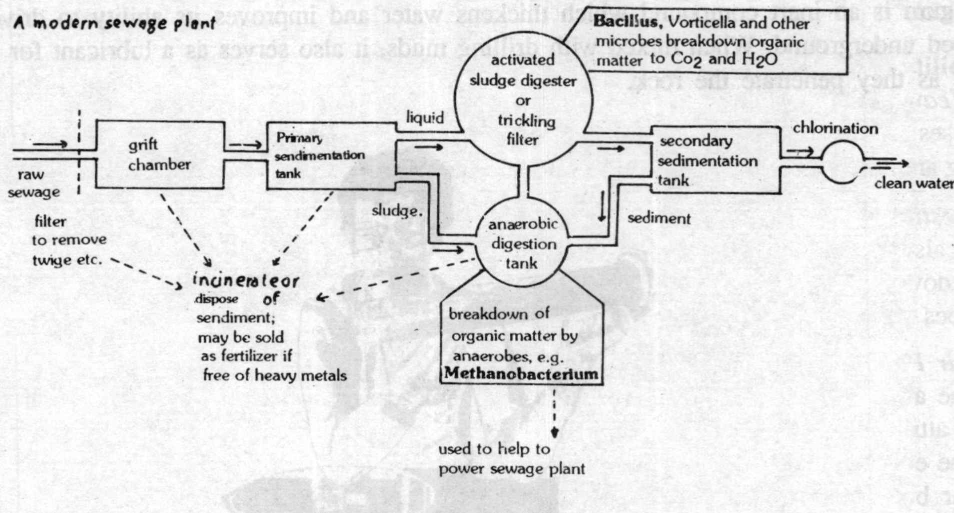

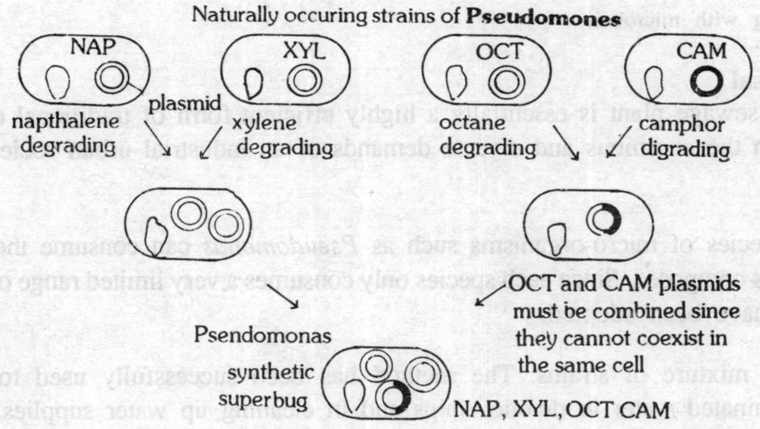

Fig. 1.4. *Waste disposal.* Waste disposal is undoubtedly the largest single application of micro-organisms by man. Its success depends upon the enormous metabolic diversity of microbes. We rely on them to dispose of sewage, farm waste, effluent from the food industry, and industrial poilutants such as organic chemicals and oil spills. The principal objectives of waste disposal are to minimise health hazards (cholera, dysentery, etc.) and to reduce the amount of material discharged into the environment which might encourage the growth of obnoxious anaerobes or otherwise upset the ecosystem.

can be employed to convert an expensive loss-making part of the production process into a profit-making sideline. Sugar-rich effluent at Bassett's sweet factory is here being converted into alcohol and carbon dioxide.

Possibility Future Developments

Heavy Metals. Heavy metals are dangerous pollutants produced by various industrial processes. Some microbes accumulate heavy metals (use is being made of this in uranium mining and the possibility of creating strains which could absorb toxic waste is being investigated.

Pesticides and Herbicides. The deliberate or accidental discharge of noxious organic chemicals into the environment is likely to continue. An ideal pesticide or herbicide should break down quickly once it has done its job. Many do not. Genetic engineering has produced microbes capable of destroying a number of these chemicals.

Air Pollution. Sulphur dioxide is an air pollutant produced by burning coal. When released into the air, it combines with water to form sulphuric acid ("acid rain"). Its harmful effects are partly attributable to a lowering of pH and partly because it releases soluble aluminium salts into the environment in toxic quantities. Research into the possibility of creating new strains of sulphur bacteria to 'clean up' SO_2 before it is discharged from chimneys is taking place.

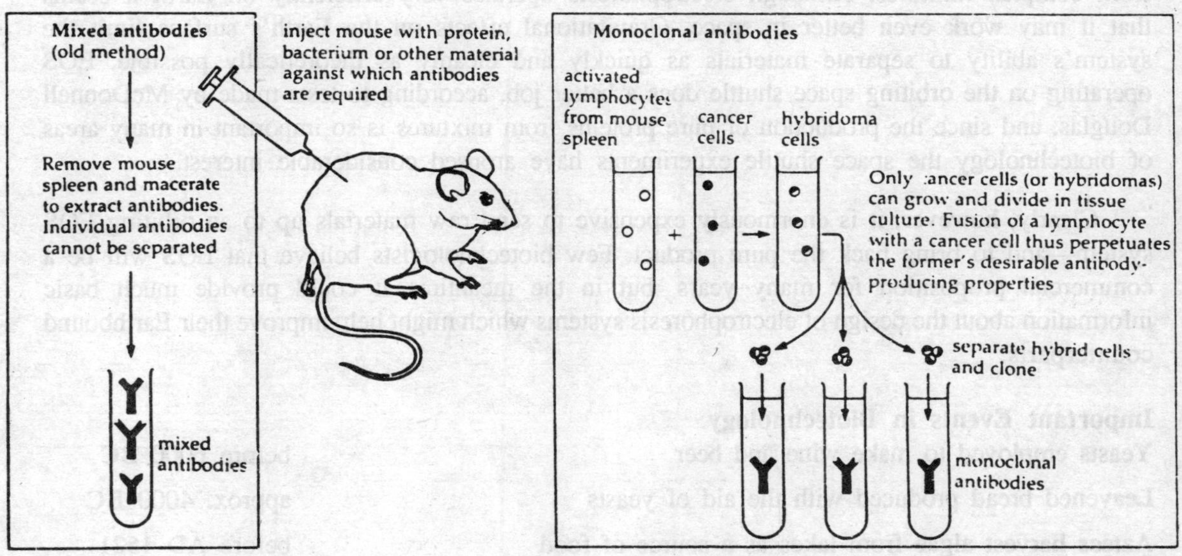

Fig. 1.5. *Monocloal antibodies: a hi tech application of biotechnology*. Kohler and Milstein (1975) devised a technique for the production of specific (monoclonal) antibodies. Manufacture of monoclonal antibodies is now under way, and the products are used to diagnose clinical conditions varying from pregnancy to cancer. Modern pregnancy-testing kits use a monoclonal antibody to react against a hormone, human chorionic gonadotrophin (HCG), which is present in the female's urine as soon as 2 days after conception. (The presence of HCG in the urine is purely fortuitous. It is secreted by the conceptus (embryo) and absorbed by the mother. It helps to maintain the corpus luteum, so preventing menstruation during the first weeks of pregnancy).

Magic Bullets at Last

The *immune system* is one of the major defence systems of the body. Its fundamental components are the antibody-producing *lymphocytes*. Each type of lymphocyte produces just one type of *antibody,* which in turn reacts with just one type of chemical (antigen). Antibodies are Y-shaped molecules with characteristic 'notches' in the arms of the Y. The binding properties of a particular antibody depend entirely upon the shape and chemical properties of the notches. In medicine, particular *(monoclonal)* antibodies are used to diagnose specific clinical conditions. In future they may also be available for the treatment of disease—in the words of Paul Ehrlich, "the physician's magic bullets."

Biotechnology in Space

Since biotechnologists are so involved in looking at the potential of very high technology systems they naturally keep an eye on sophisticated techniques being developed in many other areas. Recently biotechnologists have begun to consider how space technology might be drafted into service, and this has led to a link up between biotechnology and NASA's space shuttle programme. The giant McDonnell Douglas Corporation has been testing a novel system called electrophoresis in space (EOS). Electrophoresis itself is a well-developed technique, employed in many laboratories, which uses electric fields to separate specific materials (often proteins) from complex mixtures. Although electrophoresis operates very efficiently on Earth it seems that it may work even better in space. Gravitational effects on the Earth's surface limit the system's ability to separate materials as quickly and cleanly as theoretically possible. EOS operating on the orbiting space shuttle does a better job, according to tests made by McDonnell Douglas, and since the production of pure proteins from mixtures is so important in many areas of biotechnology the space shuttle experiments have aroused considerable interest.

Clearly, however, it is enormously expensive to send raw materials up to an orbiting EOS system—and to bring back the pure product. Few biotechnologists believe that EOS will be a commercial proposition for many years, but in the meantime it could provide much basic information about the design of electrophoresis systems which might help improve their Earthbound counterparts.

Important Events in Biotechnology

Yeasts employed to make wine and beer	before 6000 BC
Leavened bread produced with the aid of yeasts	approx. 4000 BC
Aztecs harvest algae from lakes as a source of food	before AD 1521
Copper mined with aid of microbes, Rio Tinto, Spain	before 1670
Antoni van Leeuwenhoek first sees microbes with his newly designed microscope	1680
Louis Pasteur identifies extraneous microbes as a cause of failed beer fermentations	1876

Alcohol first used to fuel motors	approx. 1890
Eduard Buchner discovers that enzymes extracted from yeast can convert sugar into alcohol	1897
Large-scale sewage purification systems employing microbes are established	approx. 1910
Three important industrial chemicals (acetone, butanol and glycerol) obtained from bacteria	1912-14
Alexander Fleming discovers penicillin	1928
Large-scale production of penicillin begins	1944
Double helix structure of DNA revealed	1953
Introduction of many new antibiotics (streptomycin, cephalosporin, etc)	nineteen-fifties
Mining of uranium with the aid of microbes begins in Canada	1962
Brazilian government initiates major fuel programme to replace oil with alcohol	1973
First successful genetic engineering experiments	1973
Hybridomas which made monoclonal antibodies first created	1975
US outline guidelines for genetic engineering	1975
US National Institute of Health introduces guidelines on genetic engineering	1976
Rank Hovis McDougall receive permission to market fungal food for human consumption in UK	1980
US Court decides that genetically engineered microbes can be patented	1980
Monoclonal antibodies receive US approval for use in diagnosis	1981
Biotechnology firm Cetus sets Wall Street record for first public offering of stock ($115 million)	1981
Genetically engineered insulin approved for use in diabetics in US and UK	1982
Animal interferons approved for protection against cattle diseases	1984

Prospectus for the Future

Genetically engineered growth hormone approved for treatment of dwarfism	mid-eighties
Interferon used to treat some viral diseases	mid-eighties
Monoclonal antibodies widely employed in diagnoses	mid-eighties
Genetically engineered hepatitis vaccine introduced	mid-eighties
New antibiotics produced by cell fusion	mid-eighties
Commercial production of dyes and industrial chemicals from algae	mid-eighties
Genetically engineered proteins used to treat heart attacks and strokes	mid-eighties
Monoclonal antibodies employed to boost the body's defences against cancer and other diseases	mid-eighties
New vaccines against foot-and-mouth disease	mid-eighties
Growth hormones used to increase yields of meat and milk from cattle	mid-eighties
Raw materials for plastics industry obtained from microbes	late-eighties
Interferon employed to treat certain types of cancer	late-eighties
More industrial chemicals produced by microbes	late-eighties
Genetically engineered microbes help extract oil from the ground	nineties
Microbes widely employed to extract metals from factory wastes	nineties
Small-scale production of hydrogen from bacteria	nineties
Monoclonal antibodies used to guide anti-cancer drugs to cancerous tissues	nineties
New crops created by genetic engineering are able to manufacture their own fertilizers and resist drought and diseases	nineties

Features of Biotechnology

Biotechnology, like many new terms, means different things to different people. Some definitions are drawn so widely that they include conventional agriculture and animal breeding. At the other extreme, some popular accounts equate biotechnology with genetic engineering, ignoring numerous other equally exciting and rewarding techniques that now exist to turn our knowledge of biology to practical ends.

The crucial feature of biotechnology, as defined here, is that it makes use of *microbes*, or *cells obtained from plants and animals,* excluding activities which involve whole plants or animals, for example, growing wheat or raising cattle. Microbes are usually grown under carefully controlled conditions in large containers, sometimes huge metal vats with a capacity of 100,00 litres (22,000 gal.) or more. When supplied with the right nutrients, they can grow extremely rapidly. The total weight of microbes in a vessel may double in as little as twenty minutes. Thus, vast amounts of microbes can be obtained very quickly from a few starting organisms. Sometimes the biotechnologist's aim is simply to generate as many as possible at the cheapest cost. This biomass is widely used as a foodstuff for farm animals and, in some parts of the world, as a food for humans.

Most of the new biotechnological processes, however, are much more sophisticated. They aim to harvest certain valuable materials manufactured by the microbes. These include antibiotics, fuels and an enormous range of chemicals for industry. There are many materials of great value that, for all their versatility, no known microbe manufactures naturally, and this is where genetic engineering steps into the picture. The potential of genetic engineering is almost boundless and when its awe-inspiring power became apparent in the seventies, serious concern arose about the wisdom and safety of 'meddling with nature' in such a fundamental manner—altering the very genetic make-up of organisms. Alarming visions of mutant 'killer bugs' were conjured up. The scientists involved in this research were the first to pose questions about its possible dangers, and an intense public debate quickly followed throughout the world. Memories of the passions raised linger on in the minds of many, but it is not the purpose of this book to go into the complex details of these arguments. The final chapter briefly examines some of the new information that has been gathered over the last few years, which has convinced the vast majority of researchers — and the public bodies set up to oversee their work—that the dangers originally proposed do not, in fact, exist.

Problems of Biotechnologists

Today, as the bioindustrial revolution gathers momentum, we are faced with other questions of immense significance concerning the routes biotechnology will follow. Will there be further improvements in the health of people in developed nations with little attention to the immeasurably more severe afflictions of the developing world? Will increased fuel supplies only be created at the expense of diminished food resources? Will some nations be left behind in the rush towards biological industries and, if so, how will their economies be affected? How can research and development best be guided towards ensuring that biotechnology yields the most beneficial results?

As in the early days of the microchip revolution, the potential social and economic impact of new technologies can only be estimated by comprehending the underlying principles and their applications. Thus, the major part of this book presents specific examples of how biotechnology has already affected our lives and how its influence will become ever more pervasive in the next two decades. The aim is to provide readers with the information they need to assess the wider ramifications of biotechnology as they unfold.

The biotechnology boom of stock-market fever and screaming headlines has been fuelled to some degree by over-optimistic speculation. Fortunately, a healthy realism is now taking hold, and the days of bio-hype seem numbered. Biotechnology is emphatically not a fashionable catchword for interesting laboratory experiments, nor is it the key to instant megaprofits. Biotechnology will survive and thrive by taking and translating knowledge drawn from many areas of science and technology into practical processes.

The problems today's biotechnologists face are sometimes complex and always challenging. The dramatic progress towards solutions to many of these problems—and the vast rewards for success—make biotechnology one of the most remarkable and fascinating endeavours of the eighties.

EXERCISES

Q. 1. Define the following terms:

(a) Antibiotic;
(b) Biomass;
(c) Downstream processing;
(d) Fermenter;
(e) Immobilisation;
(f) Monoclonal antibody.

Q. 2. What can a cow teach a fermenter technologist?

PRACTICAL WORK

Immobilisation

To make pellets for trapping yeast

(a) To 20 cm^3 of deionised distilled water (i.e.) deionised and *then* distilled), add 0.18 g of sodium alginate (BDH or Sigma Chemicals).

(b) With 0.5 g of dry yeast, make a slurry using deionised distilled water, *or* use commercially available enzyme solution.

(c) Add 1 cm^3 of (b) to 10-12 cm^3 of (a). Suck this mixture into the body of a 10 ml syringe.

(d) Slowly release (c) drop by drop from the syringe into a gas jar filled with 0.2 M $CaCl_2$ solution. If distinct 'blobs' do not form, try stirring very gently as the alginate mixture falls through the $CaCl_2$ solution.

(e) Wash encapsulated cells (or enzyme) free of $CaCl_2$ using deionised distilled water.

2

Techniques of Biotechnology*

Techniques of Biotechnology

Biotechnology is an interdisciplinary activity. It involves a close collaboration from specialists in many fields: molecular and cell biology, chemical engineering, computing and economics. Our main concern here is to look at those aspects of biotechnology which are of particular consequence to the biologist.

Organisms

Table 2.1 gives a list of raw materials that are being altered by biological means. Much of this processing could be achieved, by organic chemists. However, the non-biological alternatives may have a number of disadvantages (Table 2.2).

There may be prohibitive costs in the non-biological alternatives associated with fuel, materials, equipment and downstream processing (extracting and purifying the product). However, biotic systems do not have a monopoly of advantages. Indeed there are often stringent requirements regarding:

pH (often within pH $\pm$ 0.5 of the optimum);

temperature (often within 2-3°C of the optimum);

oxygen (or lack of it)

minor nutrients (small but precise requirements for vitamins, minerals etc.);

osmotic potential of the media (to prevent plasmolysis or lysis);

sterility (to prevent the growth of undesirable organisms).

The choice between a biological and a non-biological industrial system is rarely clear cut. It may depend on the installation and operating costs, which can vary from time to time and from place to place. There may also be powerful social and political pressures influencing the choice of system.

---

* Excerpts taken from the review published by M. Johnson Portgail.

TABLE 2.1

Applications of Biotechnology

	Applications	*Organisms involved*	*Type of organism*
1.	**Food and beverages**		
	Baking, wine	*Saccharomyces cerevisiae*	Y
	Lager	*Saccharomyces carlsbergensis*	Y
	Yoghurt	Lactobacillus bulgaricus	B
		Streptomyces thermophilus	B
	Vinegar	*Gluconobacter suboxdans*	B
	Blue-veined cheese,	*Penicillium roquefortii*	F
	Camembert	*Penicillium camembertii,*	F, B
		Streptococcus lactis,	
		Streptococcus cremoris	B
	Single-cell protein using		
	paper waste	*Candida utilis*	Y
	petroleum products	*Saccharomyces lipolytica*	Y
	methane, methanol	*Methylophilus methylotrophus*	B
2.	**Chemical Industry**		
	(i) Industrial chemicals		
	Ethanol	*Saccharomyces cerevisiae*	Y
	Acetone, butanol	*Clostridium autobutyliticum*	B
	Citric acid	*Aspergillus niger*	F
	Xanthan gum	*Xanthomonas campestris*	B
	Dextran	*Leuconostoc mesenteroides*	B
	Biodegradable plastic (polyhydroxy butyrate)	*Alcaligenes eutrophus*	B
	(ii) Enzymes		
	Proteases ('biological detergents')	*Bacillus spp.*	B
	Amylase (hydrolysis of starch	*Aspergillus oryzae*	F
	Glucamylase and maltose to glucose)	*Aspergillus niger*	F
	Rennin (cheese making)	*Bacillus spp.* (via genetic engineering)	B
	Pectinase (wine making)	*Aspergillus spp.*	F
	Cellulase (stubble digestion in agriculture)	*Trichoderma reesii*	F
	(iii) Fine Chemicals and Pharmaceuticals		
	Penicillin	*Penicillium chrysogenum*	F
	Cephalosporin	*Cephalosporium acremonium*	F
	Streptomycin, kanamycin		
	Neomycin, tetracycline	*Steptomyces spp.*	B
	Insulin, somatostatin, interferon	*Escherichia coli, Bacillus subtilis*	B
	(via genetic engineering)		
	growth hormone, etc.		
	Riboflavin	*Eremothecium ashbyi*	Y
		Pseudomonas denitrificans	B
	Vitamin B_{12}	*Proprionibacterium*	B
	Lysine, nucleotides	*Cornybacterium glutamicum*	B
	Vaccines	Various	B, V

3. Medicine	
Diagnosis of infectious disease, genetic defects, or other clinical conditions (pregnancy) by monoclonal antibodies, DNA probes and restriction mapping	B, C
4. Environmental Applications	
Disposal of pollutants (sewage, oil spills, sludge)	A
Mining of copper, uranium, and oil	B
5. Agriculture	
Breeding by protoplast fusion; genetic engineering for improved product quality, disease resistance and reduced fertiliser dependence, etc.	A
Biological control of pathogens	B, V

A, all categories of microbes and various eukaryotic cells; B, bacteria, C, eukaryotic cell cultures; F, filamentous fungi. V, viruses; Y, yeast.

TABLE 2.2

Advantages of Biotic Production Systems

	Chemical (non-biological) processing	*Biological processing*
1.	Very high or low temperatures often required	Moderate temperatures normally employed: heating and refrigeration costs are lower
2.	Organic solvents often needed; may be both expensive and toxic	Water normally employed as solvent for growth or enzymic reactions
3.	Inorganic catalysts often used; often toxic, expensive and lacking specificity, so resulting in unwanted byproducts. Separation of catalyst and product sometimes difficult	Enzymes highly specific, so less wasteful of substrate; byproducts (if any) usually less noxious. Removal of of enzymic catalyst generally easier
4.	Extremely acidic or alkaline conditions often employed, demanding expensive corrosion-resistant containers	Moderate pH usually employed
5.	Raw materials often expensive	Raw materials usually cheap
6.	Lengthy preparation time often required	Short fermentation or reaction time, yielding a higher rate of production.

Which organism?

From Table 2.1 it can be noted that all the organisms are microbes. There are several reasons for this:

(i) They show a wider variety of potentially useful metabolic pathways and biochemical reactions than higher organisms do.

(ii) Their growth rates are exceedingly high. This is due to their extremely large surface-to-volume ratio that promotes the rapid transport of materials in and out of the cell. The doubling time for bacteria, for example, is about half an hour, and for yeasts just over 2 hours. In contrast, the times needed to double the mass of a young plant or a calf are about 2 weeks and 2 months respectively.

(iii) Their growth needs are generally fairly simple and can be easily provided.

(iv) They have comparatively simple and well-documented regulatory systems. Thus, given the right conditions, they can be induced to undertake particular biochemical tasks.

(v) Their genetic systems are quite simple, and new desirable characteristics can be introduced by genetic engineering.

(vi) They have no natural death (no fixed lifespan). They, or their enzymes, can be immobilised and used over and over again. *Immobilisation* makes them easier to handle and can substantially reduce costs.

The only other living materials in Table 2.1 are 'pseudomicrobes,' i.e., cell and tissue cultures. Although they have some features in common with cultures of 'real' micro-organisms, they have much more exacting growth requirements and possess complex, poorly understood regulatory and genetic systems. However, they do possess some unique properties making them particularly valuable in certain processes.

Higher organism

Higher organisms play two major roles in biotechnology. (i) They provide *biomass,* i.e. the essential raw materials such as starch, cellulose and sugar waste on which micro-organisms can be put to work. (ii) They are the traditional sources of food, clothing and construction materials, they themselves become the subject of biotechnology, and particularly genetic engineering.

In any commercial biotechnological process the microbe or system finally chosen will depend on the answers to very specific questions concerning the yield, the ease of product purification and the cost of providing and maintaining appropriate conditions.

3

Applications of Biotechnology

Introduction

The applications of biotechnology are mainly to three contrasting industries and these will be considered in detail. Its application to pharmaceuticals, fuel production and single-cell protein production, illustrates the kind of problems which may confront a relatively new hi-tech industry.

Applications of Biotechnology to Medicines

To get a clear understanding of various biochemical processes and the applications of biotechnology in various life processes let us first acquaint ourselves fully with the chemistry of life. Understanding it will provide us the key to biotechnology and it will make our problem easier.

One of the most impressive achievements of science has been the explosion of knowledge about the chemical composition of organisms and the way these chemicals interact to create the phenomenon of life. Probably the most powerful impetus given to biological research was the final acceptance last century that it is futile to search for 'vital forces' which distinguish living organisms from inanimate matter. The special nature of living beings does not reside in unique chemical principles, but rather in the immensely sophisticated way wherein they utilize the ordinary laws of chemistry.

Organisms are sometimes compared to chemical factories. The strength of this analogy lies in its emphasis on the chemical nature of life—the fact that growth, development and reproduction all depend on chemical reactions. However, the analogy does obscure some of the most fundamental characteristics of organisms, many of which are directly relevant to biotechnology.

Perhaps the most notable feature of the living organisms is the sheer diversity of chemical processes they undertake. Most chemical factories are designed to convert specific raw materials into just a few products. Evolution has endowed organisms with the ability to take in a wide variety of raw materials (nutrients) and transform them into literally thousands of different types of materials, each with a particular biological role. Biotechnology draws its strength from these powerful chemical 'skills,' developed over billions of years of evolution.

Application of Biotechnology to the Prevention, Diagnosis and Cure of Disease

Introduction. The introduction of penicillin and a host of newer antibiotics has lifted the scourge of many infectious diseases and saved millions of lives. Now, biotechnologists are paving the way for a massive assault on many more of the world's most devastating diseases, including cancer, diabetes, hepatitis, malaria and sleeping sickness. Less common, but equally dangerous inherited diseases, such as haemophilia, will also be tackled with the aid of biotechnology.

The medical applications of biological industries are exciting, and progress is astonishingly fast. A few years ago many current techniques existed only in the realms of science fiction, and many more will become facts in recent future. Many diverse aspects of biotechnology are being drafted into the fight against disease. Naturally occurring and genetically engineered microbes can be used to manufacture drugs, vaccines, hormones and enzymes; new tools, such as monoclonal antibodies, should aid diagnosis and therapy; while cell fusion could provide novel and powerful antibiotics.

This chapter examines the contributions biotechnologists can make in the prevention, diagnosis and cure of three groups of diseases: those that are caused by an invasion of the body by viruses, bacteria and other micro-organisms; those that result from some imbalance in the body's natural chemistry; and those whose causes are less well understood, including heart disease and cancer.

Microbes—Their Role

In the eighteen-sixties and seventies, two of the greatest scientists of the time, Louis Pasteur in France and Robert Koch in Germany, were hard at work establishing a theory that has probably had more impact on modern medicine than any other. Between them, Koch and Pasteur proved that certain diseases of humans and other animals — including tuberculosis, anthrax and cholera — are caused by particular, identifiable microbes.

With the benefit of hindsight, this germ theory of disease may seem obvious, but at the time the notion was little short of revolutionary. Although Pasteur and Koch were not the first scientists to propose such a theory, their work provided the conclusive proof and dispelled the air of mystery that had previously shrouded ideas concerning the nature of disease. Before the germ theory took root, only the imagination of those who investigated diseases seemed to limit the possible 'causes'— miasmas in the air, supernatural influences, and character defects of the sick person had all been invoked to 'explain' diseases.

Once the germ theory had become entrenched in medical thinking, a new and more scientific era of diagnosis and treatment could begin. We now know that many common diseases are caused by bacteria, viruses and fungi, and that diseases spread as these are passed from person to person, either directly or indirectly. An immense amount of effort in now devoted to tracking down and identifying infectious microbes. Once the culprit has been unmasked, specific means of combating it can be sought.

The study of microbes, microbiology, revealed that by no means all of them are harmful. This is fortunate since all of us play host to literally millions of microbes, and we would not survive long if most were not benign. With the advent of biotechnology, our relationship with microbes is entering a new phase — some microbes have been put to very positive medical uses, particularly in the production of *antibiotics*.

Antibiotics

Today there are more than 100 different antibiotics available for use on humans. Large though this number is, it is only a small fraction of the 5000 or so compounds isolated from microbes which have been shown to kill or disable other microbes. The large number of antibiotics that are not used in medicine are rejected for a range of reasons: some produce too many harmful side effects; some are too expensive to manufacture on a large scale; and some simply cannot do a specific job as well as another, readily available antibiotic.

Classes of Antibiotics

The four major classes of antibiotics—the penicillins, the tetracyclines, the cephalosporins and erythromycin — are worth over US $4 billion in bulk sales each year, and all are superb examples of the art of biotechnology. Although the details of how each group came to benefit humans differ, the same general principles apply to each. The reason for the development of this range of antibiotics, and for the quest for new ones, can be seen in the story of the cephalosporins.

Story of Cephalosporins

In 1945, Guiseppe Brotzu, a Sardinian professor of bacteriology, found a micro-organism in the sea near a sewage outfall. This fungus, a species of *Cephalosporium,* produced a substance that killed a wide range of bacteria. Brotzu did not have the facilities to analyse this substance, so he sent his organism to Oxford. There, in the laboratory of Howard Florey where the early penicillin work had been done, Guy Newton and Edward Abraham discovered that the fungus manufactured a novel type of penicillin, which they named penicillin N. Then in 1953 they made another discovery of much greater importance — this organism also manufactured another antibiotic, which they called cephalosporin C.

The immediate advantage offered by cephalosporin C was that it could kill bacteria that had become resistant to penicillin. The phenomenon of resistance to antibiotics has proved to be the major driving force behind the quest for new antibiotics. When penicillin was first introduced it appeared to be a wonder drug and, indeed, many of the commonest dangerous bacterial infections were stopped in their tracks. However, within a very few years doctors found to their dismay that some infections that had previously succumbed rapidly to a course of penicillin now managed to survive — the bacteria had become resistant to penicillin.

Resistance to a particular antibiotic can appear in a population of bacteria in a variety of ways, including the transfer of antibiotic-resistance plasmids from other species or strains. The mere fact that an antibiotic comes into common use perversely (from our point of view) encourages the appearance of resistant bacteria. For example, the early penicillin-resistant bacteria appeared to have 'started' making an enzyme known as penicillinase, which attacks the antibiotic before it can do its job. In fact, there may always have been some individual cells making small amounts of the enzyme, but they were vastly outnumbered by the penicillin-sensitive type. As these latter were exterminated by penicillin, the resistant strains were able to thrive in their place.

Cephalosporin C was developed to tackle diseases, such as pneumonia, caused by penicillin-resistant *Staphylococcus,* but before cephalosporin was ready to be used on a large scale these bacteria had been largely conquered by one of the semi-synthetic penicillins, methicillin (produced by chemically modifying the basic penicillin molecule obtained directly from *Penicillium* moulds). Cephalosporins did, however, come into their own in the nineteen-sixties when they began to be used to fight other types of bacterial infection, and they now form a crucial part of the armoury of anti-microbial compounds.

Both penicillin and cephalosporin are members of the group of antibiotics known as beta-lactams, after a characteristic type of chemical ring structure they possess. They operate by preventing certain bacteria from building proper cell walls.

Streptomycin, on the other hand, belongs to the group of antibiotics known as aminoglycosides. These work by preventing certain bacteria from manufacturing their proteins. Once inside the bacterium, streptomycin disrupts its ribosomes, the small globular structures on which the genetic information carried by mRNA is translated into proteins.

This antibiotic was discovered by Selman Waksman, one of the great figures in the history of microbiology, and his colleagues at Rutgers University, New Jersey. They spent years studying microbes from the soil, and in 1944 they announced that they had found a new antibiotic and named it streptomycin, because it came from the filamentous microbe *Streptomyces* griseus. Streptomycin proved particularly valuable because it attacks microbes which are unharmed by penicillin and cephalosporin. In particular, it revolutionized the treatment of tuberculosis: patients were no longer condemned to years of slow therapy in sanitoria, since the new drug could often effect a cure within months.

It is an odd fact, which has never been properly explained, that the majority of antibiotics, including streptomycin, are produced by a rather narrow range of organisms, collectively known as the antinomycetes. In appearance, actinomycetes are similar to the moulds that make penicillin and cephalosporin. The branching network of filaments in actinomycetes are composed, however, of bacterial or prokaryotic cells, whereas moulds consist of more complex, eukaryotic cells.

The alarming spread of resistance to several kinds of antibiotics demands that the quest for new forms continues. In the short term, no great advances can be expected through the type of genetic engineering already discussed—that is, the insertion of one or two genes which instruct a cell to make some protein it would not normally produce. None of the commonest antibiotics are proteins and they are not, therefore, the *direct* products of genes. Most antibiotics are constructed inside the cell by a chain of discrete chemical reactions, each of which is catalysed by a separate enzyme and, of course, each enzyme is made according to the plan encoded in its own gene. It is a daunting task to find out exactly how a particular antibiotic is created, what enzymes are involved and which genes underlie the whole process.

However, genetic engineering might be used to create modified antibiotics. There are many examples of related antibiotics used in medicine which are chemically very similar. A classic example is the penicillin group — all have the same basic structure, but the presence of a variety of comparatively minor chemical modifications results in a range of drugs with different uses. Today, only the basic structure of penicillin is obtained directly from the moulds; purely chemical methods are then used to tinker with the molecules to obtain the many types of penicillin. This can be done with some ease for the penicillins, but modifying some other types of antibiotic is more difficult. Possibly microbes could do the job instead.

The microbes would need to be persuaded to use certain enzymes they already possess to modify the standard antibiotics they normally produce. For instance, cells use enzymes called methyl transferases to swap a methyl unit (one carbon plus three hydrogen atoms) between pre-existing molecules. These methylation reactions are vital for many chemical processes within the cell. If antibiotic-producing cells could be induced to employ their methyltransferases to tack on methyl units to their antibiotic molecules, a new antibiotic would be produced with different and perhaps even more useful properties. Although such modified antibiotics are perhaps more likely to come through genetic engineering, the technique of cell fusion offers another possible avenue.

Cell Fusion

Very few biotechnological processes employ 'natural' strains of microbes — that is, those found in the environment. Such 'wild-type' strains may be admirably adapted for survival in their normal habitats, but will probably yield relatively little of the substances we desire and may not be suited to the environment of a fermenter. Remember, for example, that natural strains of *Penicillium* mould produce only one ten-thousandth of the amount of penicillin made by the strains now employed in the pharmaceutical industry.

When seeking to produce a new strains of microbe, biotechnologists may rely on evolution by artificial rather than natural selection. For example, in the development of high-yielding strains of *Penicillium* mould, mutations were induced by treating the microbes with chemicals or radiation. This is a completely random process and there is no means of knowing what alterations will occur in the microbes' genetic material. Genetic engineering is much more specific, for it attempts to introduce a specific gene into the microbes. The technique of cell fusion lies at an intermediate position on this scale of random to specific genetic modification; its results are more predictable than random mutations but less so than genetic engineering.

The term cell fusion may be unfamiliar, but there is at least one very good reason why it is central to all our lives — each of us was created by a process of cell fusion. Sexual reproduction takes place when two cells carrying different genetic information join together, or fuse. The resulting combination of genes, some from the egg cell and some from the sperm cell, yields a new individual which is different from either parent and, indeed, is genetically unique. This constant reshuffling of genes from generation to generation produces much of the diversity among individuals of a species upon which natural selection operates.

Most microbes, however, predominantly reproduce themselves by simple division, in which a cell splits to give two identical offspring. Thus, in the absence of any random mutations, no new combinations of genes are obtained and the next generation will have exactly the same properties as the previous ones. Clearly this is a great advantage for biotechnologists once the required strain has been found, but it does not aid the search for new and better strains. The technique of cell fusion allows the generation of novel combinations of genes within microbes by joining two cells together.

The principles are quite straightforward. The outer, tough membrane surrounding bacterial cells is stripped away, usually with the aid of enzymes, to leave the cell contents packaged in the much more delicate inner membrane. These fragile forms of cells, called protoplasts, can then be induced to combine by adding certain viruses or chemicals.

Cell fusion creates hybrid or recombinant cells, which contain genetic material from two or more cells. Cells to be fused may be different strains of the same microbial species or even entirely distinct species. The great advantage is that novel mixes of genetic material can be obtained — combinations which would be found only rarely in nature or not at all. Apart from the creation of modified antibiotics, cell fusion may also produce new antibiotics by activating 'silent' genes.

It is already known that the actinomycetes group of bacteria makes many hundreds of different antibiotics and more are being discovered each year. It is quite likely that these organisms have the inherent ability to make an even wider range of antibiotic substances, some of which might be medically useful. According to this theory, the genes which instruct the cell to make these, as yet undiscovered, antibiotics are 'silent' or unexpressed so that no mRNA is made according to their instructions and, thus, no proteins are synthesized. The majority of any cell's genes are turned off at any one time. If the organism has no need for the protein encoded in a gene it does not normally waste its energy by making it. The problem is to persuade the actinomycetes' cells to turn on their silent genes so that their products can be assessed. It is reasonable to assume that the genes must be active under *some* conditions — temperature, nutrient supply or other external factors — for if the organism never used the genes they would probably have been cast aside during its evolution. One could invest a great deal of effort in trying to grow the cells under an enormous variety of conditions in the hope that some trigger for the activation of the genes will be found. However, cell fusion offers a short cut.

By no means all of a cell's DNA is used to carry codes which instruct the cell how to make certain proteins. Some stretches of DNA perform a control function, allowing the cell to turn on or off the protein-coding genes according to its changing needs. Many of the most exciting and awe-inspiring discoveries in basic biology over the last few years have been concerned with the control of gene expression. The stunning and subtle complexity of gene-control systems is now being rapidly revealed, although doubtless many more surprises await discovery. The essence, so far as this potential application of cell fusion is concerned, is that the control sequences of DNA do not 'know' what genes they are controlling. In normal cells, of course, the control systems have evolved in such a way that they turn their respective genes on and off according to the cell's needs for the protein encoded in that gene. If, however, a control sequence is moved from its natural position to a place where it controls an entirely different gene, it will switch its new companion on and off according to the control sequence's perception of the cell's need for the protein made by the gene with which it is *normally* associated.

Cell fusion is one way of bringing together unaccustomed partners—protein-coding genes and control regions. The aim is to attach one of the unexpressed antibiotic genes to a control region which will turn it on under the conditions found inside fermentation vessels. Under the invigorating influence of its new control region the previously silent gene should then become expressed—mRNA being copied from the gene and enzymes being synthesized which the microbe may use to manufacture new and useful antibiotics.

When cells are fused, particularly if they are closely related strains of microbes, pieces of DNA from each of the original cells will sometimes swap places. In particular, a control region from Cell *A* may displace one from Cell *B*, and this might lead to the expression of a previously dormant gene in Cell *B*. There is, as yet, no way of guiding this exchange of genetic material between two fused cells. However, as with all microbiological experiments, one is dealing with vast numbers of cells. So long as some of the exchanges are productive this is a protentially profitable route towards the production of novel antibiotics.

Monoclonal Antibodies

The German scientist, Paul Ehrlich, Secured his place in history by laying the foundations of immunology — the study of the body's defences against infection — and by investigating the effects of chemicals on microbes, which culminated at the turn of the century in the discovery of the first effective treatment for syphilis. He is almost equally well remembered for the words 'magic bullets', which encapsulated the ideal of all drug research. Ehrlich aspired to find drugs which would eradicate microbes while producing no ill effects in the patients. So far his dream has never been completely realized; even the best drugs occasionally provoke harmful side-effect. Perhaps such perfect drugs will never be found, but in 1975 scientists came a step nearer achieving something implicit in the notion of magic bullets — the ability to direct drugs to the exact position in the body where they will do most good. In that year Georges Kohler and Cesar Milstein, working in Cambridge, England, discovered how to make monoclonal antibodies, and in 1984 their efforts were recognized by the award of the Nobel Prize for medicine.

Of all the new biotechnologies, those that utilize monoclonal antibodies are likely to have the most rapid and widespread impact on medicine. To accept this bold claim it is necessary to look at what monoclonal antibodies are and how they are made.

All animals face a constant assault from viruses, bacteria, fungi and chemicals in their environment. If these enter the body and its individual cells, the results may be devastating. The first lines of defence are formed by the skin and the membranes which surround cells, and in many cases these can ward off the threat. However, when these defences are breached or circumvented, more subtle and powerful weapons are brought to bear on the invaders. Pre-eminent among the immune system's weapons are a group of proteins known as antibodies.

Antibodies are manufactured by specialized cells in the spleen, blood and lymph glands. These so-called B-cells release antibodies which roam the body, seeking out and latching on to microbes or other foreign materials. Once the invader has been tagged by antibodies, the rest of the immune system swings into action, culminating in the demise of the undesirable alien. A central puzzle of immunology — and one which has great relevance for biotechnology — is how antibodies recognize foreign substances and attach themselves to them. The answer is revealed by the molecular structure of antibodies, and this gives the clue to their enormous potential in biotechnology.

The shape of each antibody molecule is dictated by the sequence of amino acids used in its construction. All antibodies have the same basic form, that of the letter Y, but when examined in more detail it is found that this apparent uniformity disguises an almost incredible diversity. Each antibody has two identical pockets, one at the end of each arm of the molecule. The shape of these pockets varies subtly from one type of antibody to another. It is this variation which endows antibodies with their most important characteristic — specificity. The pockets of antibodies mesh with molecular structures, and hold on tightly to them (See Fig. 3.1). In the human body there are, quite literally, millions of different types of antibodies, each having pockets with a characteristic shape.

The surface of every substance, be it a virus, a bacterium or even smooth plastic, is studded with molecules which jut out into its surroundings. When an antibody encounters a protrusion which happens to fit its pockets, the two lock together. The structure to which an antibody binds is called its antigen, and the relationship between an antibody and its antigen is very precise; an antibody will only latch on to an antigen which has exactly the right shape. The term monoclonal antibodies is applied to a group of identical antibodies all with the same shape of pocket and, thus, recognizing exactly the same antigen.

This very specific interaction can be utilized in many ways. For example, many diseases are characterized by the presence of unusual substances in the body or excessive amounts of some normal material. Monoclonal antibodies are now commonly employed to detect the presence of viruses, bacteria and other infections, by mixing appropriate antibodies with samples of blood or other bodily fluids. By providing a precise measure of the amount of specific substances in a patient's body, it will also be possible to diagnose many other disorders, such as some forms of infertility which are characterized by a lack of particular hormones.

The major technical difficulty lies in obtaining the required antibodies. There is no practical way of picking out the desired antibody from the teeming hordes of other types, so another approach must be adopted. The cells that make antibodies can be separated fairly easily. This would not brighten the prospects if it were not for one fact of nature and one crucial scientific trick. The relevant fact is that each B cell (the cells that make antibodies) produces only *one* type of antibody. Could a B cell, placed in a glass dish with suitable nutrients, grow and divide to produce a clone of identical cells, all making the same kind of antibody? This might not be the required antibody, but it would be pure or 'monoclonal,' which is more than half the battle. By repeating the process for many individual cells, persistence, technical ingenuity and luck might turn up the right antibody-producing cell eventually. Unfortunately, this neat plan founders on a simple but profound obstacle — B cells die soon after they are removed from the body. Their brief period of life in the laboratory is insufficient to provide a large enough clone even to find out exactly what kind of antibody they are producing, let alone obtain enough antibody for practical applications.

Kohler and Milstein invented the technical trick that solves the problem. Ironically this discovery, which promises so much in terms of cancer therapy, depends on the use of cancer cells. Unlike normal cells, cancer cells can be cultured quite easily. Given the right laboratory conditions they will grow and divide virtually indefinitely. Combining the antibody-producing skills of B cells with the quasi-immortality of cancer cells should produce large quantities of monoclonal antibodies. Rather surprisingly, this can be done in a very straightforward manner. Fusing a B cell with a cancer cell produces a hybrid cell, which has the properties of both original cells— it makes antibodies and it lives for a very long time. It was an arduous task to perfect a method of making these hybridoma cells, but within a few years of Milstein and Kohler's original success, the techniques had been refined and brought into common use in hundreds of laboratories.

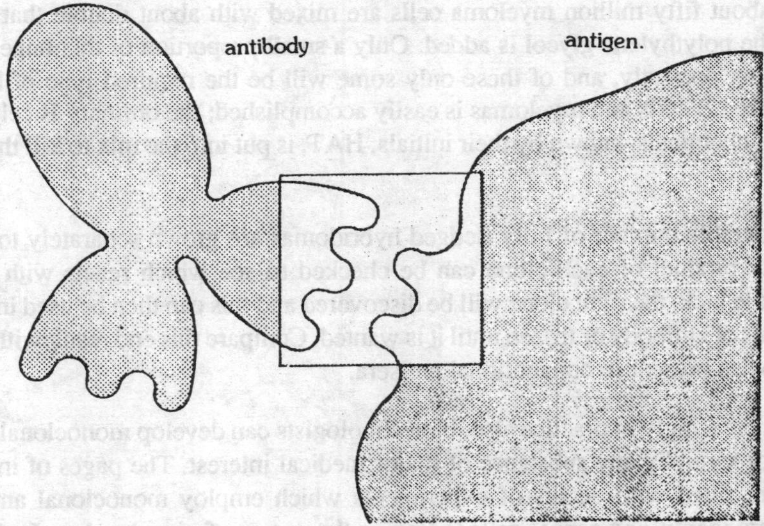

Fig. 3.1. Antibodies (also called immunoglobulins) are Y-shaped proteins. At the end of each arm are two identical pockets, the shape of which varies from one antibody molecule to another. When an antibody encounters an antigen whose shape matches the pockets, the antibody and antigen bind together.

Needless to say, these two types of cells will not join to form a hybridoma of their own accord. The biotechnologist's craft is to manipulate the cells' environment to induce them to fuse together. In the early days this was done by mixing the cells with viruses. This affects their membranes which coalesce to form a single membrane around the contents of both original cells. More recently it was found that a simple, cheap chemical, polyethylene glycol, can do the job even better.

The cells used most frequently are mouse spleen cells and mouse myeloma cells, the latter being a cancerous form of B cells. The first step is to ensure that the mouse spleen has as many B cells producing the particular required antibodies as possible. One of the most successful applications of the hybridomal technique has been to produce monoclonal antibodies which recognize interferon molecules — the natural substances that help the body fight off viral infections and may be used to treat cancer. To do this a mixture of materials known to contain interferon is injected into a mouse. (See Fig. 3.2) This mixture may be obtained, for example, from the blood of someone suffering from a viral disease. The substances in the mixture then act as antigens, stimulating the mouse to produce many spleen cells which make antibodies against these foreign materials, and among them will be anti-interferon antibodies. This is exactly the type of immune response which makes vaccination of human beings against infectious diseases so effective. In vaccination, a dead or severely weakened micro-organism is introduced into the body. The immune system makes antibodies against the foreign material, and, having 'learned' to recognize this particular threat, the immune system is prepared to combat any subsequent invasion by a virulent organism of the same kind.

Having ensured that the mouse spleen has many anti-interferon B cells, the spleen is removed and minced. Then about fifty million myeloma cells are mixed with about double that number of spleen B cells and the polythylene glycol is added. Only a small proportion of this huge number of cells will fuse together correctly, and of these only some will be the required type of hybridoma. Removing the unaltered B cells and myelomas is easily accomplished; the ordinary B cells soon die, and a combination of chemicals, known by their initials, HAT, is put into the mix to kill the unaltered myelomas.

Then the considerable number of fully fledged hybridomas are grown separately to give large clones. The antibody product of each clone can be checked to see which reacts with interferon. Eventually, with a little luck, the right clone will be discovered and this can then be used immediately to make large amounts of antibody or frozen until it is wanted. Compare this end result with the mixed population of antibodies present in conventional antisera.

In theory, and increasingly in practice also, biotechnologists can develop monoclonal antibodies which are specific for virtually any substance of major medical interest. The pages of innumerable scientific journals are filled with details of experiments which employ monoclonal antibodies in dozens of ways—from identifying cancer cells to tracing the nerves of giant leeches. In the last five years, these new tools have played an ever more important part in basic research. To meet the demand for monoclonal antibodies, exotically named companies, such as Hydrid-tech, Celltech and Allergenetics, sprang up like mushrooms in the late seventies and early eighties. By mid- '83 over 40 monoclonal antibody products had been approved for diagnostic use in the USA — including kits for pregnancy testing, cancer diagnosis and the detection of rabies viruses. Recent estimates put the value of monoclonal antibody products at over a billion dollars a year by the end of the decade.

This new technology is now poised to make a major impact in medical practice over the entire spectrum of human diseases, from cancer and kidney failure to infections caused by viruses and bacteria. The potential applications of monoclonal antibodies include boosting the natural defences of patients; improving the chances of successful organ transplantations; targeting drugs to specific parts of the body; and purifying drugs before they are administered to patients.

An example of this final group of applications, which has already been described, is the production of monoclonal antibodies which attach themselves only to interferon molecules, spurning all others. These antibodies can be fixed on to beads which are then packed inside a tube. When a mixture of materials containing a tiny percentage of interferon is poured into the tube, the antibodies pluck the interferon molecules from the soup while the unwanted materials flow away. Chemicals are then passed through the tube to loosen the antibodies' hold on interferon which is flushed out. This technique yields a solution which is 5000 times richer in interferon than the original crude mixture. This and other uses of monoclonal antibodies rely on their exquisite powers of discrimination. Looking for a needle in a haystack, and extracting it, is not such a problem if one has a magnet — and in monoclonal antibodies we can hope to find a 'magnet' which will pick up just what we are looking for.

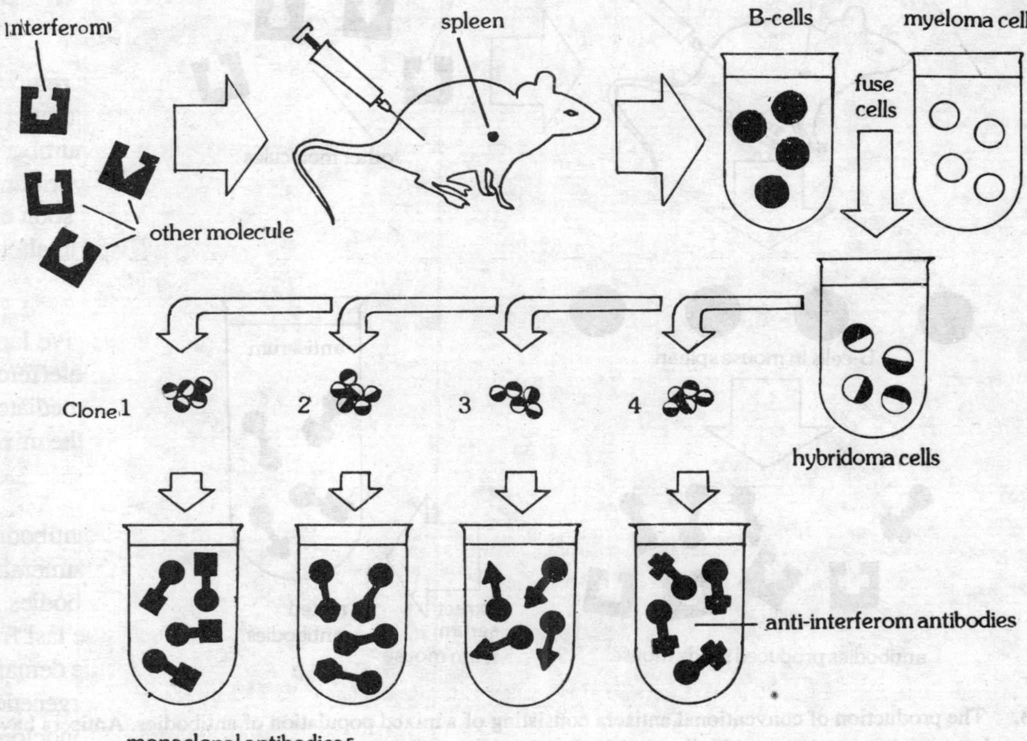

Fig. 3.2. The production of hybridoma cells which manufacture monoclonal antibodies that recognize and latch on to interferon molecules. A mouse is injected with a mixture of materials containing a small amount of interferon. A few days after the injection the mouse's spleen is removed and its B cells, some of which will be producing antibodies which recognize interferon, are fused with cancerous myeloma cells to yield hybridomas. The hybridoma clones are separated from each other and tested to see which produces anti-interferon antibodies.

Interferon

The chequered history of interferon began in 1957 in London, where Alick Isaacs and Jean Lindenmann were investigating the body's responses to viral infections. They found that a substance secreted from infected cells helped other cells resist the effects of invading viruses. They named this substance interferon because it appeared to interfere with the spread of viral infections. Since then this enigmatic material has shuttled from prominence to obscurity and back again. Until recently, all attempts to elucidate the true nature and role of interferon foundered due to a lack of sufficient pure material for investigation. Now, interferon has caught the imagination of the newly emerged band of genetic engineers who have been able to provide larger quantities of interferon, which can be purified with the aid of monoclonal antibodies. Thus the way is now clear for a detailed evaluation of interferon's potential for the treatment of viral diseases and some cancers.

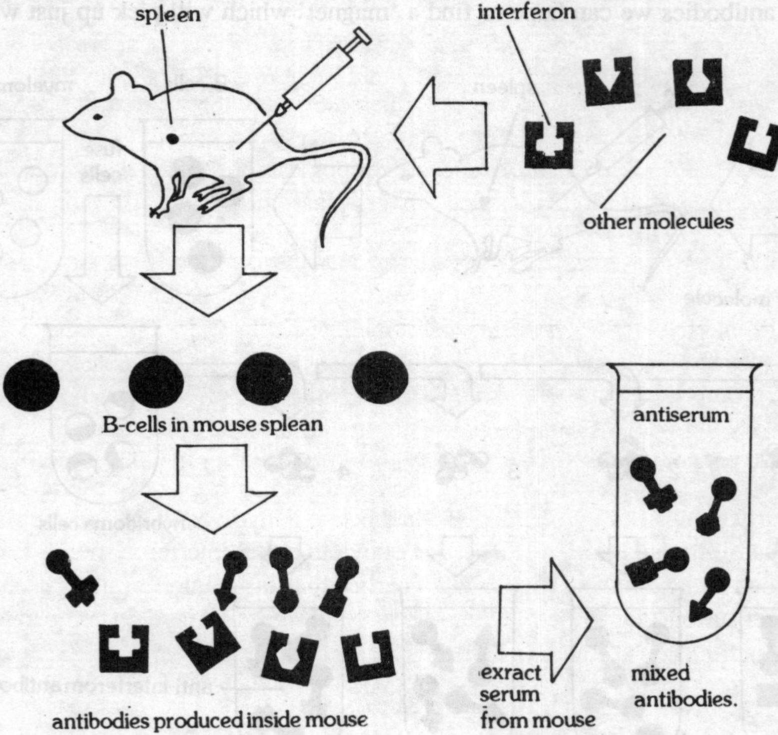

Fig. 3.3. The production of conventional antisera consisting of a mixed population of antibodies. Antisera have been used for many years in diagnosis and, sometimes, therapy. A mouse (or often a goat or a house) is injected with material that contains the substance which antibodies are needed against. Blood taken from the animal contains a certain amount of the desired antibody, mixed with many other unwanted materials. Since conventional antisera are obtained from live animals and are pure, the process is both costly and has many limitations.

Isaacs and Lindenmann were trying to find out why people suffering from an infection by one kind of virus rarely caught another type of viral disease at the same time. This fact intrigued them, for it was well known that one kind of bacterial infection often paves the way for another because the patient's defences are weakened. Why should viral infections be different? Isaacs and Lindenmann provided the clue when they discovered that, in response to an invasion by viruses, cells secrete interferon which acts as an alarm signal, alerting neighbouring cells to the invaders' presence and allowing them to prepare themselves against the imminent attack. It is now known that there are at least a dozen different kinds of interferon produced by human cells, all of which are proteins. Furthermore, the interferons produced by different species of animals are themselves different; interferon from mice has little effect on humans.

Until 1980 the sole source of human interferon was, not surprisingly, human cells. Most of the world's supply came from the laboratory of Kari Cantell in Helsinki, where vast quantities of white blood cells from donors were deliberately infected with viruses, stimulating them to produce interferon. The tiny amount of interferon manufactured by each cell and the difficulty of separating it from all the other materials meant that the blood from 90,000 donors provided only 1g of interferon in a form which was, at best, only 1 per cent pure. The scarcity of interferon and its enormous price, sometimes estimated at $50 million per g, constituted a severe brake on research. (The prices calculated for interferon were, in fact, rather specious since no-one ever possessed as much as a gram at any one time, let alone offered it for sale in such quantities. Nevertheless, the cost was extraordinarily high.)

Despite these difficulties, evidence for the value of interferon in the treatment of some viral diseases and, perhaps, some cancers did begin to accumulate. Then in January 1980 news from Switzerland sparked the interferon boom, both clinically and financially. Charles Weissmann announced that he and his colleagues in Zurich had cloned the gene for one type of human interferon and inserted it in to bacteria which promptly began manufacturing interferon. Other researchers soon followed suit and, within a year or so, the world's supply of interferon had increased dramatically, its cost had plummeted by about 90 per cent and, most importantly, very pure interferon became available at last.

By 1985 over thirty companies in the US alone had become involved in the interferon business and many countries had instigated clinical trials. The conviction that interferon merits very serious consideration has spread far beyond the small band of early enthusiasts. Interferon has been hailed as the new hope for cancer sufferers and, indeed, there are hopeful signs that it could come to play a part in cancer therapy. Much closer at hand is the use of interferon to prevent or cure several viral diseases.

Preliminary studies have revealed promising results in the treatment of rabies, hepatitis, shingles, various herpes infections and cytomegalovirus infections. The last of these is a very common virus which infects about half of the population at one time or another. It can produce fever, pneumonia and even death in new-born babies and already weakened adults. Experiments with monkeys have shown that interferon can also give protection against a lethal virus which attacks the heart muscles.

The best evidence of interferon concerns its use as a prophylatic rather than a cure. Researchers at Britain's Common Cold Research Unit have shown that interferon, given as a nasal spray, builds resistance to this all too familiar illness.

At present the cost of interferon is still far too high for it to be used against relatively trivial infections such as the common cold, even though in causes millions of lost working days. Many biotechnologists believe, however, that improved manufacturing methods may bring its price down dramatically and that it may be sold over the chemist's counter. In the USSR 'interferon' is already on sale to the public, but the dosages are so low that the preparations are almost certainly ineffective.

So far only the first few chapters of the interferon story have been written and, undoubtedly, the plot will take many twists and turns before the full potential of these intriguing substances is revealed. The challenges for the next few years are to find out more about the way they work, which types of interferon are best for each disease, and what harmful side-effects they produce. These are tasks for basic scientists and clinicians. Biotechnologists are trying to develop more effective and cheaper methods of supplying pure interferon for research and, eventually, for its widespread use in medicine.

Anti-viral Vaccines

In 1967 over 10 million people were infected with smallpox and the disease was endemic in more than thirty countries. Today, this hideous affliction has been wiped off the map. The World Health Organization's mass vaccination programme against smallpox is arguably the greatest triumph of modern medicine, and its success highlights the immense benefits which can accure from the development of effective vaccines against viral diseases. Polio, yellow fever, rabies and rubella (German measles) are just a few of the other viral diseases which can now be fought with vaccines. There remains, however, many widespread and serious viral diseases for which no cheap and effective vaccine exists, and biotechnologists are in the forefront of the crusade to bring these intractable diseases under control.

The US Office of Technology Assessment points to seven major viral diseases in humans for which vaccines are needed: hepatitis, influenza, herpes simplex, mumps, measles, common cold and varicella-zoster (shingles). The encouraging progress being made towards a vaccine against hepatitis illustrates many of the general themes of the biotechnological approach to viral vaccine production.

Three major forms of hepatitis exist: two are termed hepatitis A and B, and the third is known, rather unhelpfully, as non-A non-B hepatitis. Most attention has focused on hepatitis B, serum hepatitis. The virus that causes this disease has been identified and is responsible for thousands of serious liver infections and roughly 100 deaths a year in Britain, while in the US, the Center for Disease Control estimates that there may be as many as 150,000 cases a year, resulting in about 1,000 deaths. In other parts of the world the disease is even more common; in some African and Asian countries as much as ten per cent of the population suffers from hepatitis B. Furthermore, hepatitis is thought to be a major cause of liver cancer.

The standard method of producing anti-viral vaccines is to grow the virus either inside a suitable animal or, preferably, inside cells grown in the laboratory. The viruses are then collected and either killed or severely weakened before being injected into humans. In response to these foreign intruders, the body's immune system makes antibodies which will attack them. This takes time, but since the viruses are in a harmless form the delay does not matter; the dead or feeble viruses will do no damage while the immune system is gathering its strength. If, later, a live and fully active virus of the same type grains access to the body, the immune system is already prepared to stamp it out.

The crux of the problem is that there is often no convenient method for growing large quantities of a particular virus. Viruses are amazingly simple structures consisting of just a small piece of genetic material encased in a coat of protein. Peter Medawar, an eminent British immunologist, has summed up viruses as 'just bad news wrapped in protein.' Fortunately, as explained in the discussion of monoclonal antibodies, the immune system only recognizes foreign substances by their external features, in this case the viral protein coat. This fact is being used by genetic engineers as part of a plan to produce an anti-hepatitis vaccine. They have taken the genes that code for part of the hepatitis coat protein and introduced them into the bacterium, *E. coli*. The bacteria manufacture the viral protein without, of course, making any of the viruses themselves. If all goes well with the clinical trials now under way, this genetically engineered protein will be available as a vaccine within a year or two.

This general approach can be applied to the prevention of the spectrum of viral diseases. Apart from the human diseases mentioned earlier, some of the most rapid progress has been made in the veterinary area, most notably with a vaccine for foot-and-mouth disease. The first stage is to identify a characteristic feature of the particular virus' protein coat, and the exquisite powers of discrimination offered by monoclonal antibodies are invaluable here. Genetic engineers then clone the appropriate gene and persuade bacteria to synthesize it in large quantities. Among recent advances in this area is the cloning of a rabies virus antigen by scientists at the French firm of Transgene and several groups of researchers have turned their attention to the viruses that cause influenza, the common cold and herpes.

Genetic engineering not only promises to provide protection against diseases that cannot be fought by conventional vaccines, it should also mean safer versions of existing anti-viral vaccines. Viruses themselves will not be employed in these new vaccines but only one of their proteins made by bacteria, eliminating the risk that the vaccine could be contaminated with live viruses. This is something that happens extremely rarely now, but expensive and time-consuming procedures are needed to ensure that such an accident does not occur.

Aids

Even five years ago the term 'acquired immune deficiency syndrome' was unknown. Now AIDS is causing intense concern among doctors and the general public. The overwhelming reaction to AIDS has many causes: the unheralded emergence of an insidious and deadly disease has kindled a latent fear of the unknown; the immense social effects of the disease's spread in certain communities have no parallel in recent times; and the reaction from the press has been extreme — sometimes unreasoned.

When AIDS was first discovered in 1981 it was restricted almost entirely to a few specific groups, notably male homosexuals, intravenous drug users, haemophiliacs and Haitians; the majority of AIDS sufferes still come from these groups. The effects of AIDS very considerably, but the common theme is a drastic reduction in the patient's ability to fight off infections (immune deficiency). The debilitating effects of AIDS often lead to a previously rare type of cancer known as Kaposi's sarcoma, and to very severe pneumonia. Once the disease has taken hold, the chances of recovery from AIDS are very slim.

Against this frightening background it must be noted that AIDS is still a comparatively rare disease in most countries. By early '85 the UK, for example, had just over 100 cases. The incidence in the USA was much higher, with over 8,000 cases and nearly 4,000 deaths. The fact that the first cases of AIDS were diagnosed in the USA and that the spread of the disease in other countries seems to be following something like the US pattern, albeit with a time lag of two or three years, gives good reason to fear that it will become more common elsewhere. (It is now thought that a form of AIDS may be endemic in parts of tropical Africa, and perhaps it spread from there to the US and hence to other parts of the world.) One fact that is often overlooked in popular accounts of AIDS is that it seems that many people may be able to fight off the AIDS infections and do not develop the disease itself.

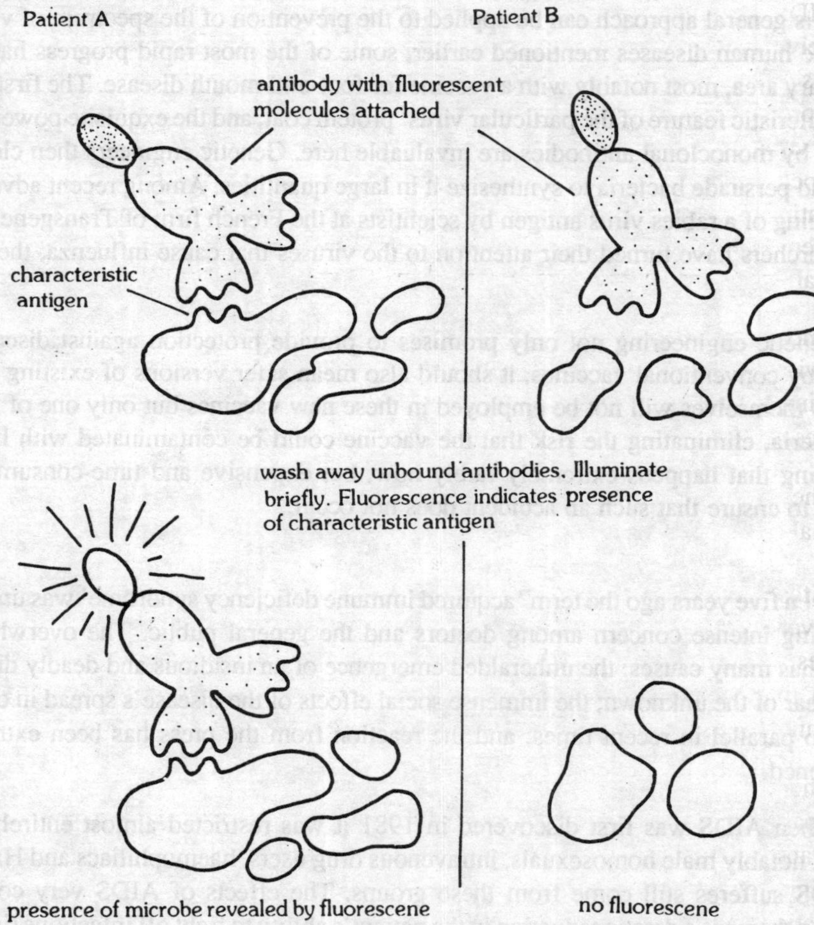

Fig. 3.4. Monoclonal antibodies in diagnosis. The presence of specific materials in samples taken from a patient will help doctors diagnose an illness. Microbes, viruses, proteins and many other substances all have characteristic antigens. Monoclonal antibodies which recognize specific antigens can be linked to fluorescent molecules which shine under ultraviolet light. These fluorescent antibodies are added to a sample of blood or other material taken from the patient. If the antigen is present, the antibodies will cling to it. The sample is then washed to remove unbound antibodies. Any light emitted after the washing process betrays the presence of the particular antigen for which the test has been devised. Fluorescent markers are just one of several ways of making visible the presence of antibodies. Radioactive labels can also be employed.

In 1984 scientists in France and the USA discovered that AIDS is probably caused by a virus, variously termed LAV or HTLV III, which can be transmitted by very intimate contact with someone carrying the disease. Although AIDS virus can be found in the saliva of sufferers there is no evidence that it can be transmitted by this means rather, it appears to pass from one person to another either by sexual contact or by contact with infected blood. Doctors and scientists have yet to find effective ways of treating or preventing AIDS, but two biotechnological approaches are receiving great attention.

AIDS has killed a number of haemophiliacs, and it is believed that they contracted the disease by receiving blood proteins contaminated with the virus. Haemophiliacs need to be given certain proteins (Factor VIII or Factor IX) to help their blood clot normally and prevent excessive bleeding. These proteins are extracted from donated blood which, if taken from someone infected with the AIDS virus, can pass on the disease. New methods for killing viruses in blood products are being developed, but biotechnologists are also planning to have clotting proteins manufactured by microbes, eliminating the possibility of viral infections. The first steps along the way to genetically engineered anti-haemophilia proteins have been made with the cloning of the genes for Factors VIII and IX and the small scale production of the proteins by bacteria.

Another potentially important advance came in 1985 — the unravelling of the genetic structure of the virus itself. As noted earlier knowledge about the protein coat of a virus (which is constructed according to the virus's genetic make-up) is a crucial part of developing genetically engineered vaccines.

These two themes of research into AIDS, along with current tests concerning the therapautic potential of interferons, are offering some hope that the disease can be contained.

Parasites

Over 1000 million people in the tropics live under the threat of painful and multilating parasitic diseases which exact a toll of millions of human death each year. Even more distressing is the fact that in many areas these diseases are becoming *more,* not less, common. The greatest benefit biotechnology could bring to humans would be to help lift the scourge of these diseases.

There is a clear possibility that genetic engineering will produce new and effective vaccines for many parasitic diseases, but in the near future monoclonal antibodies are likely to be of much greater importance. A major problem that faces all doctors, especially those working in ill-equipped hospitals and clinics, is to make accurate diagnoses of diseases. A battery of individual monoclonal antibody preparations which can be used to identify specific parasites would be of immense benefit. The production of monoclonal antibodies which recognize characteristic features of particular parasites is now beginning, and their use in simple tests should become possible within a very few years.

In 1975, the United Nations Development Programme, the World Bank and the World Health Organization set up a Special Programme for Research and Training in Tropical Diseases. Six diseases were targeted for a concerted attack, two of which, malaria and leprosy, are well-known to everyone. While the other four, filariasis, leishmaniasis, schistosomiasis and trypanosomiasis, may be less familiar, the threat they pose is just as great. The efforts of biotechnologists are showing signs of success in the treatment of three of these diseases, malaria, leprosy and leishmaniasis.

Malaria

In the nineteen-fifties it seemed that malaria would soon be conquered—perhaps not eradicated entirely, bit certainly stamped out as a major cause of death in many countries. By the seventies, such confidence had evaporated. Malaria had fought back against all that science and medicine could throw at it and the disease was re-establishing itself in 'clean' areas. It is now estimated that there are 300 million cases of malaria, and in Africa alone there could be as many as a million deaths a year.

The optimism was engendered by the initial success of a two-pronged attack on malaria: the malarial parasite itself was being killed with drugs; and DDT and other insecticides were controlling the mosquitoes which spread the disease from person to person. Increasingly, however, both the parasites and the mosquitoes have become resistant to the chemicals which previously killed them with ease. A new approach is urgently required, and the production of vaccines with the aid of biotechnology is now a real possibility.

The chills, fever and profuse sweating so characteristic of malarial infections are the body's responses to millions of tiny, single-celled organisms which have invaded the patient's red blood cells. The malarial parasites, *Plasmodia,* are not bacteria but belong to the group of organisms known as protozoans. Like our own cells, they have well-defined nuclei enveloping their DNA and are unaffected by the antibiotics which kill bacteria.

The malarial parasites are transferred from one person to another by female mosquitoes which feed on human blood. It is quite impossible to catch malaria directly from an infected person. *Plasmodia,* injected into the bloodstream by the mosquito's bite, migrate to the liver where they proliferate. Within a few days as many as 30,000 microbes burst out of each liver cell and invade red blood cells. There the *Plasmodia* grow and multiply until eventually they burst out in search of fresh cells to infect. Strangely, the millions of individual *Plasmodia* act in concert; they all stream out of infected cells at about the same time, repeating the process every two or three days. This phenomenon is reflected in the cyclical nature of malarial fever, in which periods of relative calm are punctuated with acute attacks.

One of the reasons for the malarial parasite's 'success' is that while inside blood cells it is hidden from the patient's immune system. However, it is exposed to danger during the brief movement from one cell to another. All too obviously, the human immune system often cannot meet the challenge. A good vaccine which 'teaches' the body how to make antibodies against malarial parasites before the infection occurs would make all the difference. People in high-risk areas would then be able to fight off the invading parasites as soon as they enter, while there are only a few to deal with. Without vaccination, the inevitable delay in mounting an immune defence gives the parasites time to build up their forces.

It was not until 1976 that William Trager, at New York's Rockefeller University, discovered how to grow *Plasmodia* in the laboratory. At first it appeared that this might lead to the production of anti-malarial vaccines. The standard method of growing large amounts of the parasites, killing them and injecting the remains into humans, was tried. Although initial tests on animals were encouraging a major, and possibly insuperable, obstacle intervened. The parasites were grown in the laboratory inside human blood cells. Unless all traces of the human cells are removed from the vaccine the body is likely to react against these as well as the parasites, and the results of this response could be dire.

Recently scientists — notably Ruth Nussenzweig and her colleagues at New York University — have made monoclonal antibodies which recognize a protein on the surface of *Plasmodia*. Mice and monkeys injected with these monoclonal antibodies resisted malarial infections. It is conceivable that this could form the basis of a passive immunization programme. Such immunization is termed passive because the protection depends on foreign antibodies injected into the body, rather than on antibodies made by the individual's own immune system, as in active immunization.

Far more exciting, however, is the possibility that large amounts of the parasite's protein could be made with the aid of genetic engineering. Pure samples of this protein could then be inoculated into human beings, making them immune to the malarial parasite.

Trypanosomes

The term trypanosomiasis covers several intractable, tropical diseases, each caused by a different species of the group of protozoans called trypanosomes. The two commonest forms of these diseases in humans are Chagas' disease, which afflicts 10 million people in South America, and African sleeping sickness, which claims 10,000 fresh victims each year. In Chagas' disease, *Trypanosoma cruzi* invades the heart, nervous system and gut, and it is especially fatal in children. The two species of trypanosomes responsible for sleeping sickness also attack the nervous system, and, unless treated, the disease in invariably fatal. The indirect effect of trypanosomes on human welfare is also great. Other species of trypanosomes infect cattle (killing three million a year), sheep and goats, preventing the inhabitants of vast tracts of Africa from rearing these animals for food.

It must be stated at the outset that biotechnology probably has little to offer in the short-term for the alleviation of this suffering. Nevertheless, there is one contribution biotechnologists might make, and a brief look at trypanosomes will also emphasize the huge challenge faced in attempting to overcome the natural cunning of parasitic organisms.

The work of immunologists, and, more recently, molecular biologists studying the genes of these organisms, has revealed a truly remarkable strategy for avoiding the attentions of the human immune system. Trypanosomes are masters of disguise. The antigens on the surface of most cells are relatively constant — those found one day will probably still be there the next. Trypanosomes are quite different; they can call upon a well-stocked 'wardrobe' of antigens with which to dress themselves. No sooner has the patient's immune system learned to recognize a trypanosome by one of its antigenic outfits, than the trypanosome done another set of antigens plucked from its store. The whole process of identifying the intruder must start a new, and so on almost indefinitely. Consequently, the immune system is never able to mount an effective campaign against these chameleon-like organisms.

This behaviour results from the trypanosomes' ability to switch on and off an array of different genes, each coding for different antigens. It rules out any reasonable hope of producing vaccines against the disease, for dozens of separate vaccines would be needed, one for each of the organism's many guises. Perhaps the only good thing that can be said for these detestable creatures is that through studying them a great deal has been learned about the way genes can be manipulated, and this knowledge may yet underpin new applications of genetic engineering.

However, biotechnology might help combat sleeping sickness by the production of two types of protein. Recently an anti-cancer drug, daunorubicin, was found to be active against trypanosomes if the drug is first linked to albumin or ferritin, two proteins found in human blood. The supply of these proteins is limited at present. If they were to form part of treatment for sleeping sickness, biotechnologists would undoubtedly be called upon to produce the required amounts, probably by inserting the appropriate genes into bacteria and purifying the protein thus manufactured.

Leprosy

Other diseases may have attracted more folk tales, but few are as well-entrenched as those surrounding leprosy. One is that leprosy is highly contagious; in fact, about 95 per cent of the population is probably immune to the disease and even susceptible individuals will only catch it after long-term exposure to the infection. Another myth claims, wrongly, that leprosy causes parts of the body to drop off. It does, however, produce numbness in the extremities of the body, making accidental damage more likely. Finally, while it is true that the overwhelming majority of lepers live in tropical countries, cases of the disease are by no means unknown elsewhere. For example, there are over 400 cases in the US among recent immigrants, and only a century ago leprosy was still found in northern Norway.

Leprosy is caused by a bacterium called *Mycobacterium leprae,* and about 11 million people are affected by the disease. Until recently the chemical dapsone formed the basis of the effective treatment, but the alarming rise in bacteria resistant to dapsone demands new types of therapy, at much greater costs to the poor nations where leprosy is common.

One of the principal aims of the Special Programme mentioned earlier is to develop reliable skin tests to identify people at particular risk from the disease. Very few animals can become infected with leprosy bacteria, a surprising exception being the nine-banded armadillo. Purified bacteria from armadillos are being used in small-scale tests, but this may be impractical for general use. Obviously a vaccine is needed, but there is, as yet, no way of growing the bacteria in the laboratory so that a vaccine can be manufactured. Unfortunately, there seems to be little research devoted towards this important challenge.

Liposomes

An entirely new approach towards more efficient drug delivery is now proving its worth in the treatment of one of the most dangerous parasitic diseases, leishmaniasis, and is rapidly becoming a prominent feature of many areas of biotechnology. Tiny globules of fat-like material, known as liposomes, can deliver potent poisons to cells infected with leishmania parasites. Since many millions of people are afflicted with these microbes, liposomes could clearly make a great contribution to human welfare.

The membranes which form such a vital part of all cells contain large amounts of materials called phospholipids. The many different types of phospholipids share the same basic structure—two long 'tails' composed of fatty materials and a 'head' containing nitrogen, oxygen and phosphorus atoms. Most people have seen how fatty or oily liquids tend to accumulate into droplets when powred on water, rather than spreading evenly across the surface. Substances which behave in this manner are called hydrophobic (water-fearing). The fatty tails of phospholipids (the lipid parts) are hydrophobic and shun contact with water but the phospholipid heads avidly seek out the company of water molecules and so are hydrophilic. Figure 3.5 illustrates one very efficient way in which these two opposing inclinations of a phospholipid molecule can be reconciled.

In the nineteen-sixties, scientists capitalized on this natural phenomenon by learning how to assemble phospholipids into hollow spheres called liposomes. Some consist of just one sphere, while others are made up of several phospholipid shells, one inside the other rather like Russian dolls. The space in the middle can be used to entrap all manner of materials, including drugs. In the last few years there has been much research into the advantages of wrapping drugs inside liposome packages, and the treatment of leishmaniasis is one of the most promising applications.

Leishmaniasis afflicts roughly 100 million people in South and Central America, Africa, Asia and parts of Southern Europe. The disease takes several forms, each associated with a different species of protozoans called *Leishmania*. The most devastating is kala azar in which *Leishmania donovani* invades the spleen, liver and bone marrow, causing almost certain death unless properly treated. An epidemic of kala azar spread among over 70,000 people in the Indian State of Bihar in the late seventies, killing more than 4000.

There is no effective vaccine against kala azar, and the sick are usually treated with compounds which contain antimony. Antimony compounds are similar to arsenic compounds, and so it is not surprising that there is a very real danger of poisoning the patients rather than curing them. Although modern antimony drugs are safer than those used some years ago, the basic problems remains — antimony does human cells no good at all.

Liposomes offer a major advance in terms of safety, for antimony drugs are up to 700 times more effective if they are first encapsulated inside liposomes. This does not mean that the patients are cured that much more rapidly, rather that only a fraction of the normal dose of the drug need be given to the patients, greatly reducing the risks. This trick seems to work because liposomes injected into the bloodstream tend to be most rapidly absorbed by the same types of cells that *Leishmania* parasites colonize. Thus, liposomes laden with antimony drugs deposit them predominantly in the liver and spleen where they are needed.

Obviously this method of delivering drugs might be extended to other diseases which primarily affect these organs, and current research indicates that this is a real possibility. However, the potential of liposomes does not end here. In this example liposomes tended to home in on liver and spleen cells only, and this would appear to limit their use. Fortunately, it does seem possible to make liposomes which are targeted against different types of cell. To some extent this can be achieved by altering the composition of the liposomes themselves, selecting different sorts of phospholipids for example.

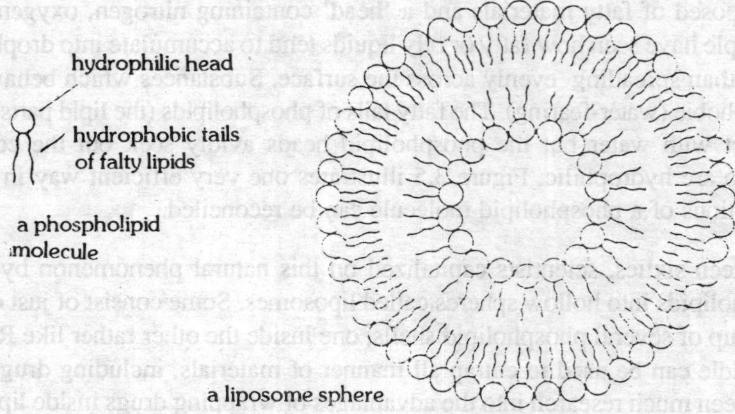

Fig. 3.5. Phospholipids and liposomes. A phospholipid molecule consists of two hydrophobic "tails" of lipids, and a hydrophilic 'head'. When large numbers of phospholipids are mixed with water they assemble themselves into spherical structures called liposomes. The cross-section of a liposome shows two concentric spheres of phospholipids with a water-filled space in the centre.

It is most unlikely that the specificity of liposomes can ever approach that offered by antibodies. It turns out, however, that these two entirely distinct tools can be combined, perhaps giving the best of both worlds. Liposomes loaded with drugs can be linked to antibodies which recognize specific microbes, and so the antibodies provide the guidance system, while the liposomes carry the 'warheads.'

Liposomes are particularly well suited to transporting substances about the body. They have been used in experiments to introduce a wide range of materials into animals, including antibiotics, enzymes, insulin and even compounds which mop up plutonium. Perhaps their most important feature is their ability to protect drugs and other materials from destruction by the hostile environment of the body. The liposome shell prevents the body's enzymes from attacking the drugs before they can do their work. Furthermore, liposomes are non-toxic and do not usually provoke a response from the immune system.

Sticky DNA and Medical Diagnosis

Quick and accurate diagnosis is the first and crucial step in combating disease. Sometimes a patient's symptoms are so clear and characteristic of a particular disease that doctors have little trouble in making the correct diagnosis. In other cases, a certain set of symptoms could have a number of different causes; identifying the disease might then depend upon a battery of slow, expensive or unreliable tests.

Monoclonal antibodies are the best developed of the new tools that biotechnologists can employ to pinpoint specific diseases but an even newer technique that has grown out of genetic engineering research may prove yet more powerful. Short pieces of DNA called probes can be designed to stick very specifically to certain other pieces of DNA. This fact underlies the many applications of DNA probes. They can be used, for example, to find out if a patient's blood contains the DNA of a particular bacterium, and hence provide a diagnostic clue. In principle and probably in practice — it will become possible to design probes which can be used to look for almost any type of DNA that interests scientists or doctors. The implications of DNA probes are profound. Not only might all manner of microbial infections be identified but probes could also indicate genetic defects, detect contamination in donated blood, help in matching organs for transplantation and even assist seed suppliers test the quality of their seeds. All this possible because DNA probes can provide a spotlight on the basic genetic structure of organisms.

The technique depends on the fact that two strands of DNA which have 'complementary' sequences of bases will stick together. To check if a sample contains a particular type of DNA (say, from a microbe) a biotechnologist will mix a DNA probe into the sample. The probe has been designed so that it is complementary to some part of the DNA which is being sought out. If the probe finds its opposite number, the two link together, and a range of methods can be used to check if this has happened. If the 'target' DNA is absent then the test gives a negative result.

DNA probes are claimed to be a thousand times more sensitive than conventional diagnostic methods which rely on mixtures of antibodies extracted from animal blood, and the market value of diagnostic DNA probes has been predicted to reach $1 billion a year in the US alone by 1992. The sexually transmitted disease gonorrhea is one of the prime targets for DNA probe manufacturers. More than ten million tests for this disease are carried out annually in the US, each costing about $25. The current tests usually involve growing the microbes in a laboratory for sometime. DNA probes might be quicker and cheaper; they may even tell doctors if the particular infection involves a strain of microbe that is resistant to certain antibiotics.

Similarly; DNA probes could be used to check that blood destined for transfusion is not contaminated with hepatitis viruses or other harmful substances, or that food is free from microbial contamination.

DNA probes are now begin developed for a multitude of purposes, including the selection of organs for transplantation and the diagnosis of some genetic diseases discussed in the next section.

Correcting Nature's Errors

As well as combating diseases caused by viruses, bacteria and protozoans, biotechnology can also help cure disorders which result from imbalances or defects in the body's chemistry, including some inherited diseases. It is now possible to manufacture an enormous range of natural compounds which can or might relieve illnesses such as diabetes, arthritis, bone disorders and nerve damage, as well as producing pain killers and materials to treat burns and wounds.

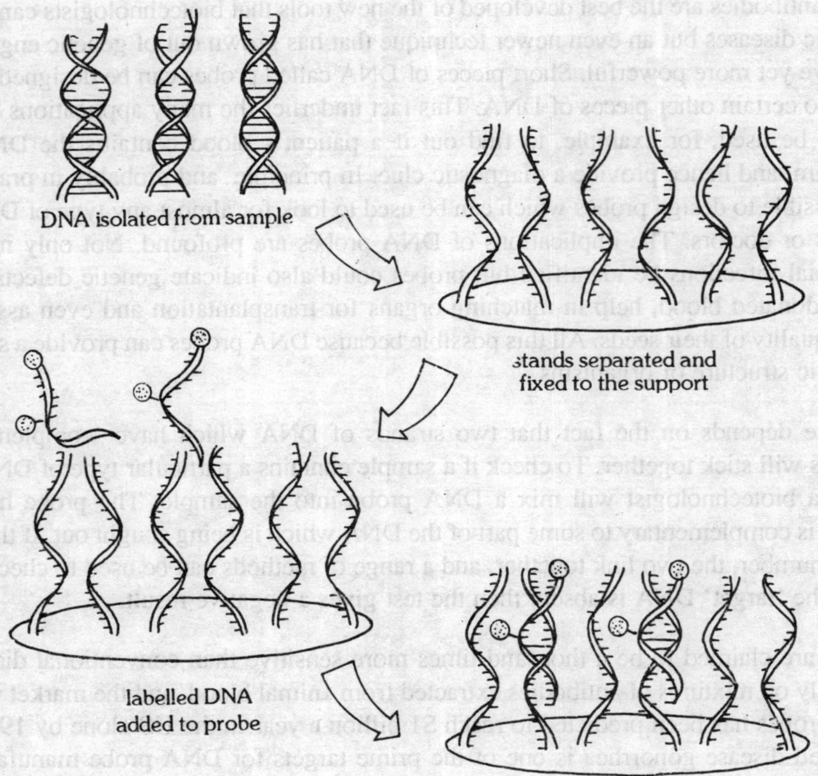

Fig. 3.6. Double stranded DNA is extracted from the sample (e.g., microbial cells) and fixed to a support. Chemicals are added to separate the strands. Single strands of DNA designed to be complementary to some characteristic part of the sample DNA are introduced. These strands (the probes) have previously been 'labelled' with some material that can be readily detected (e.g. radioactive atoms or compounds that give off light). The probes attach themselves to the sample DNA and any that remain free are then washed away and the amount of labelled probe still fixed to the support via the sample DNA is measured. If the sample DNA is not the type the probe has been designed to latch onto then no probes will remain on the support and the test gives a negative (but still informative) result.

Hormones

Our bodies contain a dazzling array of different hormones, each with a vital task in co-ordinating the activities of individual cells and tissues. Without hormones, cells would have little idea of the overall state of the body and could not know what tasks were required of them to keep the whole organism functioning properly. Hormones convey all sorts of information—for example, that food is available for digestion; that danger threatens and muscle cells must prepare for action; and that a child is developing in the womb.

Not surprisingly, a defect in one of the hormone systems can produce serious illnesses. These can sometimes be remedied by supplying the patient with the right amount of a particular hormone, and biotechnology can help in its manufacture.

Many of the body's hormones are composed of chains of amino acids. The term protein is generally reserved for chains of about twenty or more amino acids and some hormones have very short chains. Therefore, the term polypeptide hormone is usually employed for all hormones made up of amino acids (after the chemical link that joins two adjacent amino acids, called a peptide bond). The best-known polypeptide hormone is insulin and a great deal is understood about how it maintains the correct amount of sugar in the bloodstream, and how some types of diabetes can be treated by supplying patients with extra insulin. However, there are many other important polypeptide hormones, the names of which are likely to become more familiar as they come to be widely used in medicine through the efforts of the genetic engineers who will help increase their supply. They include growth hormone, a deficiency of which causes some forms of dwarfism, and nerve growth factor, which could help in the treatment of injuries and restore nerves after surgery.

Genetic engineers are turning their attention to three categories of substances with proven or possible medical applications, and polypeptide hormones provide examples of all these. First, there are substances for which we already have a source of supply, but cheaper and better materials might be made by microbes (for example, insulin). Second, there are substances with a known medical value whose supply is inadequate to meet the present demand (for example, growth hormone). Finally, there are substances which may prove to be very useful, but larger quantities are needed before their effects can be tested (for example, nerve growth factor).

Until the advent of genetic engineering all the insulin given to diabetics was extracted from the pancreas of cattle or pigs. It is estimated that there are 60 million diabetics in the world, more than half of whom live in developing countries where they are rarely diagnosed or treated. About one-third of the approximately 15 million known diabetics in the developed world are currently receiving an average of 2 mg of insulin a day. Insulin derived from animals costs about $ 75 per kg to produce, but one of the world's leading manufacturers estimates that this might be reduced by two-thirds if it could be made on a large scale by genetically engineered microbes. (This dramatic decrease in the cost of producing insulin would not, however, be reflected in an equivalent reduction in the cost of the drug to the patient, since many other factors contribute to this, such as packaging and distribution, which would remain unchanged.)

The intense interest in genetically engineered insulin rests largely on two factors. First, insulin is a test case of whether genetically engineered microbes can be efficiently incorporated into the pharmaceutical industry. Second, there are hopes that this insulin may be better and safer than animal insulins. The insulin molecules made by cattle and pigs are very slightly different from those made by humans. The insulin molecule consists of fifty-one amino acids, and at one point the pig insulin has a different amino acid compared to the human version, while cattle insulin differs at three points. These small variations mean that some patients eventually suffer allergic reactions to the animal hormones. Insulin produced by microbes is exactly the same as human insulin. It will take time to discover whether this makes a substantial difference in practice, but theoretically it should bring a benefit to some diabetics. Human insulin may also prevent or reduce some of the long-term effects of diabetes, which include damage to the retina leading to blindness and kidney damage.

The major problem in the case of growth hormone is lack of supply. A considerable proportion of children who fail to grow to a normal height do so because their pituitary glands fail to produce enough growth hormone. If the lack is diagnosed early enough, such children can be given extra growth hormone extracted from cadavers. The therapy involves an injection of about 7 mg of hormone a week for several years, while a single pituitary gland provides only about 4 mg. In 1981, 800 children in the UK could have benefited from such treatment, but there was not enough of the hormone available to give them all as much as they needed. In the US, about 1600 children are receiving growth hormone .

Human growth hormone produced by genetically engineered bacteria is currently on clinical trials with over a hundred patients in the USA, and the firms Genentech in California and Kabi-Vitrum in Sweden expect to market it in the near future. A single 500 litre (110 gal.) vessel filled with these bacteria can provide the equivalent of no less than 35,000 human pituitary glands. Once growth hormone is available in relatively large quantities it can be tested for its effects on other disorders, including senile osteoporosis (a loss of calcium from bones), some bleeding ulcers, and aid healing after burns, wounds and bone fractures.

The final group of polypeptide hormones contains less familiar but equally vital substances, including somatostatin, calcitonin, cholecystokinin, bombesin, vasopressin, parathyroid hormone, nerve growth factor, adrenocorticotropic hormone and erythropoietin. Some are well understood and for these there is already a clear medical need for better supplies to treat specific diseases. Others are more enigmatic. Only when they are all available in large quantities, in a pure form, will their true significance be revealed and their potential medical applications become apparent. The medical and scientific communities are now looking to genetic engineers to provide these materials.

Considerable progress has already been made in some, cases. For instance, somatostatin, a hormone found in the pancreas, the brain and elsewhere, was the first human molecule to be obtained from bacteria through genetic engineering — by a team of reseachers in California in 1977.

At present these hormones are obtained from four sources; animals (in which case they may not be identical with the human hormones); human organs, blood or urine; purely chemical manufacturing processes; or cells grown in laboratories (cell or tissue culture). Each method has its disadvantages, for example, cost, limited supply, purity. Genetic engineering, while not without its own problems, is highly likely to prove better than any other technique for making significant quantities of at least some of these materials. The following is a brief survey of some of the present and potential medical applications.

Bone disorders affect large numbers of people — up to 3 per cent of Western Europeans over forty, according to one estimate. Among the most important factors in keeping bones healthy is the maintenance of the correct concentration of calcium in the body. Calcitonin is one of several hormones whose job is to keep the calcium balance right, and a certain types of bone disease extra calcitonin can be beneficial. Calcitonin consists of thirty-two linked amino acids and can be manufactured by chemical means. It is, in fact, the largest polypeptide manufactured in this way for pharmaceutical use. This chemical production method might be superseded if a biotechnological process, involving genetic engineering, can be developed which makes either a cheaper or a purer product. Parathyroid hormone may also find a role in the treatment of bone disorders, either on its own or in combination with calcitonin. These are two of several hormones synthesized in the thyroid gland.

By contrast, adrenocorticotropic hormone (ACTH), a slightly larger polypeptide of thirty-nine amino acids, is not manufactured chemically ACTH is produced in the pituitary gland and is one of the most important hormones, controlling the actions of many others. It is obtained from animals and can be used to treat inflammations.

Cholecystokinin and bombesin both reduce the appetite. While the ideal solution for most overweight people is to eat less without the aid of medicaments, these two hormones would doubtless find a ready market if manufactured commercially.

Erythropoietin controls the development of blood cells and it could possibly be used to treat burns, bleeding and other conditions involving the blood.

At a more speculative level, there is some evidence that a very small polypeptide (only seven amino acids long) called MSH/ACTH 4-10 affects concentration and memory. *If* this is the case *and* there are no serious side-effects, then great interest would be shown in the material.

Endorphins

'Among the remedies which it has pleased Almighty God to give to man to relieve his sufferings, none is so universal and so efficacious as opium.' Thomas Sydenham, writing in 1680, might have added that few drugs are also so dangerous. Morphine, the major active ingredient in opium, and heroin, its close relative, are two of the most potent pain-killers known. Why should these and similar chemicals derived from poppy seeds produce such profound effects on the human body? The key to the enigma began to be revealed in 1975 when John Hughes, Hans Kosterlitz and their colleagues in Britain, discovered that two small chemicals inside pigs' brains had an effect very like opiate drugs. This initiated a flurry of research which may lead to the production of highly effective and safe pain-killers. Subsequent work in Britain, the US and elsewhere showed that there are a number of other hormones which mimic the effects of morphine and similar drugs. First to be discovered were the enkephalins, each of which consists of only five linked amino acids. Then came the endorphins, rather larger molecules, which were given their name because they are endogenous (growing from within), morphine-like substances. One of the endorphins, dynorphins, is 1200 times more potent than morphine.

The enkephalins and endorphins undoubtedly have many other effects on the body apart from deadening pain. The interaction between them is complex and subtle, but knowledge about them is accumulating rapidly. In theory, these substances should be safer than drugs manufactured chemically or extracted from plants, and this has inspired many scientists to investigate them more thoroughly.

Already at least one firm, Endorphin Inc. of Seattle, is looking at the medical potential of these substances and they have contracted researchers at University College, London, to clone the gene for pancreatic endorphin. Most interest has centred on the endorphins made by the pituitary gland, but it seems that, when injected, these endorphins do not easily enter the brain in an active form, whereas endorphins manufactured by pancreatic cells do appear to reach the brain when given by injection.

Again, it must be stressed that this kind of research is still in its very early stages and a medically approved product is probably years away. In particular, it is very possible that endorphins and enkephalins will prove addictive, just as morphine is. This problem and the need to target the drugs, might be tackled by chemically modifying them.

Steroid Hormones

While genetic engineering is the driving force behind the production of polypeptide hormones, biotechnology has already made a substantial contribution to the production of the quite different steroid hormones. Twenty years before genetic engineering became possible, a mould had brought relief to millions of people through the manufacture of cortisone.

The official birthplace of cortisone was an unusually august spot for a chemical compound — the ballroom of New York's Waldorf Astoria Hotel. In April 1949, the Mayo Clinic announced that a new medication for rheumatoid arthritis had been discovered, and it was known by the humble name of compound E. Very soon vitamin E began to be peddled as a cure for arthritis, despite the fact that its most notable connection with compound E was the presence of the same capital letter! In July, compound E was renamed cortisone and its illustrious career was launched.

The demand was immense, but supplies were extremely limited, and chemists set to work to find a way of making cortisone in bulk. They succeeded, but the process was complicated, entailing thirty-seven separate chemical reactions, and the cost of cortisone was accordingly high, \$200 per g. Help, as so often, came from a microbe. In 1952, it was found that the bread mould, *Rhizopus arrhizus*, could convert another steroid, progesterone, into a compound from which cortisone could be made much more easily. The price of cortisone dropped to \$6 per g, and by 1980, further improvements had reduced the cost to \$0.46 per g. This was possible not only because the microbes circumvented a number of steps in the synthesis of cortisone (only eleven are now required) but also because the high temperatures and pressures, and expensive chemical solvents required for the chemical method of production could be avoided.

The case of cortisone illustrates very clearly that biotechnology and the methods of the chemical industry should be not viewed as all-or-nothing alternatives. The combination of chemical and biological processes is likely to play an ever-more important part in industry over the next decade.

The basic raw materials for the manufacture of cortisone and other steroid hormones are sterols, a group of compounds which are usually obtained from soyabeans and the Mexican barbasco plant. As part of the process of making oestradiol and testosterone (both used in contraceptives) and other medically important steroids, two or more microbes are drafted in to make precise modifications to chemicals which are eventually converted into the final products.

Used with caution, steroids help alleviate inflammations, skin diseases, allergies and many other illnesses. Without microbes, many of these drugs would be totally unavailable or exceedingly expensive.

Blood Proteins

Human blood may appear to be a homogenous liquid, but in fact it has dozens of different components. There are several different sorts of cells carried by the blood; including red cells which transport oxygen and various cells which form part of the body's immune system. As well as these cells, there are many sorts of proteins. Some, such as Factors VIII and IX help maintain the blood's clotting ability, but the commonest blood protein in albumin. In 1979, about 90,000 kg of albumin were used during surgery and to treat burns and shock. This figure is expected to be more than doubled during the mid-'80s. At present, human albumin is separated from donated blood, but the gene was cloned in 1981 and before long vast quantities of albumin are expected to be produce from genetically engineered microbes. With plans to produce as much as 10 tonnes a year from microbes, albumin looks set to become the first large-scale medical product of genetic engineering.

Enzyme-replacement Therapy

Each of us carries about half a dozen defective genes. We remain blissfully unaware of this startling fact unless we, or one of our close relatives, are among the millions of people who are afflicted by a genetic disease. Roughly one in ten individuals has, or will later develop, some inherited disorder, and approximately 2800 specific conditions are now known to be caused by defects (mutations) in just one of the patient's genes. Some single-gene diseases are fairly common—cystic fibrosis is found in one out of every 2500 babies born in Britain—and, in total, diseases which can be traced to single genetic defects account for about 5 per cent of all admissions to children's hospitals.

The reason why most of us do not suffer harmful effects from our several faulty genes is that each person has two copies of nearly all their genes, one derived from their father and one from their mother. The only exceptions to this rule are the genes found on the sex chromosomes in males. Males have one X and one Y chromosome, the former from the mother and the latter from the father, so each cell has only one copy of the genes on these chromosomes. In many cases, one good gene is sufficient to avoid all symptoms of disease. If the potentially harmful gene is 'recessive' then its normal counterpart will carry out alone the tasks assigned to both. Only if we inherit from our parents two defective copies of the *same* recessive gene will a disease develop. Since our handful of gene defects are spread through thousands of different genes it is rather unlikely that you and your partner will carry the same faulty genes. So long as this unfortunate combination is avoided, your children will not suffer from recessive genetic diseases.

If the effective gene is 'dominant', it alone can produce the disease even if its counterpart is normal. Obviously only children of a parent with the disease can be affected, and even then on average half the children will be unaffected. Huntington's chorea, a severe disease of the nervous system which only becomes apparent in adulthood, is an example of a dominant genetic disease.

Finally, there are X-linked genetic diseases, in which the defective gene lies on the X chromosome. As males have only one copy of the genes on this chromosome no others can step in to fulfil the defective gene's function. Both Duchenne's muscular dystrophy and haemophilia are X-linked.

Much research has focused on the so-called genetic metabolic diseases, in which a defective gene causes a particular enzyme to be either absent or ineffective, thus upsetting the delicate balance of the body's chemistry. Biotechnology can sometimes provide the missing enzymes.

One of the best-known inherited diseases is haemophilia, a disorder in which an individual's blood fails to clot properly. Queen Victoria was the most famous carrier of the defective gene responsible for this disease, and through her it was transmitted to the royal families of Russia, Spain and Prussia. Minor cuts and bruises, which would do little harm to most people, can prove fatal to haemophiliacs because they lack substances which promote the chemical reactions by which molecules in the blood link together to prevent further bleeding. Many haemophiliacs are now treated with proteins known as Factor VIII and Factor IX which are extracted from the blood of donors; both of these materials have now been obtained from genetically engineered microbes.

Gaucher's disease is one of the ten known hereditary lipid storage diseases, most of which cause extensive damage to the nervous system. Children suffering from this disease lack the enzyme which normally breaks down one of the many forms of lipids in the body, glucocerebroside. This material accumulates in the liver, spleen and bones, producing swellings, damage to nerves and sometimes, death before the age of two. If the right enzyme could be delivered to the cells engorged with this lipid then the symptoms of Gaucher's disease could be relieved, and this is exactly what has been attempted. One of the scientists involved in this research, Roscoe Brady of the US National Institutes of Health, has said, 'I think our technology is on the verge of giving us the solutions we need. But until we can produce the required enzymes in much greater quantities and target them to specific areas, replacement therapy will have to be considered experimental." Biotechnologists have developed a number of techniques for producing quantities of biological materials such as enzymes and targeting them.

At present the enzyme needed to combat Gaucher's disease is purified from human urine or placenta where it is present in very small quantities. It is time-consuming and expensive to collect sufficient enzyme even for clinical trials, let alone the full-scale therapy of many patients. Genetic engineers could solve this problem by persuading bacteria to manufacture it.

The specific biochemical problem, usually the lack of an enzyme or other protein, has been identified for about 200 of the known metabolic diseases. It is not yet possible to say how many can be treated by enzyme-replacement therapy, but during the next decade the picture should become much clearer. However, it is virtually certain that without the aid of biotechnology, doctors would have little hope of obtaining pure enzymes in large quantities at a reasonable cost—two of the prerequisites of any successful therapy of this type. We can be confident that as soon as the effectiveness of a particular enzyme-replacement therapy is demonstrated, biotechnologists will be able to meet the demand for the desired enzyme.

Gene-replacement Therapy

Even the ambitious plans for enzyme-replacement therapy pale in comparison with gene replacement, one of the most radical notions ever put forward in medicine. Since inherited metabolic diseases result from absent or faulty genes, the ultimate cure would be to provide the patient with the correct genes which will enable the body to make the required enzymes or other proteins and so eliminate the root causes of the disease.

Despite the wide publicity gene-replacement therapy has received, the day when it becomes common is far away. Although genetic engineers can merrily stitch all manner of genes into bacteria, gene-replacement therapy-like medicine in general — aims to affect something vastly more complex and subtle than mere cells; it is concerned with healing a whole individual consisting of billions of semi-independent cells.

Each one of the cells in the body carries the same genes, but different cells make different uses of the enormous variety of genetic information available to them. For example, skin cells have insulin genes, but do not use them; that task is left to specialized cells in the pancreas. Not merely do cells from different organs take on specialized roles, but the role of a specific cell can change with time. Most notably, the many changes that take place at puberty result from a whole panoply of genes being switched on or off in the cells of several organs.

The major practical obstacle barring the way of effective gene therapy is that there is, as yet, no sure method of introducing genes into human cells in such a way that they are subjected to the body's normal control systems. Clearly, if the gene remains switched off nothing has been accomplished. Equally, if it becomes too active — instructing the cell to make excessive amounts of a particular protein — the consequences could be as damaging as those of the disease it was intended to cure.

The only attempts at gene replacement in humans were made in 1980, when Martin Cline from the University of Los Angeles went to Italy and Israel to attempt to cure two patients suffering from thalassaemia, a severe and usually fatal blood disease. There are several different types of thalassaemia, all caused by defects in the gene coding for globin, a vital part of the haemoglobin molecule which gives blood its red colour and carries oxygen around the body.

Red blood cells, which are rich in haemoglobin, are derived from bone marrow cells. Accordingly, Cline and his colleagues removed some marrow cells from their patients and tried to introduce a normal globin gene into them (using vectors analogous to the plasmids employed in the genetic engineering of microbes) in the hope that this would take over the function of their own defective genes, allowing the patients to make normal haemoglobin. As globin genes were the first human genes to be cloned, there was no difficulty in obtaining normal genes for insertion into the marrow cells. After these cells had been genetically manipulated, they were replaced into the patients' bones.

These operations, however, failed, possibly because the genes entered the wrong part of the patients' chromosomes and did not begin making mRNA, the first step in the synthesis of a protein. Indeed, the endeavour provoked fierce controversy, since most experts considered the attempt at gene replacement to be very premature. These cases do, however, illustrate the principles behind gene replacement, which may eventually find a place in medicine.

Gene replacement is an ideal subject for armchair speculation. Is it possible that human beings could be designed to precise specifications? In most important respects the answer is, almost certainly, no, and for this we should doubtless be grateful. While it is quite conceivable that single genetic defects might be remedied, it is extremely unlikely that more substantial changes could be wrought on an individual's genetic make-up. It might be possible to correct a few of nature's errors, but there is little point in thinking we can improve on her best efforts, which are exemplified by the fortunate majority of us who do not suffer from a serious inherited disease. The manipulation of 'genes for intelligence' is pure fantasy. There is not even good evidence that what we choose to call intelligence (as measured through IQ tests) depends primarily on inherited factors. Even that part of intelligence which is genetically determined will almost certainly be a result of complex interactions between a multitude of discrete genes. The task of understanding such an intricate system is fearsomely difficult; the prospects for being able to tinker with it in a specific way are vanishingly small.

It seems, however, that the simplest type of gene therapy trials may be initiated within a year or so. Several scientists, especially in the USA, believe that it will soon be feasible to introduce normal genes into patients suffering from genetic disorders. The most likely candidates for this type of treatment will be rare diseases such as Lesch-Nyhan syndrome or adenosine deaminase deficiency (ADAD).

Lesch-Nyhan syndrome patients are unable to manufacture an enzyme known as HPRT and this leads to a bizarre impulse for self-mutilation including very severe biting of the fingers and lips. ADAD has a profound effect on the patients, immune system, usually with fatal results. The normal version of the defective genes involved in both these illnesses have now been cloned.

Although these are likely to be the first diseases tackled by gene therapy, both are very rare. There are only a couple of hundred new cases of Lesch-Nyhan syndrome each year in the USA, while no more than 50 cases of ADAD are known in the world. If gene therapy does become practicable its biggest impact is likely to be in the treatment of diseases in which the normal gene needs to be introduced into only one organ—that is, in cases where even in healthy people that gene is only expressed in cells of a particular tissue. Two such diseases are phenylketonuria (PKU) and sickle-cell anaemia.

PKU affects around one in 12,000 white children, and if untreated they become severely mentally retarded. The disease is caused by a defect in a gene which normally produces an enzyme in liver cells. PKU can be diagnosed soon after birth, and its worst effects are avoided if the children are given a special, and highly unpalatable, diet for their first few years.

Sickle-cell anaemia is widespread, predominantly among the black populations of tropical Africa and the US, and also in parts of the Middle East and Mediterranean areas. The disease is named after the characteristic shape of the patient's red blood cells. Like the thalassaemias, sickle-cell anaemia is caused by a fault in globin genes. Here again the hope is that bone marrow cells can be genetically engineered to enable the patients to manufacture normal globins, thus eliminating the excruciating pain and premature death associated with the disease.

Needless to say, before therapy for a genetic disease can begin—either by gene replacement or by enzyme replacement—doctors must have an accurate diagnosis of the precise genetic defect in a patient. Here DNA probes are sure to have a major impact. Most of the present uncertainties in diagnosis should be removed by designing probes which can distinguish between defective genes and their normal counterparts.

In the next decade or so it should become clearer whether gene-replacement therapy will have a significant rule to play in twenty-first-century medicine. Until then we must continue to rely on conventional treatments and genetic counselling, in which couples who both carry the same deleterious genes are advised on their chances of producing affected children.

Scourges of the Affluent

While most of the diseases discussed so far in this chapter tend to afflict all age groups, but often the young in particular, cancer and cardiovascular diseases primarily strike mature adults and the elderly. Cardiovascular diseases, particularly heart attacks (caused by a blockage of the blood vessels that supply oxygen to the heart), and strokes (blockage of blood vessels around the brain), account for more than half of all deaths in developed countries, such as the US and UK. The National Science Foundation estimates that about one person in four now living in the US will develop some form of cancer and, on current trends, one in six will die from it.

The scale of the problem needs no emphasizing and an attack on these two groups of diseases, particularly cancer, is a prime aim of many biotechnologists. The challenge is formidable, not least because the causes of these diseases are far from clear. There are, however, some encouraging signs that new methods of diagnosis and treatment, based on biotechnology, will have an impact before too long.

Cancer and the Roots of Malignancy

Among the relatively affluent third of the world's population no disease evokes as much dread as cancer. The insidious and painful nature of most cancers amply justifies the public's concern. This disquiet has prompted many a drive to 'find a cure for cancer', most notably the US National Cancer Act of 1971 which pledged no less than $600 million to the crusade against cancer. Why, then, in the face of a massive and concerted attack from many of the world's finest scientists and doctors, has cancer remained so intractable?

In fact, cancer research has been more successful than is often recognized. For example, early diagnosis and improved therapy have made the outlook much brighter in cases of cancer of the cervix and some leukemias. However, the death rates for lung cancer, the most prevalent type of cancer in British and American males, and breast cancer, the most frequent in females, have remained obstinately high.

This slow progress has led to strident enquiries about why there is not a cure for cancer around the corner, similar to that for polio. This brings us to the nub of the problem—cancer is not a single disease. Polio is caused by identifiable viruses and this made possible the production of highly effective vaccines to combat the disease. The chances of finding an analogous anti-cancer vaccine are virtually nil. Cancer has many causes and takes many forms.

When our bodies are functioning properly, every cell and organ forms part of a precisely co-ordinated and regulated system. The whole process of growth, from a single fertilized egg cell to a mature adult, depends on cells growing, dividing and taking on specialized tasks according to an intricate pattern. Once an organ or other group of specialized cells has reached its correct size, signals are generated in a way which is not understood in detail and cell growth largely ceases. From then on new cells are normally generated at a slow rate, just sufficient to replace those that are dying. In cancerous cells, the usual regulation has gone haywire; they continue to grow and divide, invading areas of healthy tissue with disastrous results.

The obvious problem which faces cancer researchers is to determine how cancer cells escape the body's normally stringent control mechanisms. Many details have been uncovered about the molecular processes that seem to be deranged in such cells, but a coherent and comprehensive picture has yet to emerge. Rather more extensive knowledge exists about the agents responsible for transforming cells.

About 500 different chemicals are known to be carcinogenic (cancer producing) in animals, a few dozen of which have been implicated in human cancers. Obviously the search for human carcinogens is greatly impeded by the ethical impossibility of testing chemicals to see if they cause cancer in humans. Thus, most of the firm evidence about human carcinogens is gathered by looking at the incidence of specific types of cancer in people whose work or habits expose them to particular chemicals. A strong link between vinyl chloride, a chemical used to make the plastic PVC, and liver cancer has been established, and the link between cigarette smoking and lung cancer seems incontrovertible, to mention just two of the best known examples. Materials as diverse as benzene, asbestos and chromium have also been implicated in human cancer. A fierce debate rages around the importance of these environmental factors in cancers, with some experts claiming that 85 per cent of all cancers can be traced to them. In any event, 'industrial' carcinogens are unlikely to account for many cancers in the population as a whole, since most people do not come into contact with them.

It is a daunting task to try to unravel the plethora of claims relating life-style to cancer. Certainly, particular dietary habits appear to be associated with some types of cancer — for example, lack of fibre with cancer of the colon, and the high incidence of stomach cancer prevalent among those on traditional Japanese diets. It is questionable, however, that such findings can form the basis for a 'cancer-preventive diet.' In its 5-year Outlook on Science and Technology, published in 1982, the US National Science Foundation notes that the 'avoidance of one form of risk seems almost necessarily to engender another.'

Some radiations, including those produced by nuclear reactions and ultraviolet radiation in sunlight, clearly cause cancer. Sunlight is the major cause of skin cancer, but fortunately this is one of the most effectively treated forms. There is, however, a deep controversy about the amount of radiation that we can tolerate without harm. Everyone agrees that massive doses of radiation induce cancers, but opinions differ on the danger of the much smaller amounts that the general population is likely to encounter. Some experts assert that the incidence of cancer is directly proportional to the amount of radiation and that even small doses will cause some cancers; others aver that there is a threshold level, below which no harm is done. It may take many years before one side or the other has assembled enough evidence to establish its case. In the meantime, prudence seems to favour the assumption that all levels of radiation are potentially damaging.

The final important suspects in the production of cancer are viruses. Nearly every major university is engaged in the study of viruses that cause tumours in animals. Their combined efforts have revealed intricate details about the way viruses subvert a healthy cell. Dozens of specific viruses are known to transform normal cells into cancers in mice, rabbits, chickens, monkeys and other animals. However, only four forms of human cancer have been strongly linked with viruses — a type of liver cancer with hepatitis viruses, two rare forms of blood cell cancer, and a cancer of the nose and throat found mainly in Africa. A few years ago it seemed that viruses might be the key to cancer in general, perhaps even that there was a 'human-cancer virus'. That idea fell into disfavour, but with the very recent research on oncogenes it is beginning to reappear in a new guise.

Oncogenes are sections of DNA present in every cell—healthy as well as cancerous. They seem to lie dormant until activated, when they transform the cell into a cancerous form, initiating the headlong proliferation typical of the disease. Intriguingly, their structure is very similar to that of certain viruses, so perhaps there is a relatively small number of human-cancer 'viruses', although these 'viruses' are an intrinsic part or our genetic inheritance and not extraneous infections. If this theory stands up under closer scrutiny, there may at last be a unifying theme in cancer research, one which has been sorely lacking until now. The task will then be to discover just how these oncogenes are roused from their dormant state to exert their malignant influence. This may result in the apparently disparate 'causes' of cancer outlined above being linked by their shared ability to activate oncogenes.

These ideas have profound implications for cancer research. Paradoxically, they would mean that despite the crucial role of virus-like material in the induction of cancer, there would be no hope of manufacturing an all-purpose, anti-viral vaccine to protect against the disease. Vaccines can only be used to boost the body's defences against *foreign* materials, not against indigenous components of cells. On a more optimistic note, it is conceivable that ways could be developed of permanently inactivating oncogenes, thus hitting at the very roots of cancer. At present this is certainly science fiction, not science fact, but the pace of modern molecular biology is so rapid that one can rarely say that fiction will never turn into fact.

This catalogue of largely unanswered questions about the nature and causes of cancer may seem a gloomy background against which to set the current endeavours of biotechnologists. However, substantial advances can be made, even in a field so beset with theoretical and practical problems. One indication of such progress came in the summer of 1983 when scientists from Britain and the US announced that they had discovered that one of the known oncogenes contains the instructions for manufacturing a substance called platelet-derived growth factor (PDGF). This substance is normally synthesized by blood cells in response to wounds and it prompts the growth of new cells to repair the damage. Clearly, if growth-stimulating substances such as PDGF were manufactured by other types of cells they might be induced to grow uncontrollably. By 1985 over 20 oncogenes had been identified and researchers from dozens of laboratories around the world had been able to fill in more details about their actions. Many seem to encode proteins which, like PDGF, stimulate cells to grow; others seem to encode proteins which cells use to detect and respond to growth factors. The discovery of these genes and the ways they interact may well provide vital clues about the development of cancers.

More about Interferon

Few drugs have received a barrage of publicity comparable to that showered on interferon. Stories of precious vials of interferon being flown around the world to offer hope to cancer sufferers have bolstered a widespread belief that interferon can knock out cancer.

In the face of the media hype— sometimes supported by reckless quotes from scientists and doctors — it is tempting to over-react and dismiss interferon as a genuine aid in cancer therapy. The bald fact is that no-one has proved unequivocally that interferon can cure any form of cancer. Until very recently there was not enough interferon available in the world for proper clinical trials to test its efficacy. This situation is rapidly changing. Genetic engineers are supplying more and more interferon, and a number of level-headed institutions are funding research into its effects; for example, the American Cancer Society is spending $5 million and Britain's Imperial Cancer Research Fund has allocated about £1 million.

Encouraging signs are now emerging. Interferon seems to benefit patients suffering from cancers of the skin, bone, breast and blood. The next few years will see a steady stream of scientific publications presenting the results of tests with several different types of interferon in patients with many forms of cancer. Only when this detailed information has been accumulated will it be possible to say what role, if any, interferon will have in cancer therapy in the future. At present, the best guess is that it will have some part to play, not as a miracle cure, but as an extra weapon in the clinician's armoury.

One new line of research concerns the notion of producing hybrid interferons. The dozen or more types of interferon fall into three classes—alpha, beta and gamma — according to the type of cell that produces them in greatest abundance. Alpha and beta interferons are chemically similar and are derived from leucocytes (a type of white blood cell) and fibroblast cells from connective tissue of muscle and skin. Gamma interferon is rather different and is obtained from cells of the immune system. By carefully snipping the genes that code for interferons it will be possible to induce bacteria to manufacture a hybrid interferon consisting, for example, of the first half of one form joined to the second half of another. It could then be tested to see if it is more effective than either of the natural products.

One of the many surprises thrown up by research into interferon concerns the role of sugar molecules which are attached to most interferons manufactured in human cells. Like many proteins, interferons are not simple chains of amino acids, but also have sugar groups tacked on to the chain at certain points. Proteins like these are called glycoproteins. It would be difficult indeed for bacteria to make them—they would not 'know' where or how to affix the sugar groups and genetic engineers are a long way away from working out how to tell them. Fortunately, interferons which lack the sugars and can be made by suitably instructed microbes seem to act in just the same way as their natural glycoprotein counterparts.

Beyond Interferon ?

Given the number of false hopes aroused about 'cancer cures' no one should be hasty in hailing new solutions of one of mankind's most intractable problems. Too many 'highly promising' materials have fallen by the wayside.

The relatively scanty knowledge about the action of interferons appears rich indeed when set alongside our knowledge about many of the multitude of other materials involved in the immune system. Four which have received particular attention during the mid-80s are interleukin-2, tumour necrosis factor, lymphotoxin and macrophage activating factor. The genes for the first three of these have recently been cloned and increasing quantities will become available for the long and frustrating tests needed to examine their efficacy. Various trials reported up to the beginning of 1985 indicate that all are still in the running as potentially important cancer therapies, but it is unlikely that any will become firmly established within the next five years.

The scale of the task of trying to support the body's natural defences against cancer can be judged from the fact that interleukin-2 and lymphotoxin are only two of about 100 so-called 'lymphokines' produced by white blood cells. The US National Cancer Institute concluded that several of these materials may have great potential in cancer treatment.

It would be futile — and reckless — to attempt to predict which, if any, of these substances will really prove valuable. We can be sure, however, that if any do find a major place in cancer therapy they will become household words almost overnight.

Monoclonal Antibodies Home in on Cancers

Before cancer cells can be attacked with antibodies it is essential to find out what distinguishes them from normal cells. Only then can antibodies be manufactured which will home in on malignant cells and ignore all others. This is a formidable task, but substantial progress has already been made.

The root of the problem is simple to understand — cancer cells *are* human cells, albeit ones which have taken a treacherous turn. It is not surprising, therefore, that the surfaces of cancers cells (where the antigens lie) are very similar to healthy cells. Fortunately, the inner turmoil of cancer cells is reflected in subtle changes to their surface antigens. This has meant that, in recent years, monoclonal antibodies which pick out these characteristic antigens have been developed. No one has yet managed to find an antigen which is completely cancer-specific — that is, one which is found on cancer cells and absolutely nowhere else.

It is now possible to make monoclonal antibodies which recognize antigens characteristic of (but not entirely specific for) several types of cancer, including three of the major killers, cancer of the breast, lung and bowel. This list, which also includes leukemias and cancers of the skin and pancreas, will surely be extended very rapidly. The immense labour involved can be judged from the fact that researchers at the US National Cancer Institute had to examine over 15,000 different types of monoclonal antibodies to find one which is specific for a form of lung cancer that strikes about 32,000 Americans each year.

At present, monoclonal antibodies are used almost entirely for diagnosis and monitoring the effectiveness of conventional forms of cancer therapy. The success of a treatment can be measured by the decrease in the number of cancer markers detected by the appropriate antibodies in a sample of the patient's blood. The therapeutic applications of antibodies are still in their infancy. The most encouraging news so far has come from Ronald Levy and his colleagues at Stanford University School of Medicine. Their preliminary reports suggested that a cancer of, as it happens, antibody-producing cells (B-cell lymphoma) may be treated with monoclonal antibodies.

Monoclonal antibodies may, be themselves, trigger the patient's own immune system to start attacking a cancer, but it is a refinement of this idea which is creating interest among cancer specialists. This is to use antibodies as 'guided missiles' to deliver a toxic 'warhead' straight on to cancer cells. The majority of drugs distribute themselves fairly evenly throughout the patient's body, but in many cases they are really only needed at one or a few specific sites, namely where the cancer has taken hold. Many of the most powerful drugs produce harmful side-effects in the patient by affecting normal cells as well as diseased ones, and this is particularly true for anti-cancer drugs. It would clearly be desirable to give them in smaller doses, while making sure that they reach the spots where they can do most good. This is just the facility offered by monoclonal antibodies. It seems entirely possible that monoclonal antibodies targeted against antigens found only on cancer cells can be linked to anti-cancer drugs, delivering them directly to the cancer. The beauty of this technique is further enhanced by the fact that it would not even be necessary for doctors to know *where* all the cancers are — the monoclonals would find them.

Caird Edwards and Philip Thorpe of London's Chester Beatty Research Institute have called this a retiarian approach to cancer therapy. The retiarii of ancient Rome were gladiators armed only with a net and trident. Monoclonal antibodies carrying anti-cancer drugs could work in a similar manner— the antibodies would ensnare cancer cells and then the toxic drugs would deliver the *coup de grace*.

The possibility of targeting drugs very precisely would open the way to a whole new range of anti-cancer agents. Since the anti-cancer drugs cannot be sent directly towards cancers, only relatively non-toxic substances can be employed. (The word 'relatively' is significant here. All effective anti-cancer drugs cause considerable damage to other parts of the body.) If a drug linked to a monoclonal antibody could zero in on cancer cells only, then far more lethal cell poisons could be used. One such is ricin, a protein from the humble castor bean plant. Ricin is so potent that one molecule is enough to kill a whole cell.* Experiments in Dallas, Texas, showed that ricin linked to antibodies killed cancer cells in the bone marrow of mice without destroying healthy cells.

In fact, ricin, can be made less dangerous. Ricin, like many other toxins such as diptheria and cholera toxins, consist of two amino acid chains linked by a single bridge. One, the B chain, helps the ricin molecule cross the membranes that surround cells, while the other, the A chain, is the part which does the killing. By using only the toxic A chain joined to an antibody, the risk that ricin will get into healthy cells is reduced.

Obviously the effectiveness of ricin and similar poisons depends on their entering the cell under attack. With the removal of the B chain how will ricin A chain get in to carry out its destructive role? The complete answer is not known, but part of it is an exquisite example of nature's cunning — and how it can sometimes backfire. When an ordinary antibody latches on to an antigen on a cell's surface, the antibody causes no direct harm to the cell, but it does act as a signal to other parts of the immune system which destroy the cell. One way the 'tagged' cell can avoid this fate is to engulf the antibody so that it can no longer alert the immune system. This avoids trouble if an ordinary antibody is involved. However, if the antibody carries with it a toxin such as ricin, the cell, like the Trojans who welcomed the Greek horse, has ensured its own destruction.

At present nearly all the monoclonal antibodies used in research are derived from mouse cells and a few from rats. Incredibly useful though these are, they do suffer from one important disadvantage : because they come from a different organism, the immune system of a human patient may produce its own antibodies to attack the mouse monoclonals. To prevent this futile battle between antibodies, it is necessary to make human monoclonal antibodies in the laboratory. At present this cannot be done efficiently, but there is every reason to believe that it may soon be achieved. The main requirement is for a suitable human myeloma cancer to be found, one which is analogous to the mouse myeloma cells already available. This could then be fused with spleen cells to provide hybridomas manufacturing human monoclonal antibodies.

In the meantime, the use of mouse antibodies in diagnosis presents no danger, because the antibodies do not need to be introduced into a patient, but simply mixed with samples of blood or other substances from the patient. For severely ill patients, it may be worth taking the risk of injecting mouse antibodies if they offer some significant chance of a cure.

Enzymes Undermine Cancers

Monoclonal antibodies rely on differences in the *structure* of cancer cells. Another emerging biotechnology depends on differences in the way cancers actually *work* —that is the chemical reactions which go on inside the cell, or its metabolism. Again, the great similarity between cancerous and normal cells must be stressed, but a few potentially important differences have been unearthed. One serves to illustrate another way that biotechnologists can help combat cancer.

Asparagine is one of the twenty amino acids used to construct proteins, and without it no cell can live for long. Normal human cells can manufacture their own asparagine from simple and readily available chemicals in their environments. Certain types of cancer cell, however, lack this ability, and they must pick up the asparagine they need from the bloodstream. Any discrepancy of this kind immediately excites the attention of cancer researchers, and this prompted ideas of cutting off the vital supply of asparagine to cancer cells and thus starving them to death.

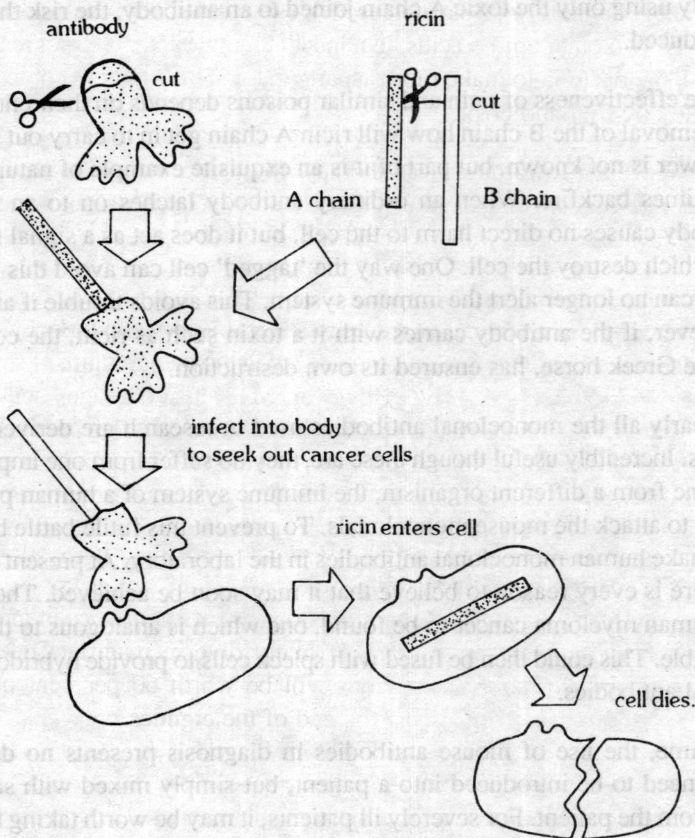

Fig. 3.7. Using monoclonal antibodies to guide poisons to cancer cells. The poison ricin is composed of two amino acid chains. The A chain, which is the part responsible for its toxic effects, is attached to a monoclonal antibody that recognizes antigens on cancer cells. When the antibody / A -chain combination is injected into the body it will seek out cancer cells and kill them.

There is an enzyme, known as asparaginase, which selectively destroys asparagine molecules. Asparaginase is found in many diverse organisms, including yeasts, bacteria and plants. Most of the medical supplies are now obtained from four types of bacteria, the most important of which is *Erwinia*. Purified asparaginase has been injected into children suffering from the blood cancer acute lymphoblastic leukemia, and about half of the children improved significantly after this treatment. Presumably the asparaginase had cleared most or all of the asparagine from their blood, with disastrous effects on cancer cells but little on normal ones. Asparaginase might also be effective against other types of cancer, but this has not yet been proved.

Obviously the biotechnologist's task is to design efficient fermentations to provide clinicians with as much asparaginase as possible at the lowest cost. In particular, the yield of asparaginase from the bacteria growing in large vats is increased by adding various simple chemicals. The most effective seem to be two other amino acids, leucine and methionine. It is not really known why these amino acids cause the bacteria to make more asparaginase. This kind of problem is biotechnology is still usually tackled by trial and error.

Clearly this approach towards cancer therapy could take many forms. The search for other 'metabolic markers' of cancer cells continues. Once discovered, drugs can be designed which exploit characteristic weaknesses of these otherwise all too vigorous cells.

Quite a different use of microbial enzymes is being developed to aid doctors in the treatment of cancer. Methotrexate (MTX) is a widely used anti-cancer drug which works by damaging an enzyme called dihydrofolate reductase. This enzyme is involved in the manufacture of several compounds that cancer cells, in particular, need to survive. However, MTX also damages healthy cells so it is vital that only the minimum effective dose is given to patients. Since different patients break down the drug at different rates it is important to know just how much MTX is in their bodies at any time. The amount of MTX in the blood can be measured by testing its effects on dihydrofolate reductase, and the enzyme used in this test is obtained comparatively cheaply from a bacterium called *Lactobacillus casei*.

Undoubtedly microbial enzymes will continue to grow in importance in medicine as the inherent advantages of microbes as excellent sources of many diverse materials become fully exploited. According to one estimate, microbial enzymes will be worth 60 per cent of the total $1 billion diagnostic enzyme market that will exist by the end of the eighties.

Organ Transplants

No recent medical advance has received more public attention than organ transplants, especially of hearts and kidneys. Admiration for the delicate skills of surgeons has rather overshadowed the crucial contributions of immunologists. Somehow the patient's body must be persuaded to accept a foreign object — the transplanted organ — and not assault it with the immune system. A number of drugs which suppress the body's immune system have been developed in recent years and one of the best is Cyclosporin A, a compound made by the fungus *Tolypocladium inflatum*. This organism, which was first discovered in a sample of Norwegian oil, is set to revolutionize organ transplantation. Without Cyclosporin A only about 35 per cent of patients who receive liver transplants survive for a year, with the drug the figure nearly doubles. Similarly, trials of Cyclosporin A have increased the success of kidney transplants from 50 per cent to over 80 per cent.

Cyclosporin A works by affecting the 'T cells' of the immune system, and its uses appear to extend far beyond organ transplantation. There are many diseases which seem to be caused when the body is attacked by its own immune system. One such 'auto-immune disease,' uveitis, an inflammation of the eye which can result in blindness, has been treated successfully with Cyclosporin. The drug has also shown considerable promise in killing schistosomiasis and malarial parasites. There are even indications that it can help control some types of diabetes. When one considers that Cyclosporin A has only been available to a few scientists and doctors for seven years or so, it is remarkable how many beneficial effects it has revealed. Cyclosporin A could clearly become one of biotechnology's greatest gifts.

Drugs which are given to suppress the immune system and so help prevent organ rejection also increase the risk that the transplant patients will catch infectious diseases, in particular, kidney transplant patients sometimes suffer from herpes viruses. Interferon could play a vital role in guarding against this.

Monoclonal antibodies could also help to supplement the patient's debilitated defences, but more importantly monoclonals can be used to select the best possible organ—that is, one which is as similar as possible for transplanting into a particular patient. Tissue typing, as the science of matching organs is known, will be revolutionized by monoclonal antibodies. These will identify the antigens on potential donor organs allow doctors to select those that are most similar to the patient's own organs. This approach goes right to the root of the problem of rejection, since it is foreign antigens which prompt the patient's immune system to fight against the new organ. Ensuring that transplants have as few foreign antigens as possible should reduce the need for drugs and, thus, keep the patient's immune system in better shape.

DNA probes could also play an important role in tissue typing. The antigens which characterize an individual's organs and cells are encoded in a set of genes known as the MHC (major histocompatibility complex). By comparing the ability of DNA probes to bind to DNA from the MHC of both the donor organ and that of the patient's own cells it should be possible to select organs which are as similar as possible to those of the patient.

Cardiovascular Disease

As we have all observed, blood around a wound quickly solidifies into a clot which prevents further blood loss. Clearly this is a vital process; if this clotting ability were absent or reduced, as it is in haemophiliacs, a major cut could prove fatal. Equally, however, it is essential that clots do not form inside the veins and arteries where they would halt the free flow of blood carrying oxygen to the tissues. If clots do form the effects can be disastrous — a heart attack if the obstruction is near the heart, and a stroke if it is near the brain.

Many substances are involved in the body's control of the clotting process. Some substances promote clotting, while others retard it. This system of checks and balances is a marvellous example of biological regulation, but sometimes, for many different reasons, blood vessels do become blocked, and two products of biotechnology can help unplug them, minimizing the damage done.

In the nineteen-thirties, intense research began into an enzyme produced by *Streptococcus* bacteria. This enzyme, streptokinase, has been used since the seventies to clear obstructions in leg and lung veins, and, in 1982, the US Food and Drug Administration approved it for treating heart attacks. Promising though streptokinase is, it suffers from a drawback common to the great majority of bacterial enzymes that are introduced into a patient's body — it can provoke an immune response which diminishes its effects. Attention is, therefore, shifting towards an enzyme which has similar properties but is made by human cells — urokinase.

Much of the world's supply of urokinase comes from Japan, where it is the seventh largest selling drug with sales of $ 150 million a year. At present, urokinase is either extracted from urine or obtained from kidney cells in laboratory cultures. Both processes are exceedingly expensive and this fact, together with the growing interest in using it to treat blood clots in Japan and Europe, has motivated genetic engineers to try to clone the relevant gene and introduce it into bacteria. At least two groups have claimed success, but their subsequent progress in developing a commercial fermentation system for producing the drug is shrouded in mystery.

In theory at least there is no reason why genetically engineered urokinase and other enzymes involved in the breakdown of blood clots should not be generally available by the mid-to-late eighties.

A group of compounds known as tissue plasminogen-activators (tPAs) have attracted even greater attention in the last four years. Urokinase seems to act by decreasing the clotting ability of blood throughout the body, with the consequent risk of internal bleeding. The tPAs, by contrast, appear to be more precise in their action, attaching themselves to blood clots and stimulating other components of the blood to break down the clot, without reducing the blood's clotting power elsewhere in the body. Some tPAs have now been cloned and are now being used in clinical trials.

An unlikely alliance of fireflies and bacteria is beginning to find a place in the diagnosis of heart and liver disease. The wakes created by ships as they plough through the oceans, especially in tropical waters, are frequently marked by a strange shimmering light. This bioluminescence comes from millions of luminous bacteria in the water. Both these bacteria and fireflies produce light by chemical reactions involving enzymes.

Increasing amounts of fatty substances, known as triglycerides, in the bloodstream warn of the risk of atherosclerosis, the narrowing of arteries caused by a build-up of fatty materials on the inside of the vessel walls. Although there are already several methods of measuring the amount of triglycerides, a newly developed bioluminescent approach is more accurate and faster, and may be cheaper.

Luminescent bacteria give off light when the enzyme luciferase catalyses a particular type of chemical reaction. This reaction can only take place in the presence of an energy-rich compound, known by its initials as NAD (P)H. Thus, the amount of NAD(P)H can be determined by the light emitted from a solution which contains luciferase extracted from bacteria. When blood containing triglycerides is mixed with another type of enzyme and NAD(P)H, the concentration of NAD(P)H in the solution changes according to the quantity of triglycerides present. This changes the amount of light given off by the solution, giving an indirect but accurate measure of the triglycerides in the blood sample. Similarly, firefly luciferase can be used to show the concentration of creatine kinase (which increases after a heart attack) and other materials.

Similar methods have been developed for measuring alcohol, certain hormones and bile acids (which indicate liver damage). furthermore this approach could be employed to analyse 200 or more substances in the body if these are found to be significant indicators of various diseases.

The prevention, diagnosis and treatment of cancers and diseases of the heart and blood vessels are certainly among the largest challenges facing biotechnologists today. Few major advances have yet been made, but encouraging progress in several areas provides the hope that the next decade will see significant improvements in our ability to combat these diseases, which now kill over half the population in the developed world.

4

From Biotechnology to Biotics*

Introduction

Technological innovations are mostly the result of the convergence of several independent paths. This was the case with genetic engineering. It is because of dramatic advances in virology (the study of bacteriophages), bacteriology (the study of the colon bacillus and plasmids), molecular biology of the gene (the genetic code), and enzymology (restriction enzymes) that it was possible to collect the knowledge, methods, and techniques that built up this new field of research.

During the next decade, "classical" biotechnology is likely to continue to develop in an important and spectacular manner. Genetic engineering will benefit from new transfer and cloning techniques. Eucaryotic cell cultivation will be improved through the wide use of synthetic growth media and a better understanding of cellular growth factors. Gene transfer and gene expression in these cells, by means of multiorganism vectors, will be widely used. Hybridoma techniques will also experience considerable development in diagnostic tests, as in the development of therapeutic agents which can be specifically aimed at target cells. Molecular hybridization probes will be used to sort and select genes, for diagnostic or production purposes. Mammalian cloning techniques will be implemented by cattle breeders, and for mass production of laboratory animals. Enzymatic engineering will be applied to an increasing number of industrial processes, either biochemical (isomerizations, specific electrodes) or chemical (hemisyntheses). Considerable developments are likely to occur in plant gene transfers.

This outstanding development of biotechnology during the last test years has been possible because of man's capability to decipher the codes of life, reprogramme the cells, amplify the control signals by turning appropriate molecular switches, and use adapted hosts capable to grow in large quantities. Most of this development has been centred around genetic engineering and hybridomas. New laboratory equipment and production apparatus has recently been born from the convergence of independent research and development paths in physics (mostly electronics, microelectronics, optics), organic chemistry (syntheis and analysis of macro-molecules), analytical chemistry, electro chemistry, immunology (antibody/antigene complex and detection), computer sciences.

* Excerpts taken from the review published by Joel De Rosnay Paus, France.

In the early fifties and more recently, fields of interest were born from cross-fertilization between biology and physics: bionics, bio-electronics, biomimetics. It appears useful to bring together such fields of research by creating a new concept resulting from the convergence of several disciplines (as well as automatic engineering, data processing, and micro-electronics), with biology. A new concept capable of dealing with such different fields as: computers in biology; micro-electronics and new devices; biochemical reactions in automatic machines and analytical apparatus; electronic properties of molecules to make transducers, probes, chips; enzymes as micromachines and micro-robots; new biomaterials.

It appears appropriate to introduce a new word necessary to identify and recognize this hybrid new research field, at the heart of the development of biotechnology for the next 50 years. By analogy to other new fields related to computer science, such as telematics or robotics, its proposed name is "biotics."[1]

Biotics is a new field of research, development, and applications resulting from the marriage of several disciplines and particularly: solid state physics, micro-electronics, organic chemistry, electrochemistry, molecular biology, and through the extensive use of information technology (data processing, automation, robots) in all areas of biology. This marriage has been made possible due to the following properties of biological and information systems:

1. The biological revolution that we have experience for nearly 20 years, is in fact a revolution in the understanding of codes, languages, memories and biological communication. Notions such as the genetic code, the control of gene expression, receptors, peptides hormones, or histocompatibility antigens, represent some of the elements of communication and network regulation of living systems. Therefore bridges can be built between biological systems and information science and engineering.
2. Biological informational macromolecules (DNA and proteins), are prone to the processing and memorization of data by computers. Biological macromolecules carry their information in a sequential and linear form. The sequential aspect of biological macromolecules means they can be treated like "printed text" on computers, automating the successive steps of their synthesis or analysis by a step by step identification or addition of the different nucleotides or amino acid sequences.
3. The last major reason that facilitates the convergence of biology and information sciences is our capability to handle molecules one by one and assemble them in any possible order. Before such micro-engineering techniques, we were only able to handle molecules, not as independent units, but statistically.[2]

THE BASIS OF BIOTICS

Historical

The advent of biotics was made possible in recent years by a better understanding of the language of life and through progresses in information sciences. The capability to decode, read, and write the language of life is turning a new page in the history of mankind, much in the same way as did the discovery of writing and printing centuries ago.

In 1953, we achieved—the complete analysis of a protein (insulin) by Frederick Sanger[3] and the structure of the double helix by J. Watson and F.H.C. Crick.[4] Then, Marshall Niremberg in 1963 deciphered the genetic code, and discovered its similarity among bacteria and man.[5] For the first time, molecular biologists were able to understand the meaning of the hereditary information, and establish the relationship between the genetic code, and the succession of amino acids in a protein sequence.

During the 60s there was an improvement of Sanger's techniques and we got Pehr Edman's methods of protein degradation.[6]

In 1976, Walter Gilbert and Alan M. Maxam[7] and Frederick Sanger,[8] proposed a new method of chemical analysis of DNA. An automatic machine able to "read" the nucleic acid sequences has recently proposed to perform such functions.[9]

A microanalyser, developed in California allows to sequence 40 to 200 amino acids per day from only 10 nanograms of protein.

Such automatic protein synthesizing machines have been developed and commercialized. They are presently used in many university research laboratories, and in the pharmaceutical industry. For nucleic acids, an important breakthrough occurred with the first synthesis of a gene by H. Gobind Khorana at the University of Wisconsin in 1964[10, 11] then at M.I.T.

In 1980, Vega Labs in California announced the first automatic gene machine. The machine had been developed by Itakura. Shortly after, a Toronto based company, Biologicals, marketed a machine developed by Kevin Ogilvie of McGill University. In early 1981, Leroy Hood, the inventor of the microanalyser, in association with Genetic Instruments, proposed a new automatic gene machine.[12]

Now that we are partially mastering the codes and languages of life, and that we know how to produce them at will. Today, it is possible to produce any sequence of a gene. This synthesis is still being made step by step, but soon biologists will use prefabricated sequences which the machine will automatically call up and link to the other elements that enter into the synthetic gene. Computerized handling of such biological information constitutes one of the major impacts of biotics on many fields of biology and biotechnology, specially in three major fields: gene libraries, analysis of sequence data, and molecular programming languages.

- The number of genetic messages deciphered and read in research laboratories throughout the world increases at the rate of 15 per cent per month. This leads to the quadrupling of the known number of gene sequences in less than one year. The world ATLAS of protein sequences contains over 400,000 gene sequences taken from international scientific publications.
- Because of the sequential nature of nucleic acids and proteins, it is possible to compare and match sequences that have been stored in computers much as we compare sentences or words in our spoken languages. These analysis and matching procedures open the way to the equivalent of applications of modern word process: searching for key words (i.e. sequences), coding data, storing information in data banks, finding homologies of sequences,

etc. This capacity will permit to check if proteins with different properties are coded by similar genes; or find in different proteins a common antigenic sequence, likely to serve as a basis for the fabrication of a synthetic vaccine. Recently, computer analysis of genetic sequences of a platelet factor involved in wound healing (PDGF), and of a simian sarcoma virus (SIS), has revealed interesting similarities, establishing important relationships between and oncogene and a naturally occurring factor.

- Computerized sequence analysis can also help find "common ancestors" of certain proteins, thus extending the genealogical tree of living species to the level of molecular evolution. Genetic "word processing" can also help to determine automatically the restriction sites of the DNA molecule. Approximately 100 types of restriction enzymes are known today. Each one breaks DNA in precise places. The computer can indicate which fragment will be obtained if one enzyme is used rather than another. It can then contribute to gene splicing and editing routines by keeping track of the different restriction sequences and their specific ends (sticky ends). New stand-alone units, combining a microcomputer and graphic display capabilities have been recently introduced on the market to provide molecular biologists with tools for the planning and execution of gene splicing experiments. One of those units, known as "BION" has been introduced by intelligenetics.[13] The future of such applications lies in the combination of large computers with numerical capabilities, sophisticated graphic displays, and artificial intelligence. Different systems can be linked through high-speed data transmission networks, like the Xerox Ethernet.

Gene synthesizing machines open the way for "gene programming machines" capable of writing programs that determine and control the fundamental reactions of life. A computer program is made up of a succession of instructions stored in a memory, and executed one by one by the central processor. These instructions are organized in "words" holding the operational code and address where data must be stored and retrieved. All computers, micro, mini, or maxi operate on a limited number of instructions such as READ, PRINT, ADD, SUBTRACT, GOTO (unconditional branch), IFTHEN (conditional branch), SHIFT, START, STOP. These instructions are expressed by the user in high level languages, such as FORTRAN or BASIC. This high level language is then translated into a series of electronic impulses of binary digits 0 and 1, the only code understood by the computer.

Life programs are also written in a language common to all living beings. As the computers uses binary digits 0 and 1, living systems use the four chemical letters of genetic code: A, T, G, C. These bases, assembled three by three form words, the "codons." Each codon specifies an amino acid position in the protein chain, or other instructions necessary for the expression of the genetic program. The genetic code is read, sequence by sequence, by the enzyme RNA polymerase which transcribes DNA's information in copies of messenger RNA. Molecular biologists and geneticists know the codons corresponding to the start signal (INITIATION), and to a certain extent, the stop sequences (TERMINATION), and punctuation. The binding sites of the repressor, a protein controlling gene expression, have also been identified as well as the sequences of the promoter gene on which the RNA polymerase is attached. One specific sequence appears to play an important role in initiation at this level, the code TATA (TATA BOX).

Our capacity of using an interdependent set of biotechnological tools, makes it possible to adopt a new integrated strategy for research and development in biotechnology: Computers are used to store and analyse biological sequence data; organic chemistry, for the synthesis of proteins and polynucleotides; small oligonucleotides, as probes to trap larger molecules through hybridation techniques (RNAs or DNAs which could then be used for cloning experiments); peptides to produce monoclonal antibodies for the isolation of proteins The coming years will see a tremendous improvement of such strategy, which is summarized by the following diagram. It shows clearly that any major progress in the biotechnology field, is now dependent on progress made in automatic macromolecular analysing and synthesizing machines, as well as in large computers for storing data, representing and handling macromolecular structures, or small desk-top computers to plan for cloning and other molecular biology experiments.

Biomolecular Programming and Microbiomachines

Today geneticists are in a position comparable to the one of computer specialist writing programs in binary language. They are at the stage computer science was during the early 50s. But already the codes and symbols they use lead them to go one step further; to the equivalent of "an assembly language." The terms they use (operons, introns, exons, transposons), represent a coded language. The next step will probably be the creation of a molecular programming language made possible by the development of gene synthesizing machines, and particularly by the progress made in

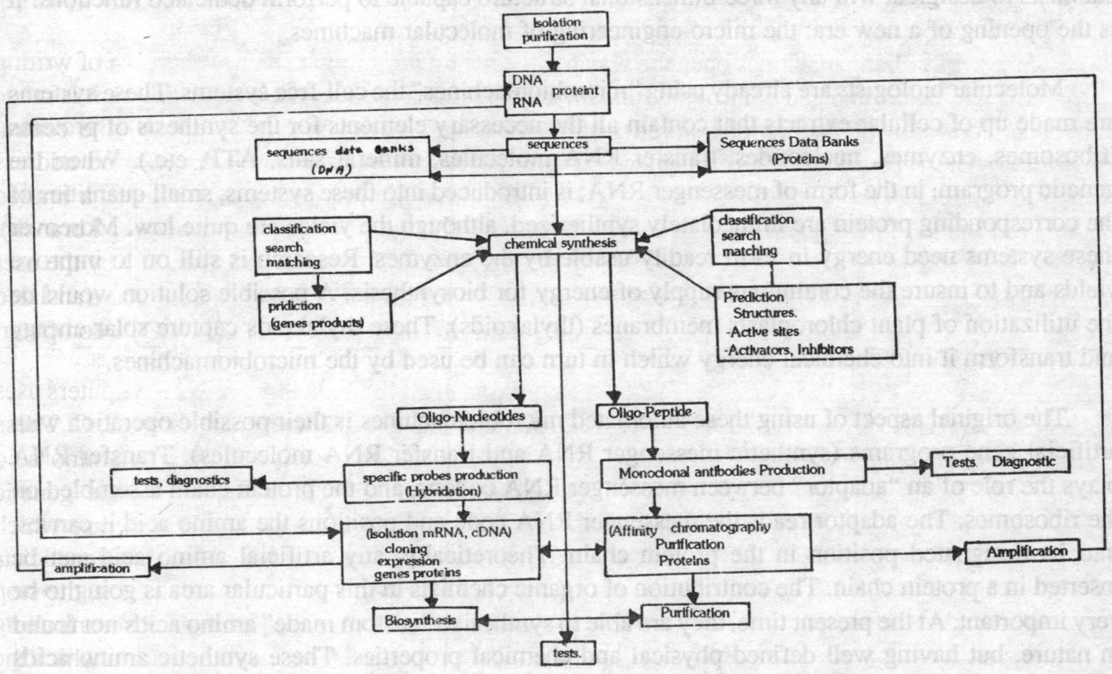

DIAGRAM . "From Biotechnology to Biotics: The Engineering of Molecular Machines,"

the automatic assembly of sequences from previously stored data. Biologists will probably never reach an equivalent level of sophisticated languages like FORTRAN or BASIC, because the mechanisms of reading, assembly, and execution in biological systems are so different than those occurring in eletronic computers. Nonetheless, in combining certain sequences, biologists already know how to write molecular programs with "loops," a repeated transcription of the same sequences (for instance, in multiple copy and thermosensitive plasmids). Will molecular programmers be able to perform "jumps analogous to the GOTO instruction, from one sequence to another, or even conditional branching? Or more, bring in subroutines to help complete the molecular program? The advent of gene machines and genetic engineering techniques is opening the way toward a novel form of molecular programming language.

If such a potential was to be achieved, biotics would follow a path similar to that of editing in book and magazine publishing. We can expect a similar procedure with gene synthesizing machines or for molecular programming. The user will compose and complexify the programs of life, calling up entire pre-assembled sequences, and different modules of amplification and control.

Molecular programming languages will become even more important through the use of dedicated small computers linked to graphic displays, which will allow the molecular biologist to plan for cloning experiments. Coupled with powerful computer graphics programs for molecular design, this technology represents a new symbiosis between man and computers, enabling biologists and organic chemists to design at will any three-dimensional structure capable to perform dedicated functions. It is the opening of a new era: the micro-engineering of molecular machines.

Molecular biologists are already using "microbiomachines" the cell-free systems. These systems are made up of cellular extracts that contain all the necessary elements for the synthesis of proteins (ribosomes, enzymes, nucleotides, transfer RNA molecules, mineral salts, ATP, etc.). When the genetic program, in the form of messenger RNA, is introduced into these systems, small quantities of the corresponding protein are immediately synthesized, although the yields are quite low. Moreover these systems need energy in form readily usable by the enzymes. Research is still on to improve yields and to insure the continuous supply of energy for biosynthesis. A possible solution would be the utilization of plant chloroplasts membranes (thylakoids). These thylakoids capture solar energy and transform it into chemical energy which in turn can be used by the microbiomachines.

The original aspect of using these automated microbiomachines is their possible operation with artificial gene programs (synthetic messenger RNA and transfer RNA molecules). Transfer RNA plays the role of an "adaptor" between messenger RNA codons, and the protein chain assembled on the ribosomes. The adaptor reads the messenger RNA code and positions the amino acid it carries, into its designated position in the protein chain. Theoretically, any artificial amino acid can be inserted in a protein chain. The contribution of organic chemists in this particular area is going to be very important. At the present time, they are able to synthesize "custom made" amino acids not found in nature, but having well defined physical and chemical properties. These synthetic amino acids could then be linked to artificial transfer RNA molecules and incorporated into artificial enzymes of predesigned properties, by a cell free system of microbiomachines. These artificial proteins could lead to numerous applications; catalytic microsurfaces, specific electrodes, ultramicrocircits, energy conversion microsystems.

Microbiomachines are in use in all living systems. Proteins and enzymes are molecular machines, able to turn shafts, transmit tension, connect parts, move molecules, hold fluids, modify workpieces, move components. They are known as flagellar motor, microtubules, ribosomes, collagen, vesicles, metallic complex, functional groups.[2] They can be used on production lines, with enzyme pools to construct complex macromolecular structures and assemblies. They can store and read programs like numerical control systems, as in the case of RNA and DNA reading and transcribing the genetic code. Cables, glue, containers, drive-shafts, motors, pipes, pumps, conveyor belts, already exist in the molecular world. Now with the knowledge gained in such fields as microbiology, enzymology, genetic engineering, and computer assisted molecular design, it is possible to engineer such micromachines, and assemble them on catalytic surfaces or arrays, transforming the micro-engineering capabilities of such machines into the macro-engineering necessities for mankind.

Structural proteins and artificial enzymes will be used for the creation of ultrastructures analogous to those observed in the living cells. The properties of such artificial ultrastructure will be determined by the preferential orientation of the macromolecules in space, and their specific functions. Biologists will be able to reproduce the electron transport chains, like the cytochromes systems; the enzyme pools located side by side in the mitochondrial membranes; or the molecular units catalyzing the photosynthetic reactions in the chloroplasts membranes.

Molecular Information Processing and Molecular Electronic Devices

The most promising areas of biotics results from the convergence of micro electronics, solid state physics, and molecular biology.[14] This hybrid technology is leading to the development of molecular electron devices (MED). Technological breakthroughs in this area are opening the way to the development of "chemical computers." Molecular electron devices could represent a significant break-through for computers in the next 30 years. From the tube era (40s to 60s), to the transistor era (60s to 80s), we are now entering into the molecular electron devices era (80s to 2000 and beyond). The challenge is enormous. We need to start from scratch and reinvent all the components of present day micro-electronics. [15]

First, engineers have to work on a switch, able to shift information in one state or another. By interrogating this component, one should be able to find out in which state the switch is. The development of a molecular switch will be equivalent to the discovery of the transistor.

The second major achievement will be the building of a true memory with different switches, that is, an array of molecules which could undergo reversible alterations, and therefore can be reused.

The third major step is to construct molecular wires, able to transport information through distances, like conjugated chains of carbon atoms on which solitons move rapidly in order to connect switches and memories.

The fourth step will be represented by the assembly of switches, memories and wires in three-dimensional structures or arrays, organized in assemblies of modules at several levels of communication and interconnections, able to perform co-ordinated functions. This is where one of the major breakthroughs of the new molecular engineering techniques will come into play; from the properties of atoms and molecules, the system will self-assemble in highly organized structures, putting atoms, molecules, and macromolecules into place, in order to perform dedicated functions. At this step, immunclogy techniques could be used to fit molecules at predetermined spots, or to deposit heavy metals which could be used as conductors at other levels of communications.

Finally, the system should be reparable. Modules which do not perform correctly will have to be detected, corrections made, and sometimes a device replaced. These types of self-repairing automata already exist in biological systems. We begin to understand how they function at the molecular level, and how they self-assemble. Other supramolecular assemblies, like the quantosome, the hotosome, the oxysome are small self-contained factories which perform important functions for the cell. These minute devices are packed with an enormous amount of molecules and information. We are slowly starting to understand their structures, and able for the first time to copy them. But we still have to learn more from nature's design, understand how nature operates, and then translate what we have learnt into the new micromachines.

In order to make ultramicrocircuits, several laboratories are trying to synthesize proteins which do not exist in nature. Such syntheses are feasible using present day genetic engineering techniques, automatic gene and protein synthesis, and computer reproductions of the bi- and tri-dimensional structures of amino acid sequences. It is likely that from this research a large variety of molecular electronic devices and ultramicrocircuits could be conceived and manufactured. They could be rendered biocompatible, thus allowing the production of implantable logic circuits, offering the prospect of direct interface between the central nervous system of animals or human beings and computers. Such biocompatible circuits are now being implanted and tested in the brains of rats.[16] Other applications of these ultramicrocircuits could be the production of prostheses for the blind, different transducers, or solar energy converters on soft plastic sheets. A general preparation technique of biomolecular ultramicrocircuits has been patented in 1978 by EMV, a small company of the Washington area.

The production and assembly of these molecular circuits can be considered from two different approaches: a "passive" one (successive depositions, etching, grafting, doping) using technologies close to those presently used in the manufacturing of microcircuits; or an "active" one resulting from the spontaneous "growth" of the molecular circuit. In fact, automatic machines used today for synthesizing and analyzing genes and proteins, offer new models which may inspire the automation of the successive operations of molecular circuit production: growth of polymers, successive washing, reactions with other active groups, blocking, and reactivation of chemical groups.

However, at the level where the presence or absence of a single chemical link can affect the performance of an entire circuit, it is virtually impossible to construct and assemble circuits with traditional macroscopic control techniques. It becomes necessary to use auto-assembly properties of biological macromolecules, observed for instance in Langmuir-Blodgett films or during the auto-organization of viruses or predissociated cellular organelles. In other words, instead of introducing the information from the outside, as we do today with most of our machinery (drill press, lathe, or even car construction robots), we will use information from the biopolymers themselves. Such information stored in the primary sequences of amino acids, allows the three dimensional folding of a protein. It is thus possible to benefit from the properties of biological macromolecules to assemble three dimensional molecular ultramicrocircuits.

Many questions still remain unanswered: will these ultramicrocircuits be repairable? Will we be able to selectively break chemical linkages or rearrange them? Is it still necessary to utilize boolean logic, presently used in all computers? Will we be inspired by the neuronal networks of the brain? These circuits presently work, in all probability, in a non-boolean fashion, using parallel processing. To build the logic of the future, the convergence between molecular neurobiology and micro-electronics is in the forefront and it holds great promise.

From now on the evolution of biotics seems to be irreversible. In the next 50 years, this evolution will perhaps lead us to a "symbiotic" man, directly connected by his own nervous system to miniature computers, able to tap through communication networks in to any giant memory from any place on earth, or to communicate directly with any individual. This man/computer interface will probably be achieved by logic circuits compatible with living tissues and operating at the request of an internal command. The molecular ultramicrocircuits open up the way towards "artificial senses" allowing, for example, to "see" in the infrared, to detect minute quantities of radioactivity or to enhance the capacities of recognition of certain odors.

While we already have great difficulty in mastering out own neuronal circuits and effectively communicating with others, is it necessary to embark upon amplification techniques of our own intellectual and sensorial capacities with the aid of a controlled symbiosis with machines? What will happen to our relations with the outside world? Will the "symbiotic man," related physically and biotechnologically to machines that he himself created, constitute a distinct living species, eventually replacing Homo sapiens? No one knows. But it is hard to satisfy the curiosity of scientists; especially when they discover at the intersection of several disciplines new unknown territories to explore and possibly conquer...

REFERENCES

1. Rosnay, Joel de, "Biologie, informatique et automatique: Pessor de la biotique," *Prospective et Sante, 18,* 21-29 (Ete 1981); et *Futuribles, 51*, 41-58 (Janvier 1982).
2. Drexel, E., "Molecular Engineering: An Approach to the Development of General Capabilities for Molecular Manipulation," *Proc. Natl. Acad. Sci. USA, 78*, 5275-5278 (1981).
3. Sanger, F. and Thompson, E.O.P., "The Amino-Acid Sequence in the Glycyl Chain of Insulin," *Biochem. J., 53*, 353-374 (1953).
4. Watson, J.D. and Crick, F.H.C., "Genetic Implications of the Structure of Nucleic Acid. A Structure of DNA," *Nature, 171*,964-967 (1953).
5. Nirenberg, M.W., "The Genetic Code II," *Sci. -American, 216,* 80-94 (1963).
6. Edman, P. and Begg, G., "A 'Protein Sequence'," *Eur. J. Biochem., 1,* 80-91 (1967).

7. Maxam, A.M. and Gilbert, W., "A New Method for Sequencing DNA," *Proc. Natl. Acad. Sci., U.S., 74,* 560-566 (1977).
8. Sanger, F., Nicklen, S., and Coulson, A.R., "DNA Sequencing with Chain-Terminating Inhibitors," *Proc. Natl. Acad. Sci., U.S., 74,* 5461-5467 (1977).
9. Selective Cutting System for Guanine Sequence Analysis," *Science and Technology in Japan,* 2 (5), 19-21 (Jan/Mars 1983).
10. Khorana, H.G., et al., (Nishimura et al.), "Studies on Polynucleotides," *J. Mol. Biol., 13,* 283-301, et 302-322 (1965).
11. Khorana, H.G., "Total Synthesis of a Gene," *Science, 203,* 614-625 (1979).
12. Stinson, S.C. "DNA Synthesizers Based on two Chemistries,"*Chemical and Engineering News,* 17-18 (February 2, 1981).
13. Sun Microsystems Inc., 2550 Carcia Ave., Mountain View, CA 94043; Intelligenetics, 124 University Ave., Palo Alto, CA 94301.
14. Rosnay, Joel de, "Les biotransistors: la micro-electronique du XXleme siecle," *La Recherche, 124,* 870-872 (Juillet/Aout 1981).
15. Haddon, R. and Lamola, A., "The Organic Computer," *The Sciences,* 40-44 (May/June, 1983).
16. MacAlear, J. H. (EMV Ass. Inc.), "Biotechnical multi-Electrode/Neuronal Interface," Project: small Business Innovation Research (1981).
17. Kuhn, H. and Mobius D., "Systems of Monomolecular Layers—Assembling and Physico-Chemical Behaviour," *Angew Chem, Intern., 10* (9), 620 (1971); Kuhn, H. "Electron Transfer in Monolayer Assemblies," *Pure Applied Chem., 51,* 341 (1979); Kuhn, H., a. "Thin Solid Films," 1983, 99, 1; b. "Self-Organizing Molecular Electronic Devices," In *Proceedings of 2nd International Workshop on Molecular Electronic Devices.* Held 13-15 April 1983, NRL., Washington, D.C., Marcel Dekker, New York, NY.

5

Cell Biotechnology*

Introduction

Many a laboratories in the world have achieved the transfer and expression of bacterial and foreign plant genes in plant cells. Increasing attention is now also paid to the use of recombinant DNA technology to isolate and transfer genes governing agriculturally important characteristics such as salivity and growth resistance. The development of large scale culture methods for plant cells made it apparent about 25 years ago that such cultures could be used to produce secondary metabolites just as in microbial and fungal fermentations. The achievements in this direction made in U.S.A. and Japan have been tremendous, particularly at the laboratory and pilot plant stage.

The Architecture of Life

The world contains an astounding diversity of life forms, yet when we delve beneath the surface we find that the millions of species that are found on our planet have many features in common. The basic unit of biological organization is the cell. Plants and animals are all constructed from cells, each of which is surrounded by one or more thin membranes. Cells are very small — it would take 5000 human red blood cells to cover the dot on the letter i.

The plants and animals familiar to us are composed of astronomical numbers of cells — the human body, for example, contains about 100 billion cells. In these multicellular organisms, there are hundreds of different types of cells, each with its specific tasks: cells in the eye sense light, muscle cells provide the power for movement, and so on. This specialization of cells is a key feature of advanced organisms. Each type of cell contributes to the well-being of the whole organism and each depends on the others for its survival.

In contrast, most of the organisms that concern biotechnologists—*the microbes* — consist of only one cell. Each cell is an independent entity that can perform all the functions required to keep it alive and allow it to reproduce. Thus, for these creatures the term 'cell' and 'organism' are synonymous (see Figure 5.1).

The world of microbes includes various types of organism. Three groups are particularly important in biotechnology: the *bacteria,* the *algae* and the *fungi.* Bacteria generally have one of

* Excerpts taken from the review published by F.E. Young, England.

three shapes, rod-like, spherical and spiral, and most are between one and ten millionths of a metre in length. Algal cells tend to be slightly larger and one of their characteristic features is that, like plants, they can obtain energy from sunlight. The cells of fungi are often organized into large groups, as in mushrooms, but each cell is still capable of surviving on its own. Yeasts and bread mould are both fungi.

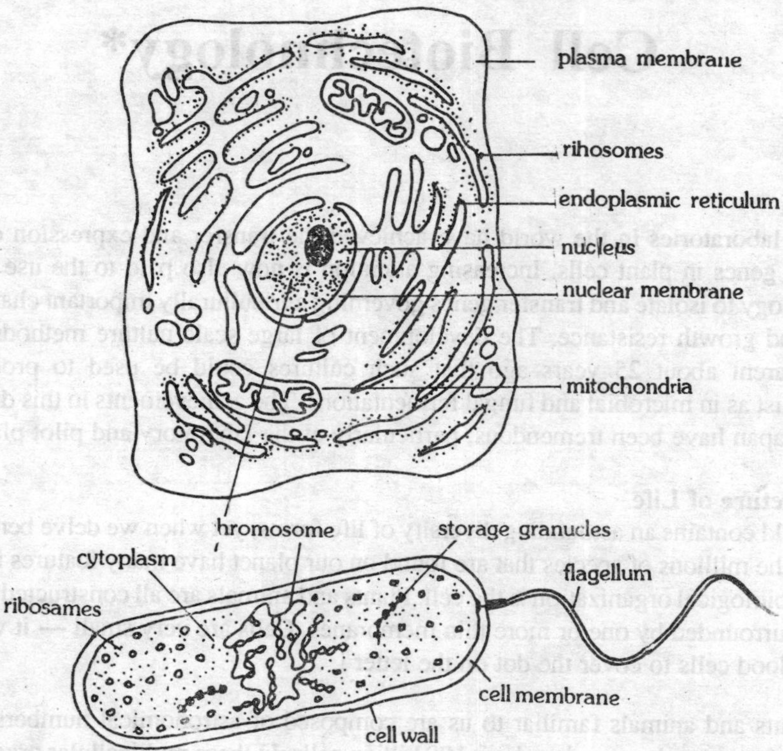

Fig. 5.1. Despite the obvious structural differences between bacterial and animal cells their basic chemistry is very similar.

(Above) A typical bacterial cell. The cell is enclosed in a thin cell membrane, outside which is a more sturdy cell wall. These help the cell retain its shape and prevent the content of the inside (the cytoplasm) from leaking away. Many bacteria have a flagellum, which is used to move the cell around in a liquid environment. Storage granules contain food, and the ribosomes are globular structures on which proteins are manufactured. Finally, the chromosome contains the cell's genes, encoded in a DNA molecule which is tightly folded, like a bundled piece of rope.

Top A typical animal cell. The cells of plants and animals are much more complex and varied in appearance than those of bacteria. The chromosome is enclosed in a nucleus with its own membrane. The ribosomes are attached to a structure called the endoplasmic reticulum. Sausage-shaped bodies — the mitochondria—provide the cell with energy. The cell is enclosed by a plasma membrane which is the equivalent of the thin cell membrane in bacteria; an animal cell has no cell wall.

Even though a bacterium may weigh only a millionth of a gram, its chemical versatility is quite dazzling. The cell consists of thousands of different types of chemicals, most of which are quite complex. All of these substances must be constructed from the relatively simple building materials which the microbe finds in its environment. Complex chemicals cannot be built up from these simple materials in one step. Instead, each cell takes in raw materials and makes a succession of small modifications to them until the final product is completed. The network of chemical reactions by which cells convert an enormous range of substances into the materials they need for life is known as their *metabolism*. All of these individual chemical reactions must be harmoniously co-ordinated, and proteins called *enzymes* play a central role at every stage.

Protein molecules consist mostly of carbon, oxygen, hydrogen and nitrogen. Proteins are constructed from twenty different sorts of simple building block known as amino acids. The staggering versatility of proteins arises from the fact that so many different shapes can be created by arranging amino acids in various ways. A chain containing only a dozen different amino acid, selected from a pool of twenty different sorts, can be strung together in billions of different permutations. Some proteins contain just a few dozen amino acid subunits, while others have over 200; most of the enzymes are composed of more than a hundred amino acid sub-units. To make a protein, the individual amino acids are linked together in a chain, and the number and order of subunits in a particular type of protein give it its individuality.

Once the cell has assembled a chain of amino acids, the chain begins to fold back on itself. Only rarely do proteins exist as a straight line of amino acids; far more commonly, the chain twists and turns to create a complex three-dimensional structure (see Figure 5.2). This folding is not random, but is dictated by chemical forces, which depend on the sequence of amino acids in a particular protein. Thus, the final shape of a protein molecule is determined by the order in which its amino acids are strung together. Some enzymes and other proteins consist of more than one chain, and the chains for these are usually manufactured separately and later assembled to form the complete structure.

The importance of proteins can hardly be overestimated. The insight of the Dutch agricultural chemist Gerardus Mulder, who in the eighteen-thirties coined the word protein from the Greek for 'primary substance', has been amply vindicated by subsequent scientific investigations. The human body contains over 30,000 distinct types of protein. Each has a very specific use — for example, some give tendons their strength and resilience, some carry oxygen about the body, and others protect against infection. It is certain that many proteins still await discovery and, even so, a list of the known proteins would fill many pages of this book. At present, biotechnology is concerned with just a small fraction of these proteins, in particular those that act as enzymes.

Role of Enzymes — As the Biological Accelerators

Enzymes are called biological catalysts. A catalyst being any substance that speeds up the rate of a chemical reaction. Some chemical reactions occur rapidly and spontaneously, without a catalyst. For example, hydrogen and fluorine molecules will combine very rapidly to form hydrogen fluoride. However, many reactions including most that are important for life, occur only very slowly in the absence of an appropriate catalyst.

Although enzymes are grouped together on the basis of their ability to speed up chemical reactions, it is crucial to note that the different types of enzymes vary greatly in their structure and function. Each type of enzyme has a particular molecular 'architecture' and most of them are able to trigger off only one particular type of chemical reaction. A typical cell may contain 1000 different types of enzymes and, between them, the many cells in a large and complex organism may have tens of thousands of distinct enzymes. So many are required because the intricate task of keeping an organism alive and ensuring that it produces healthy offspring involves a vast array of chemical reactions, nearly all of which need the services of an enzyme. There are, of course, many identical copies of each type of enzyme in a cell.

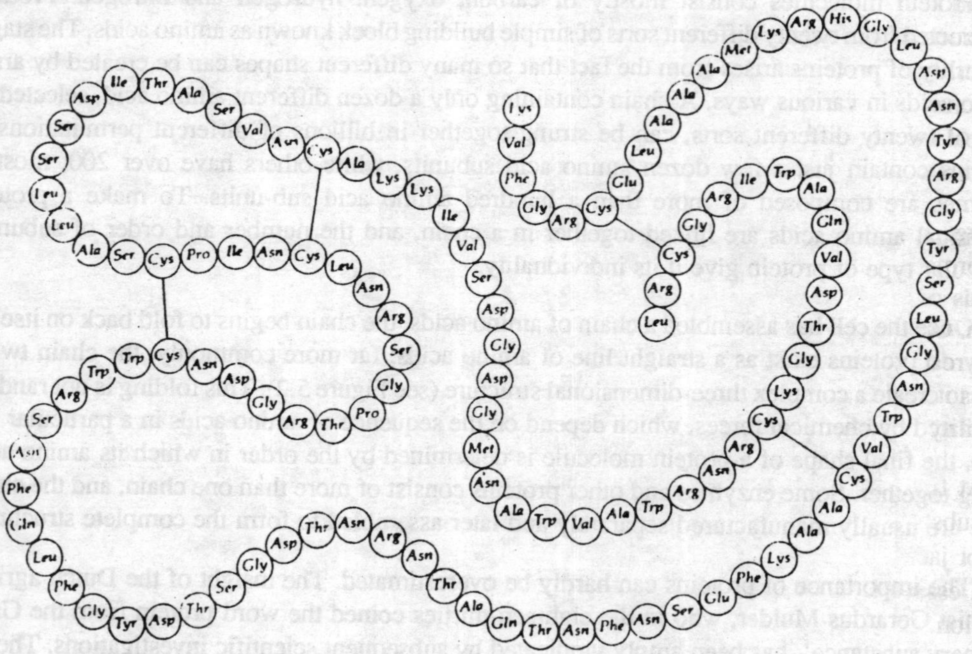

Fig. 5.2. The enzyme lysozyme is constructed from 129 amino acid building blocks, represented here by circles. Each circle is labelled with a three-letter code according to the type of amino acid it represents, and the order of these is the same in every lysozyme molecule. Obviously, the three-dimensional shape of this molecule cannot be shown on a flat page; the molecule is actually folded into a form resembling a rugby ball. Lysozyme is found in human tears and is able to break down the cell walls of some bacteria. Alexander Fleming discovered this enzyme in 1922, seven years before he discovered penicillin.

Naming an Enzyme

Most enzymes are named after the type of chemical reaction they catalyse, and normally the suffix '-ase' is added to the name. Thus, the enzyme alcohol dehydrogenase is so named because it catalyses the removal of hydrogen from a molecule of alcohol. This enzyme is partly responsible for the fact that one eventually sobers up after drinking alcohol. However, by subtracting hydrogen from alcohol it produces acetaldehyde, a compound that induces a hang-over!

Catalytic Action of Enzymes

How enzymes act as catalysts is one of the most basic problems of biochemistry. Although many of the subtleties of enzymes remain hidden, much is known about the general features that given them their marvellous powers. The key is their three-dimensional shape. Every chemical compound has a characteristic shape, and an enzyme will interact only with those chemicals whose shape it 'recognizes'. The chemicals that participate in a reaction catalysed by a particular enzyme are called the substrates of that enzyme.

The whole process whereby an enzyme picks out its substrate or substrates works rather like a lock and key. On the surface of the enzyme there are clefts and crevices that much the bumps and protrusions on the substrate. Thus, when an enzyme and its substrate encounter each other they fit together. Any other compound (that is, one that is not a substrate for that enzyme) cannot act as a key to the enzyme lock—the key is the wrong shape. Once enzyme and substrate are locked together, chemical forces come into play which break and make various chemical bonds within the substrate, and alterations of chemical bonds are the essence of reactions.

Fig. 5.3 shows, in outline, how one enzyme can recognize its two substrates and aid their transformation into two product molecules with different shapes and chemical properties. All of life depends on millions of such events occurring in an immensely complex, but highly co-ordinated system. Once the enzyme has performed its appointed task the products are released and the enzyme stands ready to repeat the whole sequence again, being in exactly the same chemical and physical state as it was at the beginning. This is characteristic of all catalysts and gives an enzyme molecule the ability to convert as many as a million substrate molecules every minute.

At first sight the extreme fastidiousness of enzymes—their rejection of all molecules except their substrates—may seem to be a disadvantage. In fact, it is crucial to their great power. Enzymes are not jacks-of-all-trades, but they are certainly masters of one. They possess both the speed of action essential for the cell and for the biotechnologist, and another equally important attribute—precision.

An enzyme not only selects its substrates from the many materials that mill around it, but it also ensures that the correct products are made. Most compounds in the cell can be altered in a number of ways—adding an atom or two here, taking a couple off there, splitting the molecule down the middle, and so on. To form complex compounds, the cell must arrange for a series of small modifications to be made, one after the other, in a consistent and predictable way. The mere accelerating effect of enzymes is not sufficient; each enzyme must deliver the correct, partly formed compound to the next enzyme in the series, so that it can perform its task. Any variation would lead to chaos. (See Fig. 5.4).

This ability of enzymes to channel reactions down particular pathways gives a high yield of the desired product and ensures that little of the precious starting materials is converted into unwanted, or even harmful, byproducts. In this way a highly complex network of reactions can go on constantly inside cells with predictable results, and predictability — at least at this molecular level — is the essence of life.

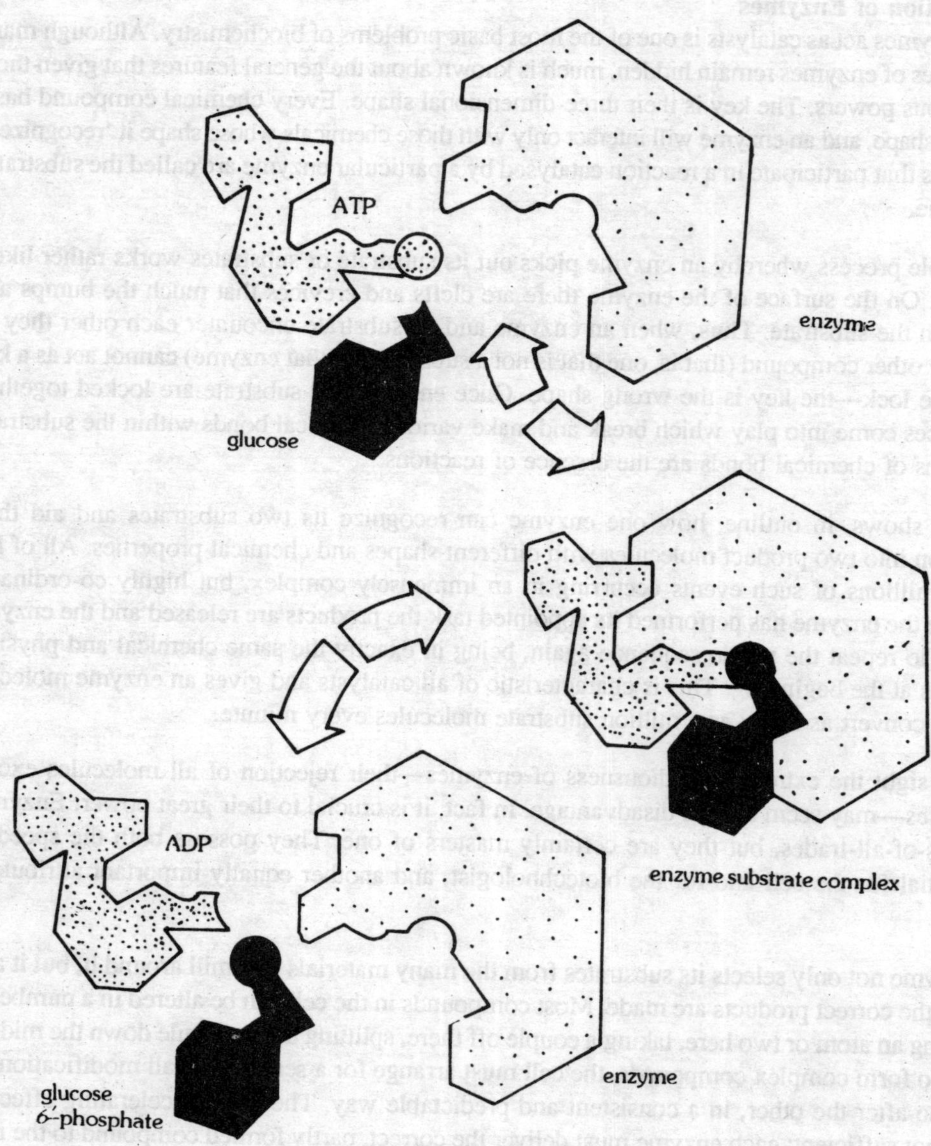

Fig. 5.3. The lock and key model of enzymes. The two molecules (ATP and glucose) which are to react with each other fit into pockets in the surface of an enzyme. When they are brought close together by the enzyme, part of the molecule (the dark circle) is transferred to the other molecule. The products of the reaction (ADP and glucose 6-phosphate) are then released.

The finely tuned metabolism of cells can present problems for biotechnologists, however. Many biotechnological processes are designed to produce large quantities of a particular substance that cells normally manufacture in only modest amounts. An example is the production of lysine, an amino acid that is widely used as an additive to animal foodstuffs.

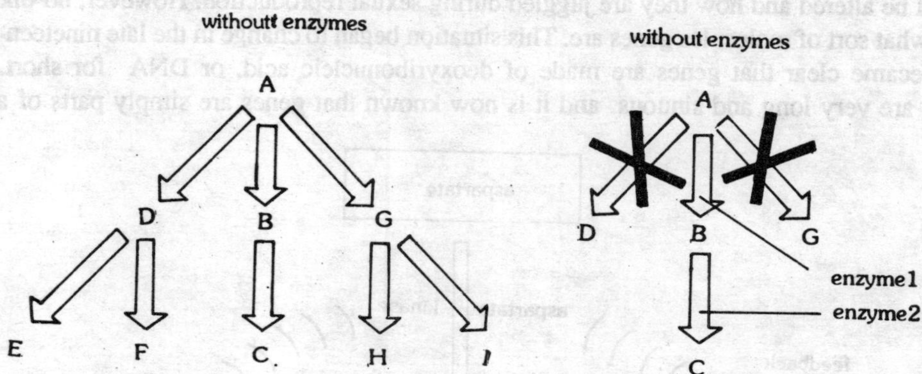

Fig. 5.4. *(Left)* How a compound might undergo chemical reactions in the absence of 'directing' enzymes. Many different products result from the several different reactions that may occur.

(Right) Two enzymes 'channel' the reaction along a single path, ensuring that only one product is made—the product the cell needs.

The vast majority of bacteria of the species *Corynebacterium glutamicum* produce only enough lysine to meet their needs. Built into the microbe's metabolic network is a system of feedback regulation whereby the organism senses how much lysine it has available. If there is too little, a series of enzymic reactions occur to replenish the supply; if there is enough lysine then the cell does not waste precious energy and raw materials in making more. Some of these bacteria, however, have regulatory systems which do not operate correctly and they manufacture much more lysine than they require (See Figure 5.5). Over-production of particular substances and the role of feedback regulation is of central importance in biotechnology.

DNA and Its Role as the Spiral of Li

According to natural law like begets like. Poppies do not grow from rose seeds, nor do sheep give birth to calves. Obvious though these facts are, it is only in the last few decades that we have begun to unravel the underlying causes of this continuity in nature.

The fundamental unit of biological inheritance is the *gene*. Experiments particularly in the early years of this century, led scientists to develop some complex ideas about how genes operate and how they are passed from generation to generation. This research eventually led to the notion that each gene is in some way responsible for making a particular type of enzyme. This 'one-gene, one-enzyme' theory was soon superseded by the more general theory that each gene is involved in the construction of one protein, that may be an enzyme or some other sort of protein. More recently, the theory has been further refined to state that each gene is responsible for making one type of amino acid chain. Some chains simply fold themselves to form an active enzyme; other enzymes consist of two or more chains.

Strange though it may seem, much of the pioneering work in genetics—the science of biological inheritance—was carried out by scientists who had only the fuzziest idea of what a gene actually is. Much was revealed about how these almost abstract entities affected physical characteristics, how they might be altered and how they are juggled during sexual reproduction. However, no-one knew precisely what sort of molecules genes are. This situation began to change in the late nineteen-forties when it became clear that genes are made of deoxyribonucleic acid, or DNA for short. DNA molecules are very long and sinuous and it is now known that genes are simply parts of a DNA molecule.

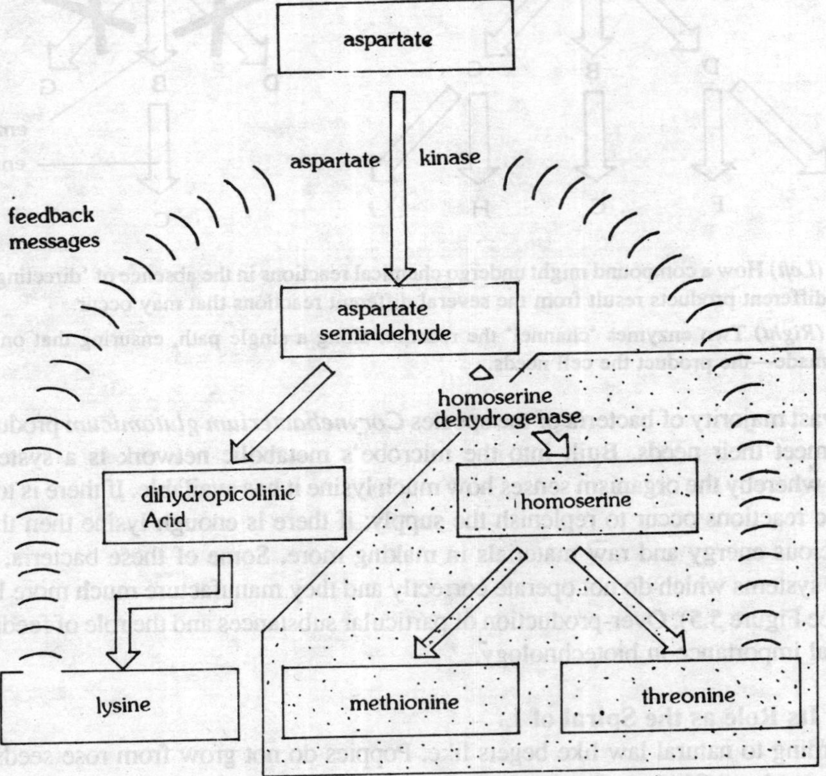

Fig. 5.5. *Corynebacterium glutamicum* manufactures lysine from aspartate, a compound which also serves as a starting material for the manufacture of another amino acid, threonine. The rate at which aspartate is converted into these amino acids, via other chemical compounds, is regulated by the combined amounts of lysine and threonine in the cell. When there are sufficient amounts of both amino acids they act to inhibit (slow down) the enzyme aspartate kinase, which is responsible for channelling aspartate down the reaction sequences that produce lysine and threonine. However, in some forms of this bacterium the enzyme homoserine dehydrogenase is missing. This enzyme plays an essential role in the manufacture of threonine, but does not contribute to the production of lysine. Since the *combined* effect of lysine and threonine shuts down aspartate kinase, the first enzyme in the series, these defective bacteria continue to make lysine in large quantities since they are 'fooled' into acting as if the cell needs more lysine. As these bacteria need a certain amount of threonine to survive, biotechnologists supply the organism with a small amount of this amino acid—enough to keep it alive, but not enough to cause it to stop making lysine.

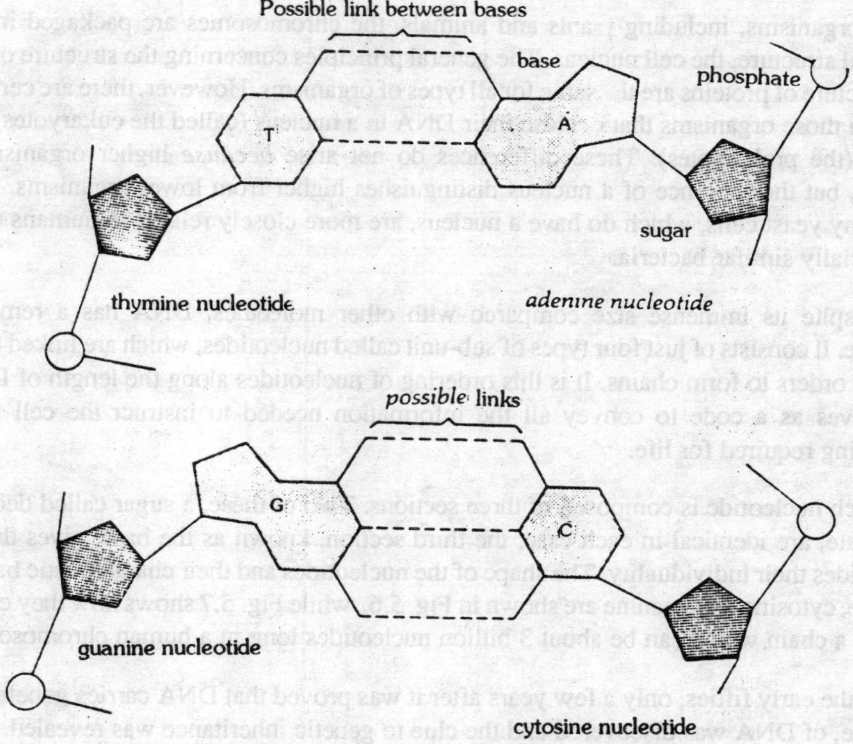

Fig. 5.6. **The four nucleotides found in DNA molecules. Each nucleotide is composed of three parts. The sugar and phosphate portions are identical in all of the DNA nucleotides. The base portions differ from one nucleotide to another. The figure also shows the potential links that each base can form—three each for G and C, and two each for A and T.**

Genes are responsible for making proteins, and genes are made of DNA, but it would be slighty misleading to imply that DNA 'makes' proteins, rather it contains the *instructions* for manufacturing proteins. The laborious assembling of amino acids into a chain is the task of other components of the cell, with DNA directing the process. The unravelling of the mysteries of DNA and protein synthesis is one of the major scientific achievements of the century, perhaps, even of any century, and has immeasurably widened the horizons of Biotechnology. Once the relationship between DNA genes and proteins (especially enzymes) was elucidated it became possible to think about ways of inducing cells to make novel proteins by inserting the right pieces of DNA.

When cells are stained with special dyes and viewed with a microscope certain features stand out, among which are the chromosomes. The most important component of a chromosome is a single, huge molecule of DNA along which many genes are ranged. The number of chromosomes in a cell depends on the species of organism from which the cell is taken. Bacteria have only one chromosome, while human cells have forty-six. The chromosomes of bacteria float freely inside the cell, but in higher organisms, including plants and animals, the chromosomes are packaged inside a roughly spherical structure, the cell nucleus. The general principles concerning the structure of DNA and the manufacture of proteins are the same for all types of organisms. However, there are certain differences between those organisms that enclose their DNA in a nucleus (called the eukaryotes) and those that do not (the prokaryotes). These differences do not arise *because* higher organisms have a cell nucleus, but the presence of a nucleus distinguishes higher from lower organisms. At a molecular level, tiny yeast cells, which do have a nucleus, are more closely related to humans than they are to superficially similar bacteria.

Despite its immense size compared with other molecules, DNA has a remarkably simple structure. It consists of just four types of sub-unit called nucleotides, which are linked together in very specific orders to form chains. It is this ordering of nucleotides along the length of DNA molecules that serves as a code to convey all the information needed to instruct the cell to manufacture everything required for life.

Each nucleotide is composed of three sections. Two of these, a sugar called deoxyribose and a phosphate, are identical in each case; the third section, known as the base, gives the four types of nucleotides their individuality. The shape of the nucleotides and their characteristic bases—sadenine, thymine, cytosine and guanine are shown in Fig. 5.6, while Fig. 5.7 shows how they can join together to form a chain which can be about 3 billion nucleotides long in a human chromosome.

In the early fifties, only a few years after it was proved that DNA carries genetic messages, the structure, of DNA was discovered and the clue to genetic inheritance was revealed. DNA takes the form of the now famous double helix (See Fig. 5.8), in which two chains of nucleotides twist around each other in a helix or spiral. The subunits in each chain are linked together as shown in Fig. 5.7, but there are also links *between* the two chains. The bases of each chain face each other across the centre of the helix and they pair up in a very specific way. Adenine (A) and thymine (T) can each form two bonds, while cytosine (C) and guanine (G) can form three bonds each. Because they match up with each other in this way, the bases A and T are called complementary bases, as are C and G. This exact and unvarying relationship between the sequences of the bases on the two strands is used by the cell for two processes: to make copies of its DNA, and to assemble its proteins in an exact and reproducible way.

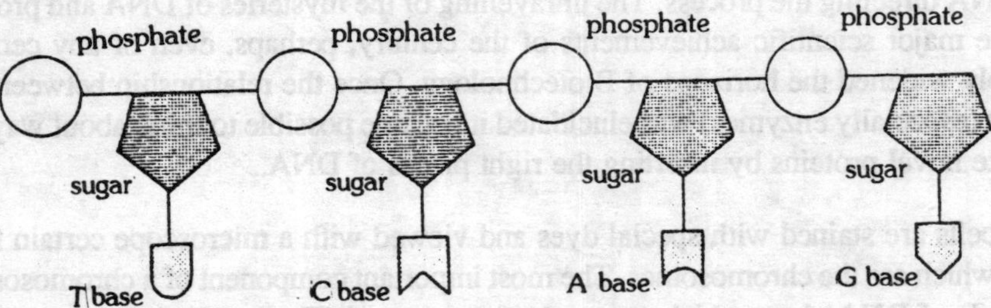

Fig. 5.7. A chain of nucleotides is formed by linking each one through the sugar and phosphate parts of the molecules. The bases of the nucleotides branch off this backbone chain.

Even the most complex multicellular organisms grow from a single cell—a fertilized egg. This cell starts out with a complete set of chromosomes, half of which are provided by its female parent and the other half by the male. When it has grown to a certain size, this first cell splits into two, both of which also grow and split into two, until, eventually, the millions of cells that make up a mature adult have been formed. If, at each division, one cell took half the chromosomes (and hence half of the genetic information in DNA) after a few stages very little DNA would be left in each cell, the material having been distributed throughout many cells, Clearly, this would rapidly destroy the complex organization of life. Instead, the DNA in a cell duplicates itself just before the cell divides and, thus the complete sets of chromosomes are available for the two new cells. This is where the double helix comes in.

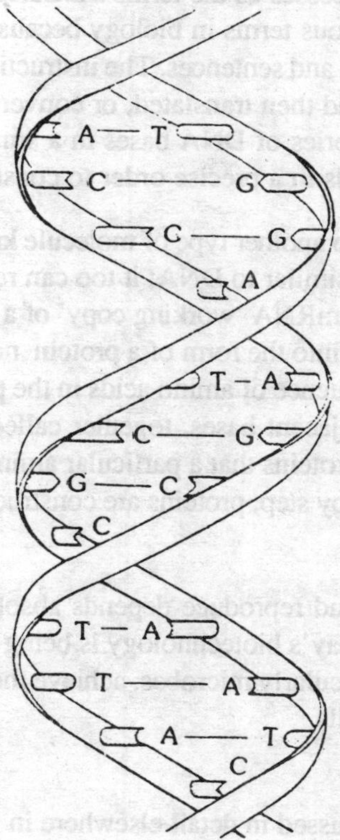

Fig. 5.8. The DNA double helix. Two strands of nucleotides twist round each other and the strands are linked by bonds between bases in each strand. Guanine (G) always pairs with cytosine(C), while adenine (A) always pairs with thymine (T). The continuous 'ribbon' (the backbone) of each strand is composed of alternating sugar and phosphate sections.

Fig. 5.9 illustrates how a molecule of DNA copies itself. The process can be viewed in terms familiar to photographers. Compare one strand of the double helix to a film negative and the other strand to a positive print. Clearly, both carry the same information or picture, but in different forms. If the positive and the negative are separated, the former can be used to make a new negative, while the original negative can be used to make a new positive. We now have two identical positive-negative pairs where only one existed before. DNA replication works much like this.

Perhaps the most elegant feature of life at the molecular level is the way in which the information for making proteins is encoded in DNA and read as instructions by other parts of the cell which assemble the proteins. A clue to the process is to be found in the emphasis placed earlier in this chapter on the specific *order* of the bases that make up DNA and the amino acids that make up proteins. When proteins are synthesized, the order in which their amino acids are assembled is dictated by the order of bases in the relevant section of a DNA molecule — the gene for that protein.

The scientists who study these processes us the terms *transcription* and *translation* to describe them. These are among the most felicitous terms in biology because they make explicit an analogy with a written language of letters, words and sentences. The instructions in DNA are first transcribed, that is, written out in a similar form, and then translated, or converted into the language of protein. Figures 5.10 and 5.11 show how the series of DNA bases in a small section of a gene are used to instruct the cell to assemble amino acids in a precise order to construct part of a protein molecule.

First the message is transcribed into another type of molecule know as messenger RNA (mRNA for short). Chemically, mRNA is very similar to DNA; it too can recognize bases in a DNA strand and through this recognition process an mRNA 'working copy' of a gene is constructed. The process of translating the instruction on mRNA into the form of a protein now begins. The sequence of base in an mRNA molecule specifies the sequence of amino acids in the protein. The words of the mRNA language consist of groups of three adjacent bases, together called a codon. Each codon tells the cellular machinery which synthesizes proteins that a particular amino acid must be incorporated into a protein at a specific place. Thus, step by step, proteins are constructed according to the instructions encoded in the DNA of genes.

The ability of organisms to live and reproduce depends absolutely on the manufacture of the correct proteins at the correct time. Today's biotechnology is being built on a deep understanding of the processes by which organisms, particularly microbes, achieve these dazzling feats of chemistry— and how we can capitalize on their skills.

Genetic Engineering

Genetic engineering has been discussed in detail elsewhere in the book. The following account provides a brief summary:

Genetic engineering means the manipulation of genes under highly controllable laboratory conditions. A central feature is the isolation and selective replication of specific genes. This is called *gene cloning*. The purpose of the exercise may be to obtain multiple copies of a gene itself or its products. The following steps are normally involved:

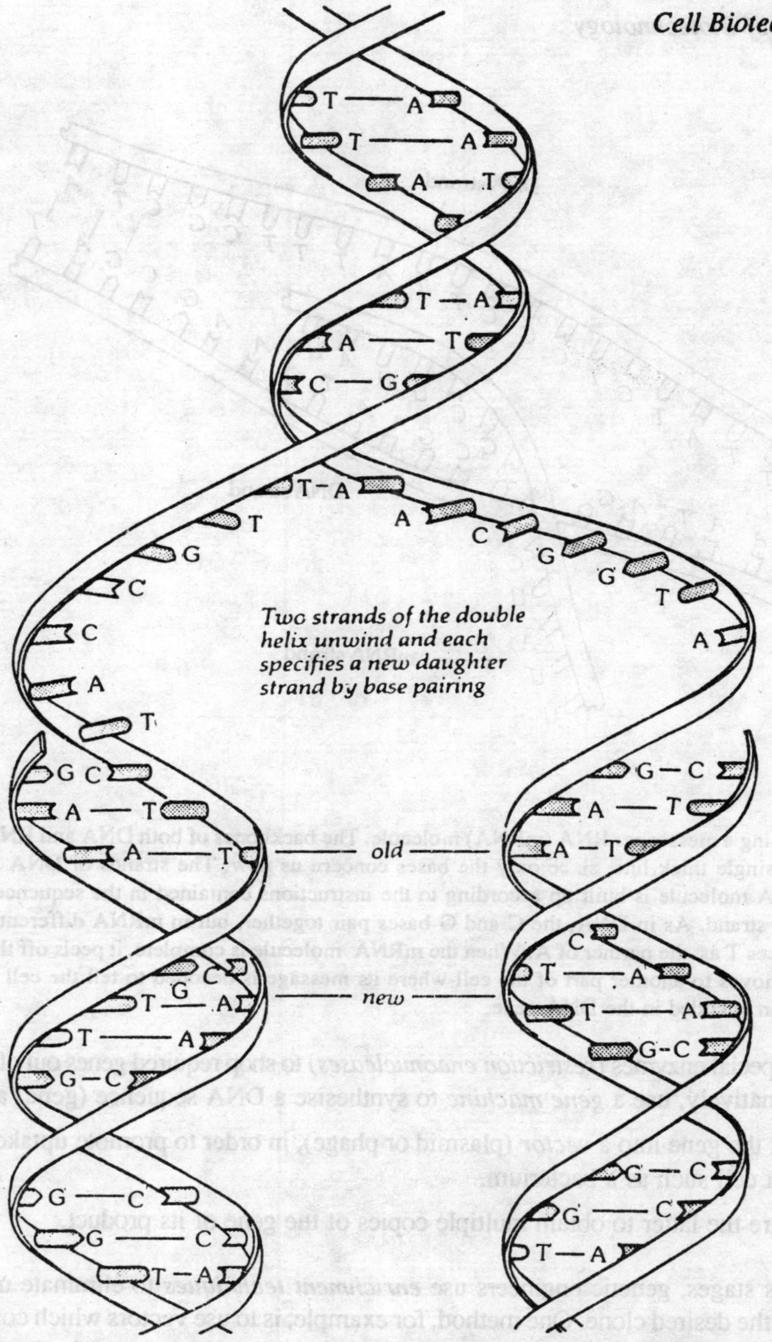

Fig. 5.9. How DNA copies itself. The two strands of the double helix begin to separate, breaking the bonds between the bases on each strand. The 'free' bases now pick up new nucleotides from a pool of such molecules provided by the cell. A pairs with T, and G pairs with C. Meanwhile, enzymes are at work joining the newly selected nucleotides into a chain. This process occurs with both of the 'parent' strands and, thus, two identical DNA molecules are created from one original. The specific pairing of the bases guarantees faithful reproduction.

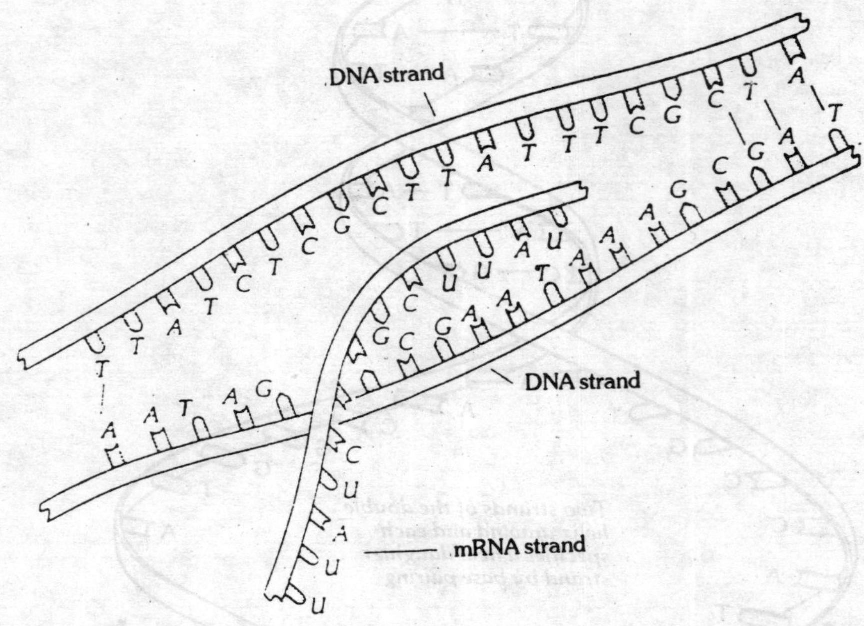

Fig. 5.10. Forming a messenger RNA (mRNA) molecule. The backbones of both DNA and RNA are represented by a single thick line, since only the bases concern us now. The strands of DNA separate and the mRNA molecule is built up according to the instructions contained in the sequence of bases on one DNA strand. As in DNA, the C and G bases pair together, but in mRNA different base, uracil (U), replaces T as the partner of A. When the mRNA molecule is complete, it peels off the DNA template and moves to another part of the cell where its message is decoded to tell the cell how to make the protein encoded in the DNA gene.

(i) Use special enzymes *(restriction enaonucleases)* to shop required genes out of a chromosome. Alternatively, use a *gene machine* to synthesise a DNA sequence (gene) artificially.

(ii) Insert the gene into a *vector* (plasmid or phage), in order to promote uptake of the gene by a host cell such as a baçterium.

(iii) Culture the latter to obtain multiple copies ot the gene of its product.

At various stages, genetic engineers use *enrichment techniques* to eliminate unwanted debris and to identify the desired clone. One method, for example, is to use vectors which confer a particular property on their hosts, such as drug resistance. If the bacteria are subsequently grown on a medium containing the antibiotic, only those which have incorporated the vector will grow. Further refinements are possible to ensure that the only bacteria which grow are those that have incorporated *hybrid vectors* (vectors into which foreign genes have been inserted).

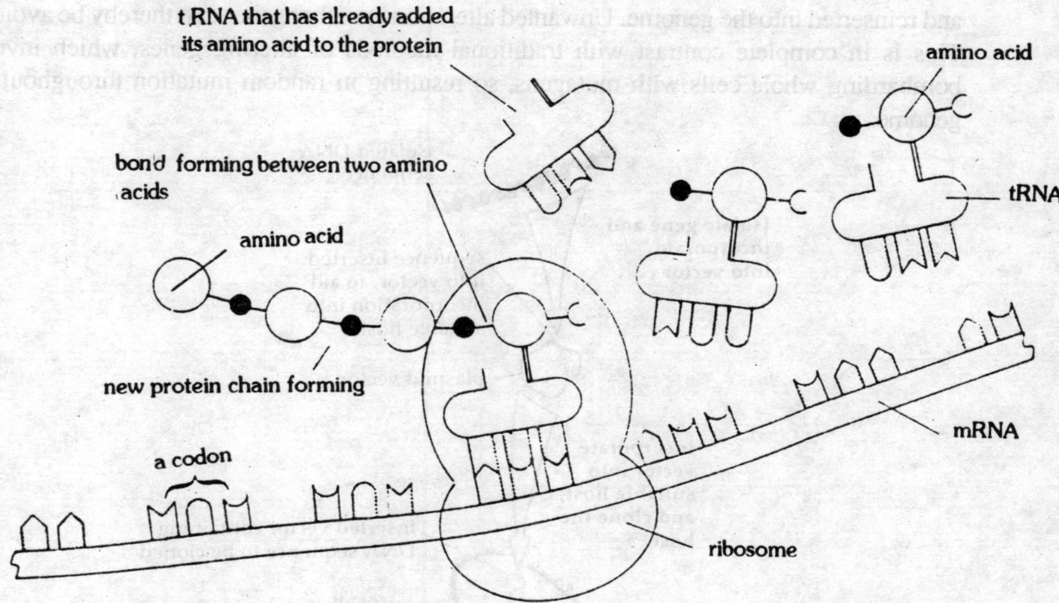

Fig. 5.11. The translation of mRNA involves dozens of types of molecules, among which are two other forms of RNA. Neither of these carries any genetic message, but they help the cell 'read' the information in mRNA. Once mRNA has been made, it leaves the nucleus (if that cell has a nucleus) and enters the main body of the cell. There, comparatively large globular structures, known as ribosomes, latch on the wandering mRNA molecules. Ribosomes, are built up from several sorts of protein plus a form of RNA called ribosomal RNA. Once the ribosomes have grasped the mRNA, the third type of RNA—transfer RNA (tRNA)—comes into action. There are many sorts of tRNA and each is able to recognize certain groups of three bases (codons) on the mRNA. Furthermore, each type of tRNA carries with it just one specific type of amino acid. The translation of the genetic code depends on the fact that one end of a tRNA molecule recognizes specific codons, while the other end of the same tRNA molecule carries a particular amino acid.

The mRNA is moved along the ribosome, much like a film moves across a ratchet in a camera. At each stage one 'frame; is made visible to the decoding centre, the 'frame' in this case being a codon. Many tRNA molecules mill around the ribosome and its mRNA. Each checks to see if the codon 'exposed' in the ribosome is one that it can match up with. If it is, the tRNA deposits the amino acid it has brought on to the last amino acid in the growing protein chain. The mRNA them moves another notch along the ribosome, exposing the next codon, and so the process continues.

Three techniques have been made possible by gene cloning which have important applications in biotechnology:

(i) *DNA sequencing*. The rapid identification of base sequences in a gene provides detailed information about the organisation of the genome and the amino acid sequences of proteins.
(ii) *Production of DNA probes*. 'Hot' gene clones can be made using radioactive nucleotides. These can then be used as tracers to identify similar sequences in intact genomes.

(iii) *Site-directed mutagenesis*. Isolated genes can be subjected to subtle and specific alterations and reinserted into the genome. Unwanted alterations to other genes can thereby be avoided. This is in complete contrast with traditional methods of altering genes, which involve bombarding whole cells with mutagens, so resulting in random mutation throughout the genome.

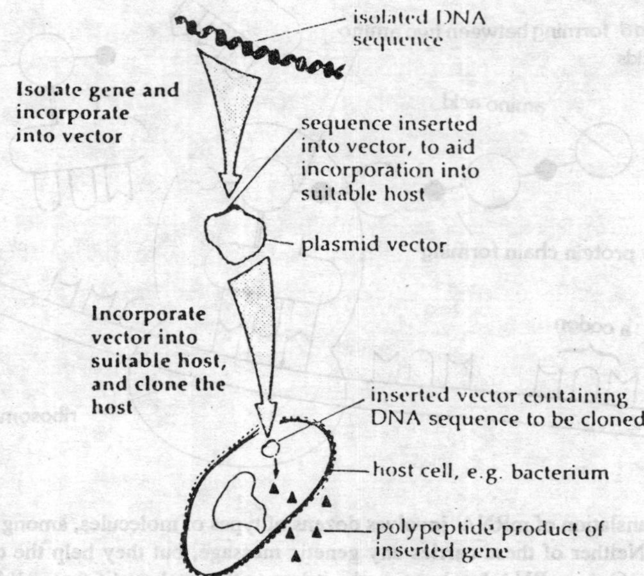

Fig. 5.12. *Gene cloning*. A specific application of gene cloning—ensulin production. The book *Genetic Mechanisms* in this series.

Applications of genetic engineering—gene cloning and associated technologies—are found in the food and drug industries, waste disposal medium and agriculture.

Somatic Cell Cultures

Animal cell cultures

The most widely used materials for animal cell cultures are young or cancerous tissues, because they can be induced to grow and divide more easily than most others. Nonetheless, all animal cells have extremely demanding requirements (e. g. 37 ± 0.5 °C; pH 7.3 ± 0.1; etc.). The cells are first gently separated using trypsin which digests the membrane proteins binding them together. They are then allowed to attach to the base of a culture vessel, to Teflon tubing or to *microcarriers* (plastic beads). This is essential because cultured animal cells only grow and divide if attached to a solid surface. Mitosis will occur until the whole surface is covered with cells, at which point growth comes to an abrupt halt *(contact inhibition)*. Animal cells are normally cultured to produce specific end products such as *monoclonal antibodies* (see somatic cell hybridisation below). Specialised animal cell can only produce similar cells; they are in no sense equivalent to zygotes and cannot develop into organisms. The situation in plants, however, is quite different.

Plant cell cultures

Two tpyes of culturing are possible with plant cells. The first, *tissue culture*, leaves the cell wall intact. The second involves preliminary disgestion of the wall with *cellulase* to produce wall-less *protplasts*. These, like animal cells, are easily ruptured unless the osmotic potential of the surrounding fluid is similar to that of the cell.

Using subcultures of plant tissue or protoplasts, clones of genetically identical plantlets can be produced. The rate at which particular strains or varieties of desirable plants are produced may therefore be greatly enhanced. Protoplasts, like isolated animal cells, may also be used for a particular type of genetic engineering called *somatic cell hybridisation.*

Somatic cell hybridisation

This technique brings together genetic material from completely different species, so overcoming the natural barriers which normally prevent hybridisation. Some the these genetic cocktails have proved very useful. In the case of plants, new species of *Petunia, Nicotiana* and other genera have been created, and plant breeders are hoping that he technique will produce improved varieties of crops. With animals, combinations such as human (or mouse) cells x cancer hybrids have been created in order to produce powerful drugs and diagnostic aids such as interferon and monoclonal antibodies.

Hybrid cells are produced by treating the parental cells with 40% polyethylene (polyethene) glycol. This results in a high frequency of fusion, and the *heterokaryons* (unfused nuclei) or synkaryons (fused nuclei) can then be cultured. A common problem with synkaryons is that they tend to be unstable and shed chromosomes from one or other set. Selective media must therefore be used to eliminate clones without the desired combinations of chromosomes.

Enzyme Technology

A number of biochemical conversions result from enzyme action. In many cases, the organisms are only used because they are needed to produce enzymes. The organism itself may even be a hindrance. Enzyme technology strives to eliminate the organism and to manipulate enzymes for their maximum industrial effect. There are several advantages over 'whole-organism' technology.

(i) *No less of substrate due to increased biomass*. For example, we may want yeast to turn sugar into alcohol. Yet even under the best conditions yeast will convert a proportion into cell wall material and protoplasm for its own growth.

(ii) *Elimination of wasteful side reactions*. Whole organisms may convert some of the substrate into irrelevant compounds or even contain enzymes for degrading the desired product into something else.

(iii) *Conditions optimal for a particular enzyme can be employed*. These conditions may not be optimal for the whole organism.

(iv) *Purification of the end product is easier*. Especially using immobilised enzymes.

Nevertheless, in some circumstances. 'whole-organism' technology may still be preferable:

(i) We may not want a 'pure' product. The difference between ethanol and an expensive wine lies in the 'impurities' in the latter. These include various aldehydes, ketones, tannins and acids, all of which enhance a wine's characteristics.

(ii) Even if we do want a pure product, the sequence of reactions leading to its synthesis may be so complex that whole-organism technology is the only realistic proposition.

(iii) Purified enzymes are extremely expensive compared with the organisms which produce them. This difficulty has been tackled from both ends: firstly, by improving the technology of enzyme production and purification; secondly, by making more efficient use of the enzyme once it becomes available.

Enzyme production

Micro-organisms are the favoured source of most industrial enzymes. Apart from the ease with which they can handled microbes produce a wide range of enzymes with distinctive and useful *properties*. A *protease* from one source might work optimally at low pH, but that from another at high pH. There is considerably more choice available than with higher organisms. However, the final choice may not depend only on the properties of the enzyme itself but also on how easily the producer organism can be grown and whether the enzyme is extracellular (cheap to purify) or intracellular (expensive to extract). Other considerations may include the yield, and what impurities are present; pure enzymes are extremely expensive, to that varying quantities of contaminants usually have to be tolerated.

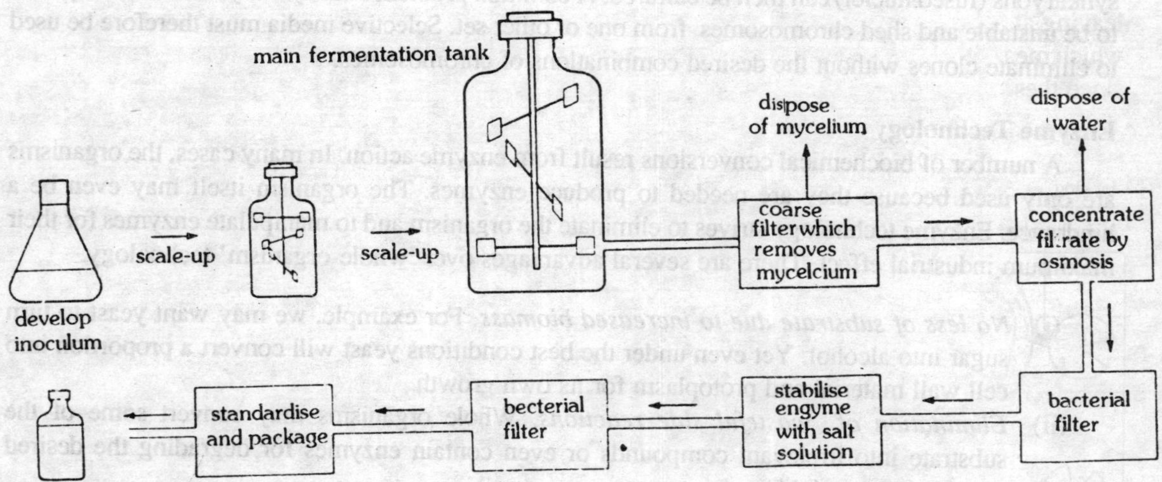

Fig. 5.13. Enzyme production.

Fig. 5.13 summarises the main stages in what has now become a multi-million pound industry. The finished product may vary from a liquid to a dry powder.

Enzyme immobilisation

Immobilisation means physically or chemically trapping enzymes or cells onto surfaces or inside fibres, gels or plastic particles. The benefits can be considerable:

(i) The same enzyme molecules can be used repeatedly, since they are not lost at the end of a production batch.
(ii) The enzyme does not contaminate the end product, neither does it pollute the environment, so simplifying purification processes (downstream processing).
(iii) Thermostability is often increased. Thus, glucose *isomerase* is stable at 65°C for almost year when immobilised. In solution in denatures within a few hours even at 45°C.
(iv) If appropriate, immobilised enzymes can be used in a non-aqueous environment. This is often convenient during the production of pharmaceuticals.
(v) Continuous ('open') production systems can be designed more easily.

Fermenter Technology

Some applications involve industrial fermentations at some point in the production process.

A *fermenter* (*bioreactor*) is a container designed to provide an optimum environment in which organisms or enzymes can interact with a substrate and form a desired product. Fermenters are so important that an entire industry has grown up around their design and construction. Some fermenters are *open*, i.e., they allow continuous processing with substrates entering at one end and products leaving at another. The ICI Pruteen fermenter is an example. However, most fermenters are *closed*, which means that processing is done in batches. This is more appropriate for antibiotics, for example, since these are only synthesised after mycelial growth.

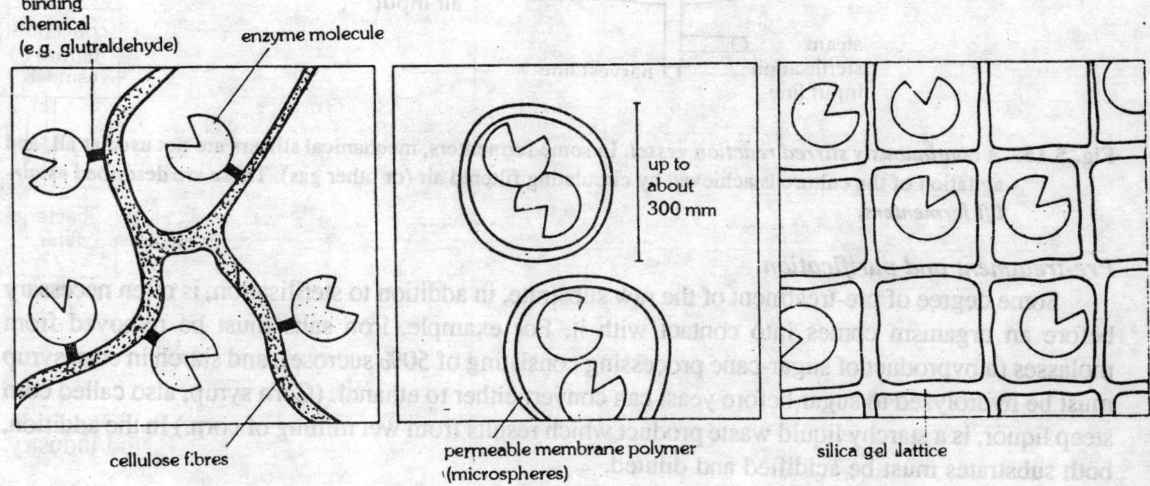

Fig. 5.14. *Enzyme mobilisation.* Materials other than those indicated may be used to adsorb or encapsulate enzymes of cells.

In a fermentation it may be necessary to regulate several factors within predetermined values: oxygen and CO_2 levels, pH, temperature and media concentration, for example. Moreover, a high degree of sterility is often needed to prevent the occurrence of an ecological succession within the fermenter. This is technically often the most difficult problem to overcome. Stainless steel or copper fermenters are usually favoured because they are highly resistant to steam sterilisation, have a low toxicity and are relatively inert to attack from most substrates, organisms and products.

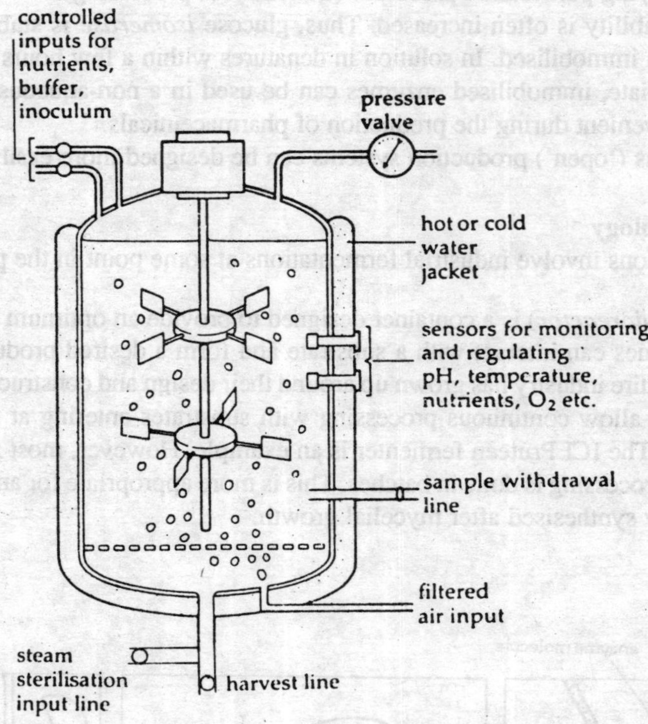

Fig. 5.15. *A continuously stirred reaction vessel.* In some fermenters, mechanical stirrers are not used at all, and agitation of the culture is achieved by circulating filtered air (or other gas). These are described as *air-lift fermenters.*

Pre-treatment and purification

Some degree of pre-treatment of the raw substrate, in addition to sterilisation, is often necessary before an organism comes into contact with it. For example, iron salts must be removed from molasses (a byproduct of sugar-cane processing consisting of 50% sucrose), and starch in corn syrup must be hydrolysed to sugar before yeast can convert either to ethanol. (Corn syrup, also called corn steep liquor, is a starchy liquid waste product which results from wet milling of corn.) In the addition, both substrates must be acidified and diluted.

Downstream processing

Desired products may be present only in very small quantities at the end of a fermentation—

perhaps just a few milligrammes per cubic decimetre in the case of pharmaceuticals. They may also be contaminated by undesirable waste products or may require some further modification. Various treatments, collectively, called downstream processing, are therefore often required.

Pharmaceuticals: Penicillin

Background

Antibodies are biochemicals secreted by micro-organisms which inhibit the growth of other micro-organisms. Only about 100 are marketed, although some 6000 are known to exist and about 200 new onces are discovered each year. Many of those which are not used have no obvious benefits over those already available, and some have adverse side effects. Half the useful antibiotics come from just three genera: *Penicillium, Cephalosporium* and *Streptomyces*. To be useful an antibiotic must cause maximum damage to a pathogen, and minimal damage to a patient. They generally work by exploiting the differences which exist between prokaryotic and eukaryotic cells. Hence, they attack murein synthesis (cell walls), bacterial membranes and bacterial protein synthesis (Fig. 5.16). Penicillin production is representative of the industry as a whole.

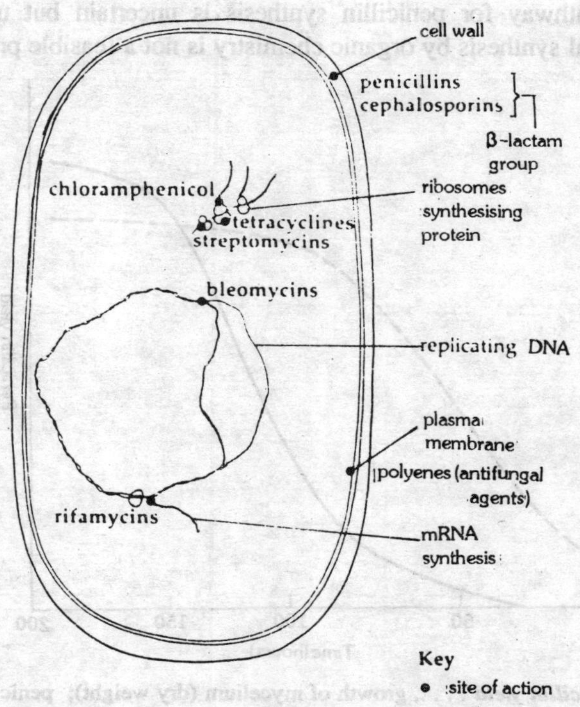

Fig. 5.16. *Sites of antibiotic action.* Out of the 6000 + antibiotics known, about 1000 come form just six genera of filamentous fungi (e.g. *Penicillium, Cephalosporium*), and 2000 come from just three genera of bacteria (e.g. *Streptomyces*). The sulphonamides are conventionally described as antibiotics, although they are entirely man made.

Fleming discovered penicillin secretion by *Penicillium notatum* in 1929, but it was World War

II, with the attendant fears for casualties and epidemics, that provided the stimulus for its commercial production. Using corn steep liquor as a substrate and an isolate of *Penicillium chrysogenum* (from a mouldy melon on a market stall!), penicillin yields of a few milligrammes per cubic decimetre were obtained. X-rays and nitrogen mustard were used to create mutants which gave higher yields and which did not excrete undesirable byproducts. Now, almost 50 years, later, strains are available yielding 10^4 times more than the original isolate. Production has also been dramatically increased by improvements to the media and to fermenter design. Phenyl ethanoic acid (phenyl acetic acid), for example, is now routinely added to the media since this induces the synthesis of a metabolic precursor of penicillin G (the most active form of penicillin). Batch fermentation is performed in sterilised 10^5 dm^3 continuously stirred reaction vessels (Fig. 5.15) the fermentation time being about 200 hours (Fig. 5.17). After fermentation is completed, the broth is passed over a filter and washed. The filtrate contains the penicillin. It is then mixed with a source of potassium ions to enhance solubility, filtered again and dried. The result is a crystalline potassium salt of penicillin G. The product is sometimes modified in order to enhance its, activity by adding new side group after first trimming it with the imobilised enzyme, *penicillin acyclase*.

The biochemical pathway for penicillin synthesis is uncertain but undoubtedly complex. Consequently, its artificial synthesis by organic chemistry is not a feasible proposition.

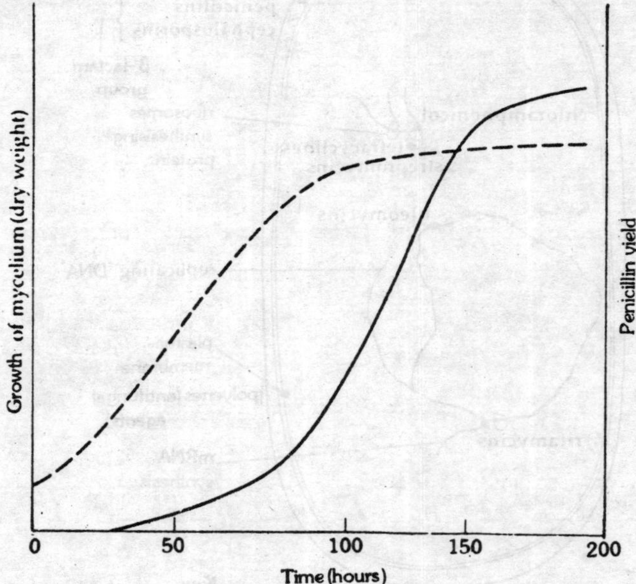

Fig. 5.17. ***Growth and penicillin yield*** **: . . ., growth of mycelium (dry weight); penicillin yield. Penicillin is a secondary metabolite, i.e. a material produced by an organism but which is not absolutely essential for its survival. Characteristically, high yields of secondary metabolites are obtained after growth has ceased and not during periods of exponential growth. In contrast, the yield of *primary metabolites* such as CO_2 and ethanol (produced by respiration) rises proportionally with the growth of the organism.**

6

Genetic Engineering*

Introduction

'Genetically synthesized human insulin seems to be safe and effective in man.' These undramatic words brought news of a landmark in science and an event that could revolutionize medicine during the next decade.

In July 1980, seventeen volunteers received injections of insulin that was made by genetic engineering at Guy's Hospital, London. In September 1982, insulin from bacteria became the first genetically engineered material to be licensed for use in humans.

The importance of insulin derived from genetically engineered microbes lies less in the value of insulin itself (though this is immense) than in the fact that it helps vindicate the claim that microbes can be persuaded to make a wide variety of substances with medical and commercial value. It has proved that microbes can manufacture foreign (for example, human) proteins, and that these proteins are safe to use. Insulin is only the first of an avalanche of valuable substances which will be manufactured with the aid of genetically engineered microbes. The list grows monthly, and it now also includes interferon, growth hormone, vaccines and blood proteins.

In the mid-nineteen-seventies an intense and fiery debate engulfed genetic engineering (sometimes known as gene cloning genetic manipulation, gene splicing or recombinant DNA research). With these entirely novel techniques, scientists can manipulate the very core of life — the DNA of which genes are composed. It is now possible to swap genes from one organism to another, inducing cells to manufacture materials they have never made before. Much of the storm over the wisdom of pursuing this awe-inspiring line of research revolved around the thorny issue of weighing possible risks (chiefly hypothetical killer 'superbugs') against the possible benefits, particularly in biological research and medicine.

Since genetic engineering is a key tool in biotechnology it is important to understand its basic principles and how these can be applied.

Reasons for Turning To Genetic Engineering

We turned to genetic engineering because of the fact that it has the possibility of transferring

* Excerpts taken from the review published by Dubhan, New York.

genes from one organism to another which is an alluring prospect since genetic engineering could reduce the cost and increase the supply of an enormous range of materials now used in medicine, agriculture and industry. Moreover, there are many substances that are found in nature in only small quantities which might well prove invaluable if they were available in sufficient quantities for their potential to be examined.

A major attraction of using microbes as the factories for making these materials is that scientists and technologists have a great deal of experience in growing such organisms cheaply and efficiently on a large scale. Brewers and bakers have been doing this for millennia and the modern pharmaceutical industry has developed a new level of sophistication which will underpin much of the new biotechnological industries.

Thus, one of the major problems genetic engineering sets out to solve is that many types of cells cannot be grown outside their normal environment (at least not without an immense amount of difficulty), but very efficient techniques for growing microbes rapidly and cheaply have already been developed. Cells cannot be cultured from, for example, the human pancreas—the natural source of insulin — but genetic engineering can create microbes which can manufacture this material.

The basic principles of genetic engineering were developed barely a decade ago, and since then there has been astounding progress which has provided us with a set of tools of quite remarkable power and sophistication. In short, genetic engineering involves inserting new genetic information into an organism — usually a bacterium — to endow it with novel capabilities. It does not follow a single fixed set of procedures. The choice of method of depends on which gene is to be transferred and what type of organism is to receive the new genetic information. The choice even depends to some extent on the personal preferences of the scientists involved.

The biotechnological applications of genetic engineering consist of the following four main stages: obtaining the gene which codes for the product the microbial factory is to manufacture, inserting the gene into the microbes; inducing the microbes to start synthesizing the foreign product; and collecting that product.

Like other branches of engineering, genetic engineering takes materials available in nature, uses specialized knowledge and tools to modify them in particular ways, and assembles the pieces to yield the final structure. Genetic engineers are faced with problems arising from the fact that generally they cannot *see* the materials they are manipulating. Today's immensely powerful microscopes can provide a lot of detail about the shape and structure of microbes, but provide little information about the parts of the cell that are of central concern to genetic engineers — DNA and protein molecules. A human cell contains thousands of genes, each of which is the blueprint for manufacturing a protein, but even with the finest microscopes it is not possible to distinguish one gene from another. Thus, many of the subtlest tricks of genetic engineers are designed to reveal in other ways just what is going on inside the world of microbes, and what effects their manipulations are producing.

In the laboratory, the first step towards creating a genetically engineered bacterium which will manufacture a human protein is to *identify and isolate* the gene from a human cell which codes for that protein. With a few exceptions — notably the cells involved in sexual reproduction — all cells of an organism contain the same genetic information, but they do not all use that information in the same way. The cells of different organs have different functions to perform, and this specialization among cells is possible because of a phenomenon called *gene expression.* A gene is said to be expressed if it is used to direct the construction of an mRNA molecule, that can then be used to manufacture the relevant protein. Genes can be turned on or off rather like a light bulb, specially one fitted with a dimmer switch. Expression is not an all-or-nothing matter; some genes will be working at full tilt, some rather lethargically, while others are idle; it all depends on what proteins the cell needs at that time.

This difference in expression between genes in specific cell types is used by genetic engineers in the most elegant method of isolating the required gene. Rather than searching through the incredibly complex mass of genes encoded in DNA, this technique focuses on the cell's in mRNA molecules. Thus, if a gene to make insulin is needed, human pancreas cells are examined to find the mRNA molecules which have been copied from that gene. There will, of course, be other types of mRNA in that cell as well, but nothing like the multitude of diverse pieces of genetic information encountered in the cell's DNA.

The mRNA molecules must first be extracted from the cells. When cells are broken open, a great assortment of materials is released, including enzymes, other proteins, DNA, pieces of cell membrane, and mRNA. All these materials are separated from each other by treating the mixture with chemicals and subjecting it to a various number of physical forces. One of the most important steps is to put the mixture inside tubes which are spun round at very high speeds in a centrifuge. This separates different kinds of molecules according to their weights and shapes. Heavy globular materials quickly fall to the bottom of the tube while lighter, sinuous molecules are found higher up the tube.

The isolated population of mRNA molecules contains the genetic information required to manufacture all the proteins that the cell was synthesizing at the time it was destroyed. This information must now be converted back into the form of the equivalent DNA sequences.

Only a few years ago, there was no known method for converting genetic information from an mRNA form to a DNA form. Much was known about the enzymes that help cells to make RNA copies of DNA genes, but information never seemed to flow the other way. Then came a revelation— some viruses can perform the trick. They do so with the help of enzymes, *reverse transcriptases*, so-called because they carry out the reverse of the operation that transcribes DNA into mRNA. Viruses, that are responsible for many diseases of microbes, plants and animals are quite unlike any other entities. They are far smaller than even bacteria, and cannot be considered as truly living organisms since they are incapable of an independent existence. Viruses consist simply of a package of genetic material wrapped inside a protein coat. They reproduce by subverting the chemical machinery of host cells and forcing them to make replicas of the invading viruses. Certain types of virus use RNA as their primary genetic material. Since all forms of life — from bacteria to humans — store their genetic information in DNA molecules, these RNA viruses have reverse transcriptase enzymes that convert their RNA genes into a DNA form, thus tricking the host cell into creating new viruses.

Mixing reverse transcriptase with the human mRNAs and the building blocks of DNA produces DNA copies of the RNA molecules. This DNA is constructed in the form of single strands, rather than the more familiar double-stranded helix. Other types of enzymes — DNA polymerases — are then used to convert single-stranded DNA into the double-stranded form by stringing together bases in a second strand according to the sequence of bases on the first strand. The double-stranded DNA so produced is known as copy, or complementary, or cDNA, because it is a copy of the genetic instructions contained in the original mRNAs (an example is given in Figure 6.1).

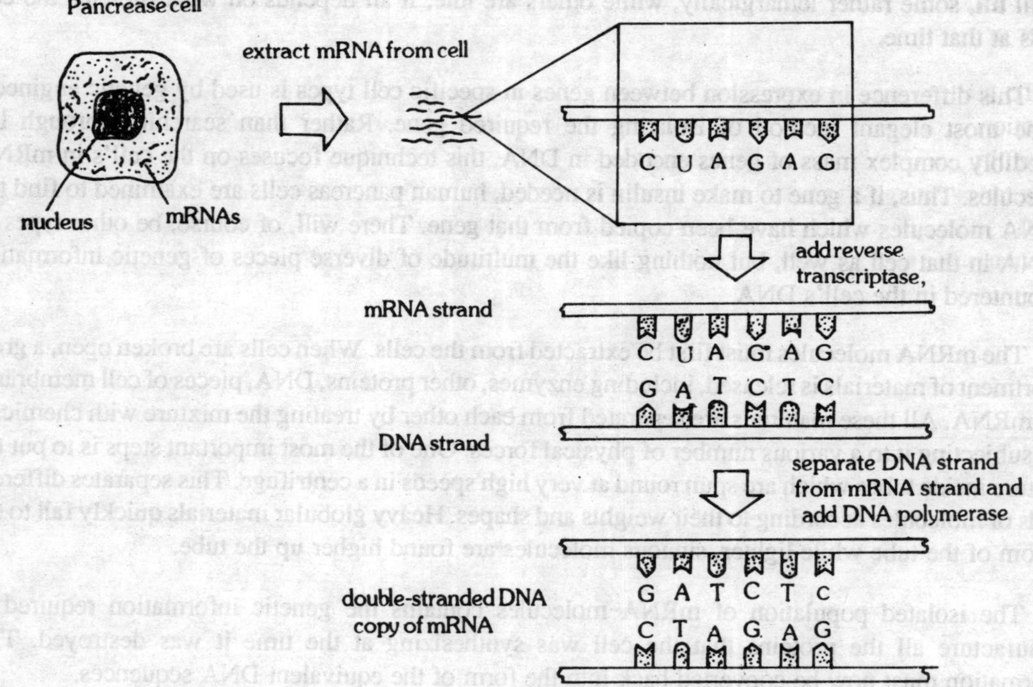

Fig. 6.1. Human insulin is one of many proteins manufactured in the pancreas. Each protein has its own gene and its own type of mRNA. Pancreas cells are broken apart, releasing many sorts of mRNA, which are then separated from the rest of the cell material. Only part of the sequence of bases of a single mRNA is shown. The enzyme, reverse transcriptase, is added to the mRNA along with the nucleotide building blocks of DNA. This enzyme assembles the nucleotides in a string according to the order of bases on the mRNA — that is, it uses mRNA as a 'template'. The newly synthesized DNA is converted from single-stranded form into a double-stranded helix with the aid of another enzyme, DNA polymerase. This cDNA contains the same genetic information as the mRNA from which it has been formed. The cDNA is, in effect, a copy of a human gene.

Since the process started with many different types of mRNA, there are now cDNA copies of many different genes, among which are copies of the insulin gene.

Plasmids

All the genes essential for a bacterium's survival are carried on its single, large, circular chromosome. There are also much smaller circles of DNA inside some bacteria and these rings are known as plasmids. Plasmids are enigmatic structures and their functions are not entirely understood, although it is clear that many of them carry genes that enable bacteria to resist antibiotics. Plasmids have an odd relationship with the rest of the cell; most importantly for genetic engineering, they will often pass from one cell to another, even if the cells are of different species.

Hence, if plasmids are taken from one set of bacteria and the human cDNA gene is 'stitched' into the plasmid ring, the plasmid's natural ability will permit it to enter bacteria and convey the human gene into its new home. A plasmid used in this way is known as a *vector*, from the Latin for carrier or bearer. Certain types of viruses can also act as vectors.

To stitch the human gene into a plasmid we need the services of another type of enzyme, a restriction enzyme. All enzymes are remarkably precise tools, and in restriction enzymes their powerful ability to distinguish between similar structures is brought to a peak of perfection. Faced with a tangled mass of DNA, restriction enzymes scan the double helix until they recognize certain specific sequences of bases, and then make a precise cut across the two DNA strands. In the case of a circular plasmid molecule, this opens the ring ready for the insertion of the human gene.

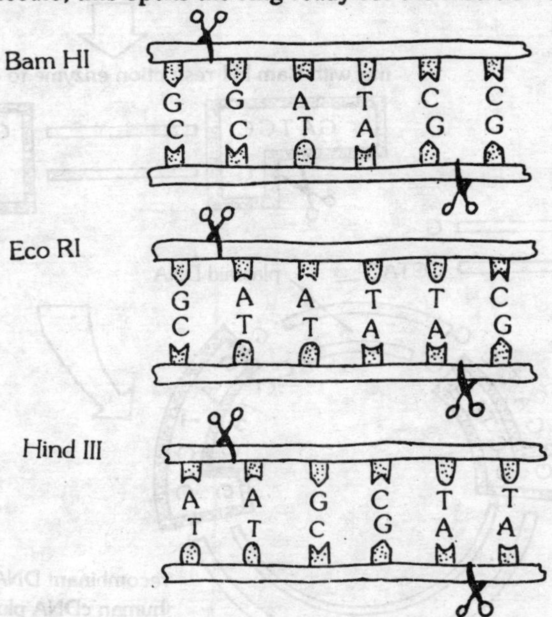

Fig. 6.2. The restriction enzyme *Bam* HI (so called because it is obtained from the bacterium, *Bacillus amyloliquefaciens* H) recognizes the sequence of six base pairs shown and cuts across the two strands to create staggered ends with hanging bases. *Eco* RI (from *E. coli* RY13) and *Hind* III (from *Haemophilus influenza* Rd) operate in a similar way.

About 300 different types of restriction enzyme have been found so far. Each recognizes a specific sequence of bases in DNA and each makes its own characteristic type of incision. Figure 6.2 shows how three such enzymes operate. A crucial point is that they do not cut straight across the two

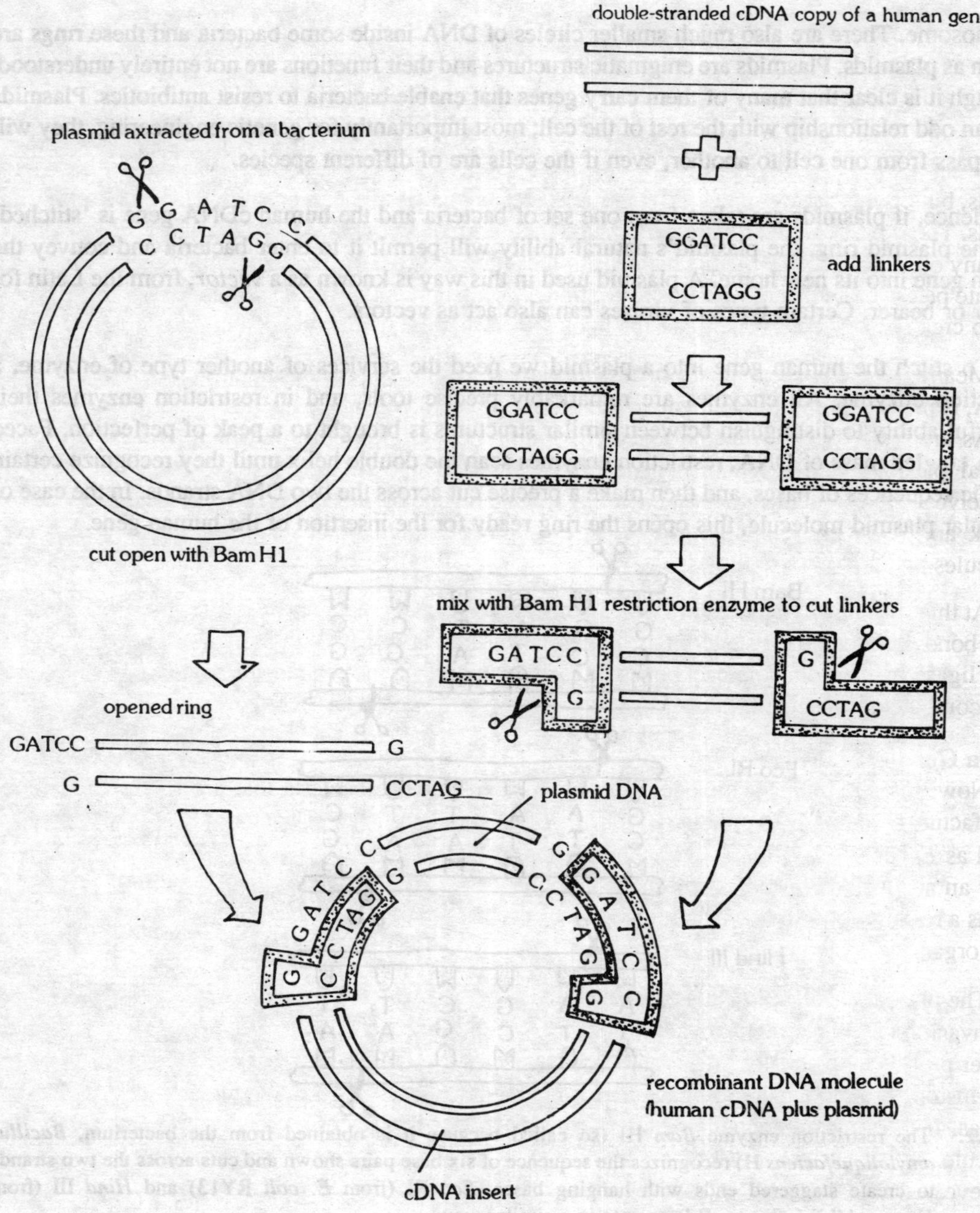

Fig. 6.3. **Creating recombinant DNA molecules. Plasmids are small circular pieces of DNA found inside some bacterial cells. The one on the left, named BR322, contains one set of the sequence of six bases recognized and cut by the restriction enzyme *Bam* HI. This enzyme splits open the plasmid ring, leaving a set of four unpaired bases at each end of the molecule.**

On the right is a cDNA molecule, to each end of which six linker bases are added. The same restriction enzyme is then used to cleave through the linkers, producing hanging (unpaired) bases which match those on the plasmid. The cDNA and the plasmids are now brought together and some will join, as shown, to produce a single circular molecule which incorporates the human gene cDNA. This is a recombinant DNA molecule. Because two pieces of DNA cut with this type of restriction enzyme possess a mutual attraction, the enzymes are said to create sticky ends

strands, but instead create staggered ends. This leaves four bases hanging from each side of both cut strands. These bases are no longer paired with their normal companions and are, thus, able to link up with any other piece of DNA that happens to have the same set of four hanging bases. Because two separate pieces of DNA cut with this restriction enzyme possess a mutual attraction, the enzymes are said to create sticky ends. These are the hooks upon that the human gene is hung (see Figure 6.3).

Meanwhile, in a separate set of test tubes and glass dishes, the human gene, in the form of cDNA, is prepared for its attachment to the opened plasmid. When the opened plasmid rings and the cDNAs are mixed some will join together forming new circles, each of which is made up of the original plasmid with the human gene cDNA inserted into it. It is also possible for plasmids themselves to join together without capturing a human gene or, indeed, for two human gene cDNAs to link and form a circle. However, it is only the plasmid/human gene hybrid or recombinant molecules that are now of interest.

At this stage the two pieces of DNA are held together rather loosely by only eight comparatively weak bonds between their sticky ends. A permanent link is forged with the aid of another enzyme, DNA ligase, which joins the backbones of each strand. Once these connections have been secured, the recombinant molecule is stable.

How a Gene is Cloned?

Now, the recombinant molecule can be inserted into the bacteria that are to act as factories manufacturing the desired protein. By far the most popular choice of bacterial host is an organism called as *Escherichia coli (E. coli)*. The predominance of this bacterium in genetic engineering is partly an accident of history. Molecular biologists have studied this particular species for decades and, as a result, more is known about the inner workings of this microscopic creature than about any other organism, including humans.

The plasmids were chosen because they have an inherent ability to enter the cells of *E. coli* and this invasion can be facilitated by adding a few simple chemicals to the mixture. Plasmids possess another property which is of great value to biotechnologists — they can make copies of themselves. Once inside a bacterial cell, a single plasmid may multiply itself to yield up to a few dozen identical replicas. If the plasmid contains a human gene, then that gene is copied along with the rest of the molecule. As the bacterium which harbours the plasmids is also growing and dividing — as often as once every twenty minutes — each daughter cell takes with it a few of the plasmids, which again reproduce themselves. Before long a single bacterium will have given rise to millions of descendants. A population of cells all derived from a single ancestor is called a *clone*, and all cells is a clone have the same genetic make-up. Thus, within a day or so a single bacterium carrying a recombinant molecule will yield millions of identical cells, all of which contain the original human gene, and the gene is then said to have been *cloned*.

Choosing the Right Bacteria

The genetic engineers now wanted to know about types of bacteria that are present in the mixture of cells growing on a glass dish. The aim being to identify the very few bacteria that contain a recombinant plasmid with its human gene, because it is these bacteria which will be used as microbial factories. They are confronted with three types of bacteria: first, there are bacteria that

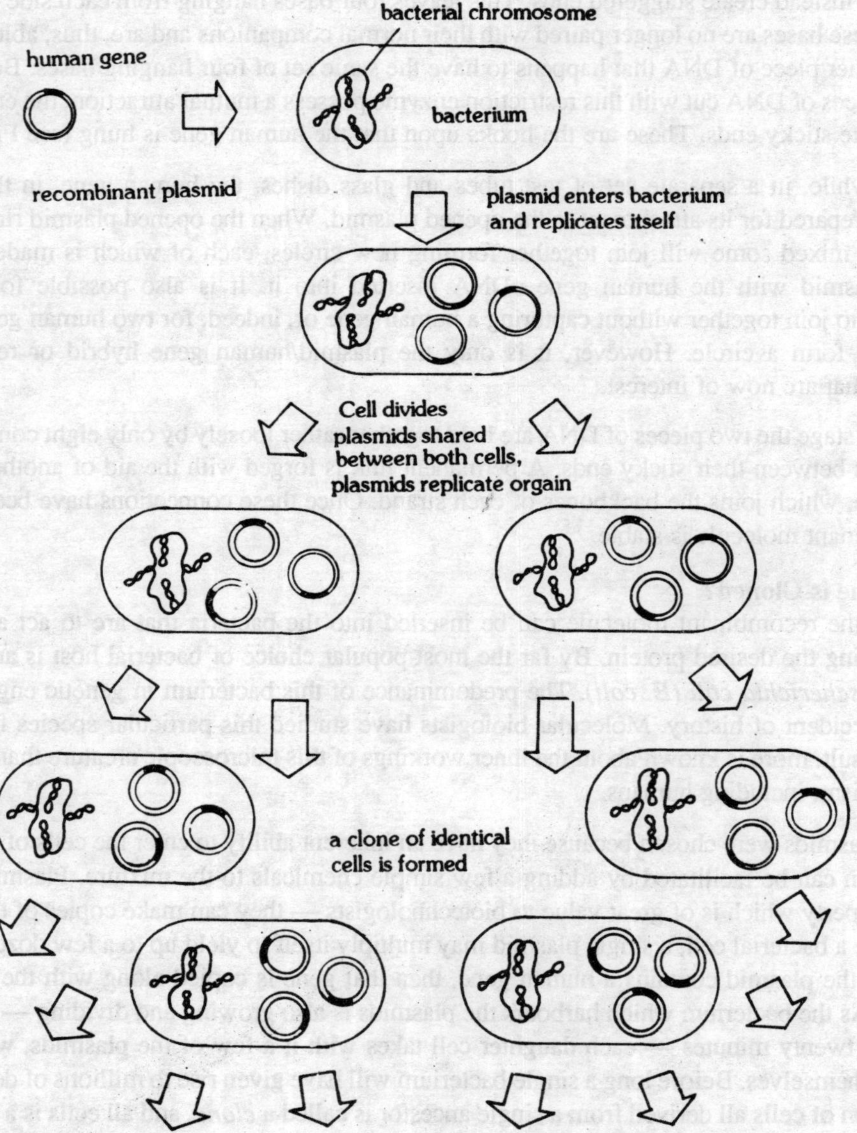

Fig. 6.4. A recombinant plasmid containing a human gene enters a living bacterium and is cloned.

have been infected by a plasmid carrying a human gene (these are the ones the scientists are seeking); second, there are many bacteria which have picked up normal plasmids (that is, those which do not contain human genes); and third, there are bacteria that have resisted invasion by any type of plasmid. Some selection process is required to identify the needed bacteria and make visible the changes occurring in this microscopic world. A simple but invaluable technique that helps during the checking process is *replic planting* (see Figure 6.5). The mixed population of bacteria is smeared across a plate of nutrients to separate the individual cells. Each cell is allowed to multiply until it has produced a visible clone of cells. Each clone is then separated into two portions. One is retained to grow on its ideal nutrients, while various experiments are carried out on the other portion. Some of these experiments will kill some of the bacteria, but because their identical twins are kept alive and well in another place, copies of every clone are available for future investigations.

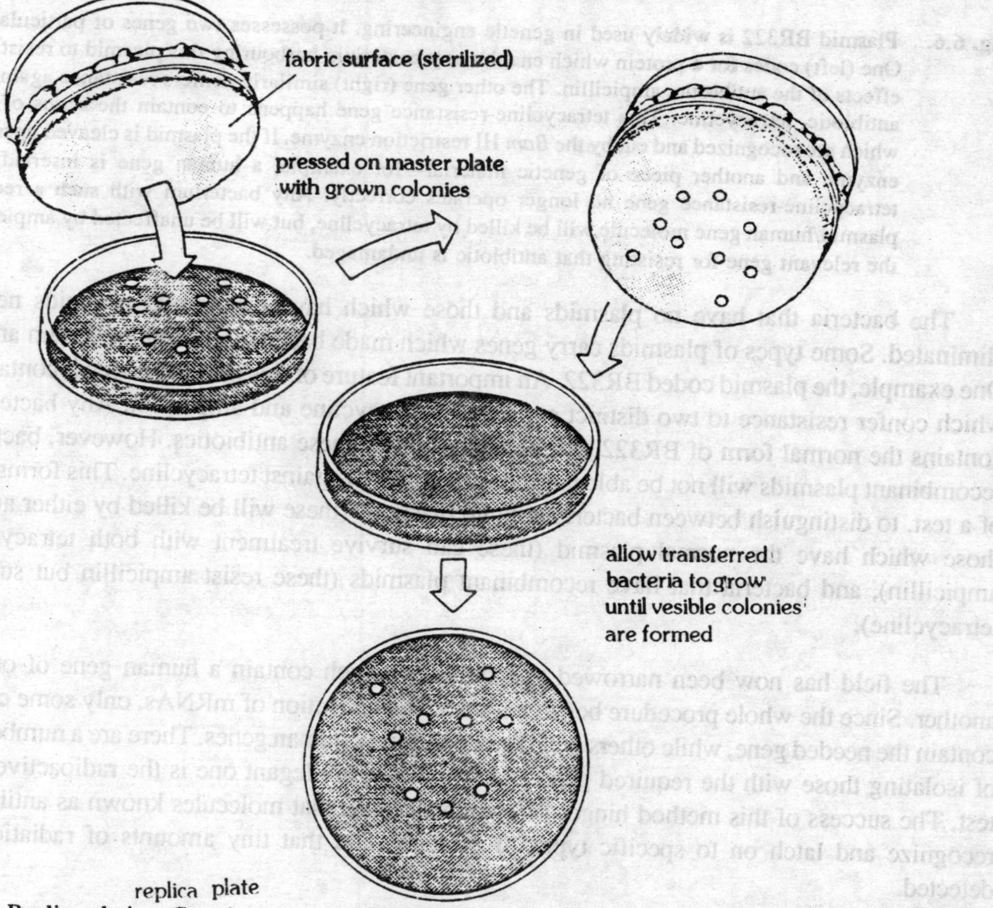

Fig. 6.5. Replica plating. Growing on a glass dish (Petri plate) of nutrients are many colonies of bacteria. All the cells in each clump are clones descended fr a single parent cell. A pad of sterilized fabric is gently placed on the plate. Some cells from each clone adhere to the fabric, and this is then lifted off and transferred to another dish which is free of bacteria. The transferred bacteria soon multiply, producing visible colonies. In this way, two plates are obtained on which identical clones appear in exactly the same positions.

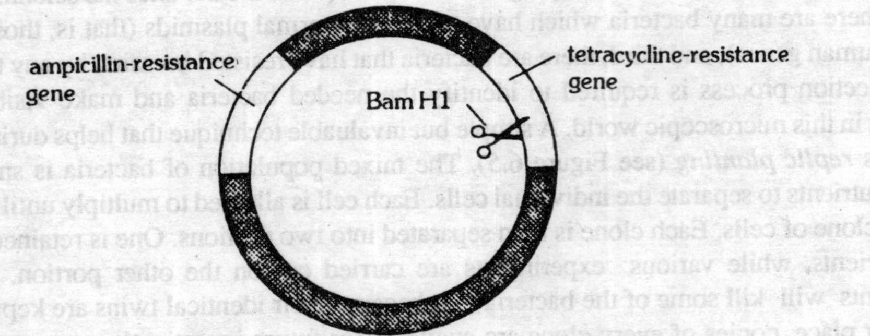

Fig. 6.6. Plasmid BR322 is widely used in genetic engineering. It possesses two genes of particular interest. One (left) codes for a protein which enables any bacterium harbouring this plasmid to resist the lethal effects of the antibiotic, ampicillin. The other gene (right) similarly confers resistance against another antibiotic, tetracycline. This tetracycline-resistance gene happens to contain the series of six bases which are recognized and cut by the *Bam* HI restriction enzyme. If the plasmid is cleaved open with this enzyme and another piece of genetic material—for example, a human gene is inserted, then the tetracycline-resistance gene no longer operates correctly. Any bacterium with such a recombinant plasmid/human gene molecule will be killed by tetracycline, but will be unaffected by ampicillin since the relevant gene for resisting that antibiotic is undamaged.

The bacteria that have no plasmids and those which have the normal plasmids need to be eliminated. Some types of plasmids carry genes which made bacteria resistant to certain antibiotics. One example, the plasmid coded BR322. An important feature of this plasmid is that it contains genes which confer resistance to two distinct antibiotics, tetracycline and ampicillin. Any bacterium that contains the normal form of BR322 will be unharmed by these antibiotics. However, bacteria with recombinant plasmids will not be able to protect themselves against tetracycline. This forms the basis of a test. to distinguish between bacteria with no plasmids (these will be killed by either antibiotic), those which have the normal plasmid (these can survive treatment with both tetracycline and ampicillin), and bacteria that have recombinant plasmids (these resist ampicillin but succumb to tetracycline).

The field has now been narrowed to the clones which contain a human gene of one sort or another. Since the whole procedure began with a mixed collection of mRNAs, only some clones will contain the needed gene, while others will contain different human genes. There are a number of ways of isolating those with the required gene and a particularly elegant one is the radioactive antibody test. The success of this method hinges on two main facts: that molecules known as antibodies can recognize and latch on to specific types of proteins; and that tiny amounts of radiation can be detected.

Antibodies feature very prominently in biotechnology and this is just the first example of their many uses. Among the many thousand types of antibody molecules, there are a few that have a very strong affinity for a specific protein molecule. To search for bacterial clones that are producing insulin, genetic engineers rely on the help of antibodies which recognize insulin molecules and stick

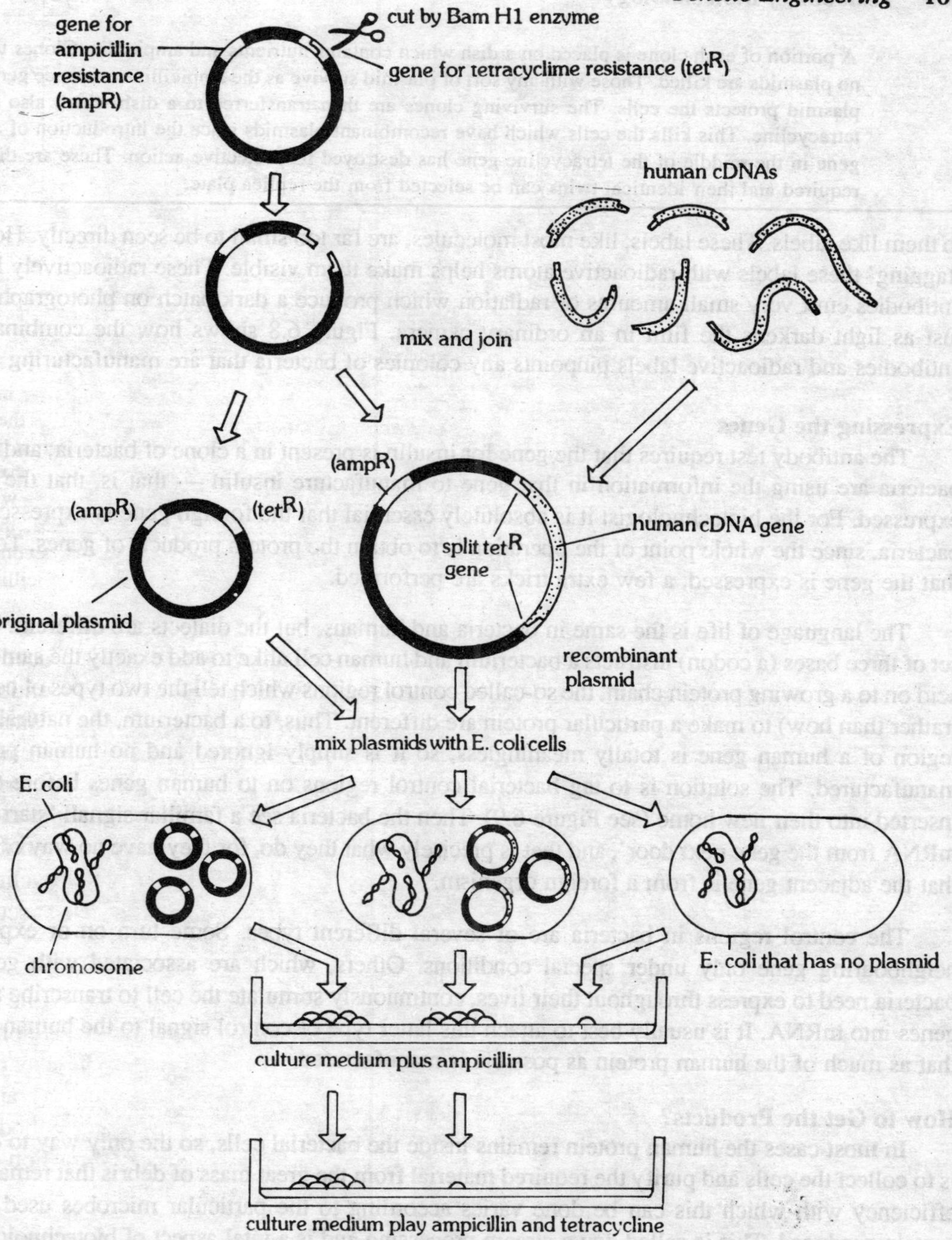

gene for ampicillin resistance (ampR)

cut by Bam H1 enzyme

gene for tetracyclime resistance (tet^R)

human cDNAs

mix and join

(amp^R)

(amp^R) (tet^R)

human cDNA gene

split tet^R gene

original plasmid

recombinant plasmid

mix plasmids with E. coli cells

E. coli

chromosome

E. coli that has no plasmid

culture medium plus ampicillin

culture medium play ampicillin and tetracycline

Fig. 6.7. Identifying the bacteria which harbour recombinant plasmid/human gene DNA molecules. Some of the bacteria that were mixed with plasmids will not have taken up a plasmid, and many of the remainder will contain unaltered plasmids which do not have a human gene. It is the third group of bacteria that are of interest — those that have been penetrated by recombinant plasmids. The clones of this last group of bacteria can be identified by a test that involves two sorts of antibiotic.

contd....

A portion of each clone is placed on a dish which contains nutrients and ampicillin. Clones that have no plasmids are killed. Those with any sort of plasmid survive as the ampicillin-resistance gene on the plasmid protects the cells. The surviving clones are then transferred to a dish which also contains tetracycline. This kills the cells which have recombinant plasmids since the introduction of a human gene in the middle of the tetracycline gene has destroyed its protective action. These are the clones required and their identical twins can be selected from the replica plate.

to them like labels. These labels, like most molecules, are far too small to be seen directly. However, 'tagging' these labels with radioactive atoms helps make them visible. These radioactively labelled antibodies emit very small amounts of radiation which produce a dark patch on photographic film, just as light darkens the film in an ordinary camera. Figure 6.8 shows how the combination of antibodies and radioactive labels pinpoints any colonies of bacteria that are manufacturing insulin.

Expressing the Genes

The antibody test requires that the gene for insulin is present in a clone of bacteria, and that the bacteria are using the information in that gene to manufacture insulin — that is, that the gene is expressed. For the biotechnologist it is absolutely essential that the foreign gene is expressed in the bacteria, since the whole point of the operation is to obtain the protein products of genes. To ensure that the gene is expressed, a few extra tricks are performed.

The language of life is the same in bacteria and humans, but the dialects are different. While a set of three bases (a codon) instructs a bacterium and human cell alike to add exactly the same amino acid on to a growing protein chain, the so-called control regions which tell the two types of cell *when* (rather than how) to make a particular protein are different. Thus, to a bacterium, the natural control region of a human gene is totally meaningless, so it is simply ignored and no human protein is manufactured. The solution is to tag bacterial control regions on to human genes before they are inserted into their new home (see Figure 6.9). Then the bacteria see a familiar signal: 'start making mRNA from the gene next door', and that is precisely what they do, for they have no way of sensing that the adjacent gene is from a foreign organism.

The control regions in bacteria are of several different types. Some turn on or express the neighbouring gene only under special conditions. Others, which are associated with genes the bacteria need to express throughout their lives, continuously stimulate the cell to transcribe adjacent genes into mRNA. It is usually best to attach this latter type of control signal to the human gene so that as much of the human protein as possible is manufactured.

How to Get the Products?

In most cases the human protein remains inside the bacterial cells, so the only way to obtain it is to collect the cells and purify the required material from the great mass of debris that remains. The efficiency with which this can be done varies according to the particular microbes used and the protein produced. This is called down-stream processing and is a vital aspect of biotechnology. The general principles are the same for genetically engineered microbes as they are for natural microbes. In essence, all the techniques make use of the particular protein's characteristic chemical and physical properties — its size, shape, electrical charge, solubility in water or other liquids, reactivity towards other chemicals, and so on.

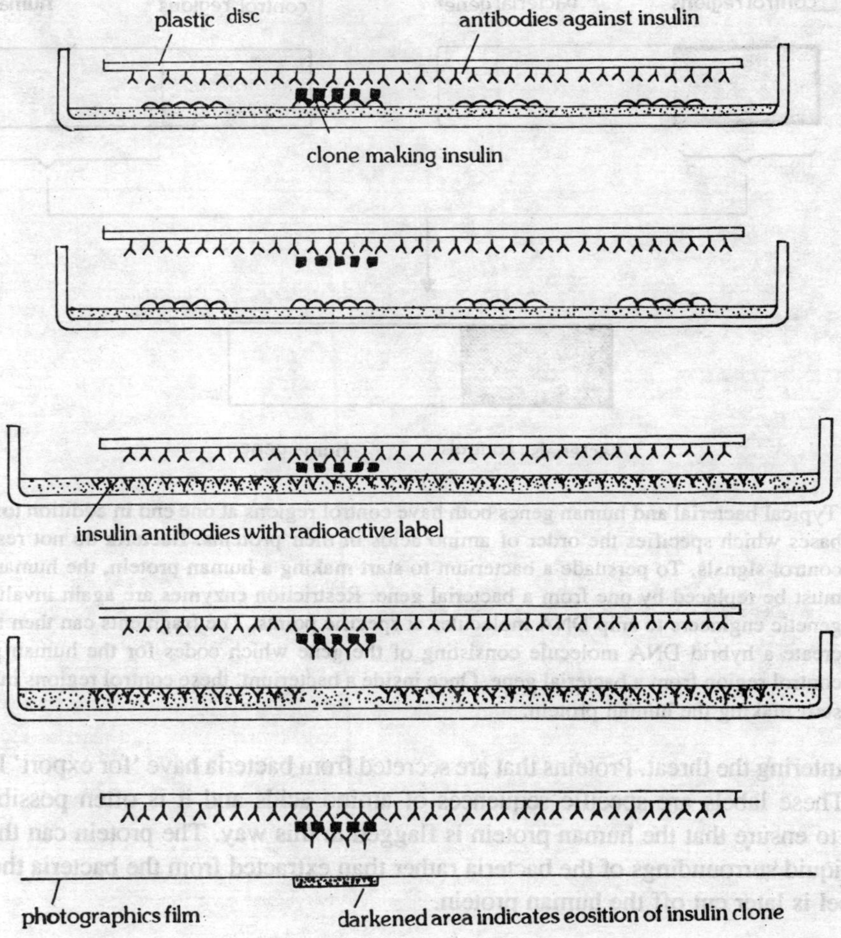

Fig. 6.8. **Pinpointing the right clone. To search for the clones that manufacture, for example, insulin antibodies which recognize insulin molecules are fixed to a plastic plate. When this plate is lowered on to a group of clones, it will pick up any insulin molecules that are present. Then the plate with its attached insulin molecules is placed on another layer of radioactively 'labelled' antibodies which can latch on to insulin molecules. These labelled antibodies are picked up *only* when they contact insulin. The whole plastic plate containing the antibody/insulin/radioactive antibody sandwiches is now placed on a photographic film. This becomes darkened wherever the radioactive antibodies are situated. By comparing the position of the dark marks on the film with the position of each clone on the original dish, the clones manufacturing insulin can be identified.**

There is a major service that genetic engineers can perform for the biotechnologists responsible for down-stream processing — bacteria can be tricked into secreting human proteins into their surroundings. Many microbes have efficient systems for transporting some of their natural proteins across their membranes and into the surroundings. This is essential for instance, in the case of enzymes that destroy antibiotics; the bacterium cannot wait until the antibiotic has penetrated the cell

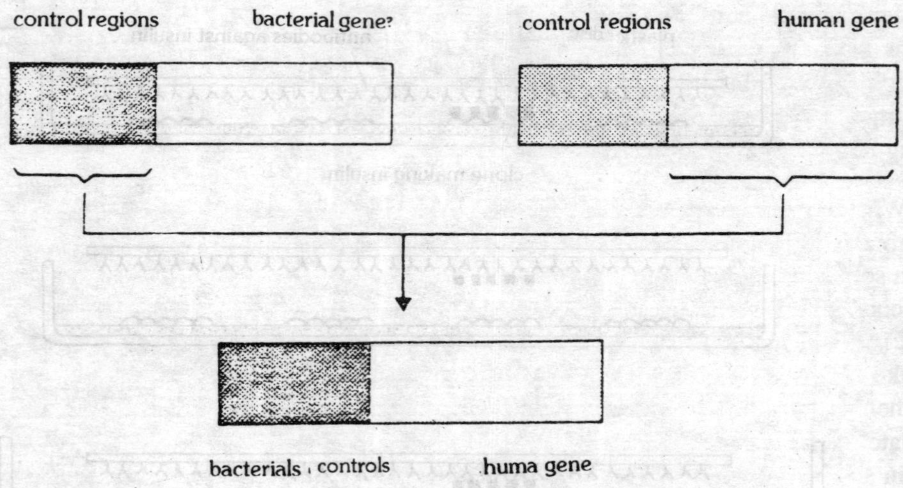

Fig. 6.9. Typical bacterial and human genes both have control regions at one end in addition to the sequence of bases which specifies the order of amino acids in their proteins. Bacteria do not respond to human control signals. To persuade a bacterium to start making a human protein, the human control region must be replaced by one from a bacterial gene. Restriction enzymes are again invaluable in helping genetic engineers to snip DNA molecules at specific points. The fragments can then be rearranged to create a hybrid DNA molecule consisting of the gene which codes for the human protein plus the control region from a bacterial gene. Once inside a bacterium, these control regions instruct the cell to start making the human protein.

before countering the threat. Proteins that are secreted from bacteria have 'for export' labels attached to them. These labels are specific sequences of amino acids and it is often possible for genetic engineers to ensure that the human protein is flagged in this way. The protein can then be isolated from the liquid surroundings of the bacteria rather than extracted from the bacteria themselves. The export label is later cut off the human protein.

The Genes and the Introns

In 1976 the summary of genetic engineering just presented would have been adequate. However, in the following year a remarkable discovery was made which bears directly on genetic engineering in the eighties — most human genes are in pieces. In bacteria and some types of algae (that is, the prokaryotic organisms) there is a direct relationship between the sequence of bases in a gene and the sequence of amino acids in its protein product. The first set of three bases indicates the first amino acid in the protein, the second set indicates the second amino acid and so on, until the last three bases signal of the end of the protein chain. The genes of eukaryotic organisms (the group which includes yeasts, plants and animals) are often split. For example, the gene which codes for collagen an important protein in tendons, bones and ligaments — is in more than fifty pieces. This does not mean that each part of the gene is on a separate DNA molecule, but that the DNA bases which code for the first stretch of amino acids are separated from the bases which code for the next stretch of amino acids by long strings of bases which do not contain information used in the construction of the collagen protein. These non-coding sequences are known variously as intervening sequences, *introns* or junk DNA. It seems that *introns* may help organisms to evolve new types of protein. By splitting the coding DNA into pieces it may be easier to juggle the basic units around to produce new proteins. This 'modular' theory is analogous to Lego building bricks; relatively few types of bricks (specific coding regions of DNA) can be rearranged to make many different structures (proteins).

Whatever the function of these introns is, they do not carry information which the cell normally uses to assemble amino acids into a protein chain. Eukaryotic organisms possess special enzymes which splice out the unwanted introns from mRNA which, when it is first transcribed from a gene, does contain introns. The resultant, shortened, messenger is called mature mRNA (see Figure 6.10) and it is this that is translated into protein. Similarly, it is mature mRNA that biotechnologists use to make cDNA copies of genes. If they were to make a cDNA copy of the mRNA including introns and then insert that into bacteria (which lack the ability to remove introns), the bacteria would translate the whole message into a string of amino acids which would be larger than the required protein and would not have the same properties.

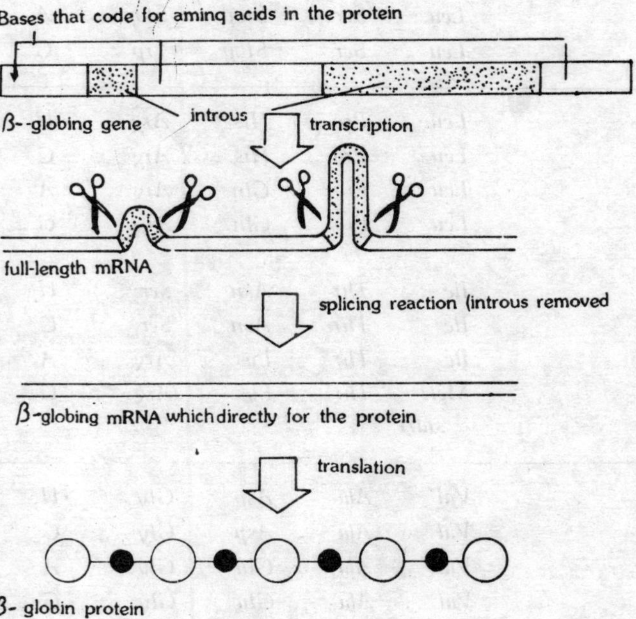

Fig. 6.10. When mRNA is first made from the gene of a higher organism it contains a copy of all the introns. Before the mRNA is employed to direct the synthesis of a protein, the cell removes the introns by a 'splicing' reaction. The shortened mRNA thus produced contains only the bases needed to direct the construction of a protein. It is this form of mRNA which is used to make cDNA.

Thus, a eukaryotic gene is like a bizarre manuscript. A few pages of exquisite prose are followed by meaningless letters and words before the next comprehensible passage. First, a messenger RNA copy of the entire text is made, then an expert editor excises the gibberish and reconnects the sections which convey useful information. The edited book is then ready to be 'read' on ribosomes and translated into a protein.

This fundamental difference between the organization of genes in animals and in bacteria severely limits the biotechnological applications of the aptly named short-gun cloning methods. In the shot-gun approach, the entire mass of DNA from human cells is cut up into thousands of pieces by means of restriction enzymes. These pieces can be inserted into plasmids and cloned as the plasmids and their bacterial hosts reproduce, but no proper human protein is produced. At best, the end product is a protein with all the right amino acids in it, but interspersed with entirely irrelevant extra amino acids.

first position	*second position*				*third position*
	U	*C*	*A*	*G*	
U	*Phe*	*Ser*	*Tyr*	*Cys*	*U*
	Phe	*Ser*	*Tyr*	*Cys*	*C*
	Leu	*Ser*	*Stop*	*Stop*	*A*
	Leu	*Ser*	*Stop*	*Trp*	*G*
C	*Leu*	*Pro*	*His*	*Arg*	*U*
	Leu	*Pro*	*His*	*Arg*	*C*
	Leu	*Pro*	*Gln*	*Arg*	*A*
	Leu	*Pro*	*Gln*	*Arg*	*G*
A	*Ile*	*Thr*	*Asn*	*Ser*	*U*
	Ile	*Thr*	*Asn*	*Ser*	*C*
	Ile	*Thr*	*Lys*	*Arg*	*A*
	Met/ start	*Thr*	*Lys*	*Arg*	*G*
G	*Val*	*Ala*	*Asp*	*Gly*	*U*
	Val	*Ala*	*Asp*	*Gly*	*C*
	Val	*Ala*	*Glu*	*Gly*	*A*
	Val	*Ala*	*Glu*	*Gly*	*G*

Fig. 6.11. The genetic code. Each codon (sequence of three bases) in mRNA specifies a particular amino acid (indicated by its three-letter abbreviation). The genetic code shows this relationship, for example, the codon UCA specifies Ser, the amino acid serine. Three codons signal 'stop', the end of a protein chain. One codon, AUG, can indicate either the amino acid methionine (Met) or the start of a protein chain.

The Gene Machines

In 1953, Frederick Sanger and his colleagues in Cambridge crowed years of labour when they announced that they had worked out the sequence of amino acids in the insulin molecule. This was the

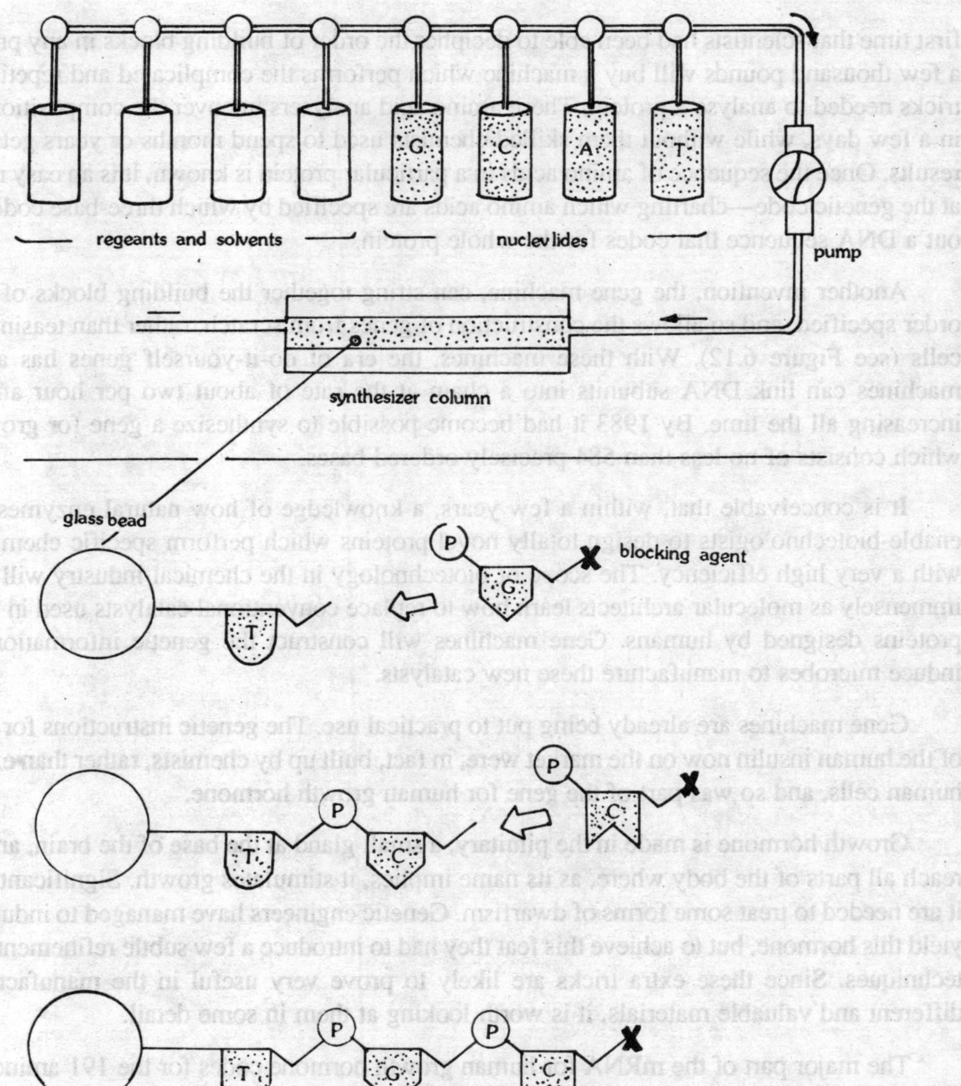

Fig. 6.12. A gene machine. Four separate reservoirs of the building blocks of DNA (the nucleotides A,T,C and G) are connected via tubes to a cylinder packed with glass beads. Suppose that the short series of nucleotides T-G-C is to by synthesized. The cylinder is first filled with beads with a single T attached, and then flooded with G bases from the reservoir. The right-hand side of each G shown in the picture is chemically modified or blocked so that it cannot join up with any of the other Gs in the surroundings. This prevents the sequence T-G-C...from building up. Any unbound Gs are now flushed from the cylinder and other chemicals added to remove the blocks. The cycle is then repeated with the introduction of C nucleotides into the cylinder and, thus the sequence T-G-C is constructed on each glass bead. Once the desired sequence is finished, the chains of nucleotides are removed from the beads with the aid of chemicals. The whole process of pumping in the necessary chemicals at the right time is controlled by a microcomputer control system.

first time that scientists had been able to decipher the order of building blocks in any protein. Today, a few thousand pounds will buy a machine which performs the complicated and repetitive chemical tricks needed to analyse a protein. These amino acid analysers uncover the composition of a protein in a few days, while without them skilled chemists used to spend months or years getting the same results. Once the sequence of amino acids in a particular protein is known, it is an easy matter to look at the genetic code—charting which amino acids are specified by which three-base codons-and work out a DNA sequence that codes for the whole protein.

Another invention, the gene machine, can string together the building blocks of DNA in any order specified, and so allows the construction of genes from scratch, rather than teasing them out of cells (see Figure 6.12). With these machines, the era of do-it-yourself genes has arrived. Gene machines can link DNA subunits into a chain at the rate of about two per hour and speeds are increasing all the time. By 1983 it had become possible to synthesize a gene for growth hormone which consists of no less than 584 precisely ordered bases.

It is conceivable that, within a few years, a knowledge of how natural enzymes operate will enable biotechnologists to design totally novel proteins which perform specific chemical reactions with a very high efficiency. The scope of biotechnology in the chemical industry will be expanded immensely as molecular architects learn how to replace conventional catalysts used in industry with proteins designed by humans. Gene machines will construct the genetic information required to induce microbes to manufacture these new catalysts.

Gene machines are already being put to practical use. The genetic instructions for the synthesis of the human insulin now on the market were, in fact, built up by chemists, rather than extracted from human cells, and so was part of the gene for human growth hormone.

Growth hormone is made in the pituitary, a small gland at the base of the brain, and secreted to reach all parts of the body where, as its name implies, it stimulates growth. Significant quantities of it are needed to treat some forms of dwarfism. Genetic engineers have managed to induce bacteria to yield this hormone, but to achieve this feat they had to introduce a few subtle refinements to the basic techniques. Since these extra tricks are likely to prove very useful in the manufacture of many different and valuable materials, it is worth looking at them in some detail.

The major part of the mRNA for human growth hormone codes for the 191 amino acids of the hormone itself. However, this hormone is secreted by pituitary cells, and a 'for export' signal is provided to ensure that this happens. This signal consists of an extra twenty-six amino acids at the beginning of the protein, and the instructions for manufacturing this signal are also encoded in the MRNA. The signal amino acids are excised from the protein when it leaves the pituitary cells, yielding the shorter active form of the hormone. This is both fine and necessary when the hormone is being made in human cells, but causes a problem when the gene is put into bacteria. The bacteria do not understand the signal's significance and do not remove the excess twenty-six amino acids. The result is a protein which is too large and would not perform its proper function when given to patients. Genetic engineers construct a partly artificial gene which will instruct bacteria to make growth hormone without the signal. This gene can then be inserted into a plasmid, cloned and human growth hormone produced.

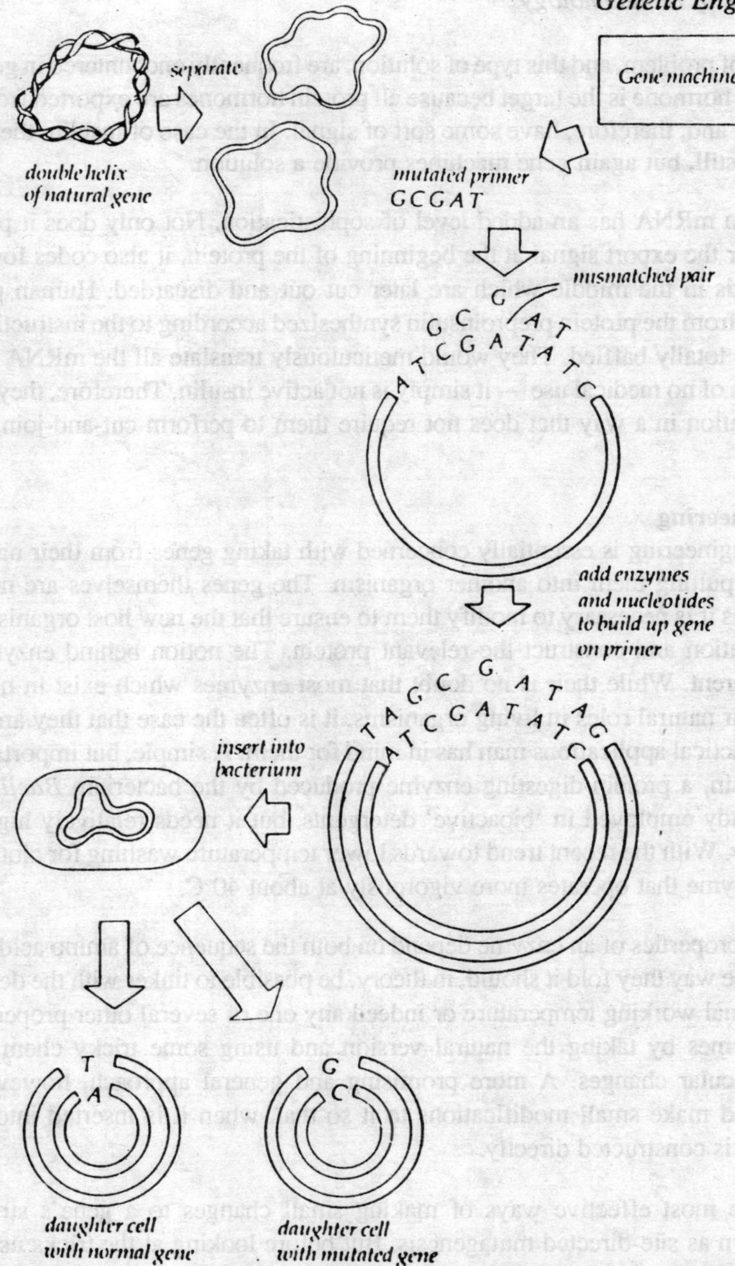

Fig. 6.13. Assembling a human growth hormone gene. First, a cDNA copy of the mRNA for growth hormone plus signal is obtained. Ideally, genetic engineers would cut this in the place which corresponds exactly with the junction between the hormone and signal, but no known restriction enzyme can slice the DNA at that point. Instead, an *Hae III* restriction enzyme is employed. This cuts fifty bases off the cDNA, corresponding to the unwanted twenty-six amino acids *plus* twenty-four amino acids of the hormone itself. The missing genetic information is then replaced by synthesizing it in a gene machine and tagging it on to the truncated gene.

This kind of problem, and this type of solution, are frequently encountered in genetic engineering, particularly if a hormone is the target because all protein hormones are exported from the human cells that make them and, therefore, have some sort of signal. In the case of insulin, the situation is a little more complex still, but again gene machines provide a solution.

The insulin mRNA has an added level of sophistication. Not only does it possess a sequence which codes for the export signal at the beginning of the protein, it also codes for stretch of thirty-five amino acids in the middle which are later cut out and discarded. Human pancreas cells can prepare insulin from the protein preproinsulin synthesized according to the instructions in the mRNA, but bacteria are totally baffled. They would meticulously translate all the mRNA into a single large protein which is of no medical use — it simply is not active insulin. Therefore, they must be given the genetic information in a way that does not require them to perform cut-and-join operations on the protein itself.

Enzyme Engineering

Genetic engineering is essentially concerned with taking genes from their natural home in one organism and putting them into another organism. The genes themselves are not usually altered, except insofar as it is necessary to modify them to ensure that the new host organism can utilize their genetic information and construct the relevant protein. The notion behind enzyme engineering is somewhat different. While their is no doubt that most enzymes which exist in nature are superbly adapted for their natural roles in living organisms, it is often the case that they are not ideally fitted to particular practical applications man has in mind for them. A simple, but important, example is the enzyme subtilisin, a protein-digesting enzyme produced by the bacterium *Bacillus subtilis*. This enzyme is already employed in 'bioactive' detergents, but it needs relatively high temperatures to work efficiently. With the recent trend towards lower temperature washing for clothes there is a clear need for an enzyme that operates more vigorously at about 40°C.

Since the properties of an enzyme depend on both the sequence of amino acids that make up the molecule and the way they fold it should, in theory, be possible to tinker with the design of an enzyme to alter its optimal working temperature or indeed any one of several other properties. It is possible to modify enzymes by taking the natural version and using some tricky chemistry to effect the necessary molecular changes. A more promising and general approach, however, is to take the natural gene and make small modifications to it so that, when it is inserted into a bacterium, the altered enzyme is constructed directly.

One of the most effective ways of making small changes to a gene's structure involves a technique known as site-directed mutagenesis. But before looking at the tricks used in this form of molecular tinkering it is useful to examine how biotechnologists first get their ideas about just *how* they want to alter the gene in question.

It is now a relatively trivial, if tedious, task to find out which amino acids are in a particular enzyme and how they are strung together. It is far from simple, however, to discover how the amino acid chain is folded into a three-dimensional shape— and it is fearsomely difficult to understand how

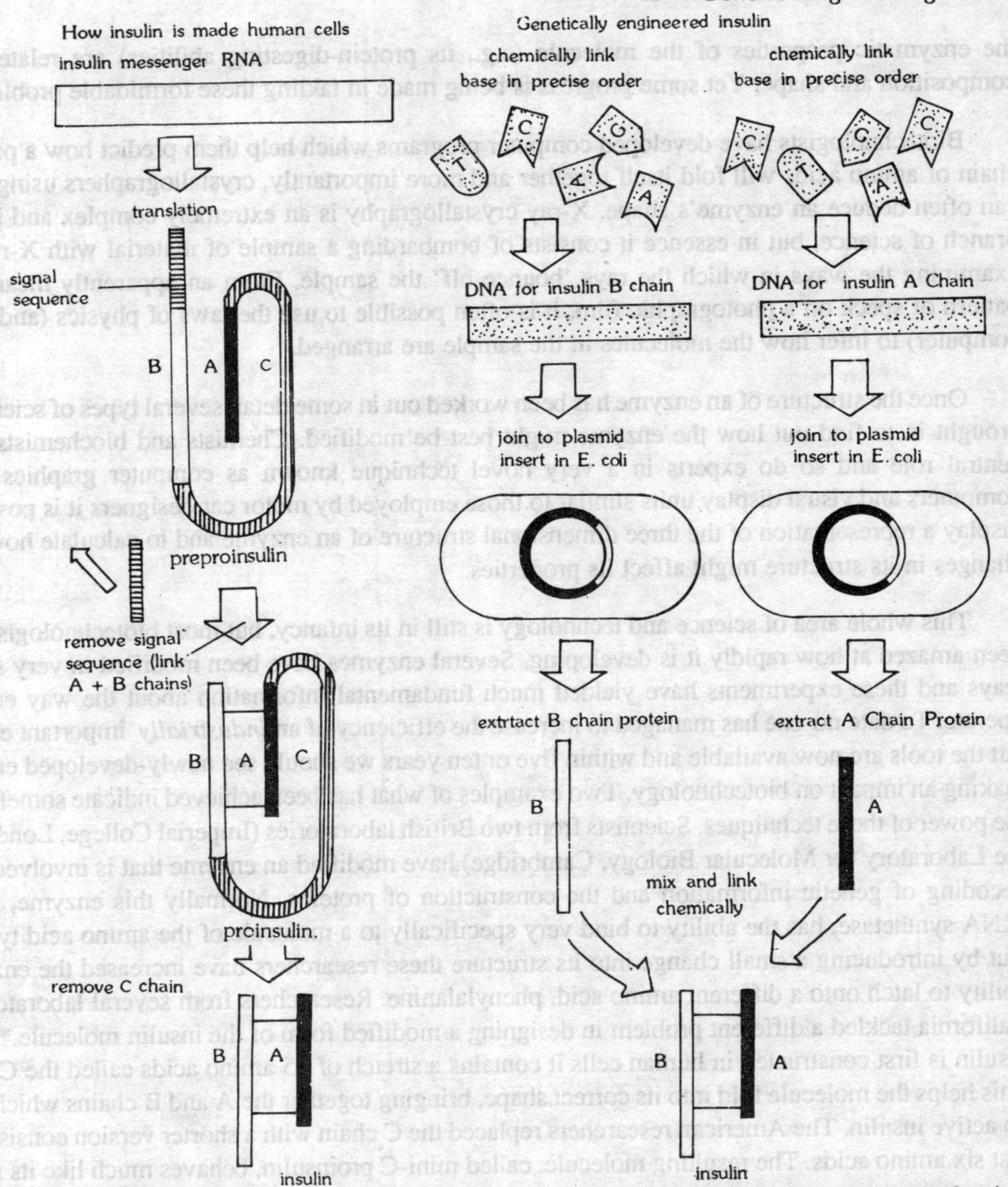

Fig. 6.14. Genetic engineering of insulin. The mRNA of human insulin codes for four distinct groups of amino acids (top left): the A, B and C chains plus the signal sequence. When mRNA is translated a protein — preproinsulin — is formed which contains all four parts. To obtain active insulin, human pancreas cells first remove the signal sequence to yield proinsulin and then cut out the C chain, leaving only the linked A and B chains, insulin itself. Bacteria cannot perform this cutting and joining process. The production of genetically engineered insulin does not start with insulin mRNA, but with but with chemically synthesized pieces of DNA which code only for the A and B chains. The synthetic gene for the A chain is inserted into one batch of bacteria, and that for the B chain is placed in another batch. The A and B chains so produced are extracted from the E. coli cells, mixed and joined chemically to yield exact copies of human insulin molecules.

the enzymatic properties of the molecule (e.g., its protein-digesting abilities) are related to its composition and shape. Yet some progress is being made in takling these formidable problems.

Biotechnologists have developed computer programs which help them predict how a particular chain of amino acids will fold itself together and more importantly, crystallographers using X-rays can often deduce an enzyme's shape. X-ray crystallography is an extremely complex and intricate branch of science, but in essence it consists of bombarding a sample of material with X-rays and examining the ways in which the rays 'bounce off' the sample. From an apparently meaningless pattern of speck on a photographic film, it is often possible to use the laws of physics (and a large computer) to infer how the molecules in the sample are arranged.

Once the structure of an enzyme has been worked out in some detail several types of scientist are brought in to find out how the enzyme might best be modified. Chemists and biochemists play a central role and so do experts in a very novel technique known as computer graphics. Using computers and visual display units similar to those employed by motor car designers it is possible to display a representation of the three dimensional structure of an enzyme and to calculate how small changes in its structure might affect its properties.

This whole area of science and technology is still in its infancy, but most biotechnologists have been amazed at how rapidly it is developing. Several enzymes have been modified in very specific ways and these experiments have yielded much fundamental information about the way enzymes operate. To date no-one has managed to increase the efficiency of an *industrially* important enzyme, but the tools are now available and within five or ten years we should see newly-developed enzymes making an impact on biotechnology. Two examples of what has been achieved indicate something of the power of these techniques. Scientists from two British laboratories (Imperial College, London and the Laboratory for Molecular Biology, Cambridge) have modified an enzyme that is involved in the decoding of genetic information and the construction of proteins. Normally this enzyme, tyrosyl tRNA synthetase, has the ability to bind very specifically to a molecule of the amino acid tyrosine. But by introducing a small change into its structure these researchers have increased the enzyme's ability to latch onto a different amino acid, phenylalanine. Researchers from several laboratories in California tackled a different problem in designing a modified form of the insulin molecule.* When insulin is first constructed in human cells it contains a stretch of 35 amino acids called the C chain. This helps the molecule fold into its correct shape, bringing together the A and B chains which make up active insulin. The American researchers replaced the C chain with a shorter version consisting of just six amino acids. The resulting molecule, called mini-C proinsulin, behaves much like its natural counterpart.

Clearly a lot of work goes into deciding how to modify a protein, but their is, of course, another equally important problem to be tackled—actually manufacturing the modified protein.

---

* Insulin is a hormone and not an enzyme, but the general principles of modifying a protein are the same, regardless of whether it is an enzyme or any other sort of protein.

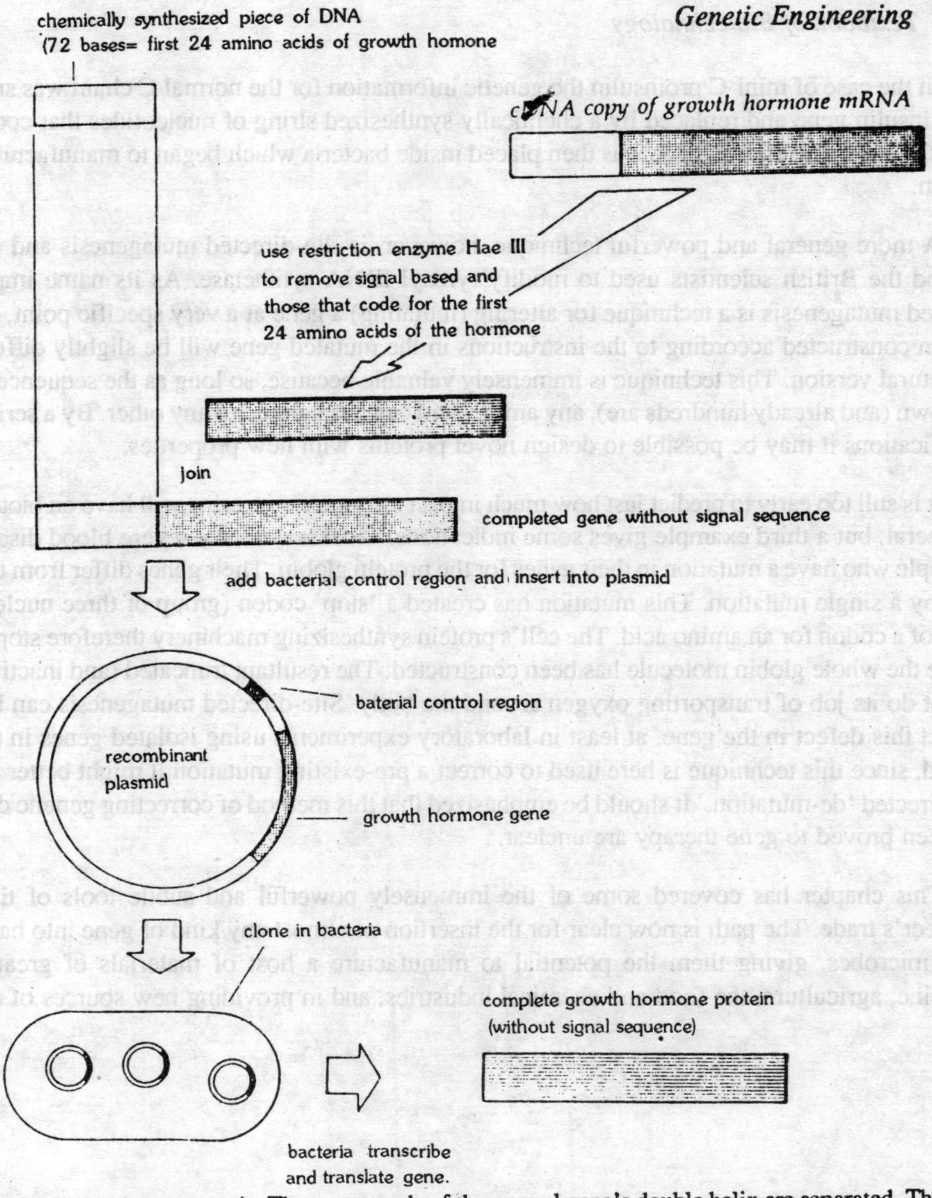

chemically synthesized piece of DNA
(72 bases= first 24 amino acids of growth homone

cDNA copy of growth hormone mRNA

use restriction enzyme Hae III
to remove signal based and
those that code for the first
24 amino acids of the hormone

join

completed gene without signal sequence

add bacterial control region and insert into plasmid

baterial control region

recombinant
plasmid

growth hormone gene

clone in bacteria

complete growth hormone protein
(without signal sequence)

bacteria transcribe
and translate gene.

Fig. 6.15. Site directed mutagenesis. The two strands of the normal gene's double helix are separated. The short sequence of nucleotides (a 'primer') that corresponds with the part of the protein chain which is to be altered is manufactured by a gene machine. This artificial sequence is complementary to the natural sequence except for a specific 'mismatch' (G replaces the T that would normally face the A in the natural gene). Because they are mostly complementary the two sequences latch onto each other. The rest of the natural gene sequence (which will not be altered) is built onto the primer with the aid of enzymes, acting on the instruction contained in the strand from the normal gene. The resultant double helix, which contains one natural gene strand and one mutated gene strand, is inserted into a bacterium. When that bacterium grows and divides it makes a double stranded gene from each of its daughter cells. One daughter will receive a gene copied from the normal strand; the other will receive a gene copied from the mutant strand and will therefore manufacture a mutated protein.

In the case of mini-C proinsulin the genetic information for the normal C chain was snipped out of an insulin gene and replaced by a chemically synthesized string of nucleotides that codes for the mini-C chain. The altered gene was then placed inside bacteria which began to manufacture the new protein.

A more general and powerful technique, however, is site-directed mutagenesis and it was this method the British scientists used to modify tyrosyl tRNA synthetase. As its name implies, site-directed mutagenesis is a technique for altering (mutating) a gene at a very specific point. Hence the protein constructed according to the instructions in the mutated gene will be slightly different from the natural version. This technique is immensely valuable because, so long as the sequence of a gene is known (and already hundreds are), any amino acid can be changed to any other. By a series of such modifications it may be possible to design novel proteins with new properties.

It is still too early to predict just how much impact enzyme engineering will have on biotechnology in general, but a third example gives some indications. Thalassemia is a severe blood disease found in people who have a mutation in their genes for the protein globin. Their genes differ from the normal ones by a single mutation. This mutation has created a 'stop' codon (group of three nucleotides) in place of a codon for an amino acid. The cell's protein synthesizing machinery therefore stops its work before the whole globin molecule has been constructed. The resultant truncated (and inactive) globin cannot do its job of transporting oxygen around the body. Site-directed mutagenesis can be used to correct this defect in the gene, at least in laboratory experiments using isolated genes in test tubes. Indeed, since this technique is here used to correct a pre-existing mutation it might better be termed site-directed 'de-mutation.' It should be emphasized that this method of correcting genetic defects has not been proved to gene therapy are unclear.

This chapter has covered some of the immensely powerful and subtle tools of the genetic engineer's trade. The path is now clear for the insertion of almost any kind of gene into bacteria and other microbes, giving them the potential to manufacture a host of materials of great value in medicine, agriculture, the food and chemical industries, and in providing new sources of energy.

7

Recombinant DNA Technology*

Introduction

In the early 1970s revolutionary developments in molecular biology made it possible to manipulate genetic material with unprecedented precision. This field of study is called *in vitro recombinant DNA technology (genetic engineering or genetic manipulation)*, because it involves manipulating genes under high controllable test-tube (*in vitro*) conditions, rather than using the classic techniques of selective breeding.

Gene Cloning

Central to this new technology is *gene cloning;* the isolation and selective replication of a specific DNA sequence. The sequence to be cloned may originate from any source (it can even be man-made). It is usual to replicate it inside a bacterium. The reasons for performing such an operation may vary from the commercial manufacture of polypeptides such as insulin, to academic research in molecular biology.

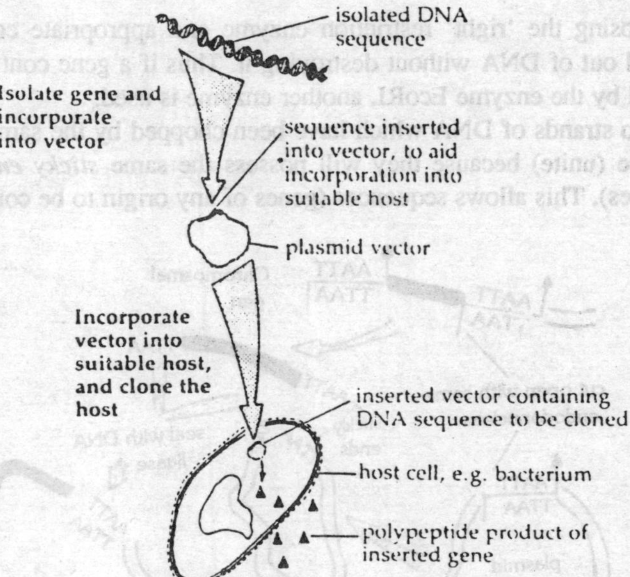

Fig.7.1. *A summary of gene cloning.* The diagram assumes that the purpose of the exercise is to obtain quantities of the gene product.

---

* Excerpts taken from the review published by Wahl, U.K.

The Isolation and Synthesis of DNA

The principal tools for the isolation and artificial synthesis of genes are enzymes. Amongst the several that are necessary for these operations, three have outstanding importance.

Restriction endonucleases

Some bacteria produce enzymes which can specifically identify and destroy the DNA of an infective virus. These enzymes, *restriction endonucleases,* only act on specific base sequences and so they chop foreign DNA into non-random sequences, usually by staggered cuts.

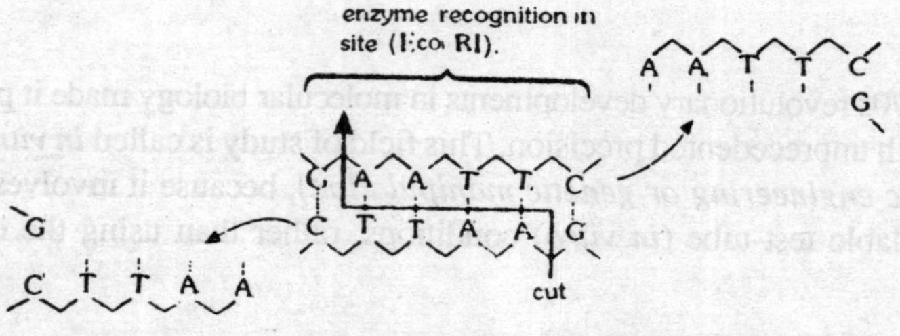

enzyme recognition in site (Eco RI).

A A T T C
G

G A A T T C
C T T A A G

G
C T T A A

cut

Fig.7.2. ***Restriction enzymes (endonucleases).*** **Some restriction enzymes, such as EcoRI (*E.coli,* restriction enzyme 1), work by staggered cuts, so creating sticky ends. Thus DNA fragments from any source ending in the sequences — *TTAA/AATT* — will hybridise with the fragment because of complementary base pairing.**

There are two important implications for the genetic engineer:

(i) By choosing the 'right' restriction enzyme and appropriate conditions, a gene can be chopped out of DNA without destroying it. Thus if a gene contains sequences which are attacked by the enzyme EcoRI, another enzyme is used.

(ii) Any two strands of DNA which have been chopped by the same restriction enzyme will hybridise (unite) because they will possess the same *sticky ends* (complementary base sequences). This allows sequences (genes of any origin to be connected together.

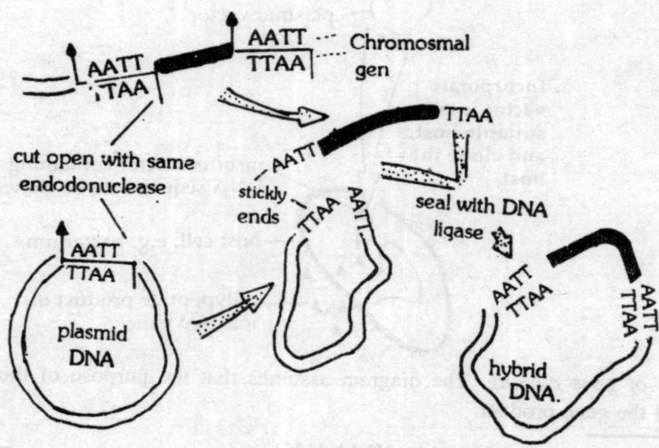

Fig. 7.3. Hybridisation following treatment with endonuclease.

Restriction Maps

By comparing DNA fragments produced by different endonucleases, a restriction map can be constructed. In the region of the α-globin gene b-m are the location of cuts produced by various endonucleases.

DNA ligase

This has already been mentioned in connection with DNA repair systems. Genetic engineers use it to glue sequences together once they have been isolated by, say, endonuclease.

Reverse transcriptase

One category of small RNA-viruses called *retroviruses* have an absolutely unique infection ('life')-cycle. Their genetic RNA codes for an enzyme called *reverse transcriptase*, which has the remarkable property of synthesising DNA from RNA.

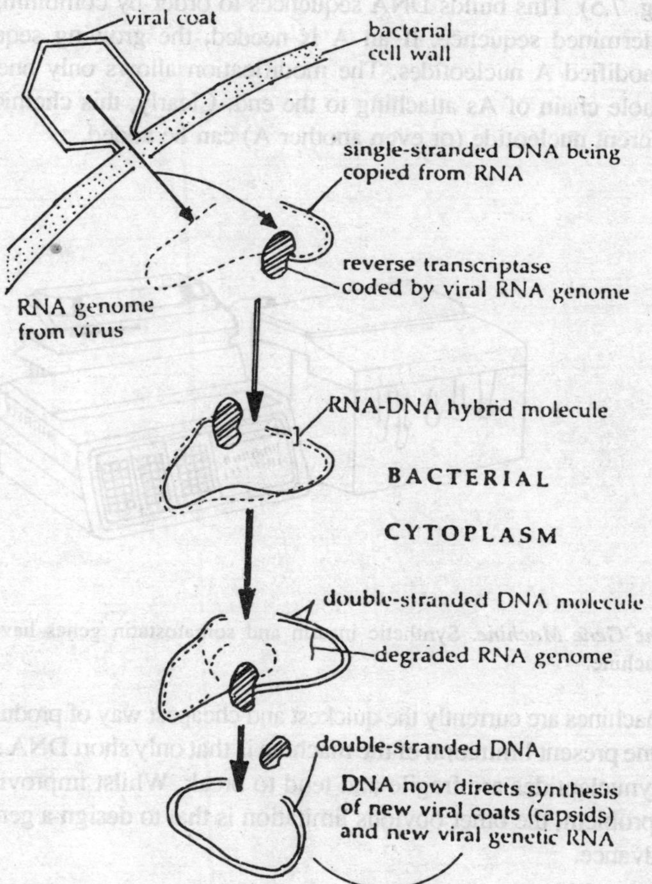

Fig. 7.4. ***The retrovirus infection cycle*. Reverse transcriptase first makes a single-stranded copy of DNA from the viral genetic material, RNA. The single-standard copy is then transformed into a double-stranded copy molecule, and the latter subsequently directs the synthesis of new viral particles.**

The importance of reverse transcriptase to genetic engineers is that it enables intron-free DNA to be made from mature mRNA. This artificial DNA can then be cloned inside a bacterium, and its product, a polypeptide, can be recovered. Such a devious route is necessary because whilst eukaryotic genes often contain introns, bacterial genes do not, nor do they have any method for removing them. Hence plugging an intron-containing eukaryotic gene into a bacterium would produce an *abnormal* polypeptide.

As Fig. 7.4. indicates, reverse transcriptase initially makes a single-stranded complementary DNA polynucleotide and subsequently converts it into a conventional double-stranded molecule. It is called cDNA (c = complementary to mature mRNA).

The Gene Machine

The most powerful tool available to the genetic engineer must surely be the computerised *gene machine* (Fig. 7.5). This builds DNA sequences to order by combining nucleotides, one at a time, into a predetermined sequence. If an A is needed, the growing sequence is put into a solution containing modified A nucleotides. The modification allows only one A to be added, in order to prevent a whole chain of As attaching to the end. Clearly, this chemical block has to be removed before a different nucleotide (or even another A) can be added.

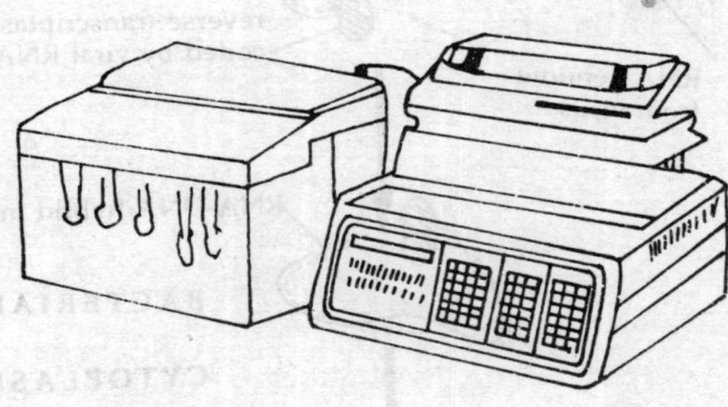

Fig. 7.5. *The Gene Machine.* Synthetic insulin and somatostatin genes have been made using this type of machine.

Gene machines are currently the quickest and cheapest way of producing predetermined nucleotide sequence. One present limitation of the machine is that only short DNA sequences can be constructed, because polynucleotides are fragile and tend to break. Whilst improving technology will doubtless reduce this problem, the other obvious limitation is that to design a gene, the base sequence must be known in advance.

Vectors and Cloning

Due to their small size and phenomenal rate of asexual reproduction, bacteria are obvious candidates for rapidly producing large quantities of genes or their products. A major problem for the genetic engineer is to plug the desired gene into a bacterial cell in such a way that it will function as required.

One method is to use temperature phages like λ (lambda) bacteriophage as *DNA vectors*. After infection temperate phages insert themselves into the bacterial chromosome, and replicate in synchrony with it for an indefinite period. Any gene which has been connected to a virus before infection will also be incorporated and replicated. At low frequency (10^{-5}, more frequently under appropriate laboratory conditions), the viral DNA breaks out from the bacterial chromosome and produces new virus particles which lyse the cell to begin a new infection cycle. Such an event may also transmit the inserted gene to new bacteria, and so assist in its multiplication.

Perhaps of even greater importance to the genetic engineer than the main bacterial chromosome are small circles of cytoplasmic DNA called *plasmids*. One category, the *R-plasmids*, confer antibiotic resistance (hence 'R'). Under selective conditions these can replicate much faster than the main chromosome, so that hundreds of copies may be present in a single cell. Consequently any gene artificially inserted into such a plasmid will simultaneously be replicated, in effect turning the cell into a 'gene factory'.

Entirely synthetic plasmids have been designed and produced, which contain a convenient arrangement of endonuclease cleavage sites, a promoter site, and antibiotic-resistant genes. They are usually named after their inventor, for example, pSC101 = Stanley Cohen's 101 plasmid. Viruses are not used to insert genes into plasmids. The latter can be taken up directly by bacterial cells using the phenomenon of transformation.

Enrichment Procedures and Identification

Producing and identifying a bacterial cell which has incorporated a functional copy of the desired gene involves overcoming some formidable problems. The first is that an endonuclease will chop the genome into thousands of fragments, only one of which will carry the desired gene. Secondly, even under the best conditions, only a small proportion of the fragments will hybridise with vectors. Thirdly, only a few bacteria will incorporate any vectors at all, let alone *hybrid vectors* (vectors which have incorporated fragments of DNA). Statistically, the chances of a complete copy of the right gene hybridising with a vector and subsequently entering a bacterial cell intact are minute. The answer to these problems is to use *enrichment procedures* in order to eliminate unwanted debris:

(i) At isolation

Electrophoresis can be used to separate the fragmented genome into sequences of different length. Only material of the right length is retained, and the rest is discarded. 'Right' means that its length corresponds to that of the desired gene.

(ii) At cloning

It is desirable to encourage the growth of only those bacteria which have incorporated recombinant vectors (i.e., vectors which have hybridised with the DNA which is to be cloned).

One method of identifying bacteria with recombinant vectors, called *insertional inactivation*, is shown in Fig. 7.6. Look carefully at plasmid pBR322 before and after hybridisation, and then answer the questions below.

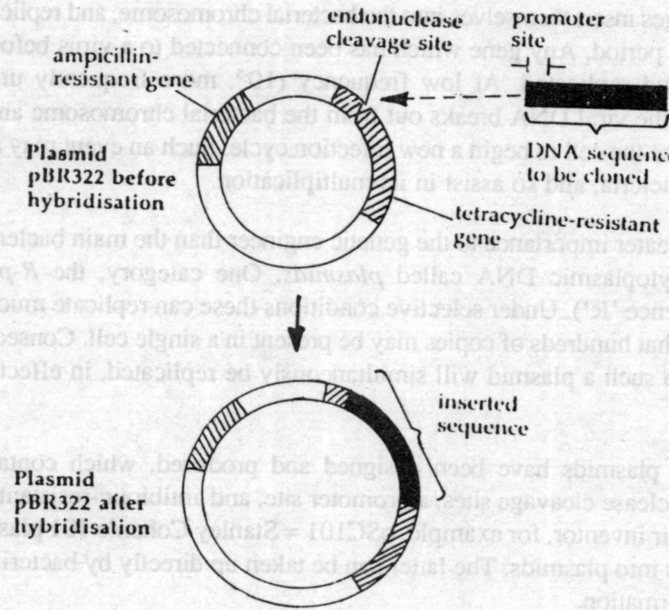

Fig. 7.6. *Insertional inactivation.* pBR322 contains two drug-resistant genes. These must be intact in order to function. One contains an endonuclease cleavage site into which a gene may be inserted.

When pBR322 is used as a vector, colonies of bacteria would be tested on two media: one containing ampicillin only, and one containing tetracycline only. Bacteria which grew on the former but failed to grow on the latter would contain recombinant plasmids, and all others could be discarded.

Identification

Identification of the colony containing the desired gene is often the most time-consuming part of the operation. The method shown in Fig. 7.7 may look simple: in fact, the desired clone is likely to be 1 in 50,000, not 1 in 5!

ASSOCIATED TEHCHNOLOGIES

Two techniques associated with genetic engineering have assumed outstanding significance. One, *DNA-sequencing* has already been mentioned in connection with the gene machine. The other is *DNA - hybridisation.*

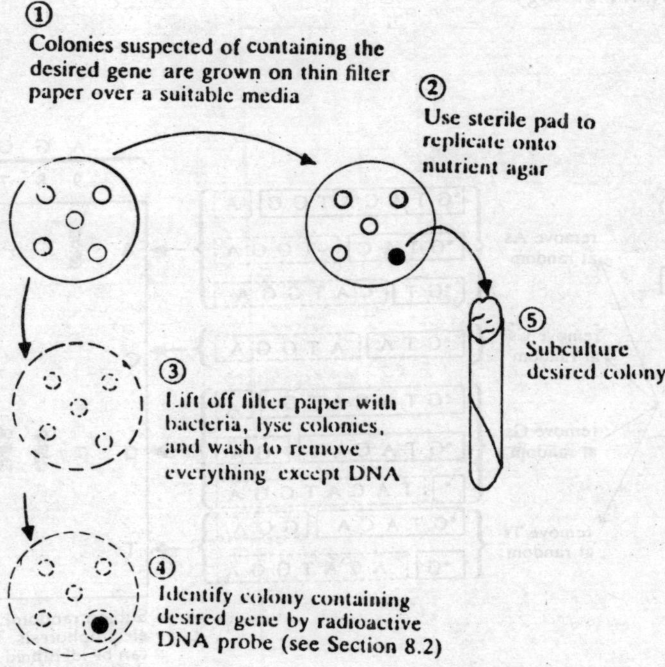

Fig. 7.7. **The replica-plate method for identifying desired colonies.**

DNA Sequencing

DNA-sequencing provides detailed information about the content and organisation of the genetic material. It is also by far the quickest method of determining the amino-acid sequence of a protein: basically all one needs is a sense-strand sequence. One technique is shown in Fig. 7.8. Essentially it involves:

(i) Cloning a DNA-sequence, as described above, then isolating and purifying it.
(ii) Labelling one end of the sequence with ^{32}P, dividing the sample into four and removing some (not all) bases of a given type from each, so that the sequences are cracked at random.
(iii) Using electrophoresis to spread out the fragments on the basis of their lengths. The smallest fragments move furthest, and their positions can be detected by the radioactive label. If the gel is calibrated, the base sequence can be read directly.

DNA hybridisation

The rate at which two polynucleotide chains hybridise (join) *in vitro* is assumed to reflect the extent to which their base sequences are complementary. Even before genetic engineering was developed, Hall and Spiegelmann used this approach to show complementarity between DNA and mRNA, and so helped to confirm the central dogma. However, the ease with which single-stranded cDNA can be produced has opened up new possibilities. Often it is made radioactive to assist its identification, and it may then be used as a *DNA probe* for identifying complementary sequences elsewhere.

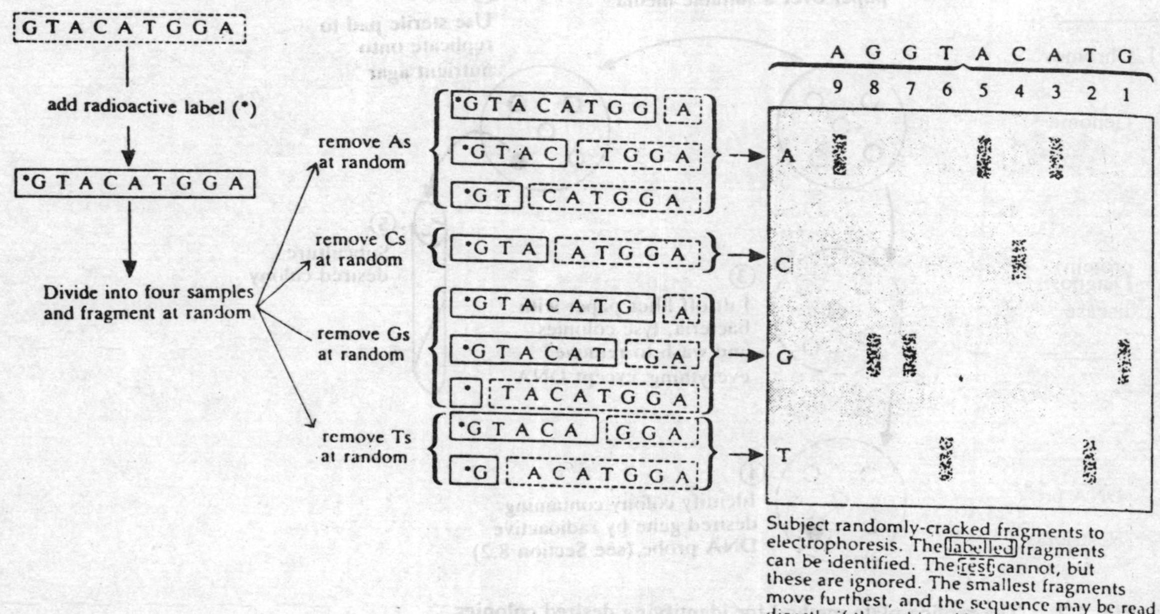

Fig. 7.8. Rapid DNA sequencing

Uses and Abuses

Recombinant DNA technology has widespread commercial and academic applications. One example is described below, others are summarised in Tables 7.1 and 7.2.

Insulin production

Pig insulin has been widely used for treating diabetes, but it is not identical to human insulin and can cause undesirable side effects. The production of human insulin by bacteria was the first major commercial achievement of genetic engineering.

(a) Reverse transcriptase may be used to generate cDNA from intron-free insulin mRNA. (The insulin gene contains two small introns and connot be used directly.) Alternatively, since the amino acid sequence of insulin is known, a synthetic gene can be constructed using a gene machine.

(b) Start and stop signals are added, together with sticky ends if necessary.

(c) The gene is inserted into a suitable plasmids, such as pBR322 (Fig. 7.6), and the plasmids are then mixed with bacteria in conditions which promote their uptake, and cloned.

(d) Enrichment procedures may be used to select bacteria containing recombinant plasmids.

(e) Isolation procedures (Fig. 7.7) are used to identify the insulin-producing clone. Finally, the clone is cultured, and the insulin is extracted and purified.

TABLE 7.1

Applications of the (Labelled) cDNA Probe

1. Chromosome mapping	cDNA will bind to the precise position on a chromosome where a gene is located
2. Genome characterisation	(i) Genes of different organisms may be compared by their ability to bind to a specific sequence of cDNA, so that the relationships between them can be determined. (ii) The number of copies of a sequence per genome can be determined by the rate at which hybridisation to cDNA occurs.
protein) 3. Diagnosis of infectious disease	Production of cDNA probes for genes known to be present in pathogenic agents can aid identification of the latter in patients suspected of harbouring them.

TABLE 7.2

Applications of Recombinant DNA Technology

1. DNA probes	See Table 7.1.
2. Restriction mapping diagnosis	The nucleotide sequence in defective genes differs from that in normal genes, and this may be reflected in different sizes/sequences of fragments when a gene is cut up with an endonuclease. May be used in pre/postnatal diagnosis of genetic defects.
3. Gene libraries (gene banks)	Genes from rare organisms may be 'stored' inside bacterial hosts until required.
4. Production of vaccines	Handling pathogenic bacteria and viruses during the manufacture of vaccines can be hazardous. A safer technique is to chop out and clone the genes that code for those proteins which trigger the immune response. Working with the products of these genes is safer than handling the pathogen itself. Alternatively, a pathogen may be rendered harmless by *site-directed mutagenesis*. This involves isolating and mutating specific genes, and then reinserting them into the pathogen. By selecting appropriate genes for mutation, an organism may lose its pathogenic affects but still trigger the immune response (and so lead to antibody production).
5. Production of foods, antibiotics and alcohols	The performance of organisms used for the production of SCP (single-cell protein) and other commercially important materials may be improved by directed alterations to the genetic material. Such techniques may replace or be used in conjunction with classical techniques of selective breeding.
6. Production of hormones and monoclonal antibodies	Hormones, e.g. insulin (Fig. 7.9), and somatostatin (growth hormone), are currently being manufactured using genetic engineering techniques. Monoclonal antibodies have extensive applications in industry and research.
7. Agriculture	Only micro-organisms can fix N_2 (g) into reduced forms which plants can use. Shortage of 'fixed' nitrogen may retard crop productivity. Attempts to incorporate nitrogen-fixing systems into cereal crops are being made. If successful, this venture could have a dramatic impact on world food production.

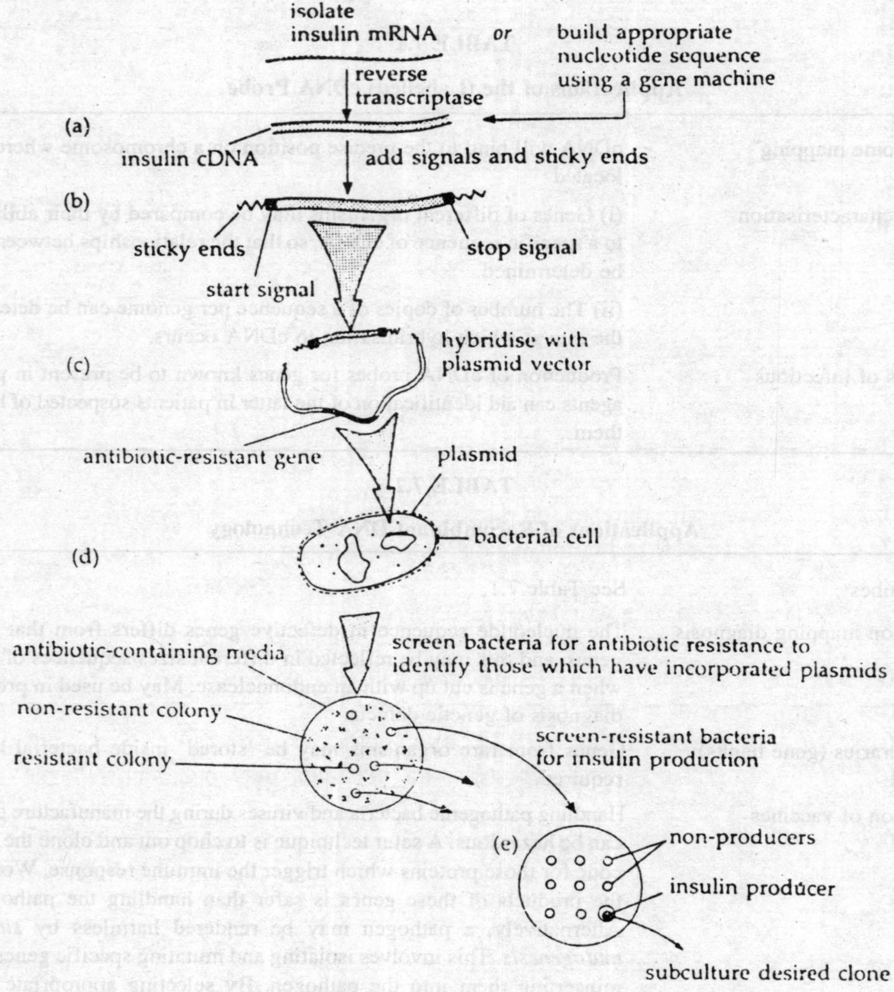

Fig. 7.9. ***Insulin production by recombinant DNA technology*** **(for explanation, see text).**

Given the varied and beneficial applications which have emerged from genetic engineering, it is nowadays difficult to imagine the considerable anxiety expressed about it by several sectors of the community in the mid-1970s. This anxiety precipitated the Ashby Report (1975) and, a year later, the Genetic Manipulation Advisory Group. As understanding and experience replaced ignorance and uncertainty, fear of the new technology receded. In 1984 GMAG was replaced by ACGM (Advisory Committee on Genetic Manipulation), whose function is to report to the government on the hazards associated with genetic manipulation. Perhaps the anxieties that were expressed a few years ago are now only of historical interest. Perhaps not. A good track record is not proof that a technology is 100% safe, and, as Ashby noted, it is always possible that an antibiotic-resistant plasmid could be accidentally incorporated into a dangerous pathogen with serious medical consequences. All microbiological work is, of course, potentially hazardous, but good microbiological practice combined with the use of *disarmed bugs* (weakened strains, incapable of living outside the comfort of a Petri dish) make the chances of an accident extremely unlikely. The question of safety is not, however, the only one. Questions which are much more difficult to answer, and refuse to go away, are the social and ethical ones. As techniques for manipulating genetic material in eukaryotes, including ourselves, continue to develop, these issues assume a more immediate significance.

REVISION QUESTIONS

Q.1. Explain more fully the advantages of '... a convenient arrangement of endonuclease cleavage sites, a promoter site, and antibiotic-resistant genes.'

Q.2. It is arguable that ever since someone consciously decided to plant a particular crop, pull up a weed, raise a specific breed of domestic animal, or kill off a parasite, the human race has been interfering with the rate and direction of evolution, i.e. manipulating genetic material. Is recombinant DNA technology really doing anything different?

Q.3. *Insulin Production* : Explain briefly:

(a) Why intron-containing DNA from the genome connot be used for cloning ?
(b) The purpose of adding 'sticky ends'.
(c) The advantages of using pBR322 as a vector.
(d) How bacteria containing recombinant plasmids may be identified and selectively grown ?
(e) How a colony containing the insulin-producing gene can be identified ?

8

Gene Cloning*

Introduction

Gene cloning involves the production *in vitro* of new DNA molecules which contain novel combinations of genes or oligonucleotides and the propagation of such recombinant DNA molecules by the exploitation *in vivo* of the replicative mechanisms of bacteria and other organisms. The developments of genetic engineering techniques have permitted the alteration of the genome of microorganisms so that it produces substances of little intrinsic value but of great medical or economic value to mankind.

(i) In Medicines

Foreign genes have been implanted into the DNA of *E. coli* to enable the production of useful proteins. Members of the antiviral family of proteins called interferons have been produced by these methods, and clinical trials to ascertain their efficacy in the treatment of certain cancers have been conducted. Human hormones such as insulin, somatostatin and somatotropin have been synthesized in *E. coli* insulin controls the level of glucose in the blood and its deficiency may result in a variety of serious diabetic conditions. To overcome problems associated with the production of insulin from pig and cow pancreases, the increasing demand for insulin and the immunological sensitization of the patient by continued injection of an animal protein, human insulin has been successfully synthesized from artificial genes on a commercial scale. The first human polypeptide hormone synthesized in *E. coli* was somatostatin which, containing 14-amino acid residues, was used to develop techniques for insulin production. Somatostatin is employed in the treatment of numerous disorders characterized by excessive growth. The hormone antagonistic to somatostatin is somatotropin (human growth hormone) which is employed in the treatment of dwarfism.

Gene cloning could potentially provide contaminant — free blood products of high purity. A coagulation factor, Factor VIII, is required by certain haemophiliacs. The risk of hepatitis and acquired immune deficiency syndrome (AIDS) due to the unintentional collection of blood from virus carriers identified this factor as a candidate for production by DNA recombinant techniques.

(ii) In Agriculture

In agriculture, techniques have been developed which permit the transfer of the characteristics

* Excerpts taken from the review published by Nuti *et al.*

of one plant to another through bacterial infection. Such techniques may create new varieties of plants with desirable characteristics, e.g. resistance to infection, the ability to withstand adverse weather conditions or the capability of nitrogen fixation.

(iii) In Pharmacy

Non-protein products may be synthesized by recombinant methods. Numerous important pharmaceuticals are small molecules, the biosyntheses of which require the sequential involvement of different enzymes. The cloning of all the relevant genes on to a single plasmid offers a means of enhanced production of the substances. The antibiotic, actinorhodin, has been synthesized in *Streptomyees* by such methods. Experimental vaccines against certain viral diseases have also been produced by recombinant DNA technology.

Gene cloning has played an important role in fundamental biological research. Its value lies in the preparation of DNA fragments containing a specific gene and has resulted in advances in knowledge of the structure, function, expression and the control of the activity of that gene. Recombinant DNA techniques have been applied to the study of the regulation of metabolism. The ascribed role of citrate synthase in metabolic regulation has resulted from the interpretation of data obtained by studies *in vitro*. However, through cloning of the *E. coli* citrate synthase can be manipulated through the regulatory unit of its gene by the non-metabolizable inducer, isopropyl β-D-thiogalactopyranoside, to permit the calculation of metabolic flux rates.

To understand fully the regulation of any metabolic pathway, the properties of the regulatory enzymes must be resolved. The manipulation of cloned genes has provided significant quantities of enzymes for their purification. Further investigation of their structure/organization including the determination of their primary structures, structure-function relationships and the analysis of the roles and expression of multigene families, e.g. protein kinase C-related genes, have expanded our knowledge of metabolic regulation. Indeed, gene technology promises enormous benefits in a wide variety of biological pursuits.

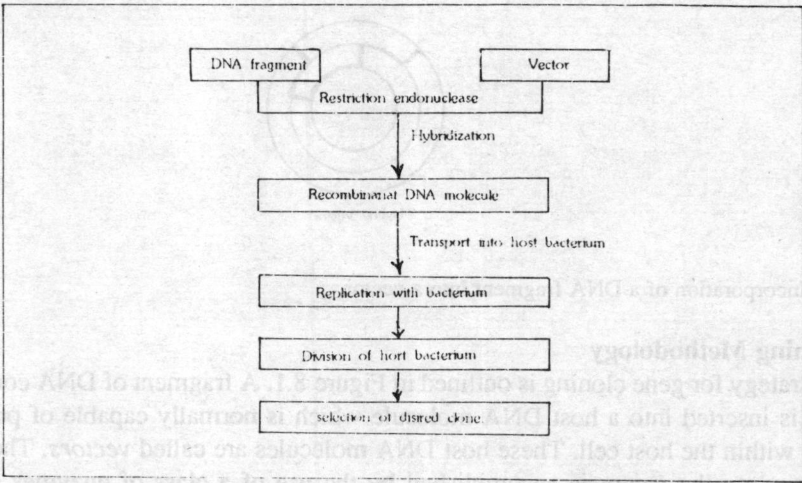

Fig. 8.1 Strategy for gene cloning in bacteria.

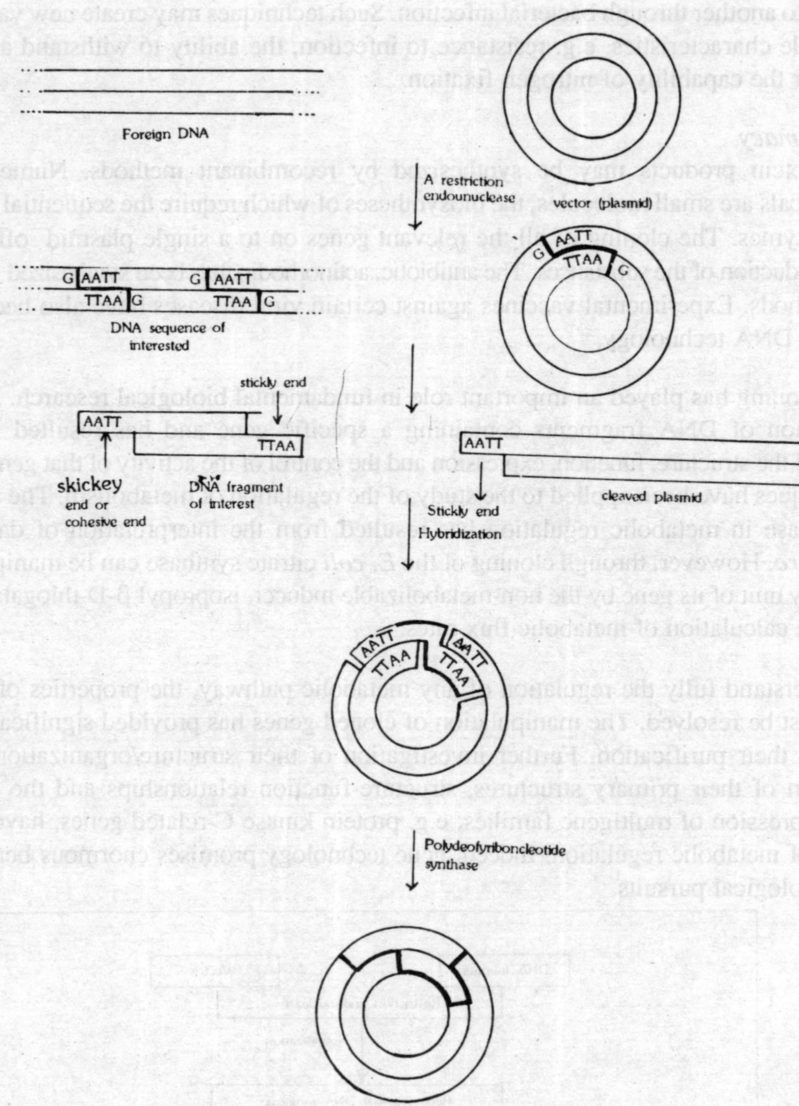

Fig. 8.2. Incorporation of a DNA fragment into a vector.

Gene Cloning Methodology

The strategy for gene cloning is outlined in Figure 8.1. A fragment of DNA containing the gene of interest is inserted into a host DNA molecule which is normally capable of promoting its own replication within the host cell. These host DNA molecules are called *vectors*. The insertion of the foreign gene into the vector is accomplished by the use of a class of enzymes called restriction endonucleases. The DNA of interest is cleaved into numerous fragments by a restriction endonuclease which results in a staggered cleavage. The vector is separately cleaved with the same enzyme to produce 'sticky' ends (or cohesive ends). Upon mixing the fragments and the treated vector, hybridization yields a selection of recombinants which are ligated by polydeoxyribonu cleotide synthase to produce vectors each carrying a different passenger DNA (Figure 8.2). Depending upon the nature of the vector, techniques are available for the transportation of the recombinants into the host cell.

The replicative machinery of the host cell is exploited by the vector to produce numerous identical copies of itself each containing the implanted gene. When the host cell replicates, copies of this gene appear in each daughter cell. Further vector replication occurs. After numerous cell divisions, a clone of identical host cells results. Each cell in the clone contains copies of the recombinant DNA molecule. The clone containing the DNA fragment of interest must be identified before the harvesting of an abundance of the desired gene.

Vectors used in *E.coli*

Vectors are DNA molecules, capable of replication in the host organism, into which a gene may be implanted to construct a recombinant DNA molecule. To be a successful vector, there are two requisites for the DNA molecule:

1. It must be capable of replication within the host cell to produce numerous copies of the inserted gene.
2. It should be relatively small.

Molecules of less than 10,000 nucleotides are more amenable to manipulation *in vitro* as lager molecules tend to shear during purification. Suitable vectors for use in bacterial systems are small plasmids and bacteriophage chromosomes. Bacteriophages are a group of viruses which specifically infect bacteria.

Plasmids are circular molecules of DNA located outside the nuclear region of prokaryotes. They usually contain one or more genes frequently responsible for the conferment of a useful property to the microorganism, e.g., antibiotic resistance (R-plasmid), promotion of sexual cunjugation between bacterial cells (F-plasmid). The plasmid DNA contains a nucleotide sequence which can act as an origin of replication to permit the independent multiplication of the plasmid. Smaller plasmid utilize exclusively the replicative enzymes of the host whereas larger plasmids (approximately 250 kilobases) may contain genes which encode enzymes specific for plasmid replication. However, some plasmids integrate with the bacterial chromosomal DNA and are replicated along with the bacterial genome. The term episome is employed to refer to these integrative plasmids.

For efficient genetic engineering, the plasmid DNA should be obtainable at high levels. Some plasmids are found only as three or four copies per cell. More useful are plasmids which have a high cellular copy number, e.g. 25-50 copies per cell. Techniques are available to amplify the copy number further. If an inhibitor of protein synthesis, e.g., chloramphenicol, is added to a culture of plasmid- containing bacteria, chromosomal, but not plasmid, DNA replication stops. The number of plasmids per cell thereby may increase to over one thousand.

Although plasmids are common features of bacterial cells, they are relatively rare in other organisms. Many strains of the yeast *Saccharomyces cerevisiae* contain a well-characterized and useful large plasmid called, because of its length, the 2 μm plasmid. Because of this plasmid this yeast is regarded as a potential host organism for exploitation by the pharmaceutical industry.

Bacteriophages, like all viruses, contain only a single nucleic acid molecule (usually DNA but occasionally RNA) which carries a relatively small number of genes enclosed by a protective protein coat called the capsid. The two main types of phage which infect *E. coli* are the head-with-tail phages, e.g., phage λ, or filamentous phages, λ M13 (Figure 8.3). The head component of the head-with-tail phages is frequently an eicosahedron (having 20 triangular faces and 12 corners) which encompasses single-or double-stranded linear or circular DNA (or single-stranded linear RNA). The tail is a complex multicomponent structure which often ends in tail fibres. In filamentous phages, the nucleic acid, which is in an extended helical form, is contained within the capsid.

The life cycles of all phages are essentially the same although many variations in the detail of the general pattern are recorded. The phage attaches to receptors on the bacterial surface. These receptors are varied and have other roles for the benefit of the bacterium. The phage injects its DNA through the bacterial cell wall. The bacterium loses its ability to either replicate or transcribe (or both) its own DNA. The phage DNA molecule is replicated usually by a phage-specific DNA polymerase encoded by phage genes. Other phage genes provide for the synthesis of protein components. The transcription of phage mRNA is usually initiated by bacterial RNA polymerase but thereafter either the bacterial enzyme is modified to recognize phage promoters or a phage-specific RNA polymerase is manufactured. New phage particles are assembled and are discharged as infective agents from the bacterium into the surrounding medium by the enzyme - catalysed (endolysin) lysis of the cell wall. A few filamentous phages including M13 utilize the process of budding in which phage particles are released through swellings in the cell wall. Since major damage to the bacterium is avoided, filamentous phage particles can be produced and released over substantial periods of time.

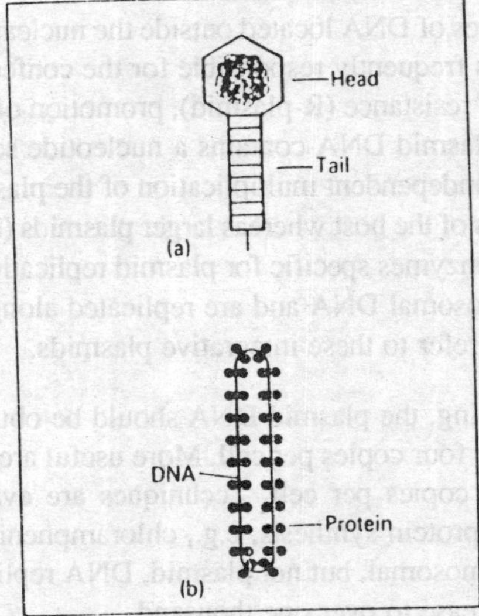

Fig. 8.3. Two types of bacterio-phage infecting *E. coli*. (a) Head-with-tail phages. (b) Filamentous phages

The life cycle processes (Figure 8.4) may be rapid (of less than 20 min duration) in which case it is called a lytic cycle. The major characteristic of a lytic cycle is that phage DNA replication if sequentially followed by the synthesis of phage proteins and packaging. The phage DNA does not exist free for any length of time in the host cell. Alternatively, the phage DNA may be implanted into the host DNA (similar to episomal insertion) and may be retained through thousands of cell divisions. However, the integrated form of the phage DNA called the prophage lies dormant until it is eventually released from the host genome. The phage DNA then directs the production of new phage particles which are released through cell lysis. This form of life cycle is described as a lysogenic infection cycle. The mechanism by which the lytic cycle or the lysogenic cycle is selected remains to be verified.

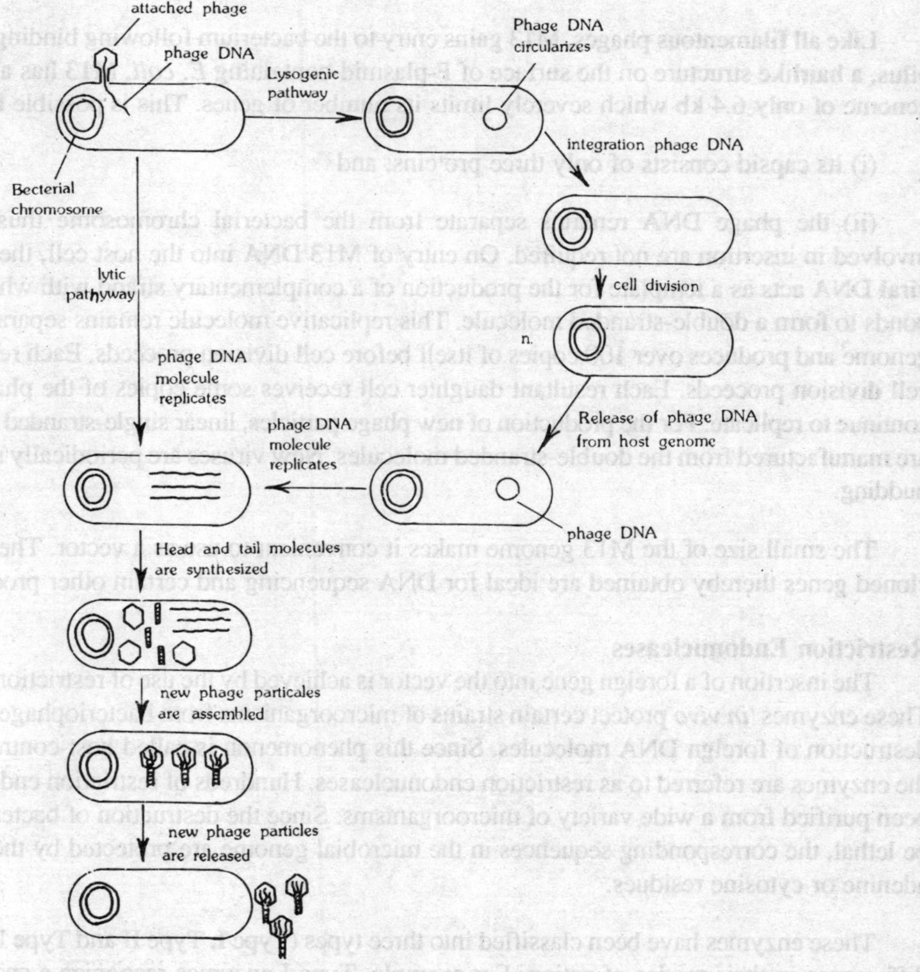

Fig. 8.4. Infection of a bacterium by a bacteriophage.

Of the many varieties of bacteriophage, only phage λ and filamentous M13 have been confirmed as major cloning vectors in *E. coli*. Phage λ has a 48.5 kb DNA molecule which has been sequenced and mapped so that the positions and functions of its genes are known. Related genes, e.g., those encoding capsid proteins or those governing integration into the host genome, are clustered so that each group may be regulated collectively. The λ DNA molecule is a double-stranded linear molecule conforming to the Watson-Crick model with the exception of a 12- nucleotide single-stranded region at both ends. These complementary regions are called 'sticky' or cohesive ends because they may hydrogen-bond with one another to circularize into an entirely double - stranded molecule (Figure 8.4) or with the ends of two different DNA molecules. The λ DNA cohesive ends are called cos sites and provide a single-stranded region to which a foreign DNA fragments, produced by a restriction endonuclease, may attach to form a recombinant molecule.

Like all filamentous phages, M13 gains entry to the bacterium following binding to the tip of the pilus, a hairlike structure on the surface of F-plasmid-containing *E. coli*. M13 has a single-stranded genome of only 6.4 kb which severely limits its number of genes. This is possible because:

(i) its capsid consists of only three proteins; and

(ii) the phage DNA remains separate from the bacterial chromosome thus gene products involved in insertion are not required. On entry of M13 DNA into the host cell, the single-stranded viral DNA acts as a template for the production of a complementary strand with which it hydrogen-bonds to form a double-stranded molecule. This replicative molecule remains separate from the host genome and produces over 100 copies of itself before cell division proceeds. Each resultant daughter cell division proceeds. Each resultant daughter cell receives some copies of the phage DNA which continue to replicate. For the production of new phage particles, linear single-stranded DNA molecules are manufactured from the double-stranded molecules. New viruses are periodically released through budding.

The small size of the M13 genome makes it convenient to use as a vector. The single-stranded cloned genes thereby obtained are ideal for DNA sequencing and certain other procedures.

Restriction Endonucleases

The insertion of a foreign gene into the vector is achieved by the use of restriction endonucleases. These enzymes *in vivo* protect certain strains of microorganisms from bacteriophage infection by the destruction of foreign DNA molecules. Since this phenomenon is called host-controlled restriction, the enzymes are referred to as restriction endonucleases. Hundreds of restriction endonucleases have been purified from a wide variety of microorganisms. Since the destruction of bacterial DNA would be lethal, the corresponding sequences in the microbial genome are protected by the methylation of adenine or cytosine residues.

These enzymes have been classified into three types (Type I, Type II and Type III), according to differences in their modes of action. For example, Type I enzymes recognize a specific nucleotide sequence but cleave variable sequences over 1000 base-pairs from the recognition site. Type II enzymes have cleavage sites which are within or close to their recognition sites whereas the cleavage sites of Type III enzymes lie 24-26 base-pairs downstream (towards the 3'-end) of their recognition sites. In addition to their nucleolytic cleavage, Type I and Type III endonucleases catalyse the methylation of DNA. Type II enzymes demonstrate only endonucleolytic cleavage; other enzymes which recognize the same sequences catalyse the methylation reactions. Restriction endonucleases are known by an acronym consisting of an abbreviation for the source organism and strain and a specific enzyme number in Roman numberals (Table 7.1).

Type II enzymes cleave their substrates in a very precise manner, a feature of major importance in gene cloning. These enzymes recognize specific short (four to six base-pairs long) palindromic sequences only in DNA and nick at least one strand. A palindrome is a word or sentence which reads alike backward and forward as exemplified in the British place — names of Glenelg and Notton. The cleavage sites are symmetrical and produce either linear (blunt-end) or staggered (cohesive-end) incisions. Since cohesive ends are valuable in the insertion of a foreign DNA through complementary hydrogen-bonding arrangements prior to ligation, blunt-end cleavages may have a single-stranded region introduced by several methods, e.g., the attachment of linkers, adaptors and homopolymer tails.

The Transport of Recombinant DNA Molecules into a Host Bacterium

The mode of entry of recombinant DNA molecules into the host cell is dependent upon the nature of the vector. The introduction of a DNA molecule into a bacterium or any other living cell is termed transformation. Certain species of bacteria, e.g. *Bacillus,* can easily acquire DNA molecules from a surrounding medium since they have developed pertinent mechanisms. However, most species, e.g. *E. coli,* can take up only small quantities of DNA. With plasmid vectors, a population of *E. coli* is bathed in cold calcium chloride (50 mmol dm^{-3}) solution containing the hybrid plasmid. This treatment enhances the DNA binding to the bacterial surface. Raising the temperature to 42°C expedites the transport of the DNA into the cell. Although this technique allows the incorporation of plasmids into the bacterium, the uptake is poor (maximum achievable uptake is approximately 1%).

Phage vectors may be introduced into a bacterial cell by two processes: transfection and packaging *in vitro*. The process of transfection is similar to transformation but the term is employed to signify the involvement of phage DNA. Transfection produces even poorer yields than plasmid transformation. The technique of packaging *in vitro* involves the preparation of two cultures of bacteria, each of which has been infected by a different defective λ strain. The strains are selected so that each carries a mutant gene for a different capsid protein. Infected bacteria synthesize and accumulate the other phage components but the lack of one protein prevents the assembly of phage particles. Upon mixing lysates from both bacterial cultures, all the necessary components are provided for the packaging of added recombinant DNA molecules into mature phage particles. These particles may infect a bacterial culture in the normal manner.

Contrived vectors called cosmids which combine some of the advantages of plasmids and phages have been constructed. A cosmid is in essence a plasmid to which the *cos* sites of the phage λ genome are attached. These sites permit the introduction of plasmid DNA into a bacterium by the *in vitro* packaging technique as long as the *cos* sites are separated by 37-52 kb of DNA. Recombinant molecules constructed using cosmids may therefore be introduced into *E. coli* with higher efficiency. A major advantage of cosmids is their capacity for cloning DNA fragments which are too large for other vectors. The recombinant DNA fragment carried by a cosmid is selected by the techniques developed for plasmid vectors.

Selection of Recombinant DNA

Low levels of DNA incorporation demand methods of detecting which bacterium contains foreign DNA. Also, since a variety of foreign DNA fragments, produced by the action of a restriction endonuclease, have been introduced into a culture of *E. coli*, some bacteria will carry DNA fragments which do not contain the gene of interest and only a few bacteria will have the appropriate fragment. A number of methods of gene detection are available, e.g., vectorial marker, hybridization and immunological detection.

To identify bacteria which have taken up the vector, a common ploy is to utilize a characteristic of the vector, i.e., a vectorial marker. Plasmids possessing an antibiotic-resistance marker will enable a micro-organism to grow in culture medium containing that antibiotic. Bacteria lacking the plasmid fail to develop colonies. For example, the plasmid pBR 322 contains antibiotic-resistance markers for tetracycline and ampicillin. Therefore, bacteria transformed by this plasmid are capable of colony formation in tetracycline-and/or ampicillin-containing medium and are easily selected.

Hybridization and immunological detection permit the identification of bacteria which carry the desired gene. The hybridization technique involves the growing of the organism on a plate of solid nutrient medium and the analysis of the DNA extracted from each resultant colony for the DNA of interest. The search is conducted through the ability of this DNA to hybridize with a complementary radioactive (^{32}P) probe. Commonly employed probes include single-stranded DNA synthesized from isolated mRNA by RNA-directed DNA polymerase (reverse transcrip ase). To reduce the enormity of the task, bacterial colonies are replica-plated from the solid medium on to a filter paper which is then treated with NaOH to rupture the cells and denature the DNA. The paper is flooded with the radioactive probe and DNA - DNA hybridization occurs with a complementary sequence. Autoradiography, in which radioactive emissions blacken the emulsion of an exposed X-ray film, is employed to locate the desired colony.

Immunological methods may be employed to detect the products of the foreign gene. After transformation, growth and replica-plating, the replica colonies are lysed to release their protein contents on to a cellulose nitrate membrane precoated with a specific antibody which interacts with the protein. Other proteins do not bind specifically to the antibody. The membrane is washed to eliminate non-specific binding of proteins and flooded with ^{125}I-labelled specific antibody or fluorescein-labelled antibody. Following further washing, the desired colony is detected by autoradiography or immunofluorescence respectively.

Upon selection of the appropriate colony, large scale cultivation of the microorganism increases enormously the number of recombinant DNA-containing cells which may be exploited for our benefit.

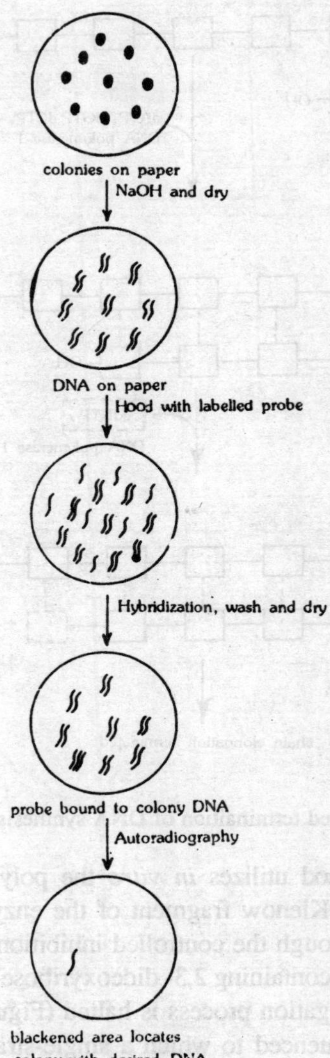

Fig. 8.5. Colony hybridization.

DNA-Sequencing Techniques

Gene cloning has made available sufficient quantities of genes to permit their primary structures to be determined. Before the advent of rapid techniques, sequence analysis was restricted to very short nucleotide sequences. Nowadays, DNA sequencing can be performed faster than protein sequencing. Rapid DNA sequencing is frequently performed by either the dideoxy method of Sanger and Coulson or the Maxam-Gilbert method.

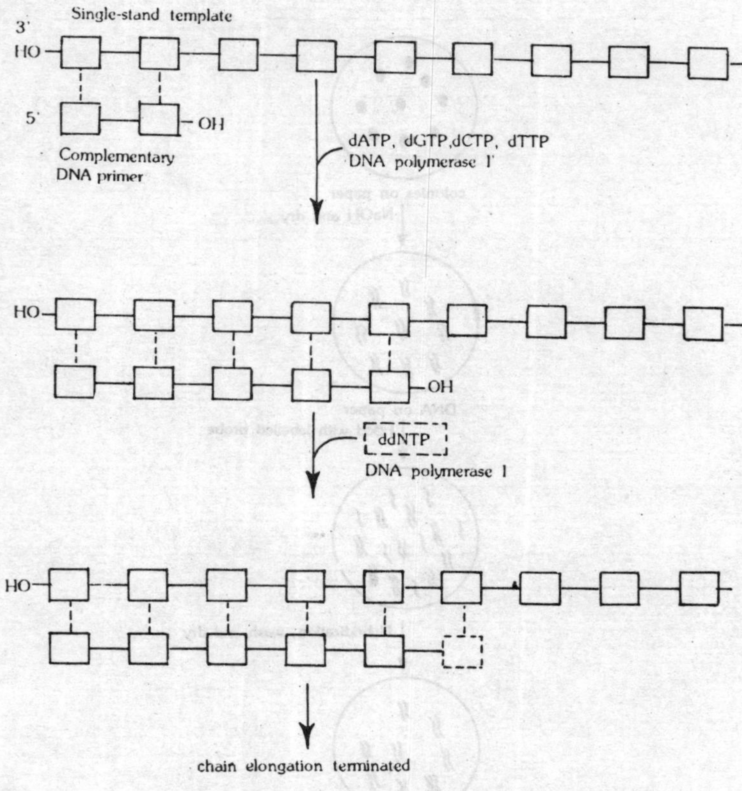

Fig 8.6. Dideoxyribonucleotide-induced terminaition of DNA synthesis.

The dideoxy sequencing method utilizes *in vitro* the polymerase activity of *E. coli* DNA polymerase I contained within the Klenow fragment of the enzyme to synthesize a collection of radioactively labelled fragments through the controlled inhibition of the replication process by the addition of dideoxyribonucleotides (containing 2,3- dideoxyribose) to the growing strand. Since a 3'-OH group is not available, the elongation process is halted (Figure 8.6). The template is a single-strand copy of the DNA to be sequenced to which a single-strand primer complementary to the preceding template bases is hybridized. For each determination four reaction mixtures are employed, each containing all four dNTPs one of which is labelled, e.g. dGTP, dCTP, dTTP and [α-^{32}P] dATP plus one different dideoxyribonucleotide (ddNTP). In each reaction mixture, oligonucleotide chains of varying length but all terminated with the same base are generated. The chains may be separated according to size by polyacrylamide gel electrophoresis. Because a radioactively labelled dNTP was present in the four reaction mixtures, all oligonucleotide chains may be detected by autoradiography. Since all four reaction mixtures are adjacently electrophoresed in the same gel, the DNA sequence can be read directly from the autoradiogram (X-ray film) starting at the fastest (smallest) fragment located nearest to the bottom of the film. A DNA segment containing 200 nucleotides may be sequenced in one experiment by this method.

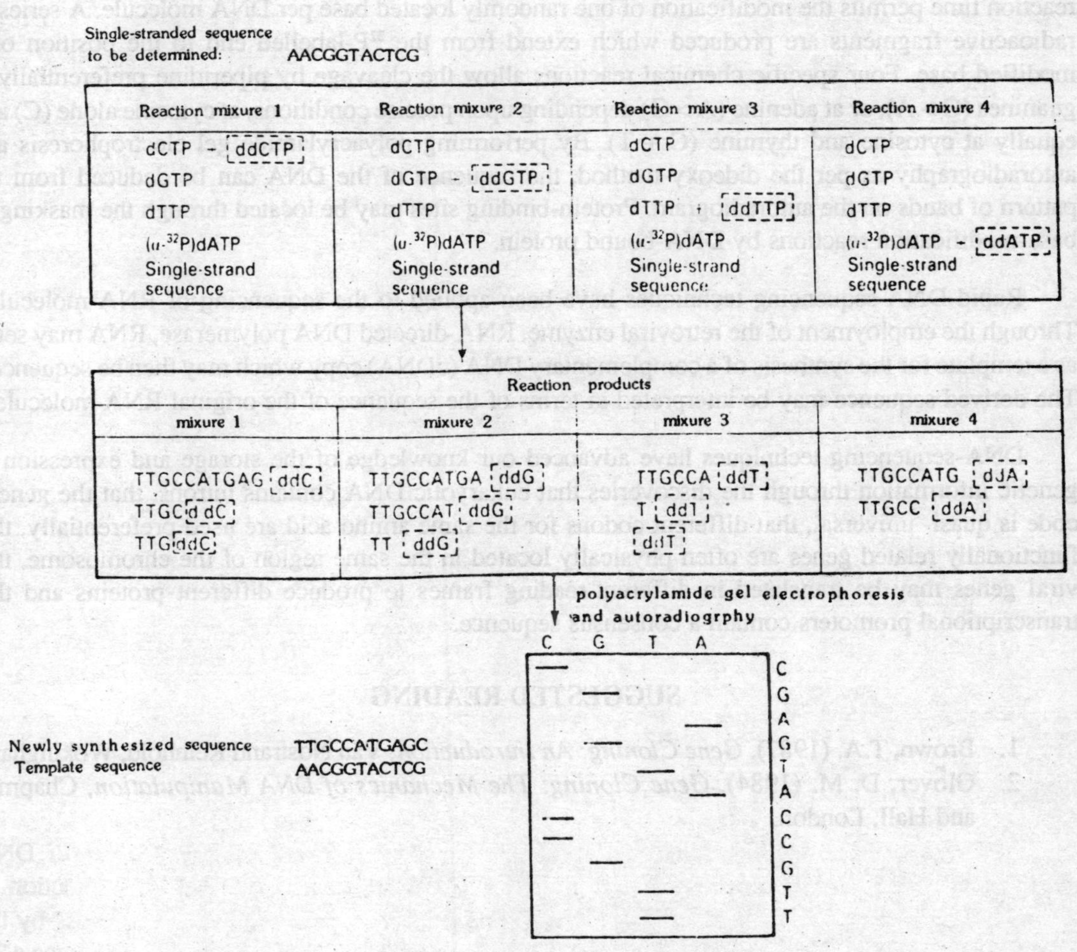

Fig. 8.7. The Sanger dideoxy method of DNA sequencing.

The Maxam-Gilbert technique provides not only an alternative to the dideoxy method but permits the sequencing of a double-stranded DNA and the identification of DNA-protein-binding sites. A specific segment of double-stranded DNA is selected by the use of an appropriate restriction endonuclease (Fig. 8.8). This segment is labelled at both 5'-ends with ^{32}P using the enzyme, polynucleotide 5'-hydroxyl-kinase. A different restriction enzyme is used to divide the segment into two fragments. Each fragment is separately processed. The complementary strands are then separated. Each labelled single-strand fragment is treated with a specific chemical reagent which reacts with one (sometimes two) of the four bases, e.g. guanosine is methylated at the N-7 position by dimethyl sulphate. The modified base renders the fragment susceptible to subsequent chemical cleavage. Piperidine at 90°C breaks firstly the glycosidic linkage between the modified base and the deoxyribose and secondly the sugar-phosphate backbone at the former position of the removed base. A pertinent reaction time permits the modification of one randomly located base per DNA molecule. A series of radioactive fragments are produced which extend from the ^{32}P-labelled end to the position of a modified base. Four specific chemical reactions allow the cleavage by piperidine preferentially at guanines (G > A), or at adenine (A > G) depending upon precise conditions, at cytosine alone (C) and equally at cytosine and thymine (C + T). By performing polyacrylamide gel electrophoresis and autoradiography as per the dideoxy method, the sequence of the DNA can be deduced from the pattern of bands on the autoradiogram. Protein-binding sites may be located through the masking of base-modification reactions by DNA-bound protein.

Rapid DNA-sequencing techniques have been applied to the sequencing of RNA molecules. Through the employment of the retroviral enzyme, RNA-directed DNA polymerase, RNA may serve as a template for the synthesis of a complementary DNA (cDNA) copy which may then be sequenced. The derived sequence may be interpreted in terms of the sequence of the original RNA molecule.

DNA-sequencing techniques have advanced our knowledge of the storage and expression of genetic information through the discoveries that eukaryotic DNA contains introns, that the genetic code is quasi- universal, that different codons for the same amino acid are used preferentially, that functionally related genes are often physically located in the same region of the chromosome, that viral genes may be translated in different-reading frames to produce different proteins and that transcriptional promoters contain a consensus sequence.

SUGGESTED READING

1. Brown, T.A. (1987). *Gene Cloning: An Introduction*, Van Nostrand Reinhold, Wokingham.
2. Glover, D. M. (1984). *Gene Cloning: The Mechanics of DNA Manipulation*, Chapman and Hall, London.

9

Microbial Relationships*

Terms and Concepts

Many microbes establish relationships with each other and with higher organisms. Usually the relationship is nutritional, although other benefits may accrue and the association can become crucial to the survival of one or both partners. In the nineteenth century, de Bary coined the term *symbiosis* to describe any situation where two different species lived together. Confusingly, some (mostly British) biologists then used the same term specifically to mean that both partners benefited. In this text the term symbiosis will be used in its original non-specific sense. *Mutualism* will be used to describe a relationship in which both partners benefit. *Parasitism* is used where only one benefits and in so doing typically reduces the fitness of the other. *Commensalism* is used where two species live together without either harming the other, but where only one partner benefits. In microbial relationships, mutualism and parasitism have been most extensively studied. In view of their enormous biological, medical and agricultural implications this chapter will concentrate exclusively on these (Fig. 9.1).

Facultatives and obligates

Associations would be easy to describe if organisms always behaved in the same way. Unfortunately, they do not. Many microbes, for example, can survive as both parasites and saprophytes. Thus *Ceratocystis ulmi,* which causes Dutch elm disease, kills the tree and then lives saprophytically on its dead remains. It is therefore described as a *facultative parasite*. In contrast, under natural conditions, *Puccinia graminis var, triticum* only grows on live wheat plants and can only be cultured on agar (saprophytic growth) with considerable difficulty. Organisms such as *P. graminis* are described as *obligate parasites*. Facultative and obligate parasites often differ in their pathogenic effects (i.e., in their ability to injure the host). Since obligates are restricted to *living* organisms, their effects on the host are often less severe, although the latter may show less vigorous growth. In contrast, facultative parasites, or parasites which have only recently 'acquired' a host, tend to be far more damaging (Table 9.1).

A question of balance

Whether a microbe exists mutualistically, commensally or parasitically with another organism is often a question of balance. We are all infected with bacteria. Most, like *E. coli,* are harmless or

---

* Excerpts taken from review published by Bull, London.

even beneficial and exist in equilibrium with the body. However, if the equilibrium is disturbed, if the body's defence systems fail to keep the microbe in check, then disease may result. For example, between 40% and 70% of adults carry the pneumonia bacterium *(Streptococcus pneumonniae)* in their throats, and so it can certainly be regarded as a normal inhabitant of the body. Yet most people do not suffer from pneumonia. Only when the defence systems fail or are weakened by, say, a viral infection, does disease occur.

TABLE 9.1

Comparison of Obligate and Facultative Plant Parasites

Characteristics	*Obligate parasite*	*Facultative parasite*
Pathogenicity	Rarely fatal	Often fatal
Growth of parasite	(i) Usually restricted to particular regions	(i) Spreads throughout plant; grows on dead remains
	(ii) Intracellular hyphae common	(ii) Almost exclusively intercellular hyphae.
Toxins	Rarely produced	Common; cause breakdown of host tissue
Specificity	Often high, especially with fungi, which may be limited to one or two species. (Generally lower with viruses)	Usually low. A single species of fungus may infect many genera of plants e.g. *Verticillium albo-arum* (wilt fungus) is pathogenic to many crops. Some exceptions, e.g. *Ceratocystis ulmi* is restricted to elm
Examples	*Puccinia graminis*, tobacco mosaic virus	*Ceratocystis ulmi*, *Pythium debaryanum*

The fact that some bacteria find their way into the body at all may be purely accidental. *Clostridium tetani* is a common soil saprophyte which continues to grow saprophytically in an open wound. However, it is pathogenic because it secretes substances *(toxins)* that affect the nervous system. The resulting disease, lock jaw, can be fatal. Hence *C. tetani* is perhaps best regarded as a well-adapted soil organism which, simply by chance, sometimes finds itself behaving as a poorly adapted parasite. The secretion of toxins by *C. tetani* is typical of the way that bacteria cause disease. Some toxins are proteins secreted by the microbial cells *exotoxins* and *aggresins* (harmful enzymes). Others *(endotoxins)* are complex polymers of carbohydrates, lipids and proteins derived from the bacterial cell wall. They are released into the body when the bacteria lyse. Both types of toxin exert their effects by modifying normal cellular function.

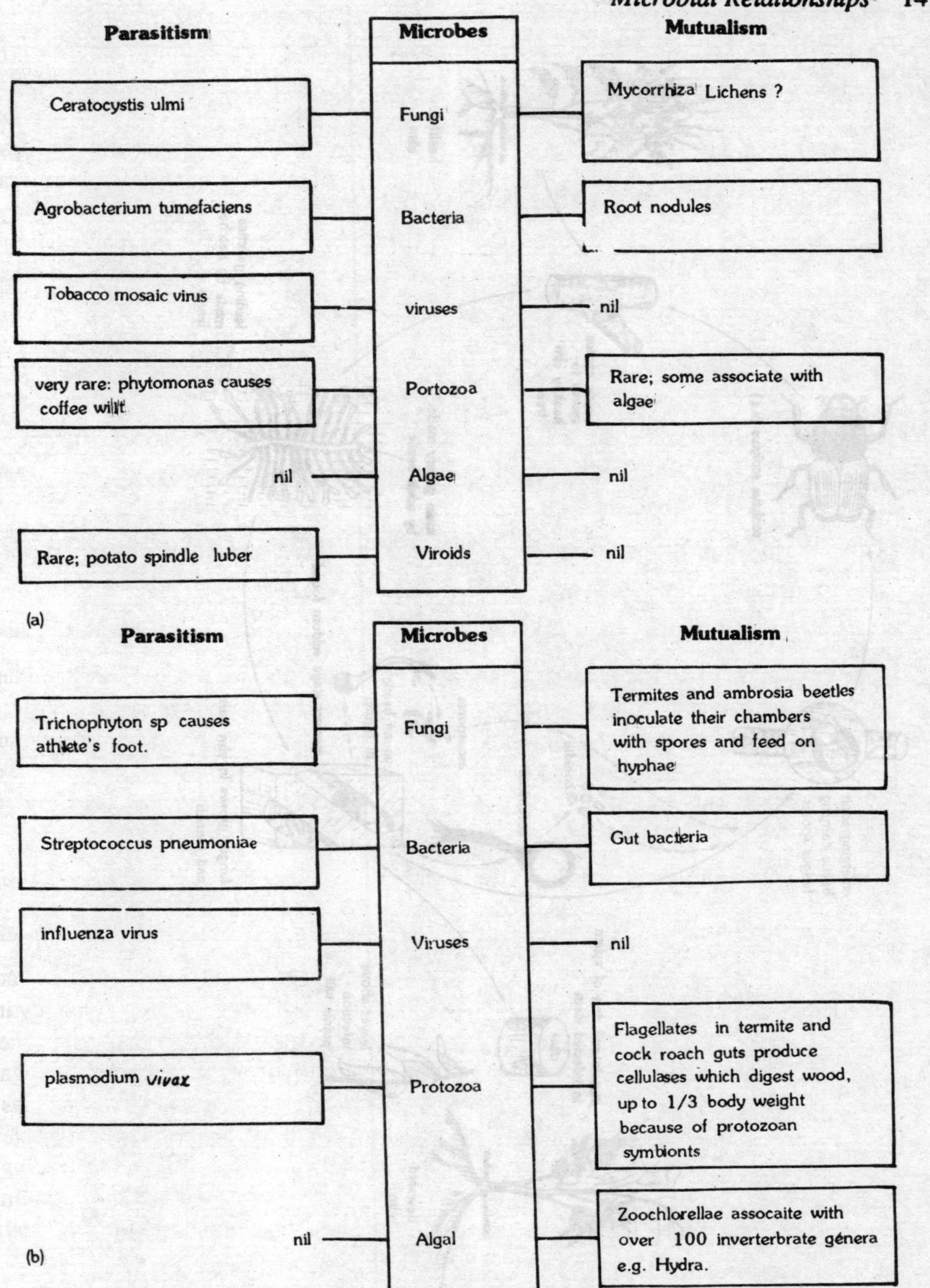

Fig. 9.1. *Microbial associations with plants and animals:* (a) *microbial associations with plants;* (b) *microbial associations with animals.*

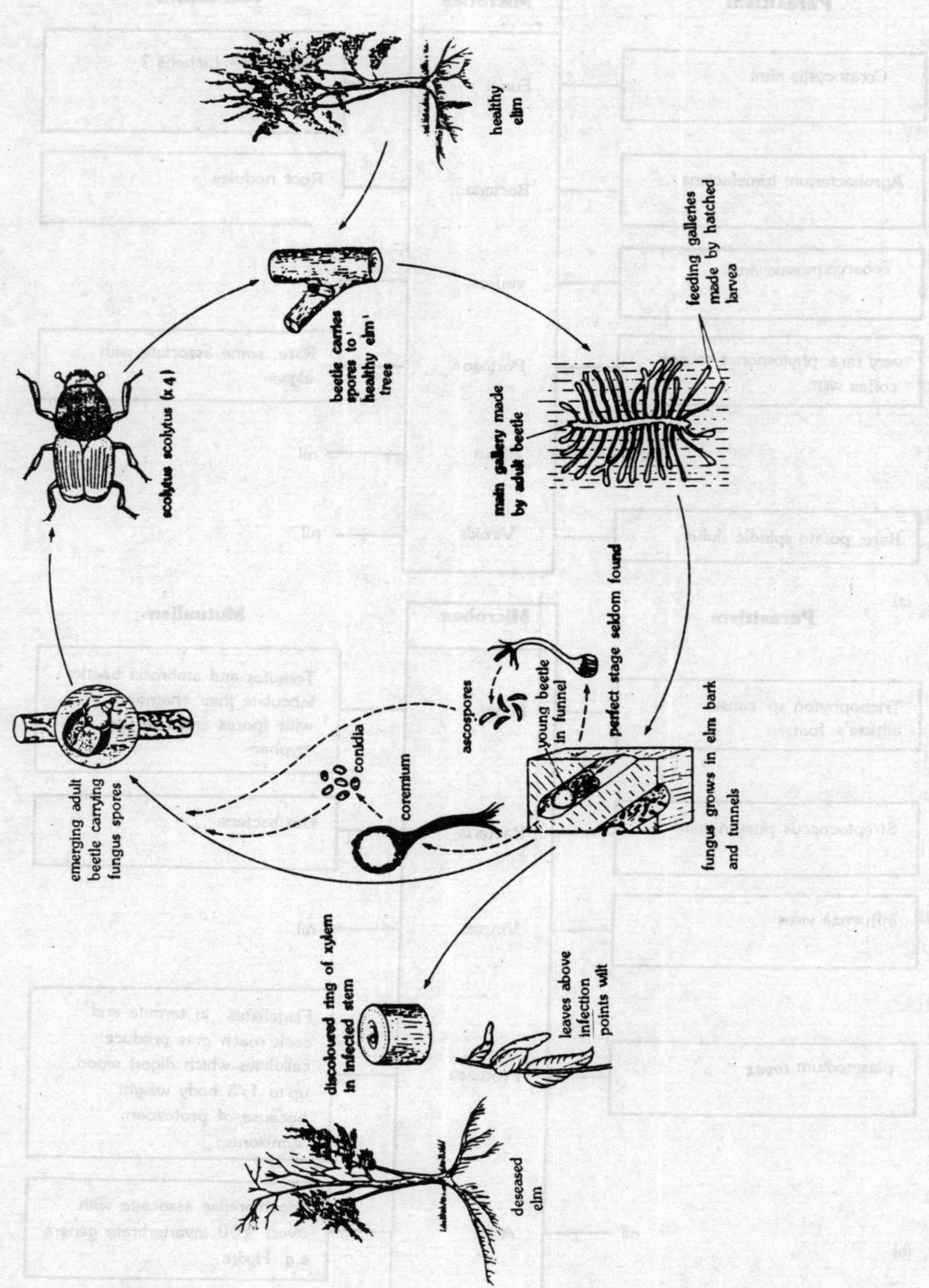

Fig. 9.2. *Dutch elm disease.*

Microbial Parasites of Plants

Fungi and viruses are by far the most serious groups of plant pathogens. Some 8000 fungi and 1000 viruses can cause disease, resulting in widespread and serious losses. Although generally less common, other plant pathogens may cause severe local problems on particular crops.

Dutch elm Disease

Ceratocystis ulmi is an Ascomycete, responsible for Dutch elm disease. It was first isolated in Holland in 1918. Heavy outbreaks occurred throughout Europe (Including the UK) during the 1920s, but then subsided. During this time the disease spread to the USA and in the late 1960s returned to the UK in an extremely virulent form. Examination of elm timbers imported from the USA showed that the mycelium of the virulent strain was present, indicating that the new epidemic had originated from diseased timber imports. Spores of the fungus are carried by the wood-boring beetle, *Scolytus* sp. The female burrows beneath the bark into the phloem and combium, usually when the elm is producing early (spring) wood in June. She releases a scent (pheromone) which attracts the male, and eggs are laid at intervals along the main gallery (Fig. 9.2). The larvae hatch and feed by chewing into the nutrient rich cambium and phloem. When mature, they burrow under the bark and pupate. The adults subsequently hatch, and emerge carrying fungal spores on their exoskeletons. Mature beetles tend to lay their eggs in diseased trees, whilst newly hatched beetles tend to feed on healthy trees. As a result the disease spreads from tree to tree. Two breeding cycles may occur in a single season.

C. ulmi possesses all the worst features of a facultative parasite (Table 10.1). Spores introduced into the elm germinate in the sap to produce a yeast-like mycelium. The xylem then becomes blocked by hyphae and tylose plugs, so choking the transport system and causing the leaves to wilt and die. Soon the disease spreads to new branches, killing the whole tree. The fungus may live saprophytically on the dead remains for many years.

Elm is highly prized as an amenity tree, but its small economic value meant that national research establishments reacted slowly to the epidemic. Before the outbreak there were about 30×10^6 elms in the UK. Roughly two-thirds of these are now dead, with near 100% losses in southern England. Attempts to regulate the disease have been made, including:

(i) felling and burning diseased wood;
(ii) using fungicides (e.g. Benlate) and insecticides;
(iii) biological control, i.e. killing the fungus with *Pseudomons* sp. (bacterium) or killing *Scolytus* using parasitic wasps;
(iv) planting disease-resistant strains, such as a Japanese—Siberian hybrid.

None of these methods has been outstandingly successful, and restocking the stricken areas will necessarily take many years. However, there is a glimmer of hope in that some regeneration of healthy shoots from the bases of killed trees is taking place. Moreover, in the north and west the disease is much less common, probably because of the presence of another elm-inhabiting fungus, *Phomopsis oblonga*. The latter adversely affects the growth of *Scolytus* larvae. Whether or not *P. oblonga* can be exploited for biological control elsewhere is not certain.

Tobacco Mosaic Virus

Viruses are second only in importance to fungi as agents of disease in plants. In common with most other obligate parasites they tend not to kill their host, although they may cause such catastrophic reductions in yield that, to the farmer, the consequences are just as bad. Table 9.2 summarises some details for tobacco mosaic virus (TMV).

TABLE 9.2

Tobacco Mosaic Virus (TMV) : Summary

Size and structure	*300 nm x 15 nm; helical ssRNA virus*
Pathogenicity	150 genera of dicots, few monocots. Examples: tobacco; tomato; potatoes. Chlorosis of leaves, stunted growth, reduced yield (Fig. 9.3). Rarely (if ever) fatal. Infects most live cells. Penetrates cuticle via stomata or cuts in leaves. Moves from cell to cell through the phloem Mostly by routine cultural manipulation of plants. Incidentally by biting insects (no specific insect vector)
Control	Sanitation; use resistant varieties; do not replant for 2 years after epidemic

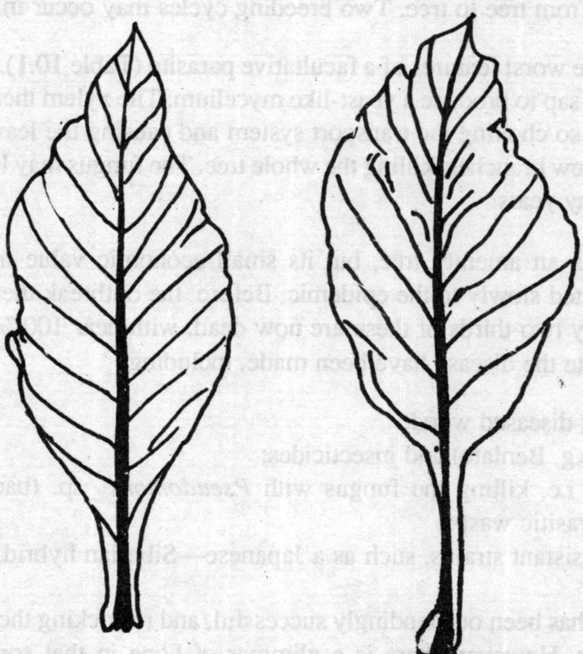

Fig. 9.3. *Tobacco mosaic disease.* Small leaves with pale achlorophyllous blotches characterise the disease. The blotches identify groups of cells which have been killed by the virus.

Insects, especially aphids, transmit many plant viruses, e.g., beet yellow virus (pathogen of sugar beet). TMV is not transmitted by aphids and is unusual in that it does not have a specific insect vector. Fungi, nematodes and routine cultivation procedures such as grafting may also transmit viruses.

Mutualistic Microbial Associations with Plants

Mycorrhiza

Probably the most universal and important mutualistic associations between micro-organisms and plants are those which develop between fungi and roots, the mycorrhizas (myco meaning fungus, and rhiza meaning root). Most, if not all, plants establish and benefit from mycorrhizal relationships, and a large but uncertain number of fungal species are involved.* There are four main types of mycorrhizas: *ectotophic, vesicular-arbuscular (V—A), orchidaceous,* and *ericaceous* (Table 10.3). The last three used to be lumped together and described as 'endotrophic,' but this is misleading as there are enormous differences between them.

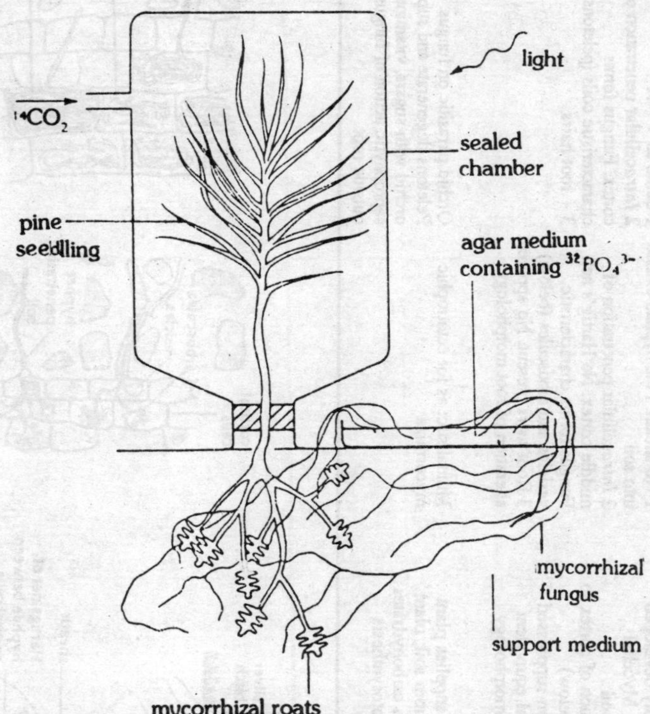

Fig. 9.4. Experiment to investigate the nutritional relationships in ectotrophic mycorrhizas *(Melin,* 1963).

Together, ectorophic and V—A mycorrhiza account for the vast majority of mycorrhizal relationships. Ectotrophic mycorrhizas are the most obvious: the roots are markedly affected, and the fungal partner often forms large fruiting bodies (toadstools) in the autumn. V—A mycorrhizas, however, are much more common. Mycorrhizas develop most successfully on nutrient-poor soils. In a forest where there is intense competition for nutrients or on a sand-dune where the soil is inherently

---

Note *** By definition, the term mycorrhiza excludes the infection of roots by pathogenic fungi such as *Pythium*.**

TABLE 9.3
Main Categories of Mycorrhiza

	Ectotrophic (sheathing) mycorrhiza	*Vesicular-arbuscular (V-A) mycorrhiza*	*Orchidaceous mycorrhiza*	*Ericaceous mycorrhiza*
Example	Typically associated with temperate trees, e.g. *Amanita muscaria* (fly agaric) with birch or pine	Most widespread type of mycorrhiza. Associated with bryophytes, ferns, spermatophytes (especially tropical trees); e.g. *Endogone* spp. with grasses	Unique to orchids, e.g. *mellea* (honey fungus) with *Gastropodia elata*	*Armillaria* Associated with heathe and related plants. Includes *Boletus* and *Monotropa* (bird's nest plant)
Characteristics	1 Fungus forms highly developed sheath around roots. Mycelial strands extend into soil 2 *Inter*cellular invasion of cortex to form Hartig's net (below) 3 Root hair formation suppressed (mycelium functional equivalent of root hairs). Root morphology altered	1 No sheath. Fine hyphae extend into soil 2 *Intra*cellular penetration of middle cortex. No Hartig's net. Fungus forms characteristic vesicles and arbuscules (below) 3 Root hairs present. No apparent alteration of root morphology	1 Same as V—A 2 *Intra*cellular penetration of inner cortex. Fungus forms characteristic coils (peletons) 3 + root hairs	1 Variable form; loose weft of hyphae surrounds root (heather) or definite sheath *(Monotropa)* 2 *Intra*cellular penetration of outer cortex. In *Monotropa* a Hartig's net may be additionally present 3 No root hairs, no epidermal cells
Association	Mutualistic: fungus supplies plant with NH_3 ad PO_4^{3-} from soil; plant supplies fungus with carbohydrates produced during photosynthesis	Mutualistic: as for ectotrophic mycorrhiza	Orchid parasitic on fungus. Peletons degenerate and supply orchid with sugars, vitamins, and saprophytic action of fungus outside root	Variable: achlorophyllous plant may be parasitic on fungus (*monotropa* type). Alternatively,

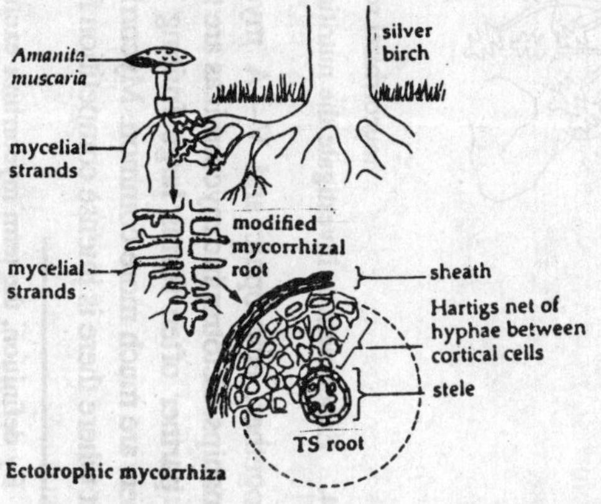

Ectotrophic mycorrhiza

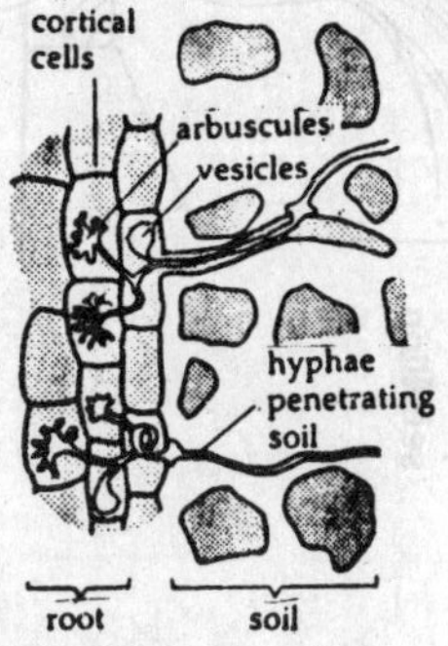

V-A mycorrhiza

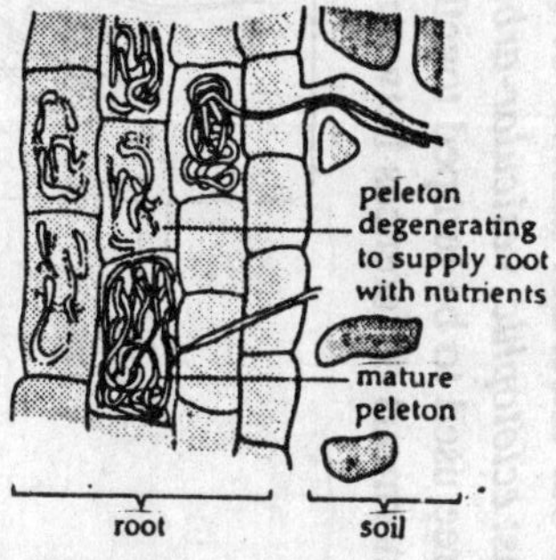

Orchidaceous mycorrhiza

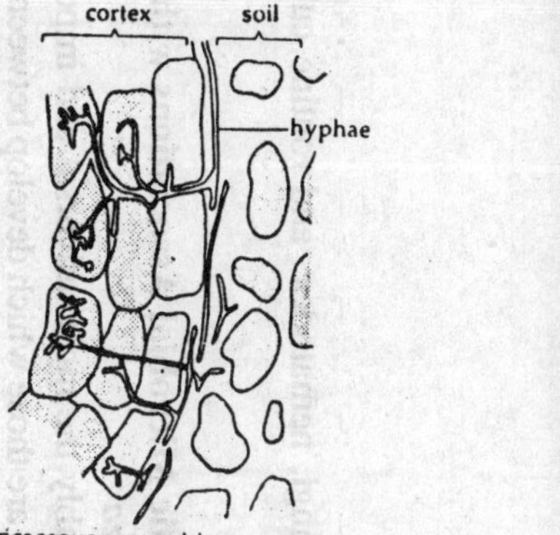

Ericaceous mycorrhiza

poor, mycorrhizas may be essential for the survival of both partners. Pine seedlings, for example, usually die within a few weeks unless they establish mycorrhizal relationships. The partners can therefore be described as *ecologically obligate symbionts (mutualists).* Functionally ectotrophic and V—A mycorrhizas are very similar, despite their obvious structural differences. Experiments indicate that in both cases the relationships are those of nutritional interdependence.

TABLE 9.4

Experiments to Investigate the Nutritional Relationships of Vesicular—arbuscular (V—A) Mycorrhiza

	Phosphate-deficient soil		*Phosphate-enriched soil*	
	Without mycorrhiza	*With mycorrhiza*	*Without mycorrhiza*	*With mycorrhiza*
Maize growing on poor soil				
Number of grains per ear	31	354	279	321
Mass of 1000 grains (g)	2.4	19.8	23.7	20.9
Legumes on poor soil, previously infected with *Rhizobium*				
Dry mass at 10 weeks (mg)	93	580	310	610
Total phosphate per plant (mg)	77	1077	383	1791
Number of *Rhizobium* nodules per plant	0	34	7	37

In mycorrhizas the fungal mycelium acts like a massive root hair system, scavenging minerals from the soil. Indeed, with ectotrophic mycorrhizas, proper root hairs never develop. Sugars formed in the leaves move down the stem as sucrose, but sucrose itself never accumulates in the fungus. The level is kept low because it is converted into isomers such as *trehalose*. The fungus therefore acts as a 'sugar sink,' the conversion enabling it to draw more easily on the carbohydrate supply available in the root. The amount of sugar passing into the fungus may be considerable, equivalent to as much as 10% of the total timber produced.

Nutritional benefits are not the only ones to accrue from mycorrhizal relationships. For the plant partner, others include:

(i) drought resistance;
(ii) tolerance to pH and temperature extremes;
(iii) greater resistance to pathogens, due to *phytoalexins* released by the fungus.

The events leading to mycorrhiza formation are poorly understood but presumably involve some form of chemical signalling. With ectotrophic mycorrhizas, definite relationships seem to exist between certain Basidiomycetes and particular trees. *Amanita muscaria* (fly agaric) and *Boletus bovinus*, for example, are common fungal partners of birch and pine, although they are not the only ones. In complete contrast, fungi forming V—A mycorrhizas are restricted to one small family of Phycomycetes, with two genera, *Endogene* and *Glomus*, forming associations with a huge variety of distantly related plants.

The orchidaceous and *Monotropa*-type mycorrhizas are very different from the above. Here the higher plant is temporarily or permanently parasitic on the fungus. Orchid seeds are minute (0.3-14 μg), without any significant food reserve. Some fail to germinate at all unless infected by a fungus. Others germinate, but development, soon ceases unless the seedling becomes infected. The fungus penetrates the cells of the cortex, bringing with it nutrients scavenged from the forest floor. The nutrient-rich hyphal coils (*peletons)* then break down, making food available to the plant. How the orchid persuades the fungus to undergo this bizarre self-sacrifice is obscure. It is certainly not because the fungus needs some essential nutrient from the orchid. This is clearly demonstrated by the hyperparasitic orchid, *Gastrodia*. This is parasitic on the honey fungus *Armillaria melea,* which is itself parasitic on trees. *A. mellea* certainly does not need the orchid for growth. Most orchids eventually become green, and so the relationship may shift from parasitism to one of mutualism. Other orchids such as *Neottia nidus-avis* (bird's nest orchid) remain achlorophyllous and parasitic on the fungus throughout their extra-ordinary existence. The relationship between fungi and orchids is completely the reverse of that between the root-infecting fungus *Pythium debaryanum* and young seedlings. In the latter case the fungus is lethally parasitic. The underlying mechanisms which determine whether a fungus-root association will develop mutualistically or parasitically (and, if the latter, which way round) are still unclear.

Other fungal associations with trees

Perhaps surprisingly, mycorrhizal fungi are rather poor saprophytes and show low levels of *cellulase* and *lignase* activity. In contrast, non-mycorrhizal saprohytic fungi (Fig. 9.5) degrade plant debris more easily and contribute substantially to nutrient recycling in the forest ecosystem. An example of a fungal parasite of trees has already been given.

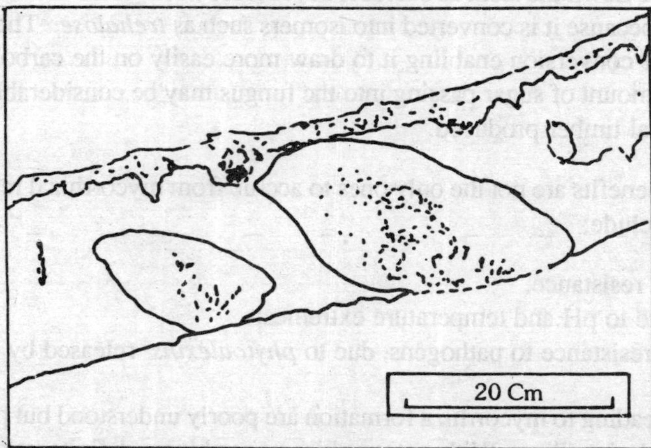

Fig. 9.5. *Polyporus (the bracket fungus)*. Many species of *Polyporus* are wholly saprophytic, although some are facultative parasites.

Bacterial pneumonia

Other mutualisms

By far the most important mutualistic relationship between bacteria and plants is the *Rhizobium*-

legume association. Blue-greens also form similar associations with specific genera. The Algae and Protozoa do not form mutualistic associations with higher plants.

Microbial Parasites of Animals

Arguably the most widespread parasites of animals are not microbes at all, but worms. However, among the microbes the viruses, bacteria and Protozoa are most serious, with the fungi playing a relatively minor role. Examples of bacterial and protozoan parasites are described below:

Streptococcus pneumonia (Fig. 9.6 & Table 9.5), the causal agent of pneumonia, was a major cause of death before the discovery of antibiotics. It is still a serious illness in infants, the elderly and infirm. The bacterium is carried by 40%-70% of adults, and about 1 in 500 individuals suffer from the disease at any one time. The symptoms are often slight, and recovery is spontaneous in about 70% of cases.

TABLE 9.5

Streptococcus Pneumoniae

Organism	Gram-positive diplococci ('double spheres') surrounded by a thick capsule Exracellular parasite. Fermentive respiration produces lactic acid by glycolysis. About 50 different serological strains
Host	Widespread in the upper respiratory tract, only invading the lungs and becoming pathogenic when the body is weakened. In lungs, exotoxins and endotoxins neutralise antibodies, enabling bacterial cells to multiply and invade host tissue. Body counter-attacks via neutrophils and macrophages (often successfully). Can infect animals, but no symptoms of disease under natural conditions
Pathogenicity	Lungs fill with fluid and blood as a result of lesions caused by bacterial growth
Transmission	Exhalation droplets
Control	Penicillin, erythromycin; isolate patients

TABLE 9.6

Plasmodium

Organism	*Plasmodium* sp.: Phylum Protozoa
Hosts	Most tropical and subtropical populations. Evidence suggests that the species which cause animal malaria are different from those which cause human malaria (and vice versa)
Paghogenicity	*See* Fig. 9.7
Transmission	*Anopheles* (mosquito); about 50% of all species are vectors
Control	(a) Prevent bites, i.e., use netting over beds, windows (not very effective) (b) Destroy insects, i.e., spray nesting sites with insecticides, drain breeding grounds (ponds, swamps) or spray with oil and insecticides

to choke and poison larvae. These strategies, although effective, are not always
feasible as paddy-fields are needed for rice growing, and ponds for watering
herds, irrigation or domestic supplies.

(c) Drugs e.g., quinine (kills erythrocyte stages only), chloroquine (kills all stages). Drugs may be used as Prophylactics (preventive agents) or for treatment. Unfortunately, many strains of *Plasmodium* are now resistant
to these drugs.

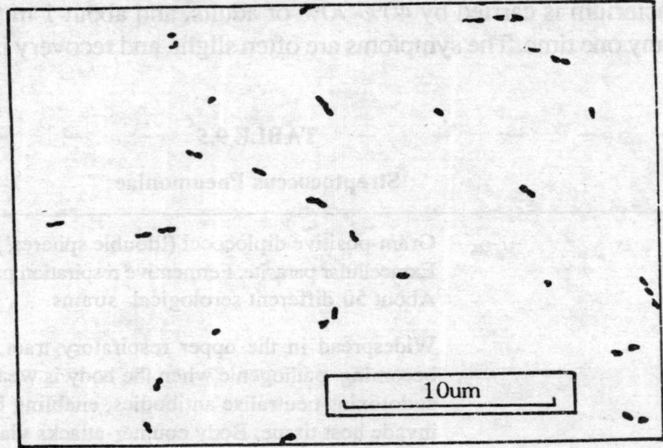

Fig. 9.6. ***Streptococcus pneumoniae.***

Malaria

25% of the world's population is at risk from malaria, and some 5% suffer from the disease at any one time, resulting in about 10^6 deaths per year. Five species of *Plasmodium* cause the disease. Fig. 9.7 applies to *Plasmodium vivax,* which results in one of the most common forms of the disease; *benign tertidry malaria.*

If the number of people suffering from malaria is any guide to success, then *Plasmodium* must be regarded as a well-adapted parasite. Several features have probably promoted its success :

(i) It is minute and therefore easily transmitted by the mosquito.
(ii) The vector introduces *Plasmodium* directly into the site of infection (blood).
(iii) As with many parasites, it has several (three) reproductive stages, so promoting both a rapid increase in population size and effective transmission.
(iv) Inside the host cells it is unobtrusive and inconspicuous, so avoiding detection (see below).
(v) In areas where the disease is endemic, the host population shows milder symptoms, and in any event the dispersive (gametocyte) stage is reached long before the host is killed.

The blood is a warm nutrient-rich isosmotic well-buffered pollution-free oxygenated environment. Unfortunately for *Plasmodium* this otherwise ideal habitat is ruined by the presence of lymphocytes

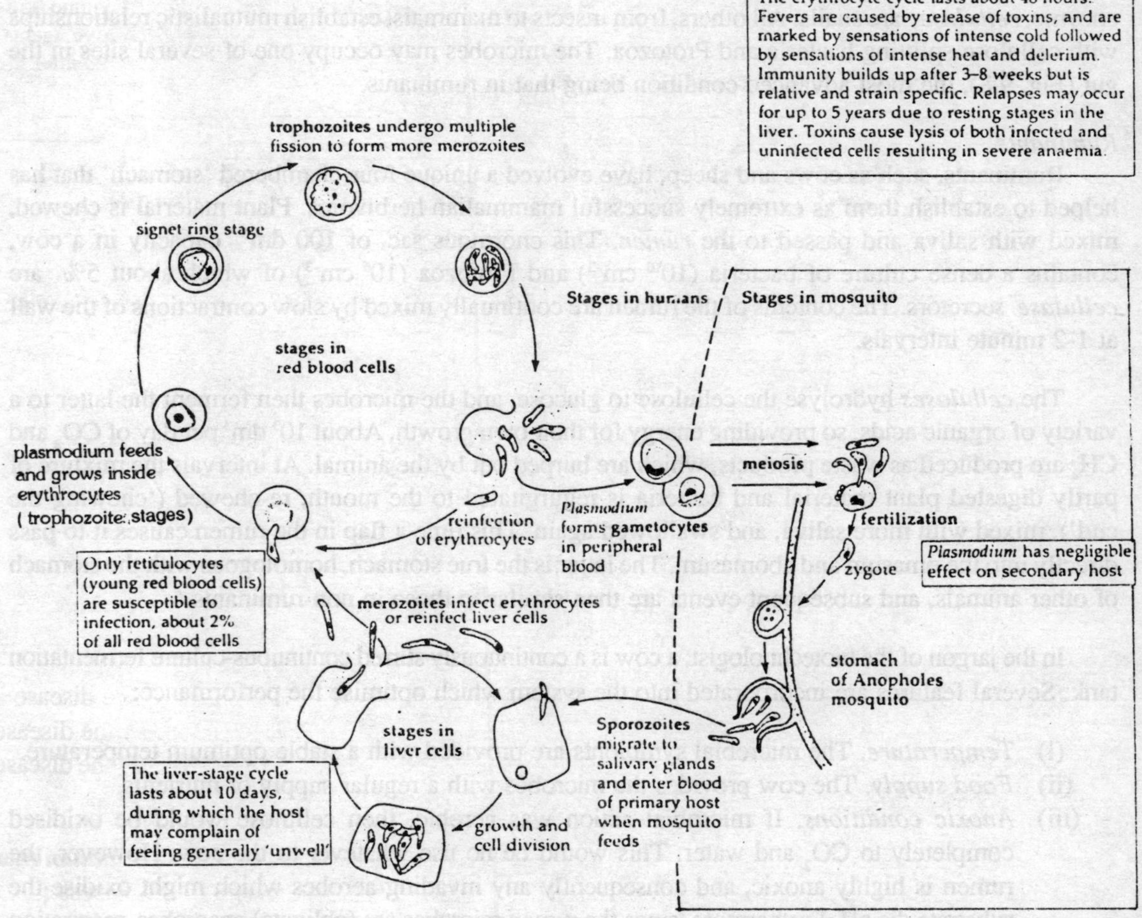

Fig. 9.7. *Life cycle of Plasmodium vivax.*

and phagocytes. However, by living inside the cells of the host, *Plasmodium* avoids detection from both. It is a very effective strategy. The only time that the lymphocytes have a chance to react against the parasite is during its quick dash from one cells to the next. Moreover, different strains of *Plasmodium* are covered in different antigens (membrane glycoproteins). The net result is that immunity is slow to build up and is only effective against one antigenic strain (*serotype*).

Mutualistic Microbial Associations With Animals

The gut flora of herbivores

Since plants are about 30% cellulose (dry weight), it would be very much to the advantage of any herbivore to digest this large insoluble inert polysaccharide (only small soluble molecules can be absorbed through the gut wall). However, the only herbivores to possess the appropriate digestive enzyme, *cellulase,* are snails. All others, from insects to mammals, establish mutualistic relationships with cellulose-splitting bacteria and Protozoa. The microbes may occupy one of several sites in the gut (Fig. 9.8), the most advanced condition being that in ruminants.

Ruminants

Ruminants, such as cows and sheep, have evolved a unique four-chambered 'stomach' that has helped to establish them as extremely successful mammalian herbivores. Plant material is chewed, mixed with saliva and passed to the *rumen*. This enormous sac, of 100 dm^3 capacity in a cow, contains a dense culture of bacteria (10^{10} cm^{-3}) and Protozoa (10^6 cm^{-3}) of which about 5% are *cellulase* secretors. The contents of the rumen are continually mixed by slow contractions of the wall at 1-2 minute intervals.

The *cellulases* hydrolyse the cellulose to glucose, and the microbes then ferment the latter to a variety of organic acids, so providing energy for their own growth. About 10^3 dm^3 per day of CO_2 and CH_4 are produced as waste products, which are burped out by the animal. At intervals the mixture of partly digested plant material and bacteria is regurgitated to the mouth, re-chewed ('chewing the cud'), mixed with more saliva, and swallowed again. This time a flap in the rumen causes it to pass directly into the omasum and abomasum. The latter is the true stomach, homologous with the stomach of other animals, and subsequent events are then similar to those in non-ruminants.

In the jargon of the biotechnologist, a cow is a continuously stirred continuous-culture fermentation tank. Several features are incorporated into the system which optimise the performance:

(i) *Temperature*. The microbial symbionts are provided with a stable optimum temperature.

(ii) *Food supply*. The cow provides the microbes with a regular supply of nutrients.

(iii) *Anoxic conditions*. If microbial action was aerobic, then cellulose would be oxidised completely to CO_2 and water. This would be no use whatever to the cow. However, the rumen is highly anoxic, and consequently any invading aerobes which might oxidise the substrate die off. Furthermore, since the rumen microbes are (obligate) anaerobes, respiration is largely *fermentive,* and the breakdown of carbohydrates is consequently incomplete. The end products of microbial respiration (fatty acids) are then used by the cow. Indeed, the cellular metabolism or ruminants is uniquely geared to fatty acid rather than to sugar metabolism, reflecting their dependence on microbially produced metabolites as a food source.

(iv) *pH*. Fermentation yields acids, but a drop in pH is prevented because the cow releases 100 dm^3 per day of alkaline saliva into the rumen. This bicarbonate-phosphate buffer stabilises the rumen contents at about pH 6.5, which is optimal for microbes. Uniquely, ruminant saliva does not contian *amylase*.

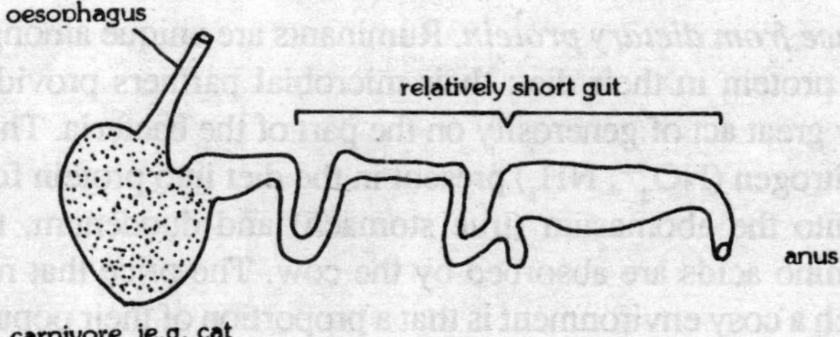

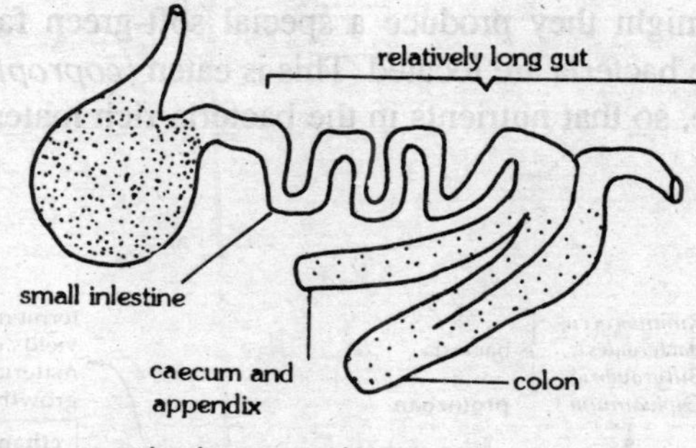

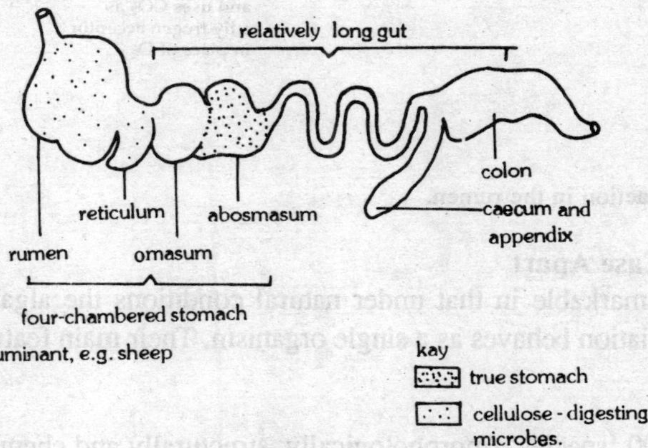

Fig. 9.8. ***Alimentary canals of mammals.*** **The first three chambers of the ruminant 'stomach' are actually modifications of the oesophagus.**

(v) *Independence from dietary protein.* Ruminants are unique among mammals in that they do not require protein in their diet; their microbial partners provide it for them. This is not through any great act of generosity on the part of the bacteria. The latter simply convert the inorganic nitrogen (NO_3^-, NH_3) present in the diet into protein for their own use but, when they pass into the abomasum (true stomach) and duodenum, they are digested and the resulting amino acids are absorbed by the cow. The price that microbes therefore pay for living in such a cosy environment is that a proportion of their population must be continually sacrificed. Independence from dietary protein gives ruminants an edge over other large herbivores. It is only possible because the microbes are positioned before the stomach and small intestine *(pregastric fermentation).* In herbivores such as horses, bacteria are located after the small intestine *(post-gastric fermentation)*, and so there is no opportunity for the microbes to be digested. However, rabbits have evolved a mechanism for reducing the problem. At night they produce a special soft-green faecal pellet derived from the caecum where the bacteria are located. This is eaten *(coprophagy)* and passes through the gut a second time, so that nutrients in the bacteria-rich material can be obtained.

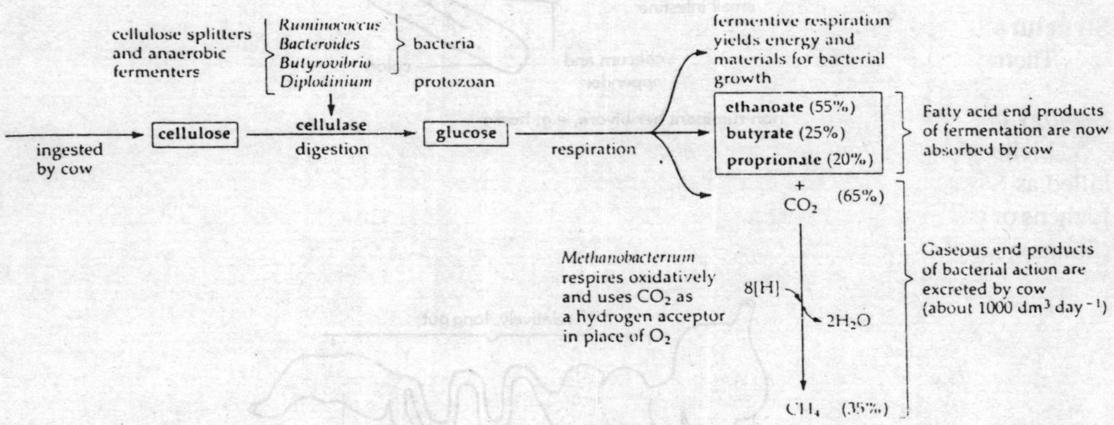

Fig. 9.9. Microbial action in the rumen.

The Lichens : A Case Apart

Lichens are remarkable in that under natural conditions the algal-fungal or bacterial (blue-green)-fungal association behaves as a single organism. Their main features are summarised below.

Organisms

There are 18000 'species' (morphologically, structurally and chemically distinct types).

Mycobiont (fungus)

These are Ascomycetes or (rarely) Basidiomycetes. They are ecologically obligate symbionts.

Phycobiont (alga or bacterium)

These are green algae (e.g. *Trebouxia,* 70% of all lichens) or blue-greens (e.g. *Nostoc*). They may be free living.

Morphology

There are three forms:

(i) *crustose* (crust like, e.g. *Xanthoria* which is the common yellowish lichen on gravestones);
(ii) *fruticose* (shrubby, e.g. *Cladonia* which is common on acid heathland.
(iii) *foliose* (leaf like, e.g. *Parmelia* which is common in woodland).

Relationship between associates

This is uncertain; the phycobiont supplies carbohydrates to the fungus, and the fungus may supply minerals to the phycobiont. There is no experimental confirmation of the latter, and the phycobiont may be able to absorb its own minerals from the substrate. 'Good' laboratory conditions cause the association to breakdown, whilst adverse conditions help to maintain it, and so the association probably enables both partners to exploit habitats which would be unsuitable for either alone.

Structure

There is often a highly organised thallus, with algae forming a definite layer.

Ecology and applications

Lichens are often *pioneer organisms*, establishing themselves on inhospitable terrain. They are killed as SO_2 levels rise, and their abundance can be used as an indicator of atmospheric pollution. Lichens or their products may be used as food (for reindeer herds), dyes (Harris tweed) and indicators (litmus).

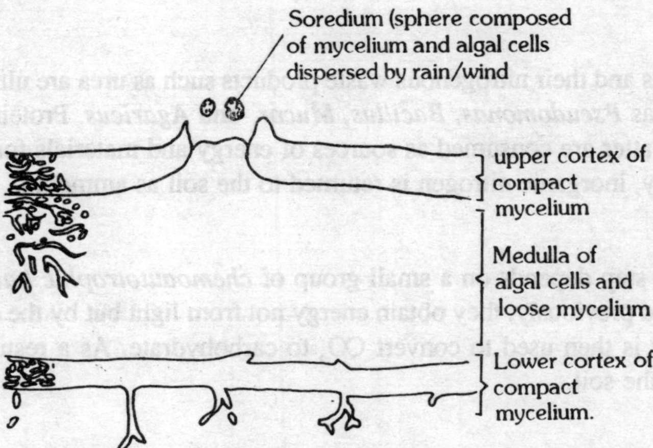

Fig. 9.10. *Generalised lichen thallus.*

Some authorities regard the lichen association as one of parasitism of the phycobiont by the fungus, and others as one of *ecological mutualism*. Whatever the exact nature of the relationship, it is extremely intimate. The reproductive structures, for example, are often unique aggregates of algal cells and hyphae, called *soredia*.

Biogeochemical Cycles

The emphasis on mammals and flowering plants in most biology courses perhaps gives the impression that the diverse nutritional and respiratory strategies of bacteria are peculiar and exceptional. Certainly the materials involved are remarkably diverse, but this only reflects the fact that the organisms concerned are exploiting the resources available in their own ecological niches. The essential nutritional and respiratory processes are in all major respects identical with those of higher organisms. Indeed, if all organisms used the same materials, then life would soon cease to exist. Nitrogen and sulphur supplies, for example, would rapidly be exhausted. As we shall see, the recycling of minerals upon which the continuance of life depends rests entirely upon the existence of a few microbial genera.

The nitrogen cycle

The main components of the nitrogen cycle are shown in Fig. 9.11 and described in more detail below.

The eukaryotes

N_2 cannot be utilised by any eukaryote. Eukaryotes depend entirely on 'fixed' nitrogen such as NO_3^-, NH_3, or —NH_2. Thus, soil nitrate is actively absorbed by plant roots and reduced to NH_3, by enzyme systems called *nitrate* and *nitrite reductases*. NH_3 is then incorporated into carboxylic acids to form amino acids and ultimately protein (see the book *Enzymes, Energy and Metabolism* in this series). In theory, plants can utilise soil NH_3 as well as NO_3^-. In practice the former is so toxic that it is of minor significance.

Ammonification

Higher organisms and their nitrogenous waste products such as urea are ultimately decomposed by saprophytes such as *Pseudomonas, Bacillus, Mucor* and *Agaricus*. Proteins are hydrolysed to amino acids and the latter are consumed as sources of energy and materials for growth. As a result of saprophytic activity, inorganic nitrogen is returned to the soil as ammonia.

Nitrification

The next crucial step depends on a small group of *chemoautotrophic bacteria*, the *nitrifying bacteria*. As described previously, they obtain energy not from light but by the oxidation of reduced nitrogen. This energy is then used to convert CO_2 to carbohydrate. As a result of their activities, nitrate is returned to the soil.

Denitrification

Nitrification is aerobic. Under anaerobic conditions a different bacterial community develops in the soil—one which results in *denitrification*. For the most part the denitrifying bacteria are saprophytic heterotrophs. They respire their food oxidatively, so the hydrogen which is removed from it has to be combined with an environmental compound. If oxygen is available, they use that. If it is not, they may turn to NO_3^-.

Nitrogen fixation

Losses due to denitrification can be offset by various prokaryotes capable of nitrogen fixation (Table 9.7). Artificial fertiliser and lighting also raise the level of fixed nitrogen in the soil.

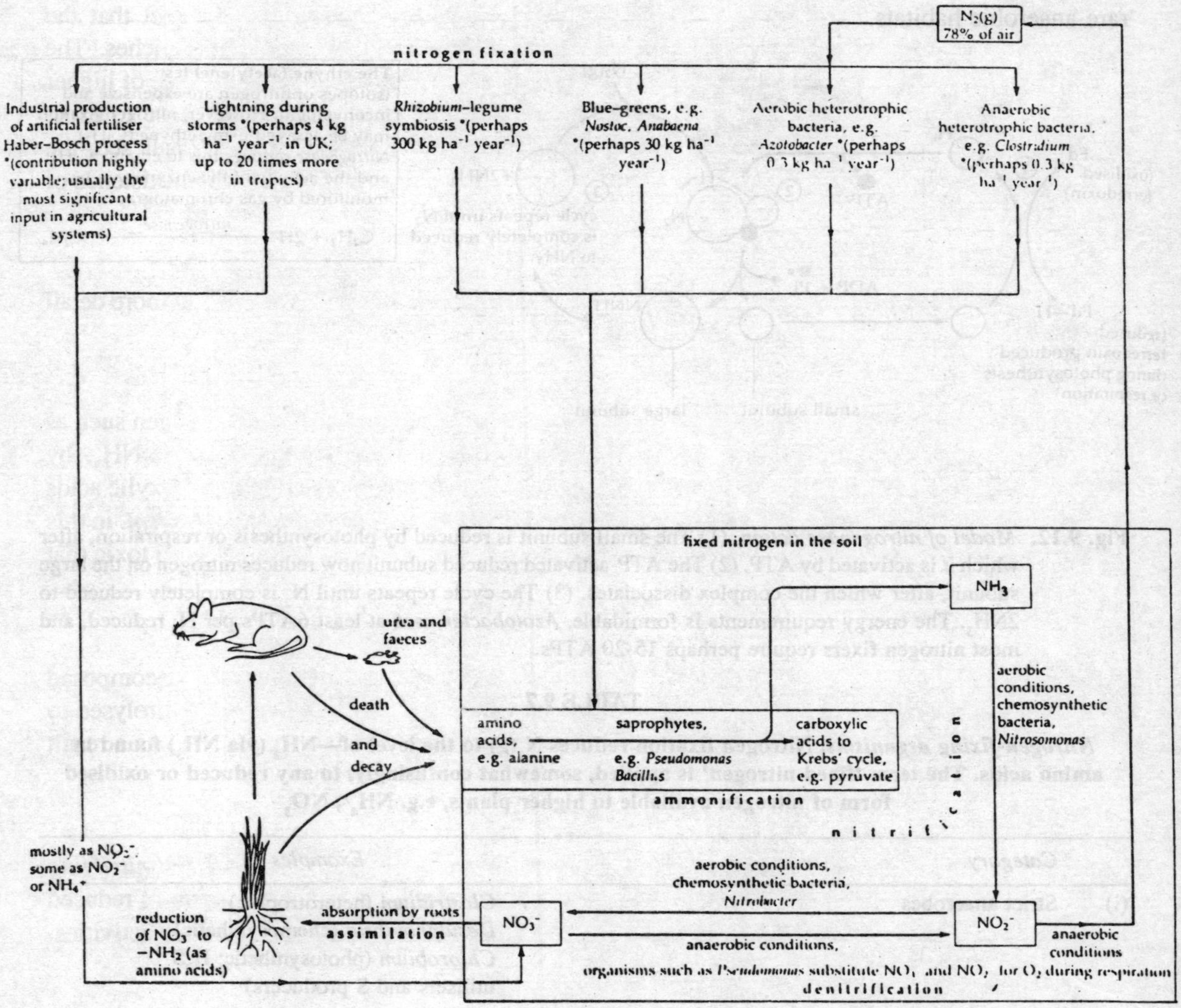

Fig. 9.11. *The nitrogen cycle.* Not all the components of the cycle operate at the same time or in the same place. The asterisks indicate that precise values for nitrogen fixation are not available. The relative importance of each contributor will vary from place to place.

Just as oxidation (nitrification and oxidative respiration) releases energy, so reduction requires it. This includes the reduction of N_2(g) to NH_3. In fact, nitrogen fixation requires a substantial amount of energy. Perhaps not surprisingly, therefore, free living N_2(g)-fixing heterotrophs make only a minor contribution to nitrogen fixation (Fig. 9.11), because the energy requirement must be met from their food supply. In contrast, the photosynthetic bacteria have an abundant source of (solar) energy available for nitrogen reduction. The photosynthetic nitrogen fixers (especially the blue-greens) thus make a correspondingly greater contribution to nitrogen fixation. (The photosynthetic sulphur and non-sulphur bacteria only make a marginal contribution, because they are restricted to comparatively rare anaerobic habitats.

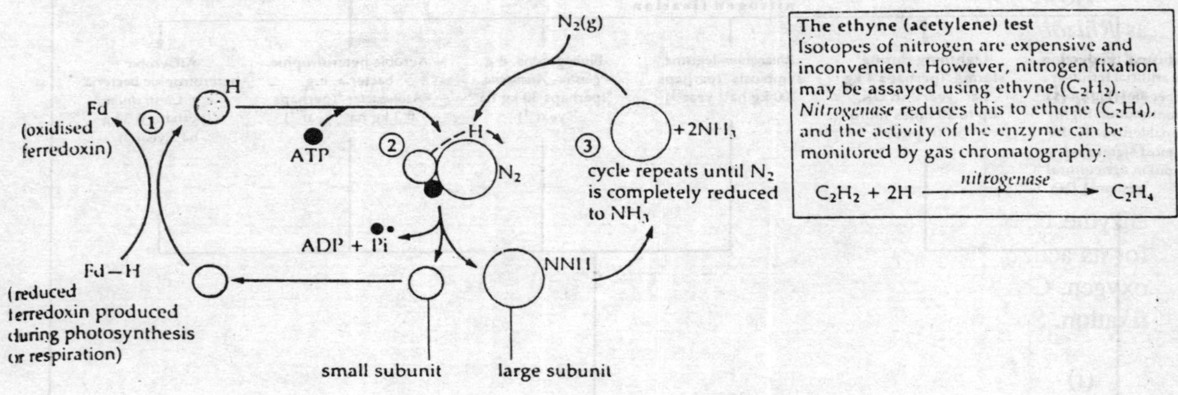

Fig. 9.12. ***Model of nitrogenase action.*** **(1) The small subunit is reduced by photosynthesis or respiration, after which it is activated by ATP. (2) The ATP-activated reduced subunit now reduces nitrogen on the large subunit, after which the complex dissociates. (3) The cycle repeats until N_2 is completely reduced to $2NH_3$. The energy requirements is formidable. *Azotobacter* uses at least 6ATPs per N_2 reduced, and most nitrogen fixers require perhaps 15-20 ATPs.**

TABLE 9.7

Nitrogen-fixing organisms. **Nitrogen fixation reduces N_2(g) to the level of—NH_2 (via NH_3) found in amino acids. The term 'fixed nitrogen' is applied, somewhat confusingly, to any reduced or oxidised form of nitrogen available to higher plants, e.g. NH_4^+, NO_3^-**

	Category	*Examples*
(i)	Strict anaerobes	*Clostridium* (heterotrophic) *Desulphovibrio* (chemosynthetic) *Chlorobium* (photosynthetic; H_2S utilisers and S producers)
(ii)	Facultative anaerobes (anaerobic when fixing nitrogen)	*Klebsiella* (heterotrophic)

(iii)	Microaerophils (anaerobic when fixing nitrogen)	*Rhizobium* (heterotrophic symbiont) *Thiobacillus* (chemosynthetic) *Spirulina* (photosynthetic; H_2O utilisers and O_2 producers)
(iv)	Aerobes	*Azotobacter* (heterotrophic) *Anabaena* (photosynthetic; H_2O utilisers and O_2 producers) *Nostoc* (photosynthetic; H_2O utilisers and O_2 producers)

However, of greater importance than free-living nitrogen-fixers are mutualistic microbes such as *Rhizobium*. The latter forms nitrogen-fixing root nodules in legumes. The legume act as a 'sink' of fixed nitrogen, and as a result the microbe is stimulated to maintain a high level of fixation. In return, the plant supplies *Rhizobium* with carbohydrates and minerals.

Physiology and ecology of nitrogen fixation

The enzyme which reduces $N_2(g)$ to NH_3 is called *nitrogenase*. It is an extremely complex enzyme, consisting of two distinct subunits, one of which contains molybdenum. A possible mechanism for its action is shown in Fig. 9.12. The enzyme is extremely sensitive to and rapidly degraded by oxygen. Consequently, oxygen must somehow be excluded from the vicinity of *nitrogenase* during fixation. Several strategies may be employed.

(i) *Avoidance*. Some nitrogen fixers live only in anoxic environments, or in places where the oxygen level is so low *(microaerophily)* that respiration effectively reduces the cellular level to zero.

(ii) *Respiratory protection*. Aerobic heterotrophs such as *Azotobacter* have greatly enhanced levels of respiration, which reduces the oxygen level of the cell. This 'pseudo-respiration' is uncoupled from ATP synthesis. Whilst permitting fixation in aerobic habitats, this strategy is extremely wasteful because considerable amounts of food are used just to create anaerobic microenvironments within the cell.

(iii) *Organelles*. In aerobic heterotrophs the *nitrogenase* is also surrounded by proteins which appear to improve its stability in aerobic conditions. It is not known how these proteins permit nitrogen to penetrate whilst still excluding oxygen.

(iv) *Heterocysts*. In aerobic photosynthesisers, nitrogen fixation is restricted to special cells called *heterocysts*. The thick wall, the enhanced level of respiration and the loss of the oxygen-generating component of photosynthesis help to keep the internal environment anoxic.

(v) *Nodules*. In the legume-*Rhizobium* association the plant root cells act as both an oxygen sink and an oxygen barrier. The most effective legume nodules produce *leghaemoglobin*, which further restricts the admission of oxygen to the bacteria.

Mutualistic associations with higher plants

The majority of the 12000 species of legumes form nitrogen-fixing root nodules, particularly if the level of fixed nitrogen in the soil is low. The formation and establishment of a nodule depend upon a sequence of chemical signals (Fig. 9.13). Special membrane proteins called *lectins* help to identify and incorporate *Rhizobium* into a root cell and, when one strain has infected a plant, no other can do so.

Nitrogen-fixing root nodules also form on alder trees. In these the bacteria are not rhizobia and belong to a different group of prokaryotes called Actinomycetes. Few details are known, but the level of nitrogen fixation may exceed even that in legumes.

Non-nodulating nitrogen fixing associations are also formed between blue-greens and various plants, e.g. fungi (as lichens), the water fern (*Azolla*), cycads and one angiosperm (*Gunnera*). The blue-green is usually *Nostoc (Anabaena* in the case of *Azolla)*. Besides these mutualistic symbioses, looser associations develop between many plants and particular strains of microbes. Maize, for example, encourages the growth of *Azotobacter* and even binds the bacterium to its roots. Efforts are being made to enhance agriculturally important nitrogen-fixing association in order to reduce the dependence of crops upon expensive fertiliser.

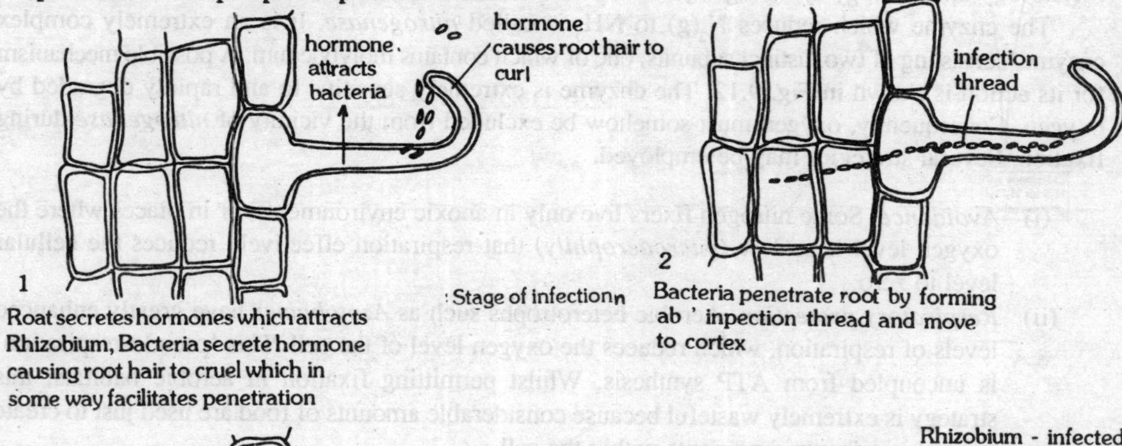

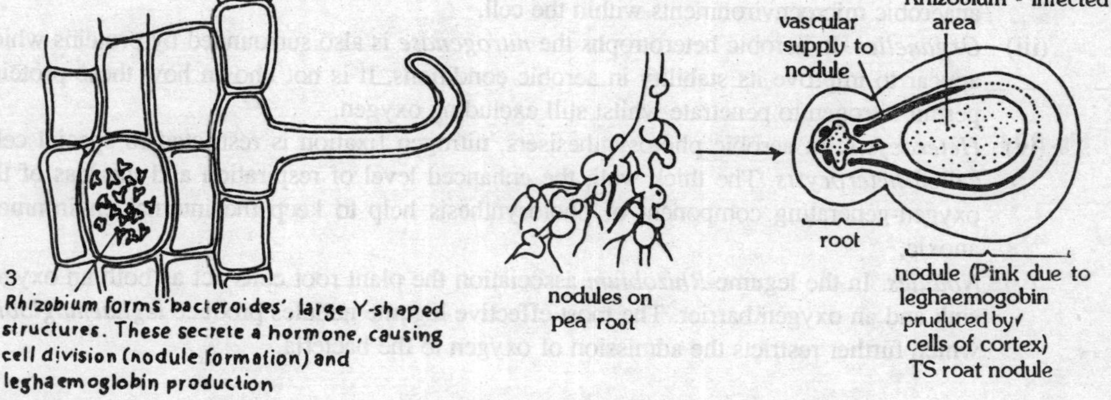

Fig. 9.13. ***Formation and establishment of a root nodule.*** **Nodules form on lateral roots. Infection occurs just behind the apices, in the root hair region.**

The sulphur cycle

Sulphur is relatively easily oxidised compared with nitrogen and as a result free elemental sulphur is rarely found in the environment. Nevertheless, the main principles of the nitrogen cycle apply equally to the sulphur cycle. As with nitrogen, it is the most oxidised form (SO_4^{2-}) that is absorbed by roots. Figure 9.14 illustrates the sulphur cycle as it might occur in a standing body of water. Study the diagram and use the questions beneath in order to deduce the main points for yourself.

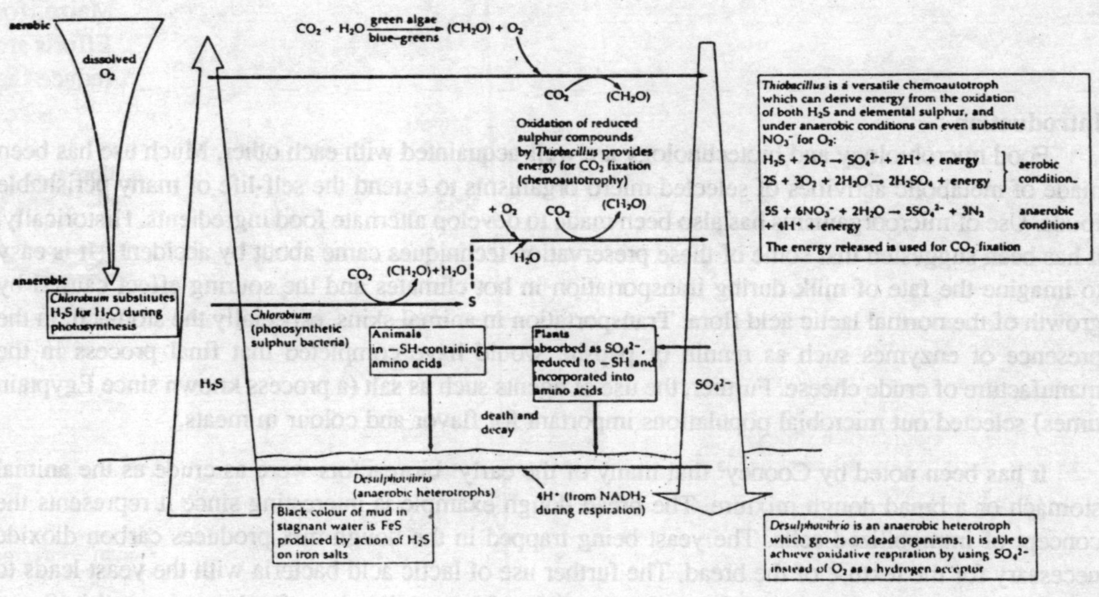

Fig. 9.14. *The sulphur cycle.* A standing column of water in, say, a gas jar will develop this cycle within about 2 months. A few centimetres of mud should be allowed to settle at the bottom, which should also be 'primed' with a little torn paper to provide nutrients for the saprophytes. Such standing columns are called *Winogradsky columns.*

10

Food Microbiology and Biotechnology*

Introduction

Food microbiology and biotechnology are well-acquainted with each other. Much use has been made of metabolic activities of selected micro-organisms to extend the self-life of many perishable foods. Use of microorganisms has also been made to develop alternate food ingredients. Historically, it has been suggested that some of these preservation techniques came about by accident.[1] It is easy to imagine the fate of milk during transportation in hot climates and the souring effect caused by growth of the normal lactic acid flora. Transportation in animal skins, especially the stomach, in the presence of enzymes such as rennin or pepsin, would have completed that final process in the manufacture of crude cheese. Further, the use of agents such as salt (a process known since Egyptain times) selected out microbial populations important for flavor and colour in meats.

It has been noted by Cooney[3] that many of the early bioreactors were as crude as the animal stomach or a bread dough mixture. The bread dough example is interesting since it represents the concept of immobilized cells. The yeast being trapped in the dough mix produces carbon dioxide necessary for the texture of the bread. The further use of lactic acid bacteria with the yeast leads to the production of a sour dough. A similar principle of immobilization of cultures is used in Swiss cheese production with *Propionibactium* supplying the holes and flavours! The reader no doubt can think of others, many of which are referenced in this book.

Unfortunately, many of these processes remained isolated from scientific principles and have been perpetuated as art forms for a long while. It is only recently that fresh concepts (or renewed vigor!)are being applied to food processes. We now live in an age where we require uniform products having strict quality control criterion. Such demands have necessitated the need for process controls and a greater understanding of the process. Extension of self-life for a centralized food system has also made demands on current technology. We now require longer shipping times and further markets than we traditionally worked with.

---

* Excerpts taken from review published by D. Collins. et al Ontario, Canada.

Finally, the need to develop alternatives or new food types has come about to counteract shortages, embargos, and the ravages of war and nature. The isolation and protection policies arising out of world recession has made us look at the need to be self-sufficient.

This chapter looks at current and future applications of biotechnology as it relates to food microbiology. The term "food microbiology" has been used in its widest sense and refers to the application of microbiology or microbiological processes to food. The depth of this application can be seen in Tables 10.1, 10.2 and 10.3 which outline some major uses of micro-organisms and enzymes in the food industry. The authors have used reviews and book references where possible so that the reader can pursue the various topics in greater depth.

Meats

The application of microbial technology to meat production lies in two main areas. These are tenderization and fermentation. In order to improve the texture of meat after slaughter, a process of conditioning or aging is required. This process is slow, measuring days after post mortem. The subtle changes that occur during this period are now becoming better understood and deal with changes in protein structure in the muscle tissues. This holding period has a number of drawbacks since there are cases where improper slaughtering practices can yield a carcass which does not follow the normal aging process and little tenderization occurs. The texture of meat depends on amount and properties of collagen of the tissue and on the state of the contractile protein. Another drawback is that undesirable spoilage can occur during conditioning as a result of microbial growth and create off-flavours with mold or slime formation on the surface of the meat.

TABLE 10.1

Major Use of Microorganisms in Foods (Meats and Poultry Products)

Microorganism	*Food Type*
Pediococcus cerevisiae	Semidry fermented sausage
	Dry fermented sausage
	Country-style ham
	Turkey sausage
	Deboned poultry meat
	Eggs
Lactobacillus species	Dry and semidry fermented sausage
	Bacon
	Fresh meat
	Poultry meat
	Eggs
Micrococcus species	Dry and semidry fermented sausage
	Processed meats (e.g., frankfurters)
Penicillium species	Dry fermented sausage
Thammdium	Fresh meat

The introduction of plant or microbial proteases has been tried to rapidly improve texture after processing. Proteases for this purpose are listed in Table 10.4. Enzymes from fungal and bacterial sources have been used but they are not the proteases of choice. The most active enzyme source is from plants (e.g. papain, ficin and bromelain).

Fungal proteases cannot be controlled with respect to their action on meats. Over application can lead to a mushy end-product. Limited breakdown of meat tissues by bacterial or fungal protease requires controlled amounts, uniform distribution, and reasonable penetration. Some of the problems have been overcome by use of carriers and ante mortem injection.[2] It is also important to note the microbial proteases are often mixtures of proteolytic enzymes with different pH and temperature characteristics.

TABLE 10.2

Major Uses of Microorganisms in Foods (Dairy Products)

Microorganism	*Food Type*
Lactobacillus species	Cheeses including Parmesan, Provolone, Muenster, Mozzarella, Brick, Buttermilk, Acidophilus milk, Yogurt
Streptococcus species	Cheeses including Romano, Parmesan, Cheddar, Colby, Blue, Brick, Cottage Cheese, Edam, Yogurt, Fresh milk, Sour cream, Butter, Buttermilk
Penicillium species	Cheeses including Camembert, Roquefort, Blue
Propionibacterium species	Cheeses, such as Swiss

Home use of microbial tenderizers is by far the biggest market since application can be controlled prior to the cooking. These proteases are activated by the cooking process. Similar applications can apply to such meats as hams or chickens. These have shown the same potential with microbial proteases. Mohr has suggested that using the natural autolytic process in meats may provide a better route for exploration for commercial tenderization of meats. He also suggests that the development of collagenases from microbial sources might be an alternative. Such enzymes have been reported in *Achromobacter*.[16] By breaking down the collagen in meat tissue, control over the rate of tenderization may be achieved.

TABLE 10.3

Major Groups of Microbial Enzymes Used in Food Preparation

Enzyme Group	*Source*	*Function*	*Industrial Application*
Amylases	fungal	Starch hydrolysed to maltose	Brewing, Fruit and vegetable juices, Starch products
Lipases	fungal	Triglycerides to glycerol and fatty acids	Cheese flavor, Milk products

Pectinases	fungal/ bacerial	Hydrolysis of pectin to peptides and amino acids	Clarification and modification of fruit and vegetable juices
Catalase	fungal/ bacterial	Hydrogen peroxide breakdown	Peroxide removal in eggs and milk products
Glucose oxidase	fungal	Converts glucose to gluconic acid	Dried egg production

TABLE 10.4

Possible Source of Microbial Enzymes for Meat and Fish Processing

Organism	*Enzyme*	*pH optimum*
Aspergillus niger	Proteases	4
Aspergillus oryzae	Proteases	4-8
Aspergillus sartoi	Proteases	3-4
Penicillium sp.	Proteases, Lipases	4-5
Clostridium perfringens	Collagenase	7-8
Clostridium histolyticum	Collagenase	7-8
Achromobacter sp.	Collagenase	7
Bacillus subtilis	Proteases	7

In the recovery of meat wastes, proteases can serve a useful function both for the hydrolysis of blood proteins as well as meat stripping. The recovered product might be a useful source of animal food supplement.

The deliberate use of microbial cultures for meat products on a commercial scale started in the 1950s and 1960s. Prior to this point in the preparation of fermented meats, a "natural" fermentation allowed the normal lactic acid producing population along the inherent quality control problems associated with such a system.

The principle of starter culture addition to meats for fermentation provides the necessary levels of lactic acid bacteria to produce the lactic acid required for taste and low pH. The other function is to antagonize the natural flora including some pathogens such as *Salmonella* and *Staphylococcus aureus* and control their growth. Organisms used in starter cultures such as *Lactobacillus plantarum* and *Pediococcus cerevisiae* provide both the acidity and anitimicrobial agents to inhibit undesirable organisms (Table 10.1). It is possible to reduce nitrite levels required in fermented sausages as the pH is rapidly reduced during fermentation. Reduction of nitrite levels can minimize nitrosamines formation and limit the concern over this proven group of carcinogens. Starter culture addition to meats has other attributes including lower histamine levels and extension of shelf life[1]. The success of starter culture, especially with the extension of shelf life, has also led to its application to non-fermented meats such as frankfurter and poultry[11]. Further extension into seafoods as a commercial enterprise is inevitable.

The future of starter culture technology presents many possibilities. It will be possible to select specific organisms for colour fixation and flavors. Fermented sausages rely on microbial sources for flavor development in the form of lypolysis and deamination of amino acids.[14] The potential for microbial lipases and proteases as flavor components is now being explored. The understanding of plasmids and their role in lactic acid production and nitrite reduction will provide further impetus for their use in meats. It may be possible to reduce nitrite levels enzymatically without destroying colour and tlavor characteristic provided by nitrite. The development of inhibitors provided during fermentation by the lactic acid bacteria similar to the antimicrobial agent nisin might replace the more traditional food preservatives. The final payoff as suggested by Bacus and Brown[1] will be the development of a more rapid aging process, greater microbial safety with respect to pathogens, and a "greater utilization of perishable raw meat supplies."

Fish

One area of application of microbial enzymes in fish processing is the use of proteases. In such processes, the marketing of protein concentrate for a prime food source from species of fish and fish offal is not acceptable (as yet!) to the consumer. The treatment of ground fish with protease enzymes for fishmeal preparation does raise a number of problems related to fishy flavors and bitterness. Such products are not particularly attractive to the human food market. The bitterness is caused by formation of bitter peptides with a high content of hydrophobic amino acids. Mohr also notes problems related to the incomplete solubilization of the fish protein (only 60 per cent soluble fraction) by microbial proteases. Research by Rutman and Heimlick suggested that an increase in solubility of fish protein may be possible by suing lower initial temperatures during the enzyme addition (cited by Mohr). Increase in yields result with this technique.

The problem of bitterness and fishy flavor may be resolved by using specific proteolytic enzymes rather than a general protease system associated with most fungal enzymes.

Dairy

Nowhere in the joint area of food and biotechnology has the success of these two disciplines been more evident than in the dairy industry. This is seen in the wide exploitation of latic acid bacteria used in the manufacture of dairy products (Table 10.2). It has also been achieved in the use of microbial sources of rennet and in the mold ripening of cheese. Cheese manufacture relies on such organisms as *Streptococcus lactis* and *S. cremoris* for lactic acid production. Rapid acid production is not only necessary for texture and taste but also to control competing contaminating organisms such as *Staphylococcus aureus* or *Salmonella* species. Any intrusion by these organisms can give rise to food infection or intoxications. In order to obtain the desired acid production as quickly as possible, starter cultures consisting of selected strains of *S. lactis* are used. Very often, flavor components such as diacetyl are required to give a characteristic taste to the product. These aroma and flavor producers are found in the main members of the *Leuconostoc* species. Some of these will only produce diacetyl at low pH values. The versatility of the lactic acid bacteria is also demonstrated by the fact that specific strains of *Lactobacillus bulgaricus, L. acidophilus* and *L. lactis* are used for specific fermented milk products such as acidophilus milk and yogurt. Combinations of such organisms will often produce the desired flavor and acidity associated with these products.

During the manufacture of cheese, a clotting enzyme rennet is added after a short period of fermentation. The source of this enzyme in the past has been from a calf stomach. In recent years, this source has become limited and so new sources have been found. Molds have provided this source and a clotting enzyme from *Mucor michei* or *M. pusillus* is now used commercially. These remarkable enzymes are suited to the dairy industry because of their clotting power and low protease activity. "Rennet like" enzymes have also been found in organisms such as *Bacillus, Penicillum* and *Endothica.*

Micro-organisms play a major role in the curing and aging of cheeses. The presence of specific amino acids, esters, alcohols, and fatty acids are influenced by specific micro-organisms. The most outward evidence of microbial activity and cheese characteristics are the mold ripened cheeses such as Camembert and Roque fort. These cheeses rely on *Penicillium* species to produce such flavor compounds as ethanol, methyl ketones and butanol. Production of these compounds rely on lipolytic (esterase) activity (Table 10.5). Application of such enzymes has been suggested to enhance cheese flavors.[6] *Aspergillus* strain seem particularly suited for application in such cheese as Italian, American, and process cheeses. Flavor in cheeses as a result of microbial action has also been reviewed by Dwivedi.[4]

Further exploration with lipase enzymes using an immobilized system has been proposed by Kilara. In order to overcome the expense of using enzymes, they should be reused. The process of immobilization achieves this end. The enzyme is placed on an inert carrier by a number of processes including microencapsulation, entrapment, and adsorption. In the system developed by Kilara,[18] succinylaminoethyl agarose was used as the carrier for preparation of partially lipolysed butter and cream. Using such systems may be useful for dietetic foods preparation. Another immobilized system with possibilities was reviewed by Greenberg and Mahoney. Using the enzyme lactase, these authors listed a number of possibilities for this system, one application being low-lactose milk. By giving milk an initial lactase treatment, it may have application for people which low lactose tolerance. The main drawback to this system is the protein-protein reaction affecting the process and the efficiency of the immobilized system. Another application for the lactase system is in the modification of whey from cheese wastes. Such a system could become a potential source of alternate sweeteners. Microbial lactases currently being considered include sources from *Aspergillus (A. niger* and *A. oryzae), Lactobacillus thermophilus,* and *Leuconostoc citrovorium.* The development of lactases from the lactic acid bacteria (or any enzymes) is an attractive idea since these organisms have a long and safe record of use in the dairy industry. Lactases from this source have a pH optimum of 6.2-6.5 and a temperature optimum of 55-60°C. Because of the central role the lactic acid bacteria play in food production, especially the dairy industry, a great deal of interest has been shown in the genetics of these organisms.

TABLE 10.5
Enzymes Used in the Dairy Industry

Enzyme	*Product*	*Purpose*
Proteases	Cheese	Flavor texture
Proteases	Cheese	Coagulation
Lipase	Cheese	Flavor, texture dietetic products
Glucose oxidase	Cheese, milk powder	Oxygen scavenger
Catalase	Milk and milk products	Pasteurization texture, flavor
Lactase	Milk	Low lactose milk

See texts for references.

In a recent review concentrating on gene transfer system and plasmids, some insight into genetics of the lactic acid bacteria has been revealed. Plasmids are common both in the streptococci and lactobacilli although most are cryptic. The most documented function as it relates to the dairy industry has been the endowment of particular plasmids to lactose metabolism and proteinase production. In the studies on lactose metabolism, plasmids have been implicated in certain lactose specific enzymes such as the phosphotransferase lactose uptake mechanism. The size of these plasmids range from 30 to 41 Md. Work done by Chassy *et al.* with *Lactobacillus* has also shown phosphotransferase activity associated with particular plasmids. Further work by this group with *Lactobacillus casei* showed by a cloning experiment with *E. coli* that lactose phosphoenol pyruvate-phosphotransferase system and P-β galactosidase are controlled by plasmids. One of the major limitations in such genetic studies with the lactobacilli is their inability to be transformable or transducible.

The other interesting aspect of plasmids is the possible link with proteinase enzyme production. In preliminary studies, cheese made with proteinase deficient *Streptococcus* mutants showed less bitter characteristics than normal cheese. This aspect of cheese production makes genetic manipulation of starter cultures of practical significance. One problem, however, in reported studies is the close link between lactose metabolism and proteinase activity. Loss of one property leads to the loss of the other function. Further research is needed to verify to see if the same plasmid controls both these functions.

As noted by Chassy et al., the genus *Lactobacillus* and *Streptococcus* are "worthy of genetic interest" because of their established use in foods, absence of toxins, and growth characteristics. It is a research growth area of the 80s.

Vegetables

In food preparation, both enzymes (Table 10.3) and organisms (Table 10.6) are used. In many of the fermentations using vegetables as substrates, the predominant group causing the fermentation is the lactic acid bacteria. Such organisms are used because of their ability to produce lactic acid rapidly and also their tolerance to low pH. Principal vegetables that are preserved by fermentation are cucumbers and cabbages. Such fermentation rely on a sequence of growth of different organisms, i.e., homofermentative lactics converting carbohydrate to lactic acid and the heterofermentors producing acetic acid carbon dioxide, ethyl alcohol, and traces of other volatile compounds. Very often during the fermentation, the success of these operations also depends on natural constituents in the substrate to inhibit competing population. By combining these natural inhibitors with the use of sodium chloride, anaerobic conditions and suitable pH, the correct organism usually will dominate. In the production of sauerkraut using cabbage as the substrate, the sodium chloride concentration is maintained at about 2.5 per cent. The sequence of lactic acid bacteria which dominates is *Leuconostoc mesenteroides, Pediococcus cerevisiae, Lactobacillus brevis* and finally *Lactobacillus plantarum*. A similar sequence occur in pickle and olive manufacture. In olive fermentation, higher levels of salt are used. As with red meats, starter cultures consisting of *L. plantarum* are used to control the natural fermentation and reduce the loss from defects due to pectinolytic enzymes produced by molds. Enzymes have an application in preparing vegetable juices both in maceration stages or liquifaction.[12] Enzymes such as amylases, glucose oxidase, proteases, and pectolytic enzymes are used. The most important of these enzymes with respect to vegetable juice production is the pectolytic ones. The pectin molecule is an important substance in maintaining the structure and integrity of vegetables and fruits. It consists of partially methylated molecules of galactose linked together (1, 4 α-D galacturonan). This molecule is broken down by two enzyme systems, pectinesterases and pectin depolymerases.[12] Sources of these enzymes are widespread in the microbial world but preferred sources are usually derived from molds such as molds such as *Aspergillus*.

TABLE 10.6
Major Use of Microorganisms in Foods
(Fruit and Vegetables and Grains)

Lectobacillis plantarum	**Carrots, cucumber, peppers, olives, cabbage**
Lactobacillus brevis	**Cucumbers, carrots**
Pediococcus cerevisiae	**Mixed vegetables, carrots, cucumbers, peppers**
Leuconostoc mesenteroides	**Carrots, beans**
***Saccharomyces* spp.**	**Wines**
Rhizopus oligosporus	**Tempeh, Miso, Tauco**
***Aspergillus* spp.**	**Tempeh, Soy**

Vegetable juices or macerates can be used for the preparation of purees for baby foods. Such purees using commercial enzyme systems have an advantage over mechanical methods by producing products with better stability and quality. By avoiding the thermal effects inherent in normal commercial methods, increased content of vitamins and pigments are obtained and less undesirable flavors are avoided.

Solid state fermentations of soyabeans for enrichment and food production have been extensively developed in many asiatic countries for Tempeh (Indonesia), Sufu (China), and Koji (Japan). Classical Tempeh is made by aerobic fermentation of treated soyabeans by *Rhizopus* sp., constituting the development of a white, mould-covered cake. *Rhizopus oryzae* and *Aspergillus oryzae* and related species have been largely responsible for the Tempeh fermentation in countries such as Indonesia, New Guinea, and other regions of the Orient. Greater digestibility is attributed to the mould fermentation.

In other countries, rice and black gram mungo legumes have been employed as a batter with leavening produced by *Leuconostoc mesenteroides* sp., *Streptococcus faecalis* and *Pediococcus cerevisaie*. Heterofermentation and low acid development occur with leavening due to carbon dioxide production. Hesseltine published a review on Tempeh and an extensive listing of Asian fermented foods, including Ragi, Sufu, Shoyu and Miso. Some of these fermented foods have been introduced into the Western part of the world, particularly with respect to Shouy or soy sauce which can be prepared from field crops.

Excellent reference texts by Pederson,[10] and Steinkraus[15] are recommended for review of the major food fermentations produced throughout the world.

Fruit

A major industry with respect to fruit is the juice industry. Enzymes derived from microbial sources have been successfully applied to juice production. Such enzymes play a role in the maceration, solubilization, and clarification of fruit pulps. The major enzymes involved are the pectinases. These enzymes can be divided into two classes.[8] The first group consists of the polymethyl galacturonases and pectinlyases, and the second pectin esterases.

The microbial sources are from molds such as *Aspergillus niger*, *A. oryzae* and *A. flavus*. The pectinases from these micro-organisms have the suitable pH optimum of use in fruit processing. The process for manufacture of pectinolytic enzymes from solid cultures or submerged cultures is given by Kilara.[8]

Crushed juices contain a high level of pulp (e.g. apple) and enzymes aid in the extraction and clarification process. The clarification process is usually completed by centrifugation of the resulting floc. It is important to note that enzyme concentration, pH, temperature, and enzyme contact time are important considerations in the clarification process. Although the largest user of microbial enzymes in fruit processing is the apple industry, soft fruits such as the berries can also be extracted and clarified by means of enzyme treatments.[12] One interesting application of microbial technology is the enzymatic debittering of citrus products. The bitter compounds, flavanoids and limonoids in citrus fruit are mainly restricted to the peel component. Navel oranges have the limonoid problem caused by the compound limonin. Enzymes for eliminating bitterness from liminoid have been isolated from *Arthrobacter* and *Pseudomonas*. Immobilized cells of *Arthrobacter globiforms* have been used to remove limonin from naval oranges. The principal bitter citrus flavanoid is naringin and is predominant in such fruits as grape-fruit. Fungal naringinase formulations have been isolated and catalyze the conversion of naringin to the non-bitter naringenin.

The use of pectin enzymes from fungal sources can also be used for clarification of grape juice. In production of juice from Concord grapes, the use of pectic enzymes is desirable in order to get a good yield.

Eggs

Fermentation of glucose from egg white by yeasts or bacteria prior to drying of the produce has been a well documented process.[5] Reasons for this are to prevent non-enzymatic browning reaction occurring during storage. As this browning reaction proceeds, discolouration, loss in protein solubility, and off-flavours result. By removing the glucose from egg products, the glucose-protein reaction associated with browning is prevented. Egg white fermentation by yeast or bacteria, however, gives rise to yeasty flavors and contamination problems. The alternative now being used is glucose oxidase-catalase enzyme preparations. In this system, the glucose is converted to gluconic acid and hydrogen peroxide by glucose oxidase and the peroxide is broken down to water and O_2 by catalase. A review of this process and the extent of its use is given in a review by Scott.[13] Enzymes like glucose oxidase/catalase belong to the oxidoreductase group and also have been used in other food products. Such uses as antioxidants, or in preservation and flavoring have been reviewed by Schmid.

In sensitive food preparations, the use of glucose oxidase/catalase has found applications in packaging as an oxygen scavenger for such foods as milk powder or coffee. Catalase has been used alone as an agent for sterilization in milk production and cheese making. Oxidoreductases such as lipoxygenase has been used in flour treatment and as a flavoring component. It has been also shown to contribute to flavor development in fruits or vegetables. Catalase has also been shown to improve the creamy flavor in dairy products. As well as flavor development, some oxidoreductases such as sulphydryl oxide may have applications for off-flavour removal in heat sensitive products like longlife milk. The status and future potential of oxidoreductase enzymes in biotechnology has been reviewed by May and Padgette.[9]

Cereal Products

Leavening of dough or breadmaking is perhaps one of the oldest microbial technologies known. Yeasts play a major role in the production of flavour, texture and volume. Few yeasts are capable of starch utilization, hence sources of amylolytic enzymes are necessary. Starch, the nutritional reservoir in plants, which is found in cereals, cassava, potatoes, legumes, and other vegetables, and must be gelatinized, solubilized, and converted into simpler sugars before yeasts and certain bacteria are able to ferment the dough. Microbial sources of amylases, and glucoamylases are available from such organisms as *Aspergillus niger*, *A. oryzae*, and *Rhizopus* species. Amylases reduced starch to reducing dextrins, and maltose glucoamylase is used to convert corn starch to corn syrup with 90 per cent or more glucose. Amylases can be divided into two enzymes, and usually they are used in combination for a number of foods, e.g., bread, fruit juice clarification, and the brewing industry (Table 10.3).

Glucose isomerase has fast become one of the most important enzymes in starch technology. The conversion of glucose to fructose has practical significance in that it increases the sweetness of corn syrups from corn starch. About 65 different micro-organisms have been identified as producers of this enzyme. Only a few organisms such as *Streptomyces* and *Arthrobacter*, produce the enzyme with the suitable temperature characteristics and activity. While both immobilized cells and enzymes have been investigated, only the immobilized cells have completely satisfied the commercial conditions. Enhancement of the enzyme activity appears to be a good task for genetic engineering, especially with the *Bacillus* species. The production of high fructose corn syrup (HFCS) has several commercial uses in the food industry. These include ketchup, salad dressings, syrups and beverages.

Microbial Biomass and Food

In retrospect, man has unknowingly consumed micro-organisms in many fermented or naturally ripened products. Fermentation technology has since progressed to such an extent in the development of fermented foods that future biotechnological applications warrant serious consideration with respect to Single Cell Protein or Microbial Protein Biomass for both human food and animal fodder. Approximately two-thirds of the world's population suffer from inadequate protein nutrition and by the year 2000, world food requirements will have doubled. In 1968, the first Single Cell Protein Conference was held at Rutgers University, New Jersey, U.S.A. However, due to the world petroleum and energy crisis, an unprecedented development in research for alternative fuels from renewable sources has been undertaken, and successful enzymatic or acid hydrolysis and economical fermentation of cellulosic materials to ethanol remains to be seen.

Developing countries have shown an interest in yeast production, and more attention to biotechnology should be directed towards the production of Bakers Yeast, Yeast Extract, and Microbial Protein as fermented products for food supplementation. Edible protein from yeast *(Saccharomyces, Candida)* has been accepted for food supplementations by the former Protein Advisory Group of the United Nations, with limits of two grams of nucleic acid per day from Single Cell protein.

Agro-industrial substrates such as bagasse, citrus, fruit, whey, wood pulp/cellulosic wastes, in addition to molasses, canning and sugar industry wastes and surplus grain offer ideal opportunities for development of single cell protein in various countries where such materials are available. Yeasts, *Spirulina, Scenedesmus,* and certain microfungi offer man an ideal source of microbial protein, and coupled with advancements in enzyme technology and genetic engineering, S.C.P. along with fermented foods will be one priority in advances in Biotechnology.

FUTURE APPLICATIONS

Aroma Chemicals

Many microorganisms synthesize compounds which have a value as flavor or fragrance components (e.g., diacetyl). Although chemicals have been proposed for cosmetics and components in household detergents, some application to food is conceivable. In a recent review by Schindler and Schmid, the commercial feasibility of volatile compounds especially from fungi have been proposed. One can visualize the candy industry as an important user of these compounds. These aroma chemicals can be obtained by direct synthesis or enzymatic transformation by fermentation. Fragrances from molds have been described for several flavors (Table 11.7). The value of commercial preparation such as these remove the vagaries of nature and seasonal variations in supplies. Although such proposals are only in the preliminary stages, the push in the biotechnology era could make such proposals feasible.

TABLE 10.7
Flavor Characteristics Produced by Molds for Possible Use in the Food Industry

Organism	*Flavor Characteristic*
Coprinius micaceus	Mushroom
Fomes scutellatus	Cherry
Pleurotus ostreatus	Mushroom
Polyporus dyrus	Quince
Polyporus betulinus	Apples
Poria xantha	Lemon
Stereum illudens	Banana
Stereum murrayi	Vanilla
Trametes odorata	Fruity
Trichoderma viride	Coconut

Monoclonal Antibodies

Monoclonal antibodies may have applications in the detection of pathogens associated with foods. In fact, there are several aspects of food quality and safety in which monoclonal antibodies may find some use. One major area would be in the field of aflatoxin detection. Using a kit developed for this purpose, detection at the processing and storage level would be feasible.

REFERENCES

1. Bacus, J.N. and Brown, W. "Use of Microbial Cultures in Meat Products," *Food Technol., 35*, 74-83 (1981).
2. Bernholt, H.F., "Meat and Other Proteinaceous Foods," in *Enzymes in Food Processing*, G. Reed, ed., Academic Press, New York, p. 473 (1975).
3. Cooney, C.L., "Bioreactors: Design and Operation," *Science, 219*, 728-730 (1983).
4. Dwivedi, B.K., "The Role of Enzymes in Food Flavors. Part I. Dairy Products," *CRC Crit. Rev. Food Technol., 3*, 457-478 (1973).
5. Grampp, E.G., "Modification of Certain Foodstuffs by Enzymes," *Process Biochem., 17* (1), 2-6 (1982).
6. Huang, H.T. and Dooley, J.G. "Enhancement of Cheese Flavors with Microbial Esterases," *Biochem, and Bioeng., 18*, 909-919 (1976).
7. Hurst, A. and Collins D.L., Thompson, "Food as a Bacterial Habitat," in *Advances in Microbial Ecology*, M. Alexander, ed., Volume 3, pp. 79-134, Plenum Press, NY (1980).
8. Kilara, A., "Enzymes and Their Uses in Processed Apple Industry—A Review," *Process. Biochem., 17* (4), 36-41 (1982).
9. May, S. W. and Padgette, S. R. "Oxidoreductase Enzymes: Current Status and Future Potential," *Biotechnol., 1* (8), 677-686 (1983).
10. Pederson, C.S. *Microbiology of Food Fermentations*, 2nd Ed. Avi Publishing Co. Inc., Westport, CT (1979).
11. Raccach, M. and Baker, R.C. "Lactic Acid Bacteria as an Antispoilage and Safety Factor in Cooked, Mechanically Deboned Poultry Meat," *J. Food Protection, 41* (9), 703-706 (1978).

12. Rombouts, F.M. and Pilnik, W. "Enzymes in Fruit and Vegetable Juices Technology. Use of Enzymes as a Processing Aid," *Process Biochem., 13* (8), 9-13 (1978).
13. Scott, D., "Dairy Industry—Glucose Oxidase," in *Enzymes in Food Processing*, G. Reed, ed., Academic Press, pp. 219-254 (1979).
14. Shahani, K.M., Arnold, R. G., Kilara, A., and Dwivedi, B.K., "Role of Microbial Enzymes in the Flavor Development in Foods," *Biotech. Bioeng., 18*, 891-907 (1976).
15. Stenkraus, K.H. *Handbook of Indigenous Fermentable Foods*, Microbiology series 9. Marcel Dekker Inc., New York (1983).
16. Welton, R.L. and Woods, D.R. "Collagenase Production by *Achromobacter iophagus*," *Biochem. Biophys. Acta., 384*, 228-234 (1975).

11

Ecology and Food Microbiology*

The absence of a common theme impeded the development of food microbiology as a discipline in its own right. The classic review by Mossel and Ingram (1955) provided a conceptual framework for the development of food microbiology in the past two decades. In practice the study of food preservation is the study of the means used to destroy or to inhibit the growth of those micro-organisms which have the potential to cause deterioration of food. Three factors can be considered to be involved (Fig. 11.1).

1. *Intrinsic factors*—the physicochemical properties of the food;
2. *Extrinsic factors*—the physical and chemical properties of the environment, either bulk or micro, surrounding the food;
3. *Implicit factors*—the physiological properties that enable particular organisms to flourish because of the interaction of (1) and (2).

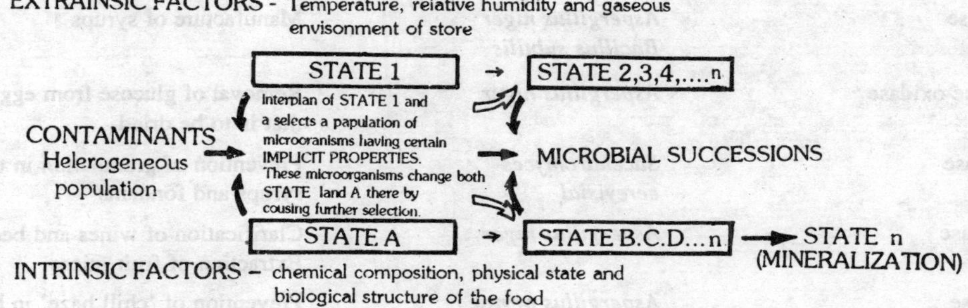

Fig. 11.1. Food spoilage and, eventually, its mineralization results from the selection of microorganisms having *implicit* properties which allow them to acquire numerical dominance (form a 'bloom') in the niches that develop from an interplay of the physicochemical properties of the food (*intrinsic factors*) and storage conditions (*extrinsic factors*).

There are two major but opposing strategies in the preservation of food.

(a) *The inhibition/destruction of micro-organisms in or their removal from a food.* It has to be

* Excerpts taken from review published by Bushell, London.

accepted that sterilization or aseptic production is applicable to particular foods only. The food microbiologist, therefore, becomes involved with the problems of hindering the growth and activities of organisms associated with the degradation and, ultimately, the mineralization of foods. It is important also to ensure that potentially pathogenic microorganisms do not survive in large numbers, multiply or leave toxic products in a food. A complete inhibition of microbial activity need not ensure stability of a food; chemical attack, e.g., fat oxidation, and degradation by enzymes, either of food or microbial origin, can lead to undesirable changes in flavour and texture, processes which can be slowed by blanching, chilling, freezing, or the addition of antioxidants, for example, the addition of sulphite to apple slices in order to prevent browning due to polyphenol oxidases.

(b) *The fostering of those organisms whose activities enhance and/or endow flavour, texture and colour; they may also inhibit the growth of potentially harmful organisms*. Today the successful outcome of many processes is helped by the use of specially selected micro-organisms ('starter' cultures) or enzymes of microbial origin (Table 11.1.).

TABLE 11.1

Microbial Enzymes used in food Manufacture

Enzyme	*Source*	*Use*
Amylase	*Aspergillus niger* *Bacillus subtilis*	Manufacture of syrups
Glucose oxidase	*Aspergillus niger*	Removal of glucose from egg white that is to be dried
Invertase	*Saccharomyces cerevisial*	Prevention of granulation in thick syrups and fondants
Pectinase	*Aspergillus niger*	Clarification of wines and beers. Extraction of fruit juices
Protease	*Aspergillus oryzae*	Prevention of 'chill haze' in beers
	Janthinobacterium lividum	Meat tenderization*
	Mucor pusillus	Curdling of casein in cheese production

*Used only in trials to date.

Setting aside the organization of the efforts of production workers, many of whom will be semi-skilled, by management, the well-being of a food in a developed country in determined by the application of principles by a relatively few technologists who factor the 'food' at some or all of the stages from harvest through production and storage until offered to the consumer. Indeed it is becoming more and more evident that little of value is achieved in food microbiology unless a complete production and distribution chain is considered.

To apply either of the strategies of food preservation noted earlier, the conditions required for microbial growth must be understood. In the following section, some features of microbial growth will therefore be considered.

Environmental Gradients

In plant ecology, the concepts of environmental gradients are used to account for the distribution of species or communities along a latitude or longitude or from sea level to a mountain top. The emphasis is given to space because of its importance relative to geological time and evolution. With some foods, such as meat having easily defined portions of fat and lean, different communities or associations of organisms are selected because of the intrinsic, chemical properties of the two substrates. A layer of dilute syrup on the top of thick syrup is an example of a practical situation which would select facultative osmophiles (at the top) and obligate ones in the depths of the syrup. Thus, with fast-growing moulds and bacteria, there are situations in food microbiology when it is useful to consider the environmental gradient concept in the sense of 'a defined space for unlimited time', when attempting to predict or interpret the distribution of different micro-organisms in a solid food or viscous liquid.

In many other situations, emphasis can be given equally to *time* and *space*. In other words, attention is directed at the *rate* at which a change occurs at a point within a food. This concept can be applied to: a food gaining or losing heat; a product in which a change in redox controls microbial selection and growth; the diffusion of salt, syrup or preservatives through a solid food—as with the diffusion of brine into meat; the production of acid by fermentative bacteria in products such as sauerkraut. Thus, for practical purposes, the organism introduced, either intentionally or unintentionally, to a food during the preliminary stages of processing can be considered to be 'stuck' at a point and, providing that the natural, implicit factors of the food will support its growth, it will be capable of growth for only as long as the imposed environmental stress is within its tolerance; for example, if temperature was being used as the environmental stress, the sluggish loss of heat could, say, result in a mesophile being provided with a temperature decreasing from 30 to 15°C over five hours, during which time it might well achieve upwards of ten generations, whereas if the drift from 30 to 15°C was achieved within three minutes, then no growth would occur.

Heat and Milk

When attempting to achieve rapid heating or chilling, the bulk and form of the food to be treated is important as well as changes caused by processing. Thus with milk, the original method of pasteurization in a large tank (at 62.8°C for 30 min) was replaced by the application of a thin film of milk to an efficient heat exchanger (71-73°C for 15-17 s). Latterly, milk has been rendered germ free by treatment at 135-150°C for upto 4 s, a process referred to as ultra-high temperature treatment (UHT). In practice UHT can be considered to be nothing other than a special case of a general approach to the heat treatment of food (High Temperature Short Time (HTST), where the objective is to achieve a germ-free state by the application of high temperatures (130°C and above) for short times, i.e., a few seconds to 6 minutes.

Pasteurization of milk was adopted initially in the United Kingdom to kill *Mycobacterium tuberculosis*, but was subsequently used for short-term preservation. The UHT method gives 'commercial sterility' without inducing a cooked flavour in the milk because the temperature coefficient (Q_{10}) for change in reaction rate is greater for the destruction of the bacterial spore (Q_{10}, 8-30) than for chemical change (Q_{10}, 2-3). Although such heat treatments will kill the psychrotrophic flora that may develop in refrigerated milk, they will not inactivate completely thermostable enzymes secreted by the organisms. On storage, the processed milk can acquire a rancid flavour, due to lipases hydrolysing triglycerides, or bitterness, due to breakdown of casein by proteases, particularly if the numbers of micro-organisms in the raw milk were large. Indeed should such pasteurized milk be used for cheese production, the carry-over of thermostable lipases of microbial origin can cause rancidity during the ripening process. Thus although technological innovation allows the production of milk with a good shelf-like and no 'cooked flavour', an undesirable feature of milk in glass bottles sterilized in tanks of boiling water, it is successful only when the production and storage of the raw material have been rigorously controlled. Moreover, the success of a process such as UHT is only assured by aseptic dispensing into germ-free containers. As the latter are commonly formed from heat labile materials, novel methods of sterilization have been introduced so that the container is rendered germ-free without leaving a residue that could cause chemical taint. H_2O_2, UV light, Cl and hydrogen chloride have been selected for these reasons. Indeed, recent studies have indicated that when H_2O_2 and UV light are used in combination, the rate at which micro-organisms are killed is greater than when either is used alone.

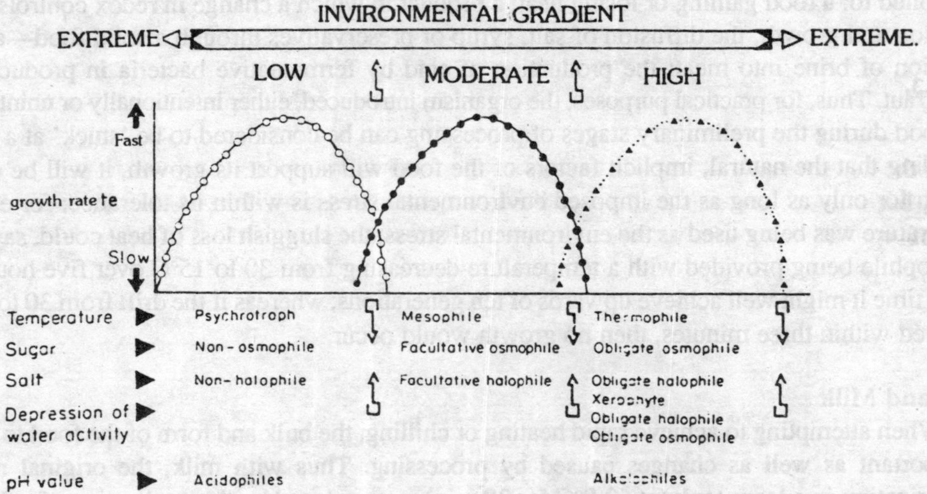

Fig. 11.2. The profiles obtained in experiments in which pure cultures of microorganisms are used to determine growth rate in a medium containing a range of concentration of H^+, salt, sugar, etc., or incubated over the temperature range from c. -5-55°C.

Tolerance Profiles

Fig. 11.2 illustrates the response of an organism to temperature, pH, salt or sugar concentration. Since it has been composed from data obtained from separate experiments using, for example, a psychrophile, a mesophile and a thermophile, the figure is probably spurious in a natural situation. From a theoretical viewpoint, it is probable that with foods containing a heterogeneous flora, competition or synergism between organisms could give profiles of the type shown in Fig. 11.3. The growth of an organism can be controlled by the qualitative or quantitative deficiency or excess of one or more factors which approach the limits of tolerance of the organism. An organisms may have a wide range of tolerance of one factor, but a narrow range for another. When conditions for a species with respect to one extrinsic or intrinsic factor are approaching the limits of tolerance, the limits of tolerance may be reduced with respect to another ecological factor. In nature, and probably in foods also, an organism rarely lies at the optimum range as determined by laboratory studies of a particular factor, thus indicating that another, possibly unsuspected, component is growth limiting.

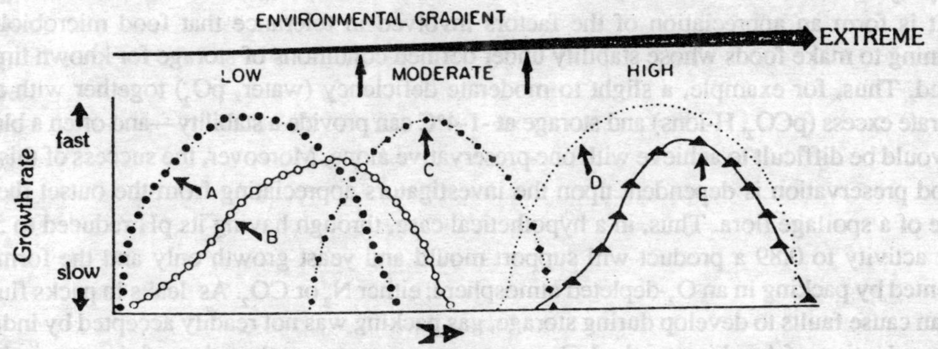

Fig. 11.3. Hypothetical growth rates of contaminants in a food subjected to temperatures over the range c. - 5-55°C or changing concentrations of an additive such as sugar or salt.

In the discussions of methods of preservation examples are given of traditional foods (butter, cheese, bacon, etc.) whose stability has been shown to be dependent upon an interplay of factors, none of which was known to those who contributed to the evolution of the craft. Although current

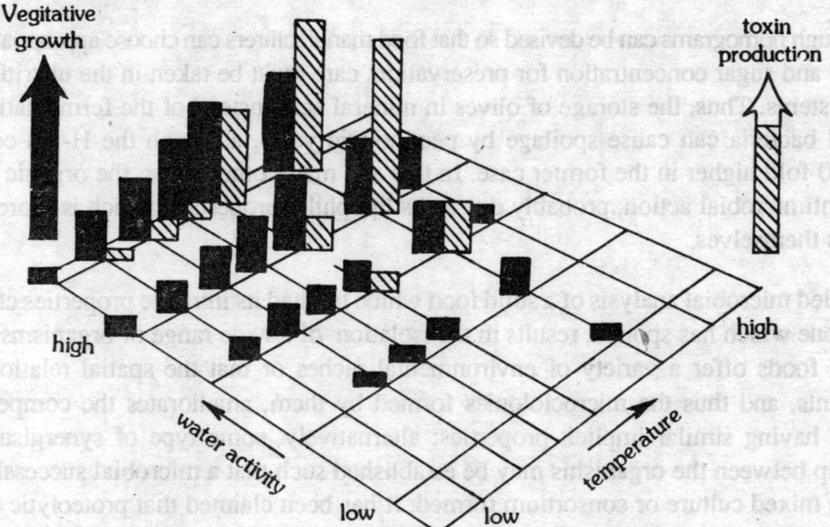

Fig. 11.4. The interplay of two factor on microbial growth in a food or a model system simulating a food are commonly displayed as 'sky-scraper' figures. This figure shows the influence of temperature and water activity on growth and toxin production by *Penicillium cyclopium* (redrawn from *Nestle Research News* (1980/81) Safety aspects of food fermentation, by M. von Schothorst).

research is demonstrating the subtle nature of the interplay of pH, NO_2^-, NaCl, pO_2, etc., in the preservation of ham and bacon, the complexities of the interactions are such that mathematical modelling is only just beginning to replace the presentation of results as models based on three-dimensional histograms (Fig. 11.4).

Interplay of Factors

It is form an appreciation of the factors involved in tolerance that food microbiologists are beginning to make foods whose stability under defined conditions of storage for known times can be assured. Thus, for example, a slight to moderate deficiency (water, pO_2) together with a slight to moderate excess (pCO_2, H-ions) and storage at -1-4°C can provide a stability—and often a blandness—that would be difficult to achieve with one preservative alone. Moreover, the success of this approach to food preservation is dependent upon the investigators appreciating from the outset the probable nature of a spoilage flora. Thus, in a hypothetical case, through having its pH reduced to 5.5 and its water activity to 0.89 a product will support mould and yeast growth only and the former can be prevented by packing in an O_2-depleted atmosphere, either N_2 or CO_2. As leaks in packs flushed with gas can cause faults to develop during storage, gas packing was not readily accepted by industry until the introduction of 'cushion packs.' Gases at pressures greater than that of the atmosphere cause inflation of the packing materials; punctured packs are therefore recognized easily. The stability of grain having more than its alarm were content of assured if the pO_2 in the void spaces is reduced either by sparging with CO_2 or by being denied exchange with the bulk atmosphere, the O_2 present originally being depleted by the activity of the grain and microbial contaminants. In the case of hay, drying in the field to low levels of water can be avoided and stability assured by adding fatty acids, e.g., butyric acid.

Although nomograms can be devised so that food manufacturers can choose appropriate combinations of low pH and sugar concentration for preservation, care must be taken in the uncritical application of such systems. Thus, the storage of olives in mineral acid instead of the fermentation products of lactic acid bacteria can cause spoilage by pectinolytic yeast, although the H-ion concentration is about 1000-fold higher in the former case. In this and many other cases, the organic acid anion has itself an antimicrobial action, probably due to its lipophilic properties, which is more effective than the H-ions themselves.

Detailed microbial analysis of a solid food which has had its intrinsic properties changed slightly and even one which has spoiled, results in the isolation of a wide range of organisms. This suggests that such foods offer a variety of environmental niches or that the spatial relationship between contaminants, and thus the microclolonies formed by them, ameliorates the competition between organisms having similar implicit properties; alternatively, some type of synergism or symbiotic relationship between the organisms may be established such that a microbial succession is favoured or a stable mixed culture or consortium formed. It has been claimed that proteolytic pseudomondas may be essential pioneers in the succession leading to the dominance of non-proteolytic pseudomonads on meat and poultry. A complex example of synergism is provided by the fermented milk product, kefir, in which fermentation is brought about by lactic streptococci, leuconostocs and yeasts associated in large aggregates or 'grains.' If an environmental extreme, such as that resulting from relatively high quantities of salt or sugar, is used to preserve a food, the resident flora consists of a very few species, namely the obligative halophiles, osmophiles, or xerophytes. In other words, preservation can be considered to be a form of enrichment culture unless sterility is achieved.

Microbiological Analysis

When attempting to interpret the information gathered from a detailed examination of a spoiled food which had not been exposed to some extreme of an environmental gradient, it is difficult to provide an explanation for those organisms which have not achieved dominance. Is their failure to grow merely a function of the inadequacy of the nutrients—either in variety or amount—in the food, due to antagonism by another organism or do they need to live synergistically with some other organism in the food? In recent years, the demonstration in the laboratory of one isolate from a food inhibiting another from the same food has been often reported—but do the laboratory observations really apply in the food in question? It would seem that this evidence can be accepted in the case of liquids offering little impediment to the flow or diffusion of mutabilities. Caution should be exercise with solid foods unless it can be demonstrated that the proximity of one organism to another is such that inhibition or enhancement of growth is feasible.

TABLE 11.2
The Food-poisoning Bacteria

Organism	*Vehicles of infection*
Salmonella spp.	Meat, particularly chicken meat; milk and cream, eggs
Staphylococcus aureus	Meat pies and meat products containing salt; confectionery
Clostridium perfringens	Cooked and reheated meats and meat products
Vibrio parahaemolyticus	Sea foods
Bacillus cereus	Rice and other starchy foods
Clostridium botulinum	Home-canned vegetables, etc.
Campylobacter jejuni	Chicken and milk
Yersinia enterocolitica	Pork and milk
Escherichia coli	Meat products

When considering the diversity of species in a food, it is essential to realize that many of the organisms may be simply remnants of associations present on an ingredient of the food or representatives of associations of habitats with which the food has had contact. In the latter case, they may be taken as indicators of the contamination of foods with members of an association in an environment with which the consumer ought not be linked by the food processing chain. Thus, by analogy to water, it has been customary to draw an unfavourable conclusion from the presence of *E. coli* in food. Its presence indicates either direct or indirect contamination with the microbial associations found in the gut of man or animals, and hence their faces. From a public health viewpoint, there has been the traditional and perhaps justifiable inference that *E. coli* may be associated with the presence of food-poisoning organisms, particularly *Salmonella*. In the past few years, the studies of the ecology of plasmids has directed attention to their involvement in the transfer of drug resistance, for example between bacteria, and has given another cause for the food microbiologist being concerned about faecal contamination. There is the fear that through serving as a vehicle in the spread of R-factors, a food may take antibiotic resistance and perhaps attributes of pathogenicity from one population to another.

TABLE 11.3

Factors Contributing to *Pseudomonas* Contamination of Curing Brines

Pseudomonas spp.	*Contributing factors*
Grow on pork during storage prior to curing	Length of storage and the temperature and relative humidity of the store will determine the rate and extent of their growth
Grow during the defrosting of frozen pork	Length of the defrosting period and the temperature will influence the development of the population of these organisms
Survive in heavily contaminated brines	A 90% reduction in the viable counts in brines (salt content, c. 26%) at 4°C can take from 25-46 days

A recent study by Gardner provides an interesting example of a situation in which the monitoring of a stage in the manufacture of bacon provides clues about the microbiological status of the meat being processed. The presence of the salt-intolerant pseudomonads in bacon-curing brine (>20% w/v NaCl) is evidence of the use of meat on which the organisms have grown because of poor hygiene or poor temperature control, especially during the thawing of frozen pork.

The influence of the environment in which the plant or animal is raised or held before processing must also be considered. Thus psychrotrophic micro-organisms make only a small contribution to the initial contamination of meat from animals raised in warm climates or fish caught in warm water. It has been noted that different types of acetic acid bacteria occur at different times of the year and, more particularly, at different stages in cider production. Thus species which preferentially oxidize sugars are found during the early stages of apple processing, the more acid-tolerant, alcohol-oxidizers at the time when yeast fermentation is almost complete. It has been noted recently that nearly all *Gluconobacter* and *Acetobacter* cause a type of rot and browning in apples and pears. Thus the changing pattern of isolations of acetic acid bacteria in fermenting apple juice may reflect in part changes in the incidence and level of post-harvest rots of apples due to infection with these organisms.

With gamebirds (order Galliformes), Changes in the flavour of undamaged muscle of uneviscerated carcasses is due to autolysis, providing the conditions of storage and the diet of the bird do not favour growth of microorganisms in the gut. Thus it was noted that a gamey flavour developed during storage (9 days at 10° C) of shot pheasants that had fed naturally on acorns and grain or had been fed grain supplemented occasionally with cabbage. There was no appreciable change in the size or composition of the microflora of the gut. Extensive greening of germ-free muscle was associated with abundant growth of *Escherichia coli* and *Clostridium* spp. in the duodenum and other parts of the gut of shot pheasants. These organisms produced copious amounts of H_2S *in vitro*. A similar discolouration of muscle has been noted also in pheasants fed proprietary turkey rations and killed by neck dislocation. Indeed the appearance of the muscle resembled that occurring in the spoilage of uneviscerated chicken carcasses where circumstantial evidence suggests that discolouration is caused by H_2S of microbial origin diffusing from the gut to the muscle. The carcasses of rock ptarmigan (*Lagopus mutus*) and willow ptarmigan (*L. lagopus*) are notable for their resistance to putrefaction when hung for long periods at temperatures up to 15° C. This has been attributed to the former ingesting the preservative benzoic acid in cranberries (*Empetrum nigrum*) and the latter antimicrobial compounds present in the buds of willow and birch.

TABLE 11.4

Isolation Pattern of Acid-tolerant, Gram-negative Bacteria

Source	*Month*	*No. of strains*	*Acetomonas spp.*	*A. mesoxdans*	*A. aceti*	*A. zylinum*	*A. rancens*	*A.ascendens*	*Others*
1	2	3	4	5	6	7	8	9	10
Cider manufacture									
Orchard soils	1	4						4	
Apple tree blossom	4	0							
Wild flowers	6	2	2						
Mummified apples	7	1	1						
Immature apples	8	0							
Fallen apples	8	8	3					5	
Factory equipment	8	0							
Apple tree leaves, twigs	9	4						4	
Mature apples	9	1	1						
Orchard grass, soil	9	0							
Factory equipment	9	8	4	2		1		1	
Harvested apples	10	63	47		8	3		5	
Harvested apples	11	3						3	
Pressed pomace (1st)	10,11,2	21	11	3		7			**Pressing season**
Expressed juice (1st)	10,11,2	28	13	5	2	5		3	
Expressed pomace (2nd)	10,11,2	17	4	4		8		1	
Expessed juice (2nd)	10,11	22	2	9	1	10			
Fermenting juice	12	2	1				1		
Farm cider	1	6	4				2		
Spoiled cider	1	12	5	2		1	4		
Dry cider	4	7	1	1		1	4		

1	2	3	4	5	6	7	8	9	10
Fermenting juice	5	1					1		
Factory sewer	6	14	2	3	1	1	5	2	
Other									
Cider vinegar generator	10	36		3	1	14	18		
Cider vinegar generator	11	10		5	1		4		
Mummified grapes	5	2	2						
Vines	6	1	1						
Strawberries	6	4	1				3		
Mature grapes	10	1	1						
Total		278	106	37	14	54	42	2	23

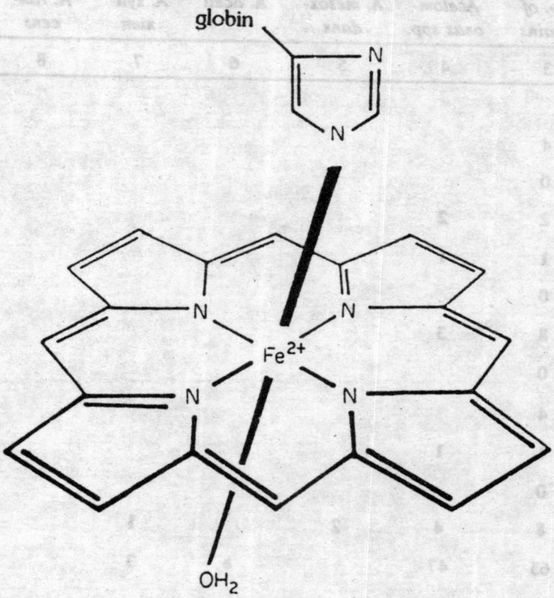

Fig. 11.5. The haem complex of myoglobin endows the meat of cows, sheep and pigs with its characteristic colour. The gaseous composition of a store or the addition of NO_2^- can change, either temporarily or permanently, the colour of meat.

Not only has attention to be given, if possible, to identifying remnants of any associations present on ingredients or derived from contact with depots of infection it has to be considered whether or not such organisms have brought about, or their enzymes will bring about, chemical changes which may adversely influence some stage in the processing or storage of food. Thus the growth of acetic acid bacteria on apples before processing results in the production of 5-ketofructose and 2-ketogluconic acid, for example, which combine with, and thus neutralize, the preservative, SO_2. The growth of *Lactobacillus viridescens* in minced meat can result in the formation of peroxides which react with myoglobin (Fig. 11.5) to give green or colourless compounds and thus an obvious fault in products such as frankfurters or ham. Similarly, pectinases produced by moulds growing on decaying flowers of cucumbers can lead to maceration during the fermentation or storage of pickles. It was noted previously that lipases and proteases in pasteurized milk can cause changes in flavour and the latter can cause sterilized milk to gel during storage.

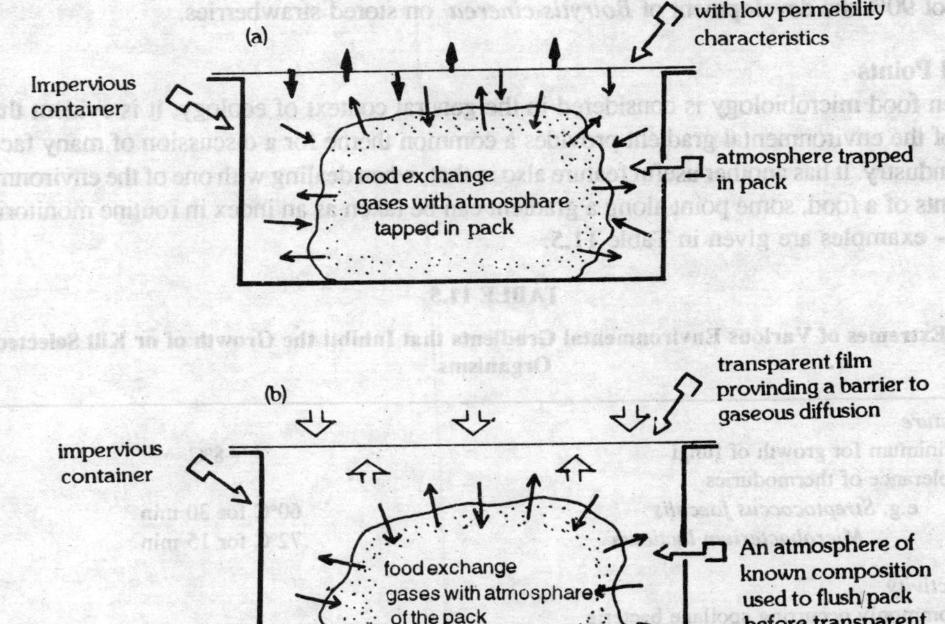

Fig. 11.6. The atmosphere around a packaged food can be *modified* by : (a) using a transparent film which impedes the diffusion of gas into or out from the pack, or (b) introducing a gas the mixture into a pack covered with a gas-impermeable film. In either case an interaction between the packaged food and the trapped atmosphere can modify the latter.

With the examples discussed above, opportunist saprophytes acquired during harvesting or during the initial stages of production of a food were the causative organisms of spoilage. In order situations, plant pathogens—the remnants of the flora on material infected in the field—can be the cause, the risk increasing when technology allows long-term storage of fruits or vegetables. In this context the plant hormone, ethylene (C_2H_4), is of importance because it can initiate the ripening of fruits picked in the green state. Thus the premature ripening of bananas—the skin changes from a green to yellow colour without a concomitant sweetening of the pulp—has been attributed to ethylene production in fruit infected with *Pseudomonas solanacearum*. If the rate of respiration of plant materials is retarded, then ripening or other physiological changes in the tissues are delayed and long-term storage is possible. A reduction of the pO_2 and an increase in the pCO_2 of storage (modified or controlled) atmospheres is one means of achieving this end. Alternatively a pressure lower than atmospheric (hypobaric storage) can be used because, in addition to reducing pO_2, it accentuates the rate of diffusion of C_2H_4 away from plant material. Neither approach can be expected to influence adversely all plant pathogenic micro-organisms; indeed the risk of plant material being damaged by such organisms is increased because of the extension in the storage life of the product. Recent research has indicated that the growth of pathogens be controlled by including CO in a modified atmosphere. Thus the addition of CO to an atmosphere containing 2.3% O_2 and 5% CO_2 reduced by upwards of 90% the development of *Botrytis cinerea* on stored strawberries.

Cardinal Points

When food microbiology is considered in the general context of ecology, it is evident that the concept of the environmental gradient provides a common theme for a discussion of many facets of the food industry. It has another useful feature also in that, when dealing with one of the environmental components of a food, some point along a gradient can be taken as an index in routine monitoring or control — examples are given in Table 11.5.

TABLE 11.5

The Extremes of Various Environmental Gradients that Inhibit the Growth of or Kill Selected Organisms

Temperature	
Minimum for growth of fungi	−8°C
Tolerance of thermodurics	
e.g. *Streptococcus faecalis*	60°C for 30 min
Microbacterium lacticum	72°C for 15 min
Water activity	
Commonly occurring spoilage bacteria	
e.g. *Pseudomonas fluorescens*	0.97
Food-poisoning bacteria	
e.g. *Clostridium botulinum*	0.95
Staphyloccus aureus	0.86
Commonly occurring spoilage yeasts	
e.g. *Cardida spp.*	0.88
Commonly occuring spoilage fungi	
e.g. *Alternaria citri*	0.80
Halophilic bacteria	
e.g. *Halobacterium*	0.75
Xerophilic fungi	
e.g. *Aspergillus echinulatus*	0.65
Osmophilic yeasts	
e.g. *Saccharomyces rouxii*	0.62
pH	
Clostridium botulinum	pH 4.6*
Zymomonas mobilis subsp. *pomaceae*	pH 3.6

Some of these have been derived empirically and others arrived at by experiment; for example, experience over the past twenty years or so has shown that few organisms grow on food stored at −7°C and that none will grow at −10°C. It will be appreciated that such statements tend to become 'a rule of thumb' in that they have a ultilitarian purpose when applied generally—they rarely, if ever, link environmental gradients and ideas of tolerance.

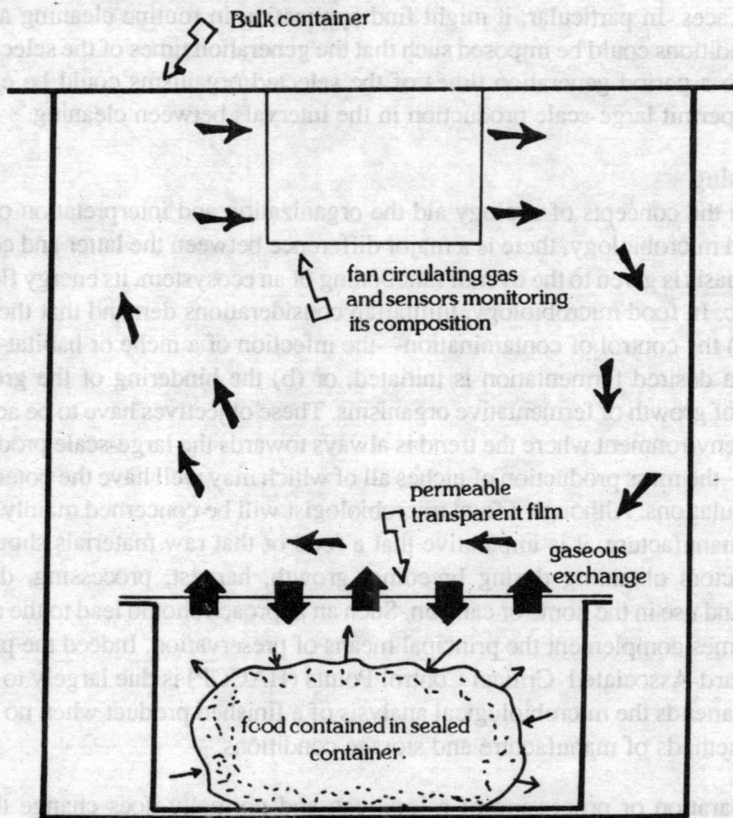

Fig. 11.7. The atmosphere in a packaged food can be *controlled* by using a transparent film which is freely permeable to gas and storing the pack in a closed container in which sensors assure that an atmosphere of a specific composition is maintained throughout storage.

In the preceding discussions, it was assumed that growth of organisms in food is essentially going to be under batch conditions, and for the majority of cases this is so. However, under certain conditions, a 'steady state' level of a limiting growth parameter can be reached when the rate or removal by microbial growth is equal to the rate of addition from some source, either by the action of another organism or by the involvement of a continuous process. Such conditions might appear to have limited application in food microbiology except in continuous fermentation systems such as the manufacture of single-cell protein and certain beers. It may be practical, however, with liquid foods such as soups, particularly if they need to be hot when put into containers, to establish a flow rate through a tank or tubes such that the flow rate is greater than the generation time of the organism so that it is unable to build up a population even though contained in a medium and at a temperature suitable for growth. Thus, through the design of equipment and planning of a process, the need for frequent stops for cleansing might be avoided by control of the flow rate. There would seem to be much scope for the application of this principle in food microbiology providing, of course, that the surface of equipment did not trap organisms or the process did not select sessile organisms which colonized surfaces. In particular, it might find application in routine cleaning adopted for working surfaces if conditions could be imposed such that the generation times of the selected organisms could be extended to a period generation times of the selected organisms could be extended to a period which would permit large-scale production in the intervals between cleaning.

Food Processing

Although the concepts of ecology aid the organization and interpretation of data derived from studies in food microbiology, there is a major difference between the latter and ecology *per se*. With ecology, emphasis is given to the overall functioning of an ecosystem, its energy flow, the composition of its flora, etc. In food microbiology, utilitarian considerations demand that the major emphasis be placed on : (a) the control of contamination—the infection of a niche or habitat—so that spoilage is prevented or a desired fermentation is initiated, or (b) the hindering of the growth of spoilage or enhancement of growth of fermentative organisms. These objectives have to be achieved *routinely* in the industrial environment where the trend is always towards the large-scale production of individual units of food—the mass production of niches all of which may well have the potential to support large microbial populations. Although a food microbiologist will be concerned mainly with the day-to-day problems of manufacture, it is imperative that a food or that raw materials should be considered in relation to factors obtaining during breeding, growth, harvest, processing, distribution, storage, manufacture and use in the home or canteen. Such an approach should lead to the adoption of methods which at all times complement the principal means of preservation. Indeed the present debate on the merits of Hazard-Associated Critical Control Points (HACCP) is due largely to the recognition that little of value attends the microbiological analysis of a finished product when no attention is given to ingredients, methods of manufacture and storage conditions.

The preparation or processing of a food can and normally does change the physicochemical (intrinsic) properties of its main ingredients and perhaps the extrinsic properties of the bulk to such an extent that the spoilage association on the processed food differs markedly from that on the major raw materials. This is exemplified by the information given in Table 12.2. During the slaughter of animals—pigs, sheep, cows, chickens—a few mesophilic organisms of gut origin may be translocated to the tissues when the constitutive antimicrobial defence systems are negated. If the meat is not chilled rapidly, organisms such as *Clostridium* spp. may grow and cause putrefaction. In modern slaughter houses the eviscerated carcasses of domestic animals are cooled at such a rate that the growth of mesophiles and the tainting of the meat around the bones are prevented. The cooling rate needs critical control otherwise the advantages gained by the microbiologist will be offset by the problems facing the food technologist. As will be seen from Table 11.2 butchering leads to the surface of meat becoming contaminated with a wide range of organisms. With storage at 1—4°C, the *Pseudomonas*/ *'Achromobacter' (Acinetobacter)* complex, of which *Pseudomonas fragi* is the main organism, forms large populations on beef, poultry, pork and fish held in a normal atmosphere at a relative humidity which prevents drying of the surface. If, however, these organisms are prevented from growing, other contaminants will form large populations. Thus the addition of tetracyclines results in yeasts becoming dominant and drying out of the surface leads to the growth of moulds, e.g. *Cladosporium herbarum*. Wrapping beef, lamb, etc., in gas-impermeable film causes an accumulation within the pack of CO_2 of microbial and tissue origin. This limits the extent of growth of *Pseudomonas* and *Acinetobacter* spp. and *Brochothrix thermosphacta* makes a major contribution to the flora. This organism becomes the principal spoilage organism when, for example, beef is held in an atmosphere containing at the outset 20% (v/v) CO_2, providing storage is at 4°C. With CO_2 storage at temperatures less than 0°C lactic acid bacteria are the principal organisms in the populations developing on the meat. In this example, an extrinsic factor temperature, is easily identified as the major elective agent. The information given in Table 12.2 indicates that at temperatures between 0-37°C, various combinations of organisms (e.g., the pseudomonads and micrococci at 10°C) dominate the spoilage flora. To account for these differences in the composition of the spoilage flora, it seems reasonable to assume that growth rate is the main implicit factor which determines whether or not an organism can contribute to a spoilage association. Such a view has received support from studies in which there was an obvious correlation between the growth rates of pseudomonads, *Br. thermosphacta* and lactic acid bacteria and the rates at which each of them spoiled meat.

Pseudomonads dominate the flora developing in minced beef held at 4°C, thus indicating that grinding the meat does not lead to the redox falling to levels sufficiently low for the growth of clostridia. The addition of sulphites to minced meat elects a population dominated by *Br. thermosphacta* and this organism is the numerically dominant contaminant in British fresh sausages to which sulphite is added at levels of 4500 p/10^6. Lactobacilli are the main contributors to the flora developing in minced meat to which glucose has been added and this situation is exploited by the manufacturers of 'continental' sausage, the stability of which is dependent in part upon acid accumulating during the fermentation process. Likewise, the stability of cured meats is dependent mainly upon NaCl electing the non-putrefying organisms : micrococci, lactobacilli, staphylococci, vibrios, enterococci and yeasts. Thus, the major factor inhibiting or encouraging the growth of particular micro-organisms in foods can rightly and valuably be considered in the context of 'microbial ecology.'

12

Microbial Food Spoilage*

A study of the literature on food microbiology and a cursory survey of the literature provides us a list of more than 50 of the approved genera of bacteria that have been associated with (1) food spoilage, (2) food poisoning, (3) food-borne disease, or (4) general contamination. An equally large number of genera of fungi and yeasts could be assembled for categories (1) and (4). Since the review by Mossel and Ingram (1955) attention has been given to the definition of the association of micro-organisms developing on, rather than a cataloguing of all the organisms recovered from spoiled foods. Generally, Gram-negative, non-aciduric, nutritionally non-fastidious bacteria are commonly associated with the spoilage of moist proteinaceous foods, related aciduric genera with acid fruit products, lactic acid bacteria with foods rich in nutrients and fermentable carbohydrates, and fungi with plant materials. For example see Table 12.1.

In case of food spoilage, the gaseous environment, temperature and the addition of sodium chloride, nitrite or glucose to meat can influence profoundly the composition of the microbial association. The use of the concept of a microbiological association by food microbiologists has a utilitarian rather than a fundamental significance simply because it is applied to the consortium of organisms present at the time that a food is deemed to be unacceptable for human use. Thus, the association may, and often does, represent only the stage at which the pioneering organisms of what would be a succession under 'normal' conditions have achieved or are approaching climax. Indeed, though this initial climax is associated with macroscopic and/or chemical changes of such a magnitude that a food is rejected the bulk of the food need not have changed to any appreciable extent. Observations of the changes occurring in milk provide a good example of this situation. Milk would be considered as sour should it curdle on addition to a hot beverage or if it had gelled in the container. These manifestations of spoilage are because of almost entirely to microbial fermentation of lactose and curdling results from the pH dropping below the pH (5.4) of casein. With prolonged storage, a succession of organisms will use lactate arising from lactose fermentation, the casein and fats and thereby bring about mineralization of the milk; all changes after the curdling phase could be of academic interest only to a food technologist. A succession in the types of micro-organisms and chemical changes can be observed when frozen peas are thawed and stored at room temperature.

---

*, Excerpts taken from the review published by Palva, London.

TABLE 12.1

A Broad Grouping of Microbial Associations and Spoiled Foods

General characteristic of food	*Examples of food*	*Microorganisms*
A. 'Moist' foods		
Rich in proteins, vitamins but deficient in fermentable carbohydrates	Meat, fish, eggs and poultry	*Pseudomonas, Acinetobacter, Alternomonas, Micrococcus, Aeromonas, Proteus*
Rich in proteins, vitamins and fermentable carbohydrate	Milk	*Streptococcus, Lactobacillus, Pseudomonas, Microbacterium lacticum*
Rich in fermentable carbohydrates and H-ions	Fruit and plant juices	*Streptococcus, Lactobacillus, Pediococcus,* Acetic acid bacteria, yeasts and moulds
Rich in proteins and vitamins but deficient in fermentable carbohydrates, salt and NO_2^- added	Bacon and ham	*Micrococcus, Lactobacillus, Vibrio, Streptococcus,* yeasts and moulds
Rich in carbohydrates, water and H-ions	Fruits	*Lactobacillus, Acetobacter, Rhizopus, Penicillium, Botrytis, Kloeckera, Hanseniaspora, Pichia, Torulopsis*
A general range of nutrients, a tendency to dry-out during storage; biological organization including an integument part of which may be damaged during harvesting and grading	Vegetables such as lettuce, legumes, carrots and brassicas	*Erwinia, Pseudomonas, Sclerotinia, Rhizopus, Fusarium, Phytophthora, Penicillium, Phoma, Pythium, Peronospora, Alternaria*
Rich in proteins and carbohydrates; acetic acid as an ingredient	Mayonnaise-type sauces and coleslaw	*Lactobacillus, Saccharomyces*
B. 'Dry' Foods		
Rich in carbohydrates	Cereals, flour	*Aspergillus, Fusarium, Monilia, Rhizopus, Penicillium, Erwinia herbicola (Enterobacter Agglomerans)*

It is now recognized that the adoption of the term 'association' signalled an important change in food microbiology because it directed attention at the few microorganisms among the initial contaminants that brought about spoilage of particular foods. In practice, 'association' has connotations of interaction/

interdependence between organisms in an environment. As food-spoilage micro-organisms can be considered to be little more than opportunist saprophytes, it may well be that in future the term 'association' will be replaced by 'bloom.' The latter is of common use in general microbiology to describe the temporal dominance of microorganisms in an environment. In the meantime, however, the term 'association' will be used in this chapter.

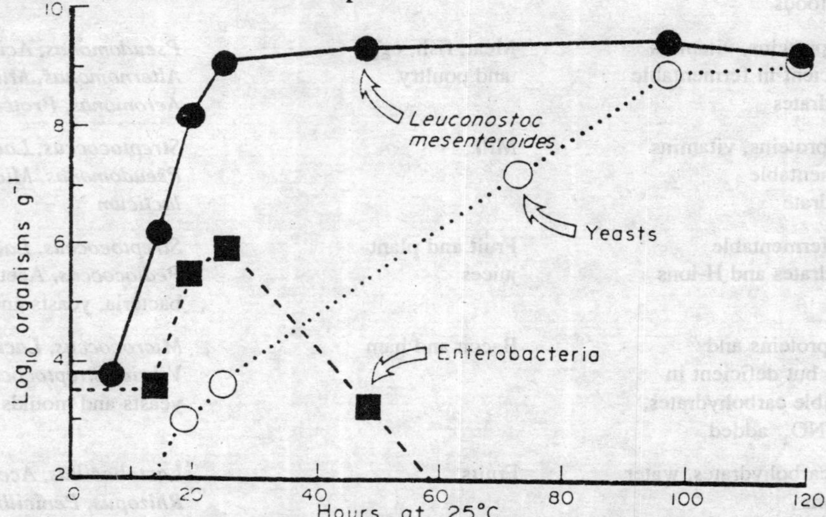

Fig. 12.1. When frozen peas are thawed and stored at room temperature their surfaces become clothed with a large population of dextran-forming *Leuconostoc mesenteroides*. The graph shows the growth of the organism and also the short period of growth of members of the Enterobacteriaceae and the subsequent emergence of yeasts as dominant organisms. Some chemical changes occurring with these changes in the size of the microbial populations.

Studies of the composition of the microbial association (Table 13.1) developing on foods of known intrinsic properties and stored under reasonably well-controlled conditions have lead to a better understanding of the implicit characteristics of spoilage organisms and, in some cases, the identification of key properties that fit an organism to a niche when it offers an environmental factor in an extreme form. Such studies show that in those foods that have no intrinsic property far removed from what is considered to be optical for the growth of a wide range of micro-organisms and a buffering system which prevents an extreme form developing, the spoilage organisms tend to be those, such as *Pseudomonas* spp., which appear to be well edapted for a scavenging/ pioneering existence in the degradative stages of the carbon and nitrogen cycle. Such foods commonly harbour a diverse flora at the time of production and the success of the few species that form an association need not result in the death of the other contaminants. As was noted previously, extrinsic factors such as the composition of the atmosphere and temperature can have a profound elective influence on the initial contaminants and ultimately it is probably the rate of growth that determines whether or not a contaminant contributes to the association present at the time of spoilage of foods such as meat. With milk the poor buffering capacity does not prevent a rapid drift in pH resulting from lactose fermentation and conditions obtain which select organisms tolerant of an acid environment, viz. the lactic acid bacteria. The many roles played by these organisms in the food industry highlight the problems faced by anyone who attempts the definition of a food-spoilage organism. Indeed as with weeds, such an organism is, in practice, the wrong organism in the wrong place at the wrong time.

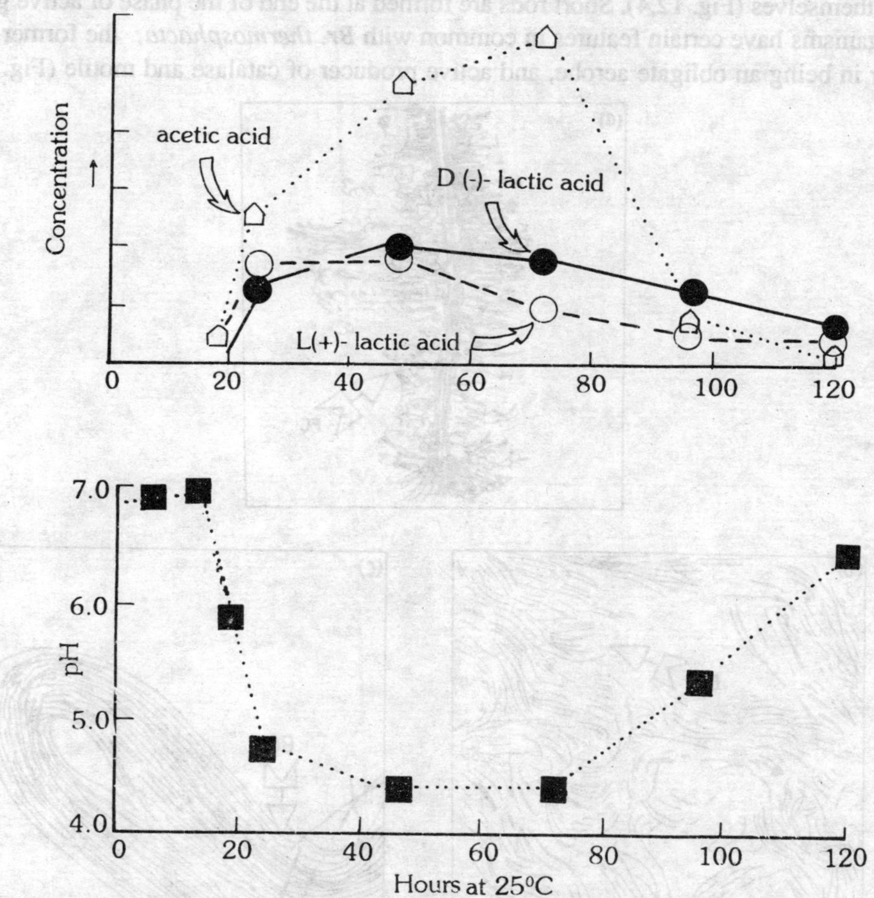

Fig. 12.2. When frozen peas are thawed and stored at room temperature, changes in the pH and organic acid content occur. The progressive increase in the concentration of acetate and lactate shown in the graph is associated with the growth phase of *Leuconostoc mesenteroides*. It is probable, moreover, that the rapid decline of Enterobacteriaceae is associated with the accumulation of acetate and a declining pH. The ultimate collapse in the concentration of acetate is associated with the latter part of the growth phase of yeasts.

The methods and ingredients used in the production of certain foods lead to the enrichment of micro-organisms that become well-known to the food microbiologist but of concern only to a relatively few general microbiologists. Such an organism in *Brochothris (Microbacterium) thermosphacta*. It was isolated originally (1951) from American pork sausages and it is now recognized as the dominant bacterial contaminant of British fresh sausages. Although it is a general contaminant of red meats, no one has identified its niche in nature. A similar situation obtains with *Kurtchia* spp., casual contaminants of red meats and areas in which meat is processed. These organisms are notable for the feather-like colonies formed on the surface of nutrient agar (Fig. 12.3) and the growth cycle in which short rods grow into long rods associated in long chains, some of which twist on themselves (Fig. 12.4). Short rods are formed at the end of the phase of active growth. Thus these organisms have certain features in common with *Br. thermosphacta;* the former differs from the latter in being an obligate aerobe, and active producer of catalase and motile (Fig. 12.5).

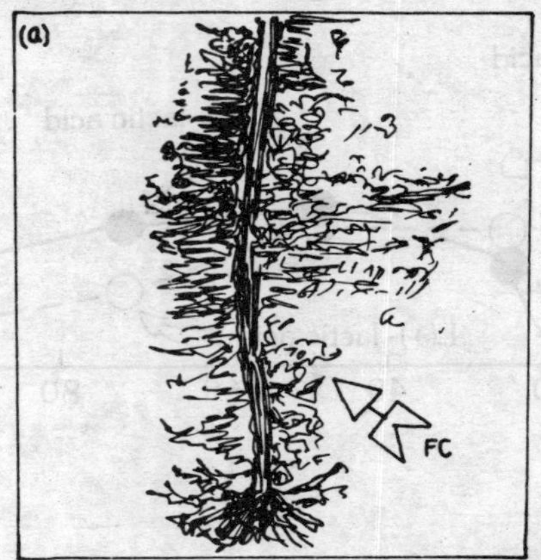

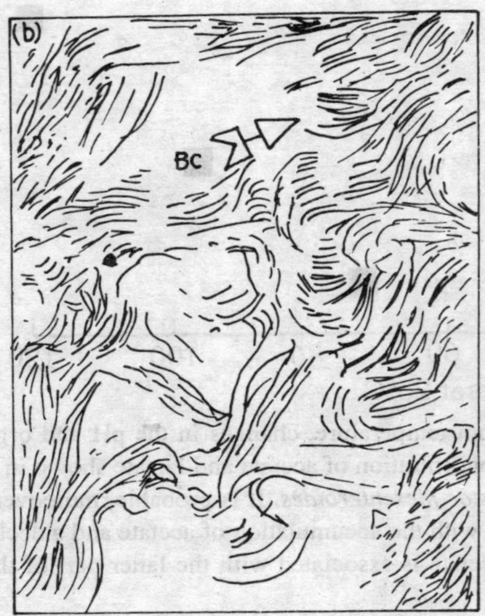

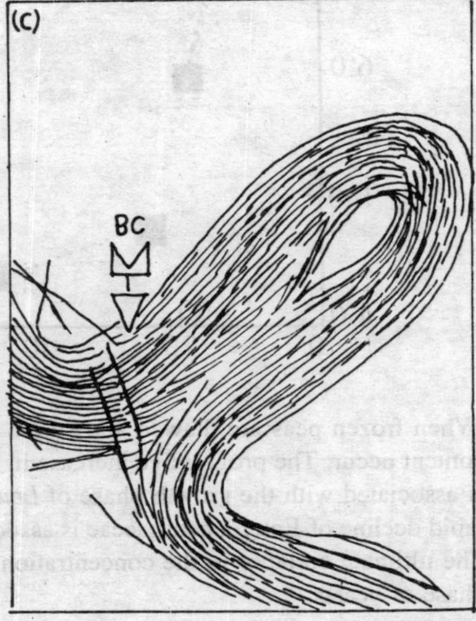

Fig. 12.3. *Kurthia zopfii*, a casual contaminant of meat and meat processing equipment, forms a feather-like colony (FC) on the surface of nutrient agar (a). Examination of the growth on nutrient agar (b and c) with the light microscopes reveals long chains of cells in bundles (BC) which snake-out over the surface of the medium (Dee Gascoyne, unpublished observations).

In general, a diverse flora is a common feature of foods offering little in the way of extremes of Environmental gradients. When extreme environmental factors obtain, either for a short period or throughout the life of the food, then selection favours organisms that have particular physiological or

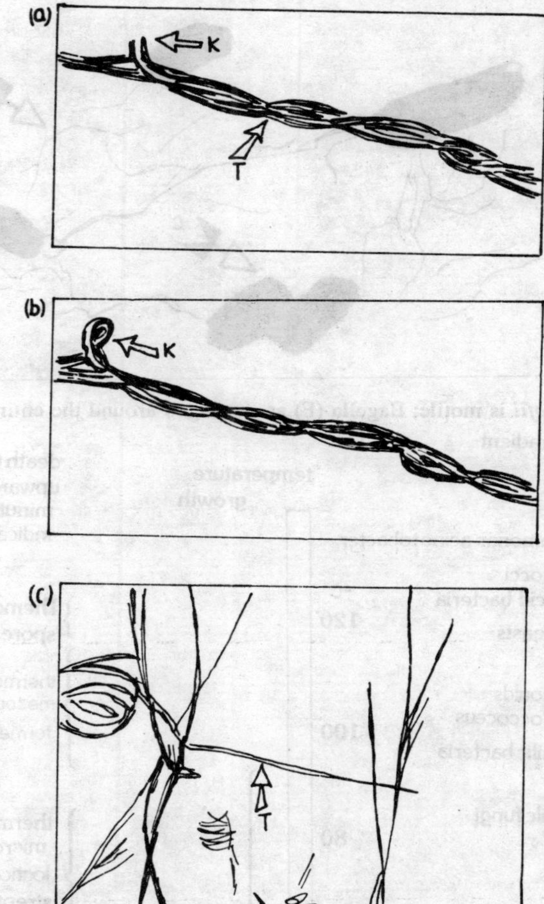

Fig. 12.4. Twisted filaments [T in (a)] are common features in the feather-like growth of *Kurthia zopfii* on the surface of nutrient agar (Fig. 12.3). The origin of such a pair of twisted filaments (K) is evident in (a) and (b) and the final form in (c).

morphological adaptations. The selective influences of temperature and water activity are shown in Fig. 12.6.. In practice, the more extreme such elective factors become, the smaller the range of micro-organisms having physiological adaptations that permit them to colonize a food. In other words, the food acts as an enrichment medium.

The presence of specific chemicals in a food also has an elective action. It was noted in Table 12.1 that an acid environment is elective for yeasts, lactic acid bacteria and aciduric Gram-negative bacteria such as *Acetobacter*, *Gluconobacter*, *Frateuria* and *Zymomonas*.

If acetic acid is an important component of a food, then lactic acid bacteria and yeasts are enriched and form the spoilage associations in products such as mayonnaise and coleslaw (Table 13.1). When this acid is used as the principal preservative, then spoilage of a food can result from the

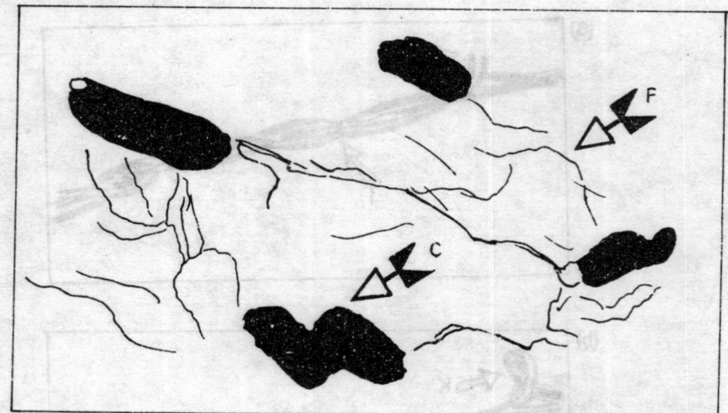

Fig. 12.5. *Kurthia zopfii* is motile; flagella (F) are arranged around the entire surface of the cell (C).

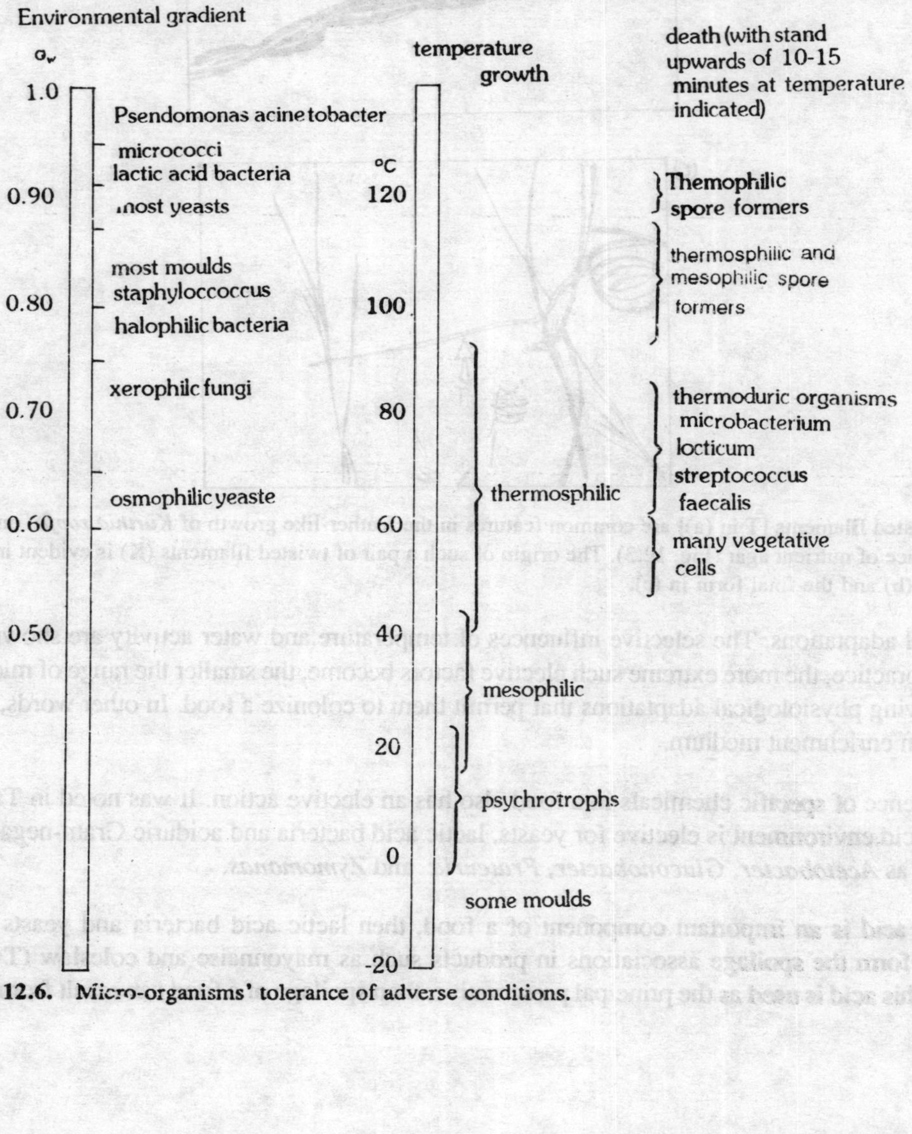

Environmental gradient

a_w

1.0
Psendomonas acinetobacter
micrococci
lactic acid bacteria
0.90
most yeasts
most moulds
staphyloccoccus
0.80
halophilic bacteria
xerophilc fungi
0.70
osmophilic yeaste
0.60
0.50

temperature
growth
°C
120
100
80
60
40
20
0
-20
thermosphilic
mesophilic
psychrotrophs
some moulds

death (with stand upwards of 10-15 minutes at temperature indicated)
Themophilic spore formers
thermosphilic and mesophilic spore formers
thermoduric organisms microbacterium locticum streptococcus faecalis
many vegetative cells

Fig. 12.6. **Micro-organisms' tolerance of adverse conditions.**

growth of a few, very highly adapted micro-organisms. The list of organisms in Table 13.5 demonstrates that the extreme environments resulting from the preparation or preservation of particular foods have led to the isolation of microorganisms notable for their tolerance of some inimical feature. *Saccharomyces bailii* is the epitome of such an organism.

TABLE 12.2

Gram-negative Bacteria Associated with Fermented Products

Bacteria	*Associated with*
Aerobes	
Acetobacter	Acetification of fruit juices, wines and beers
Gluconobacter	Commercial production of vinegar
Frateuria	
Facultative anaerobes	
Hafmia protea (*Obesumbacterium proteus*)	Minor defects in beer flavour and yeast performance
Anaerobes (at least when first isolated)	
Zymomonas mobilis subsp. *pomaceae*	Cider sickness—unpleasant flavour due to accumulation of acetaldehyde
Zymomonas mobilis subsp. *mobilis*	Common in fermenting fruit juices in the tropics—viz. pulque, the fermenting sap of *Agave americana*
Anaerobe—strict	
Pectinatus cerevisiiphilus	Isolated from spoiled beer

TABLE 12.3

Some Exceptional Implicit Properties of Micro-organisms

Organism	*Found in*	*Implicit properties*
Lactobacillus trichodes	Wines containing 20% alcohol	Grows vigorously in wine 20% (v/v) ethanol; optimum initial pH range, 4.5—5.5
Lactobacillus fructiviorans	Vinegar preserves	Maximum levels of the following growth-limiting factors given in parenthesis : acetic acid (4.2%); salt (12%), ethanol (18%); minimum pH 2.9
Streptococcus thermophilus	Cheese (a starter culture)	Optimum growth temperature, 40—50°C, will grow at 50 but not 53°C; survives 65° C for 30 min

*Leuconostoc oenos**	Wine	Best growth in medium at pH 4.2—4.8; grows slowly in 18% ethanol at pH 4.8
Pediococcus halophilus	Soy mash	Best growth in 6—8% NaCl; tolerates 15% NaCl
Saccharomyces bailii	Acid beverages and fruit juices	Growth in acid medium containing 500 p/10^6 SO_2

*This organism is probably a species of another genus.

The application of temperatures above that of the surroundings to foods also has an elective action, either by selecting organisms capable of growth at elevated temperatures or survival at relatively high temperatures. In the latter instance, resistance may be the result of an unusual feature of the vegetative cell or of a morphological adaptation. Thus, the pasteurization of milk can select vegetative cells having an above-average resistance to thermal denaturation, for example *Microbacterium lacticum, Streptococcus faecalis* and *Micrococcus* spp. Of course endospore-forming bacteria will survive also. Indeed when poor standards obtain in the cleaning of churns and storage tanks, milk can become heavily contaminated with *Bacillus cereus*. This organism's endospores survive pasteurization and, if storage conditions allow germination and outgrowth, outbreaks of 'bitty cream' occur — 'rafts' of cream float to the surface of hot coffee or tea to which contaminated milk has been added. These 'rafts' are formed because lecithinase produced by the vegetative cells modify the boundary of the milk fat micelle.

In appertization, the time/temperature relationship is such that only organisms with particular morphological adaptations survive; for example, the ascospores of years such as *Saccharomyces bailii* and *Kluyveromyces bulgaricus* can survive the heat process given to acid fruit products. The ascospores of *Byssochlamys fulva* and *Bys. nivea* can survive the processing of canned strawberries—the fruits in infected cans are reduced to a puree during storage by pectinase produced by the vegetative cells that arise from the ascospores. With extreme thermal processing, the selective pressures favour the survival of endospores having exceptional heat-resistance, viz. *B. stearothermophilus*. The storage behaviour of a canned product will be determined by storage temperature. Thus endospores of *B. stearother-mophilus* and *Desulfotomaculum nigrificans* will not germinate unless the storage temperature exceeds 40° C.

It may well be asked what value there is in identifying spoilage organisms or spoilage associations. There are at least two benefits to be gained.

1. Information is acquired about the spoilage of particular foods and thus an opportunity arises for devising specific media and methods to study the occurrence and dispersal of spoilage organisms in the factory and by ingredients, as well as for monitoring their behaviour in a commodity both during preparation and storage.

2. It enables the definition of the constellation of implicit properties that fits an organism to a niche.

Not only will (I) offer the opportunities noted above but it will provide also a body of information that permits prediction of the most likely cause of spoilage of particular foods. In practice such information will aid the food technologist in the day-to-day management of food-processing operations and it may be expected also to provide an early warning should something unusual have occurred in the preparation of an ingredient or the food itself. With (2), a thorough understanding of an organism's properties may offer the opportunity for a rational approach to the choice of methods and processes for preserving particular foods.

Another facet of the spoilage story is the study of the chemical or physical changes which the micro-organisms cause during the spoilage of a food. Again this may appear to be an academic exercise only. As with the characterization and identification of spoilage organisms, there are two possible benefits to be gained from a study of spoiled foods and the events leading to this state. A study of the events leading to spoilage may well lead in the future, when sophisticated systems of analysis are readily available, to methods which detect some early changes associated with the outgrowth of the spoilage organisms and thus a means of predicting quickly the possible interval before overt symptoms of spoilage are manifest. Such an analysis may also provide information about the nutrients used and modifications of intrinsic properties brought about by the spoilage organisms.

A study of spoiled foods will not only permit a definition of the chemical and physical changes associated with spoilage but it will enable the microbiologist to recognize those features which cause the would-be consumer to reject a food.

It is possible with may spoiled foods to identify a major change which has been brought about by the spoilage organisms. Such changes may be one or more of the following:

(a) *polymer* formation by the spoilage organisms — the polymers may be 'organized' as cells or occur as extracellular slime;
(b) *depolymerization* by the spoilage organisms—carbohydrates (starch and pectin), proteins and fats may be the substrate and through their hydrolysis the biological organization of the food may lost;
(c) *dextruction* of an emulsifier;
(d) *fermentation* of a carbohydrate leading to the accumulation of fermentation products and scouring or merely a subtle but undesirable change in flavour;
(e) *oxidation* of a product leading to a change in flavour or appearance;
(f) *reduction* of a chemical resulting in an off-flavour or a change in colour;
(g) *production* of a pigment;
(h) *production* of an odour or off-flavour.

Examples of spoilage due to processes (a)—(g) are given in table 13.6.

Little progress was possible with the chemical identification of off-odours until sophisticated systems of analysis were available. Indeed the older literature offers descriptions that can confuse rather than help the inexperience microbiologist. What use is there, for example, in a spoilage odour being referred to as 'boiled cabbage' if this dish is not a part of the cuisine of the reader? Likewise there are perhaps several opinions on what constitutes a 'dirty dish cloth' smell and a certain botanical knowledge is presumably required in order to identify a 'may-apple' odour.

TABLE 12.4

Chemical Changes—Their Spoilage Symptoms and Caustive Orgnisms.

Chemical change	*Symptoms*	*Causative organism*
(a) *Polymerization*		
Biological organization in cells	Slime on meat due to microbial growth at chill temperatures	*Pseudomonas fragi*
	Turbidity (haze) in wines, beers, ciders and beverages	Lactic acid bacteria and yeasts
	Yeast growth ('chalk moulds') on surface of bread	e.g. *Hypopychia burtonii*
Extracellular slime	'Ropey' milk	*'Alcaligenes visicosum'*
	'Ropey' cider	*Pediococcus cerevisiae*
	'Ropey' bread	*Bacillus subtilis*
	'Ropey' sugar products *mesenteroides*	*Leuconostoc*
(b) *Depolymerization*		
Destruction of biological organization	Maceration of canned strawberries	*Byssochlamys fulva*
	Maceration of carrots	*Rhizopus stolonifer*
(c) *Destruction* of an emulsifier e.g. the breakdown of lecithin by lecithinase	Bitty cream in pasteurized milk	*Bacillus cereus*
	'Custard yolk' in egg products	
(d) *Fermentation* of carbohydrate		
Accumulation of general fermentation products	Souring of milk	*Streptococcus cremoris*
	Souring of sausages	*Lactobacillus* spp.
Accumulation of a particular fermentation product (given in parenthesis)	Cider sickness *(Acetaldehyde)*	*Zymomonas mobilis* subsp. *pomaceae*
	Off-flavour in beer (Diacetyl)	*Pediococcus cerevisiae*
	Holes in hard cheese (H_2CO_2)	Coliform organisms
(e) *Oxidation* of a product e.g. ethanol	Acetification of wines and beers	*Acetobacter* and *Gluconbacter* spp.
nitrosylmyoglobin	Greening of cured meat	*Lactobacillus viridescens*

(f) *Reduction* of a chemical NO_3^-	Blown cans of hams resulting from NO_3^- being used as a terminal electron acceptor	*Bacillus* spp.
SO^{2-}_4	Blackening (formation of ferric sulphide) of pickled vegetables	*Desulfovibrio* sp.
(g) *Production* of a pigment A phenotypic property of the organism	Red spot in cheese	*Lactobacilus plantarum*
	Pink spot on salted fish	*Halobacterium*
	Fluorescent-green Whites in hens' eggs	*Pseudomonas putida*
	Purple spots in bread	*Bacillus subtilis*

TABLE 12.5

Chemicals that Taint or Flavour Food

Compound	*Contributes to*
Methane thiol (CH_3SH)	Off-odours of bacon, ham and fish Smell of some cheeses, e.g., the 'Cheddar aroma'
Trimethylamine [$(CH_3)_2$ NH]	Through using the odourless trimethylamine oxide [$(CH_3)_2$ N = O} as an electron acceptor in the absence of O_2, some micro-organisms produce trimethylamine which has a fish odour
3-methylbutanal	Malty 'off flavour' in cheddar cheese.
[$(H_3C)_2CHCH_2CHO_1$)]	Some strains of *Str. lactis* contain transaminases and decarboxylases; these convert amino acids to aldehydes. Leucine is converted to the compound shown on the left.

Although many of the terms used to describe spoiled foods may as yet defy sensible definitions, the odours may well be the major reason for a would-be consumer rejecting a product, and the reason for rejection may be based on a *general* association within a community of people of a particular *odour* and spoilage. This is a feature which has to be considered by those who attempt novel methods of preservation of foods; such methods should have A 'fail-safe' attribute so that if spoilage has occurred it will be recognized by the would-be consumer.

13

Inhibition the Growth of Micro-Organisms*

The principles of food preservation can be easily listed as under :

1. Prevention of damage of a food by mechanical causes, animals, insects, etc.;
2. Retardation of chemical changes in a food through enzyme action or purely chemical reactions such as fat oxidation;
3. Prevention or delay of microbial decomposition of a food;
4. The enhancement of growth of specific organisms which aid preservation through transformation of component(s) of the food.

Practical Approaches

There are three broad practical approaches to the problem of preventing or delaying microbial decomposition of food:

(a) inhibiting the growth of organisms;
(b) the removal of organisms;
(c) killing organisms.

The removal of organisms is a commercial possibility only with liquids such as wine, beer, cider, etc., where filtration is easily achieved.

If a traditional food or a range of ingredients is considered as a medium for microbial growth (Table 13.1), then in theory many strategies could be adopted to achieve preservation. In practice, of course, the choice is governed by economics, feasibility under commercial conditions and the need to cause minimum changes in the palatability or nutritional quality of a food. Moreover, with preservatives, the quantities used must be low so that they are non-toxic to man even when small amounts are consumed regularly over a lifetime. In practice many methods of preservation of cellular foods will result in damage to the cytoplasmic membranes; the chaos that can result from such damage is shwon in Fig. 13.1. It has to be recognized that tradition, fads and prejudices impose restraints over and above those

* Excerpts taken from the review published by Rodriguez, R.L.

noted above. Indeed a food microbiologist has to attempt the application of fundamental knowledge in an area where at times quite arbitrary restrictions operate. Thus attempts to improve the microbiological stability of tradition food will be unsuccessful if they modify, even slightly, one or other of the accepted organoleptic qualities of that food. The influence of tradition is seen in another context also; the terms drying, salting, smoking, pickling, for example, are used to classify a product that has received a particular treatment. In the following discussion it will be shown that in many instances the overall effectiveness of what on casual observation appears to be the main method of preservation is aided by the contributions of one or more minor factors. Indeed efforts tc make certain foods such as pickles and salt meat more bland or to produce new systems have centred largely on modifications of the interactions that have been identified in traditional foods. The discussion will give examples also of foods in which a preservative contributes to the preservation of organoleptic properties by influencing chemical as well as microbiological changes.

TABLE 13.1

Preservation Methods Related to Important Components of a Food or Extrinsic Factors

	Method of preservation
Component	
Carbohydrate	Extraction and purification, e.g. sucrose. Fermentation to acidulate a food
Proteins	Extraction and purification, e.g. texturized vegetable protein
Amino acids	No commercial method available
Vitamins	Avidin's binding of biotin considered by some to be an important component of avian eggs' antimicrobial defence
Fats	Extraction, purification and emulsification, e.g. butter, lord and margarine
Trace metals	The binding of Fe^{3+} by ovotransferrin appears to be the major antimicrobial agent in the hen's egg.
Water	Removal by drying or combination with sugars, salt and other humectants
pH	Deliberate acidulation. Fermentation of a carbohydrate
Extrinsic factor	
Gaseous environment	Removal of O_2; addition of CO_2
Temperature	Storage at — 1-4° C or at — 18°C
Irradiation	Ultraviolet light

Nutrient Limitation

It will become evident when reading this chapter that rarely, if ever, it is possible to identify one component as the sole cause of preservation of a food; normally it is the outcome of a subtle interplay between a major and several minor components. Thus, the choice of sub-headings in this chapter is intended to aid the cataloguing of information relating to the major components of preservation. It must be appreciated that anomalies will arise in such an approach. Thus in a discussion of food preservation through nutrient limitation it would be logical to consider water as a 'nutrient' and include dried foods or Intermediate Moisture Foods under this sub-heading. In practice such a course has not been followed because of the intention to direct the reader's attention to biological systems that have evclved to thwart microbial growth in fluids which contain abundant water and nutrients. The discussion of the antimicrobial defence of the hen's egg deals with such a system. Indeed it is now becoming widely recognized that inhibition of microbial growth through iron deprivation, the major antimicrobial defence system in egg albumen, is a ccmmon strategy in biological fluids and medical scientists (e.g. Kochan 1973) have coined the term, nutritional immunity, for this form of defence.

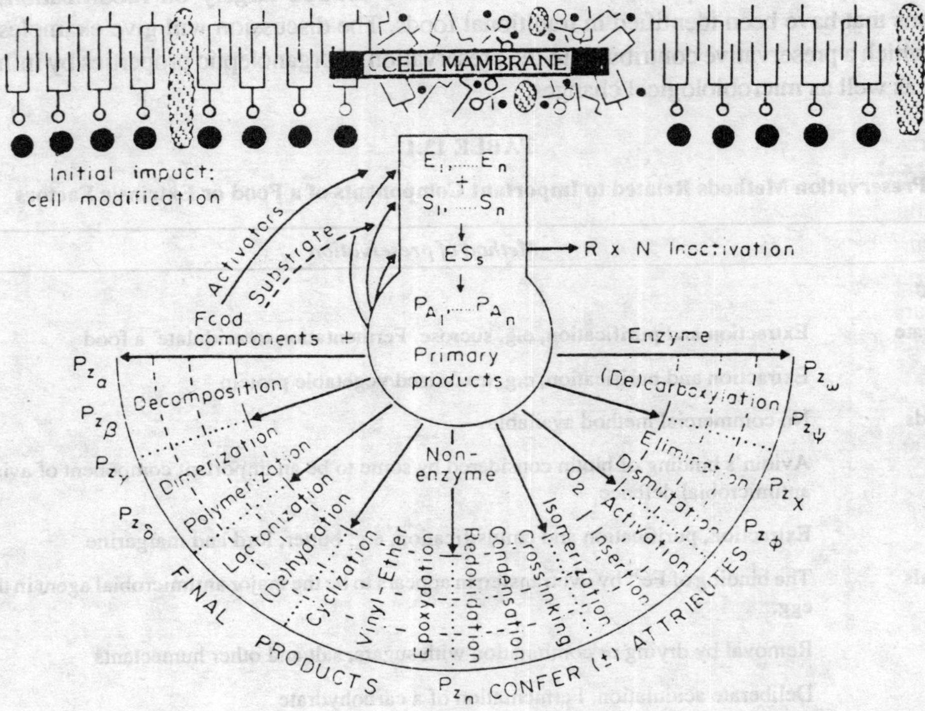

Fig. 13.1 Processing of cellular foods can cause major perturbations in nzyme activity resulting in damage to the cytoplamic mmbranes.

Butter

With commodities such as sugar, edible fats of plant and animal origin, butter, etc. (Table 13.2), microbiological stability—but not necessarily chemical stability is achieved mainly by reducing to a minimum extraneous substances, including water, which support microbial growth. With edible fats and sugars, extraction and purification is the task of chemical technologists. Butter, on the other hand, is a food whose production remains a part of the traditional food industry. Its microbiological stability is dependent mainly on the residual non-fat components of the milk together with water used in processing being dispersed throughout the fat, the latter contributing about 80% and water about 17%, by weight, of the product. Churning and processing cause the water to be broken up into very small droplets. Some estimates suggest that there are 1—2 x 10^9 water droplets g^{-1} of butter. This implies that a large proportion of the droplets will have dimensions less than those of the bacterial cell and that raw materials would have to be excessively contaminated if the majority of droplets of more than 1 μm diameter were to be infected. It has been deduced, for example, that 88% and 99.9% of the droplets would be sterile in butter contaminated with 10^5 and 10^3 bacteria g^{-1}, respectively, If, as would be the normal situation, the contaminating flora was heterogeneous, then the number of droplets containing organism capable of growth would be but a fraction of the total number of contaminated droplets. Even the largest droplets will contain only small amounts of nutrients and, as water constitutes the dispersed phase, the inward diffusion of nutrients and outward movement of wastes would be impeded. Although microbiological stability is dependent mainly upon processing removing most of the non-fat components of milk and dispersing the remnants throughout the butter, this cannot be relied upon when the product has to be stored for long periods. The control of contamination during manufacture and the use of pasteurized milk will obviously minimize the level of contamination of droplets with spoilage organisms. The environment afforded by the droplets can be made unfavourable by the inclusion of NaCl and/or lowering its pH by fermenting the cream before manufacture. Even these refinements will not ensure stability during extended storage and recourse has to be made to refrigerated storage, 0-4°C for short and — -17 — -20°C for long periods. It is common practice to store large blocks of butter and to prepare 250 g packs immediately before distribution at which time butters from various sources and of different grades may be mixed (blended). This can trigger off bacterial growth and hence spoilage. Why should this be when additional nutrients are not added? It would appear that blending causes both a redistribution and aggregation of droplets with the result that nutrients are supplied to organisms which had exhausted those available in the water surrounding them at the time of manufacture.

TABLE 13.2

Foods preserved by Nutrietnt Limitation

Fats of plant and animal origin
Butter
Margarin
Sugar*
Starches and gums

*Permissible water content of sucrose :

$$x = \frac{\text{weight of water}}{100 - \text{non-sucrose solids}} \times 100$$

In very refined sugar with about 1% of material other than sucrose, *x* should not exceed 0.03%.

The Hen's Egg

The well-being of he embryo in a cleidoic egg such as that of hens is dependent upon a non-specific defence against micro-organisms; the better understood defence based on antibodies and phagocytes would not work in the absence of a vascular system and neural control. The white makes an important contribution in terms of chemical defence (Table 13.3). In addition to lysozyme, for which no convincing case can be made for its contribution to the egg's defence, the white is an inadequate medium for microbial growth because of : (i) its low content of non-protein nitrogen, (ii) the unavailability of biotin and riboflavin through combination with proteins, and (iii) the avid chelation of Fe^{3+} by ovotransferrin. Of these, ovotransferrin working in association with the high pH (9.6) of the albumen, is of primary importance and its effectiveness is temperature dependent. At low temperatures (4-20°C), micro-organisms are prevented from growth by iron deprivation; at 38—40° C), they are killed. Suggestions that organisms, such as coliforms, would be able to overcome iron deprivation by synthesizing their own chelates, enterobactin, have not been supported by laboratory studies. Indeed this was to be expected because enterobactin, even if formed in response to the iron-poor conditions of albumen, would be quickly hydrolysed and the resulting monomers have poor chelating potential. This defence plays a cardinal role in protecting table eggs from microbial attack providing that good storage conditions (controlled temperature and humidity) limit the rate of decay of the biological structure of eggs (Fig. 13.2). Indeed with storage at room temperature, the defence of the egg is dependent upon the yolk being kept away from the shell membrane, otherwise a niche is provided at the place of contact of the yolk and membranes for microbial growth in the absence of albumen. At chill temperatures (0—4°C) rotting may well occur through chance collision of contaminants of the white with the yolk, the opportunity for collision being influenced by the size of the load of contaminants in the white. This load can be increased markedly if Fe^{3+} is deposited along with micro-organisms on to the shell membranes when eggs are washed. As this element remains localized, extensive microbial growth at the original site of infection results in heavy contamination of the albumen and raid rotting.

TABLE 13.3

Chemical Defences in the Albumen

Substance	*Action*
Lysozyme	Hydrolysis of β 1—4 linkages in peptidoglycan of the cell walls of sensitive bacteria
Ovotransferrin	Chelation of Fe^{3+} leads to stasis or death of micro-organisms
Avidin	Combines with biotin thereby making this unavailable to micro-organisms
Riboflavin-binding protein	Combines with roboflavin thereby making this unavailable to microorganisms
Non-protein nitrogen	Only very small amounts available thus restricting microbial growth
Hydrogen and hydroxyl ions	pH 9.6—10.0; not only toxic *per se* but an adjunct in Fe^{3+} -deprivation by ovotransferrin

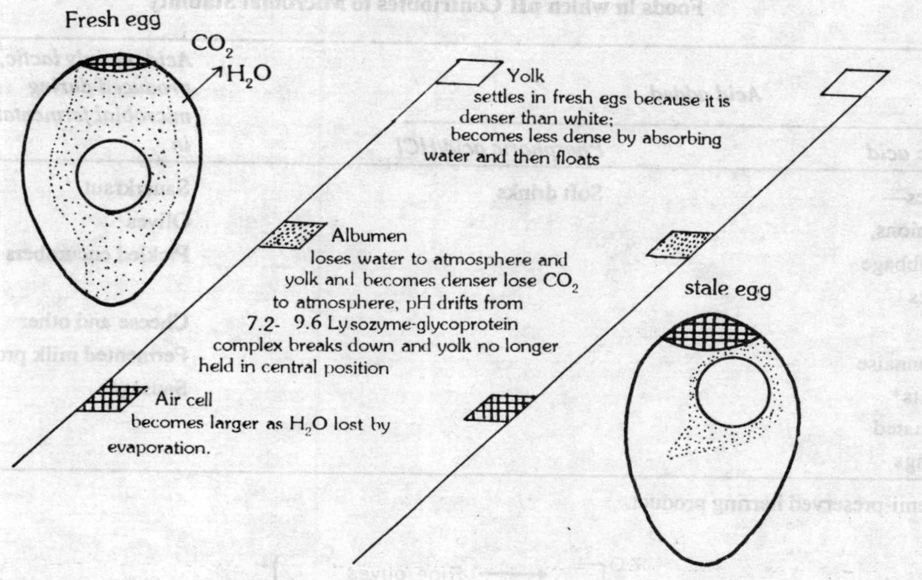

Fig. 13.2 Changes in the hen's egg during storage. The rates of change are directly related to storage temperature, the fastest changes occurring at high storage temperatures.

Cows' Milk

Freshly drawn milk is an unfavourable medium for microbial growth. In addition to a few phagocytes, it contains transferrin/ lactoferrin that chelate Fe^{3+}, agglutinins and the lactoperoxidase system which kills micro-organisms through the oxidation of enzyme systems, with H_2O_2 as the substrate and thiocyanate an essential co-factor.

pH

When it is recalled that certain heterotrophic bacteria can initiate growth in nutrient media poised at pH 1.0 (e.g. *Sarcina ventriculi*) or 9.6 (*Streptoccus faecalis)* and that there are micro-organisms which have pH optima somewhere between these extremes, it is obvious that pH *per se* is unlikely to be a completely effective preservative. Nevertheless, pH is of importance as an adjunct to other methods of preservation (Table 13.4) or processing such as canning (Fig.13.3). Moreover, pH is of particular importance from the viewpoint of food poisoning; until recently it was presumed that growth or toxin production by *Clostridium botulinum* was inhibited in a milieu having a pH of 4.6 or less. It has been shown that growth and toxin production can occur at pH 4.2, the critical pH value being determined in part by the acid used.

TABLE 13.4

Foods in which pH Contributes to Microbial Stability

Acid added		*Acid, mainly lactic, produced during microbial fermentations in*
Acetic acid	*Phosphoric acid/HCl*	
Pickles— onions, cabbage	Soft drinks	Sauerkraut
		Olives
		Pickled cucumbers
Sauces		
		Cheese and other
Mayonnaise		Fermented milk products
Tit-bits*		Sausages
Marinated herrings		

*A semi-preserved herring product.

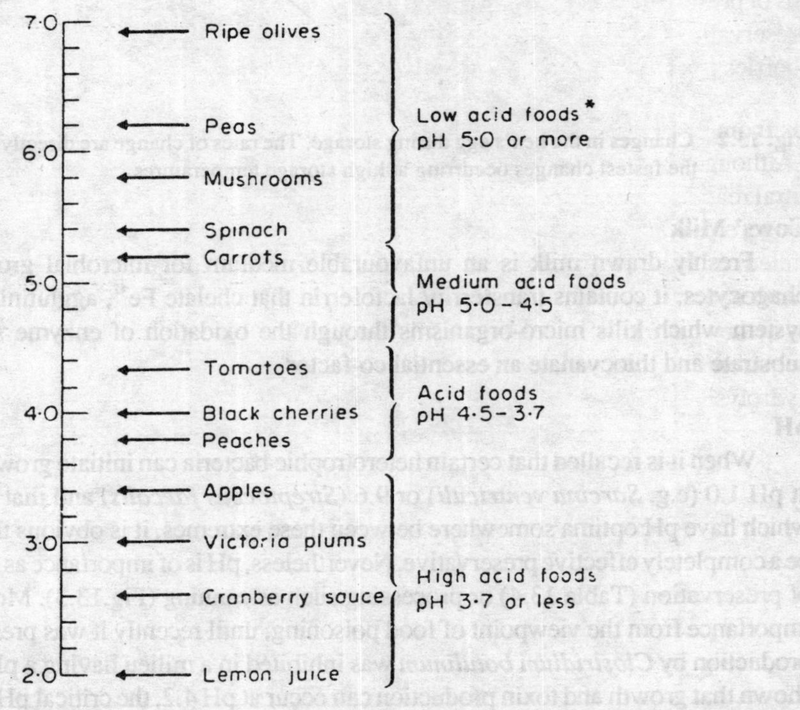

Fig. 13.3. The pH of vegetables and fruits used in canning. * The four levels of acidity that determine the choice of temperature/time relationships for appertization of such materials.

Fruits and Fruit Juices

Many fruits have acid juices and, except for syconus ones such as figs, their contents can be regarded as germ free providing the integument is whole and healthy. When this barrier is destroyed, the juice may, and normally does, act as a slelective agent in the enrichment of acid-tolerant organisms present on the fruit or processing equipment. Thus with apples used in cider production, few lactobacilli are present on the skin of hand-picked fruit, many can be recovered from fruit which have been collected from the ground on spiked rollers and enormous numbers are present in juice expressed by dirty processing equipment. In certain cases, the lower level of pH tolerance of a spoilage organism may be higher than the commercially acceptable acidity of a product. Thus with *Zymomonas mobilis* subsp. *pomaceae*, the causative organism of cider sickness, growth is inhibited at pH 3.6—3.8. Cider manufacturers can, therefore, attempt to control this organism by blending apples so that bulked juice has a pH below these values. In addition, zymonads can be competitively inhibited by encouraging an active yeast fermentation and rendered quiescent in stored cider by ensuring that essential energy sources, glucose and fructose, are depleted by fermentation. The procedures adopted for controlling this organism must not, of course, favour the growth of other acid-tolerant spoilage organisms, such as the acetic acid bacteria.

In general, however, the acids occurring in a food or raw material will not ensure stability and other means of preservation have to be adopted, viz. appertization, pasteurization, gas storage, or the addition of preservatives (Table 13.10). The naturally occurring content of benzoic acid in the juice of cranberries is such that fruit submerged in water will keep for weeks.

Acids from Microbial Fermentations

Although acid foods are not necessarily microbiologically stable, acidulation of a product having a neutral reaction may aid preservation (see examples in Table 13.4). Thus with certain plant materials (cabbage, olives, cucumber, etc.) which have juices rice in carbohydrate, poor in buffering capacity and of limited capacity to poise the redox potential, fermentation is used to increase the H ion concentration. In practice, the plant materials are held in tanks of brine, the salt acting both as a selective agent and an aid in the withdrawal of plant juices. With some material, such as olives, there may be and normally is a requirement to breakdown the antibacterial agents, aglycone and elenolic acid, which arise from the hydrolysis of oleuropein.

H^+, H_2O → + + Glucose

oleuropein

β-3,4-dihydroxy phenylethyl alcohol – elenolic acid

oleuropein aglycone

β-Glucosidase.

Figure

During the fermentation of a product such as sauerkraut, there is a succession from populations of low (*Streptococcus faecalis* and *Leuconostoc mesenteroides*) through moderate (*Lactobacillus brevis*) to high-acid tolerant organisms (*Lactobacillus plantarum*). The succession of micro-organisms in the depth of the tanks is interrupted at this point by a combination of acid, salt and anaerobic conditions. The combination of salt and low pH is perhaps the main reason for the failure of *Clostridium* spp. to contribute to the fermentation. Of course stability of the product will not be assured if acid tolerant aerobic micro-organisms grow on the surface of the brine at the expense of the acids formed by the fermenting flora. In practice, the growth of such organisms is controlled by UV-irradiation or by the exclusion of O_2 through sealing the top of the vats with water-filled plastic bags. When the fermented material is packed in small containers, the exclusion of O_2 is obviously important and stability can be ensured by pasteurization.

With the growing concern about environmental pollution and the consequent difficulties in the disposal of concentrated brine solutions, some manufacturers of pickled olives have attempted the bulk storage of material in water containing small amounts of organic acids and salt. Successful preservation can be achieved by storing olives in a solution of lactic acid (pH 3.8-4.10) containing 3-4% instead of the customary 6-7% (w/v) of NaCl. Stored olives spoiled when lactic acid was replaced by a mineral acid even though the pH was lower (pH 2.6) than that in the lactic acid solution; fermenting, pectinolytic yeasts caused softening of the lives and the integuments were distended in places by gas formation. Olives can be stored in a solution of acetic and lactic acid containing sodium benzoate providing the immersing fluid is covered so that aerobic organisms cannot breakdown the organic acids. These observations indicate that the acid radical rather tan pH *per sé* plays an important role in the storage of fermented plant material. Indeed this solution is now becoming widely recognized in food microbiology and not only in the context of food spoilage. Thus lactic and acetic acid, either singly or in combination, are more effective than hydrochloric or citric acid in inhibiting the growth of *Clostridium botulinum* and *Staphylococcus aureus*.

Deliberate Acidification

The deliberate acidification of food represents a very old method of preservation, it having been applied to fish (marinated herring, soused mackerel), vegetables (onions and cabbages), pickles, sauces and mayonnaise. Acetic acid has been the most commonly used acid. It is of interest to note that relatively few bacteria are tolerant of the acetate ion as witnessed by the successful application of acetate media in the isolation of *Lactobacillus*. The effectiveness of acetic acid as a preservative is dependent upon concentration as well as the pH of the food. Thus with pickles and sauces, acetic acid in a concentration of 3.5% of the volatiles ensures stability. Such a concentration gives a tartness which is becoming less acceptable to the consumer and, in a search for blandness and microbial stability, various combinations of acetic acid, humectants and preservatives have been tested. The admixture of sub-optimal concentrations of acetic acid and lactic acid have not proved successful in commerce. Adding sucrose and/or benzoic acid provides both stability and blandness. Moreover, the pasteurization of such products may at first sight appear to be merely additional insurance of stability. In practice, however, pasteurization may well be of particular importance, apart from destroying pectinases, in breaking a route of infection. Thus it can be assumed that the environment of factories in which sauces and pickles are produced will contain a flora characterized by the acetate tolerance of its members. Even if they did not spoil a sauce during storage and distribution, they might well do so at some time during use. As the general environment in the home or canteen is unlikely to be selective for acetate-tolerant organisms, the product would be at risk mainly to infection of factory origin. It is obvious that pasteurization is likely to play an important role in preventing such infection.

Cheese

Cheese provides an example of a product whose stability is dependent in part upon the acid reaction resulting from the fermentation of lactose by streptococci and/or lactobacilli. It differs, however, from the fermented vegetables discussed earlier in that 'ripening' is associated with the biochemical activities of micro-organisms, bacteria, moulds and yeasts, which succeed the 'starter' or pioneering organisms. Factors which enhance the succession are discussed subsequently. With cheese of the cheddar type, moulds can grow on the surface thus producing a fault as far as the consumer is concerned. This can be prevented by excluding oxygen, either by waxing the surface or covering it with gas-impermeable material, or by applying sorbic acid or one of its salts.

Sausages

The microbiological stability of 'continental' sausages is in part dependent upon an acid reaction developing as a result of bacterial fermentation of carbohydrate added to comminuted meat. As with cheese, the 'ripening' of some such products is due to the growth of micro-organisms which succeed the fermentative organisms, they may grow on the surface of the sausage, as with the moulds—mainly members of the genera *Aspergillus* and *Penicillium*—or within the sausage. In addition to the acids derived from the fermentation of carbohydrates, the preservation of fermented sausages is aided by an interplay of pH, the NO_2^- ion, salt, drying, smoking, etc.

Red Meat and pH

Red meats are considered to be less prone to microbial attack if they have an acid reaction. The latter results from the accumulation of lactic acid, the end-product of the glycolytic breakdown of glucose derived from the glycogen stored in the tissues. Glycolysis becomes the principal method of carbohydrate degradation immediately following the tissues' deprivation of O_2 due to exsanguination. The method of handling animals before slaughter has an important bearing on the pH attained in the tissues post-mortem. Excitement or fatigue will deplete the glycogen reserves; rest and feeding with carbohydrate will ensure that there is sufficient glycogen in the tissues for a reaction of c. pH 5.5 to be achieved post-mortem. In theory, an acid reaction in a cured meat such as ham or bacon would accentuate the toxicity of NO_2^- and, similarly, acid in a sulphited meat would enhance the bactericidal action of sulphite. In practice, however, modern methods of processing tend to produce products having neutral reactions due primarily to the use of 'polyphosphates' which improve the water-holding capacity of meat and meat products. Although claims have been made that phosphates, through acting as chelating agents, contribute to the preservation of foods, the available evidence does not permit an assessment of their contribution under commercial conditions.

Although the above discussion assumed that glycolysis results in a pH of 5.5 in all tissues, there are, in practice, differences in the pH obtaining in different muscles. This can lead to the localization of spoilage. With bacon stored under reduced pO_2, '*Pseudomonas mephitica*' causes greening/

blackening through H_2S production in those muscles with a reaction near to neutrality. Recent studies have shown that pH plays an important role in controlling the growth of *Brochothrix thermosphacta* on meat (Table 13.5).

Water Activity

The amount of water in a food will obviously have an important bearing on its chemical and microbiological stability. Although this has been recognized for centuries and drying is one of the oldest methods of food preservation, little headway in understanding the principles of preservation was possible whilst attention was focused on the water content of a dried food. It was established that there is a fairly broad range in the alarm water content (i.e., the highest water content at which microbial spoilage will not occur) of traditional dried foods, viz. (% water in parenthesis): dehydrated whole egg (10-11%); wheat flour (13-15%); dehydrated fat-free meat (15%); dehydrated vegetables (14-20%); and dehydrated fruits (18-25%). Progress was rapid once the concept of water activity (a_w) — the ratio of the water vapour pressure over a food to that over pure water at the same temperature—was adopted. Although fundamentalists argue that it is the relative water vapour pressure (p_w) rather than a_w which is important, the latter has achieved common acceptance and will be used in the following discussion.

TABLE 13.5

The Influence of pH and Permeability of Wrapping Films on the Growth of *Brochothrix thermosphacta* on Meat at 5°C (from Campbell R.J., Egan A.F., Grau, F.H. & Shay, B.J. (1979) *J. Appl. Bact.* 47, 505-9)

Conditions of storage	*Populations (No. g^{-1}) formd at pH*	
	5.4—5.7	6.0—6.4
Meat in closed tubes		
Containing { oxygen	**2.0 x 10^9**	**> 1.0 x 10^9**
Containing { no oxygen	**No growth**	**>1.0 x 10^7**
Meat wrappd in transparent films of varying oxygen permeability		
low (1 ml)*	**No growth**	**1.0 x 1.0^7**
intermediate (150 ml)	***c.* 1.0 x 10^7**	**1.0 x 10^7**
high (> 1000 ml)	**1.0 x 10^9**	**1.0 x 10^9**

*** Oxygen permeability : value shown m^{-2} d^{-1} atm^{-1}.**

Some, perhaps the majority of bacteria, grow well in water containing small amounts of dissolved organic and inorganic substances. Others have a requirement for low concentrations of NaCl, for example, the alteromonads; a facultative capacity to grow in high concentrations of salt, for example *Pediococcus halophilus* (growth in media containing c. 18% (w/v) NaCl), or an obligate requirement for solutions approaching saturation with NaCl, the halophilic bacteria. With yeasts, a large proportion of the known species will grow well in solutions containing 40% sugar; above this level osmophilic yeasts are selècted and at sugar concentrations of 65-70% only a few yeasts can grow and then only slowly. There are some yeasts, for example *Debaromyces*, which grow in materials containing high concentrations of salt. It is generally accepted that moulds can grow under conditions which are too 'dry' for the majority of bacteria and yeasts, and xerophilic fungi are notable because of their capacity to grow under relatively dry conditions.

When solutes are dissolved in water, some of the water molecules become arranged around solute molecules and there is an increase in the molecular forces between water molecules. This is reflected in a depression of the freezing point, a depression of water vapour pressure and the elevation of the boiling point. According to Raoult's law, the decrease in vapour pressure of the solvent of an ideal solution is relative to the mole fraction of the solute:

$$\frac{po-p}{po} = \frac{n_1}{n_1 + n_2}$$

or that the vapour pressure of the solution relative to that of a pure solvent is equal to the mole fraction of solvent:

$$\frac{p}{po} = \frac{n_1}{n_1 + n_2}$$

when p and p_o are the relative vapour pressures of the solution and solvent respectively, and n_1 and n_2 number of moles of solute and solvent respectively.

From this derives the concept of the *equilibrium relative humidity* (ERH), the unique humidity at which the rate of evaporation and condensation are equal, viz. ERH(%) = $a_w \times 100$. Moreover, in such a state the solution has a water activity *(aw)* defined as :

$$a_w = \frac{p}{po}$$

Osmotic pressure is related to a_w, viz.

$$\text{osmotic pressure} = \frac{-Rt \log_e a_w}{V}$$

where R = gas constant, t = absolute temperature, $\log_e a_w$ = natural log of a_w, and V = the partial molar volume of water. As reduction of a_w results from the increase in the concentration of a solution, it may be achieved by adding solutes—as in the preservation of foods by salting or syruping—or by removing water (Table 13.6). The refrigeration of a food will change its a_w, the formation of ice crystals leading to a progressive increase in the solutes concentration. The relationship between water or moisture content *(m)* and a_w at equilibrium may be represented graphically by a moisture sorption (m_s) isotherm. Such an isotherm is normally sigmoid in shape. There are reasons for believing that, in practice, the sigmoid curve is formed from 3 'local isotherms' (li), shown as A, B and C, which reflect three types of bound water :

1. monolayer—bound or oriented water;
2. multilayer—chemiabsorbed water;
3. mobile—capillary, solution.

TABLE 13.6

Methods of Drying Foods

Method	*Foods to which applied*
Insolation	Meat, fish and fruits
Smoking	Meat and fish
Accelerated freeze drying	Whole egg
Spray drying	Milk, egg albumen
Vacuum drying	Egg products

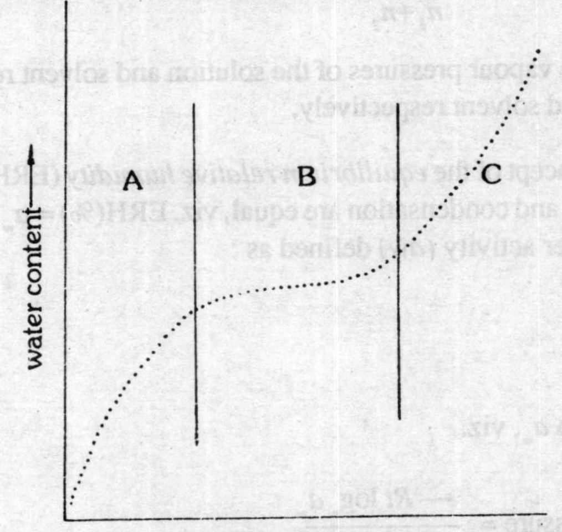

Fig. 13.4. The water sorption isotherm of a dried food.

From a microbiological viewpoint the water activity of a substrate having the local isotherm A, will not support the growth of micro-organisms and, unless protected by being in the spore state or through being clothed with extraneous proteins, fats, etc., they may well die. In terms of chemical stability, the food will deteriorate due to autoxidation, (Fig.13.5) because the decay of free radicals (Fig.13.6) is retarded by low moisture content. In the local isotherm, B, the a_w is such that it will be selective for xerophilic fungi (e.g. *Xeromyces bisporus*—minimum a_w for growth, c. 0.60/25°C), osmophilic yeasts (e.g. *Saicharomyces rouxii*—minimum a_w, c. 0.62) and obligately halophilic bacteria (e.g. *Halobacterium*—minimum a_w, c. 0.75).

With the local isotherm, B, chemical changes (Fig.13.5) will be due mainly to non-enzymic browning reactions (Fig.13.7) or to enzymes (Fig.13.8), activity of the latter increasing as increased moisture content aids the diffusion of substrates, providing diffusion is not impeded by biological structure. Thus with a 'natural storage product' such as grain having a moisture content of c. 13%, there is negligible chemical deterioration even when it is stored for years. With flour produced from such grain, chemical deterioration can assume serious proportions in a matter of weeks even though the moisture content of the flour is equal to that of the grain prior to milling. In the local isotherm, C, the conditions will become less selective as the a_w moves from 0.80-0.98. In practice, the selective pressure will move from an a_w of 0.80 which will inhibit the growth of organisms such as staphylococci to an a_w of 0.98 which will allow the growth of pseudomonads, etc. Over and above its elective property, water activity will influence the rate at which microorganisms grow. Thus a lowering of water activity away from the value which is optimal for a particular organism will lead to a progressive reduction in an organism's growth rate.

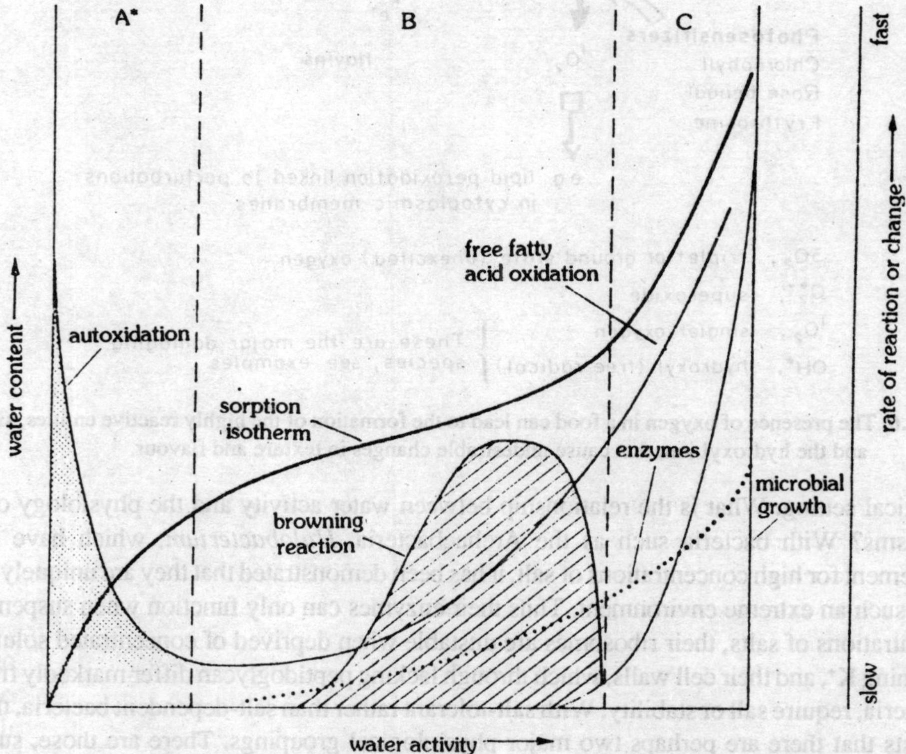

Fig. 13.5. Chemical changes and the rate of microbial growth in a food are determined by water activity.

So far, we have been concerned with water activity from the viewpoint of a selective agent in the

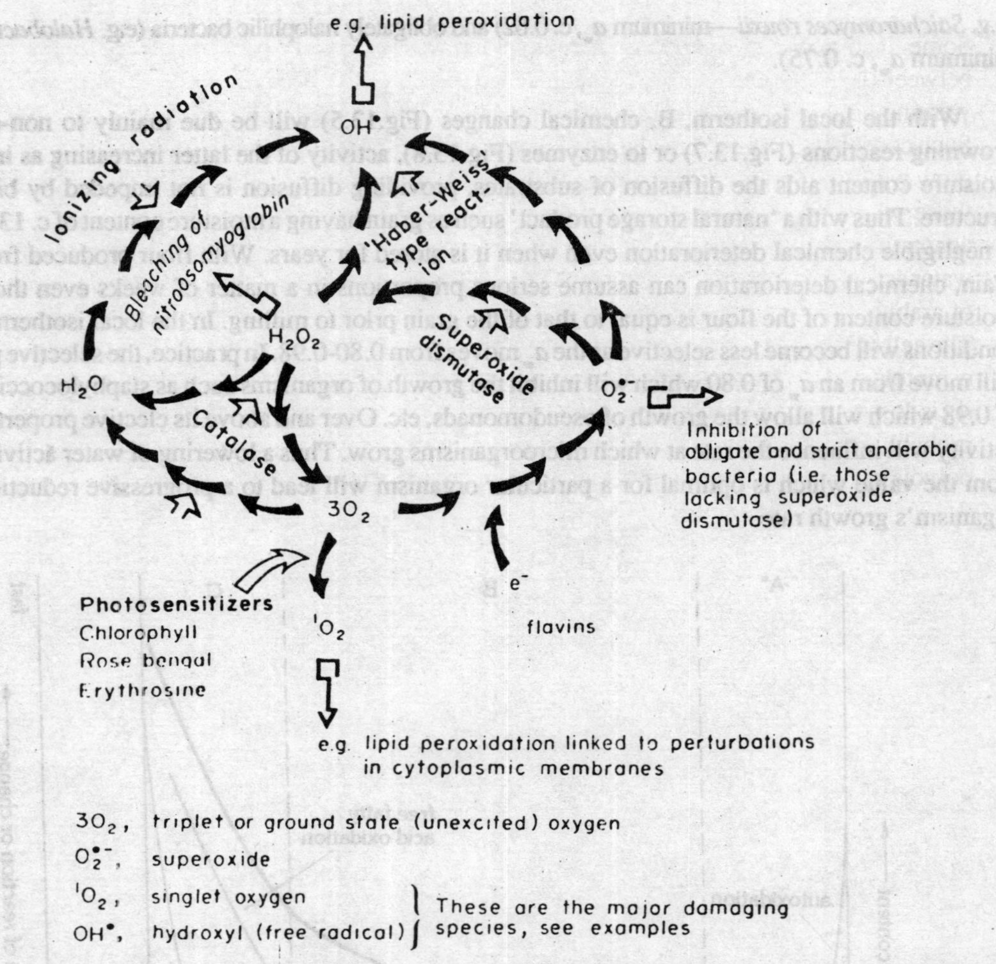

Fig. 13.6. The presence of oxygen in a food can lead to the formation of the highly reactive entities, singlet oxygen and the hydroxyl ion, that cause undesirable changes in texture and flavour.

ecological setting. What is the relationship between water activity and the physiology of particular organisms? With bacteria such as the Archaebacteria, *Halobacterium,* which have an obligate requirement for high concentrations of salt, it has been demonstrated that they are uniquely adapted for life in such an extreme environment. Thus their enzymes can only function when suspended in high concentrations of salts, their ribosomes are unstable when deprived of concentrated solutions of salt containing K^+, and their cell walls, which through lacking peptidoglycan differ markedly from those of eubacteria, require salt or stability. With salt-tolerant rather than salt-dependent bacteria, the evidence suggests that there are perhaps two major physiological groupings. There are those, such as some alteromonads, which have apparently an obligate requirement for a certain minimum level of Na^+, K^+, etc. There are others which have a marked tolerance of NaCl solutions giving a_w of c. 0.80-0.90 but which can grow equally well in a medium in which the a_w is poised by a mixture of salts other than NaCl.

Some facultative halophiles or osmotolerant bacteria, e.g., *Bacillus* spp., achieve compatibility between the osmotic pressure of a medium and their cytoplasm by synthesizing compounds, proline and γ-amino butyric acid, that confer osmoregulation without disturbing the basic metabolism of the cell. Such compounds are secreted rapidly when a medium is diluted. With yeasts, the majority of which can grow with a minimum a_w level of 0.88, there are those which can tolerate sugar solutions of 40% and those which grow, albeit slowly, in sugar solutions of 65-70% (a_w c. 0.60). There are two main adaptations that fit certain yeasts, for example, *Saccharomyces rouxii,* to growth in a medium of low water activity. In response to diminished a_w they synthesize polyols, mainly glycerol and a small amount of arabitol, which serve not only as osmoregulators but protect enzymes from damage due to low a_w. The cell membranes of the osmotolerant yeasts retain the polyols within the cell, whereas those of non-osmotolerant yeasts are leaky and much of the osmo-regulators produced in response to a diminished a_w is lost to the surrounding medium. Indeed the failure of some yeasts to become osmotolerant is probably linked to their inability to produce sufficient energy for cell maintenance and the synthesis of polyols in quantities sufficient to act as intracellular osmoregulators.

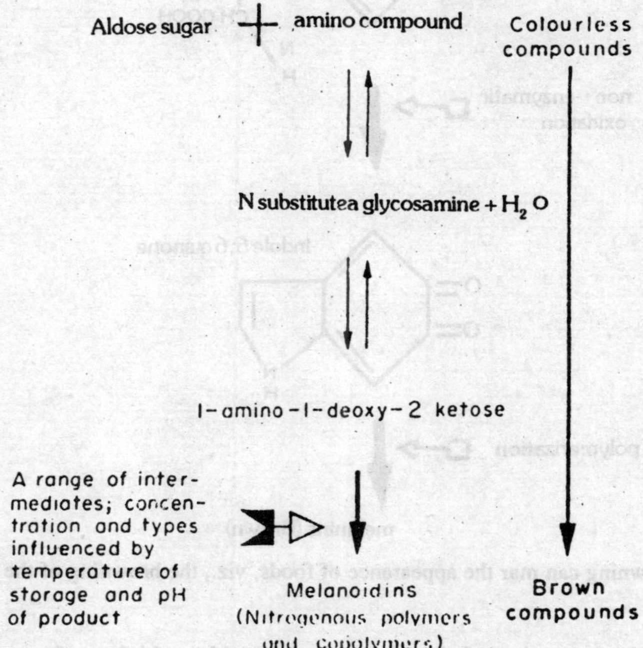

Fig. 13.7. Non-ezymic browning causes undesirable changes in the appearance, texture and flavour of dried foods.

Although reduced a_w inhibits microbial growth it can confer thermal stability on micro-organisms, a factor that needs to be considered if pasteurization of a dried product is contemplated.

Salt

When considering the use of salt (NaCl) in a food, there is a need to define the role that it is playing. It may be merely a condiment; it may be bringing about some essential change in the physiochemical

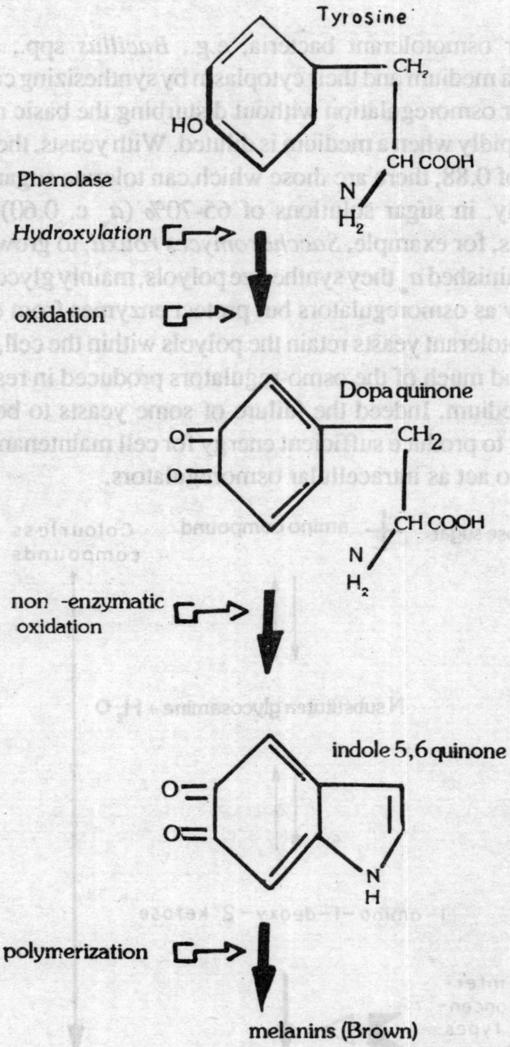

Tyrosine
CH_2
HO
CH COOH
Phenolase
NH_2
Hydroxylation
oxidation
Dopa quinone
CH_2
O=
O=
CH COOH
NH_2
non -enzymatic
oxidation
indole 5,6 quinone
O=
O=
N
H
polymerization
melanins (Brown)

Fig. 13.8. Enzymic browning can mar the appearance of foods, viz., the browning of the cut face of apples and potatoes.

properties of a food—the extraction of plant juices in vegetables which are fermented with lactic acid bacteria, or the solubilization of the myofibrillar proteins of meat so that they may act as emulsifying agents during the preparation of commodity such as Frankfurters—or it may be acting exclusively as a preservative, as in the case of dried salt fish (Table 13.7). With the last mentioned there is a need to consider the actual concentration of NaCl in the liquid phase rather than in the food as a whole. Thus with an example discussed previously, salt will be concentrated in the water droplets in butter and it can be present at levels of 10% (w/v). The term 'brine' is used to denote the percentage of NaCl in the water phase of a food, viz.

$$\% \text{ Brine} = \frac{\%\text{NaCl}}{\%\text{ NaCl} + \%\text{ water}} \times 100$$

When considering the preservative action of NaCl, attention needs to be given also to the possible synergistic action of NaCl and other intrinsic factors, such as pH, or extrinsic factors, temperature, pO_2, etc. Thus, the interplay of NaCl and pH is considered to be an important preservative system in cheese. Similarly, the salt in fermented vegetables is probably the principal agent in controlling the growth of clostridia.

TABLE 13.7

Foods in Which NaCl Contributes to Preservation

Contribution as preservative	*Food*
As principal preservative	Salted fish Salted meat
As an adjunct to	
pH	Butter Cheese Fermented vegetables
NO_2	Bacon

Cured Meats

From a historical standpoint, salting, or the curing the meat can be considered as a means whereby our forefathers preserved meat obtained from animals which had to be slaughtered in the autumn when fodder was no longer available. The Dutch used to call November, slaght-maand (slaughter month). In the evolution of salt curing, it was observed that contamination of salt with saltpetre led to the production of a heat-stable pigment, nitro-sylmyoglobin (Fig.13.9), which imparted a desirable colour to the product. The development of methods of husbandry which provide a continuous supply of animals to the meat industry and the general adoption of refrigeration have reduced the need for a fully preserved product.

At the outset, salt would presumably have been rubbed into the surface of meat and thus an inward diffusion of NaCl established. Although crystalline salt would provide an inimical environment at the surface, there would still be an opportunity for salt-intolerant bacteria of gut origin to grow within the tissues, particularly if there was inadequate chilling post-mortem or storage was at a high temperature. In such a case, the growth of *Clostridium* and *Streptococcus* spp. around the bone in a ham—i.e., at the point farthest away from the salted surface—can produce the fault, 'bone taint.'

Once sufficient salt has diffused into the tissues, the hams or sides of bacon were smoked thereby drying the outer surface and distilling upon it chemicals, such as phenols, which retard fat oxidation, or formaldehyde, methanol, cresols, etc., which inhibit microbial growth. With this example, the 'case-hardened' surface will provide an environment inimical to most micro-organisms other than moulds and a barrier against microbial invasion of the underlying tissues. A similar situation obtains with smoked fish and the efficacy of the external barrier is enhanced because proteins extracted by salt are denatured.

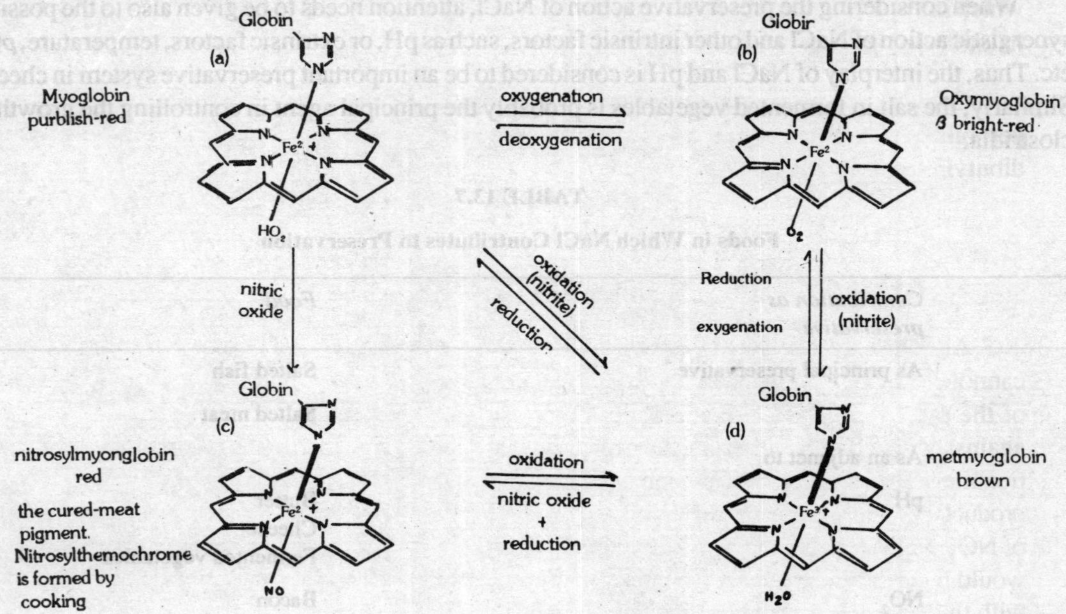

Fig. 13.9. The colour of fresh red meat products is determined by the compounds outlined above. See Dryden & Birdsall (1980) for illustrations.

Instead of applying salt to the surface, the meat can be immersed in chilled brine (20-25% NaCl) containing NO_3^- and NO_2^- as in traditional methods of Wiltshire curing. The brines may be used for many years. Thus there is an opportunity for the enrichment of halophilic and halotolerant bacteria; halophilic Gram-negative rods which lyse rapidly in distilled water can form populations of 10^8 organisms ml^{-1} and the halotolerant cocci upwards of 10^6-10^7 organisms ml^{-1}. In addition, such brines contain lactobacilli, which are unusual in that they are intolerant of the acetate ion, and causal contaminants such as *Escherichia coli* and *Pseudomonas* spp. which are introduced with the meat. The coliforms are of interest because they can retain viability for longer period in chilled brine than they do in that held at ambient temperature. The microflora developing in brines are also of interest because of their negligible capacity to digest polymers of meat, thus they do not cause putrefaction. With halophiles such as *Vibrio costicolos*, NO_3^- is reduced to NO_2^-; not only is the latter toxic to many bacteria, it is involved in the reactions leading to nitrosylmyoglobin formation, hence the cured meat colour (Fig.13.9). It also contributes to the flavour of bacon and ham. In fact, until the role of NO_2^- was recognized at the beginning of this century, curing was dependent upon the reduction of nitrate by meat and microbial enzymes and even today the 'cured' colour of certain fermented sausages results from microbial reduction of NO_3^-.

With the recognition of the important role of NO_2^- the requirement for NO_3^- reducing bacteria could be avoided simply by adding NO_2^- to the brine or meat to be cured. Likewise, the impregnation of tissues with salt and NO_2^- can be achieved by direct injection and a short period of storage in a 'cover brine which may be used for a limited period only—a practice which may lead to a population dominated by halotolerant cocci rather than halophilic rods.

Cooking NO_2^- = cured meat products may result in the formation of extremely small amounts (parts per billion) of nitrosamines (e.g. dimethyl-nitrosamine, diethylnitrosamine, dipropylnitrosamine, dibutylnitrosamine, etc.).

$$(H_3C)_2N\text{—}H + NO_2 \longrightarrow (H_3C)_2N\text{—}N{=}O$$

Dimethylamine ⟶ N-nitrosodimethylamine

Although the levels are low, the production of carcinogens during food manufacture or preparation cannot be ignored. As the derivation of nitrosamines is due to NO_2^- in meat, a reduction in concentration of the preservative offers one practical approach to the problem. Such a policy has to be balanced against the good public health record of cured meat products, especially those that rely upon a low heat treatment for stability. It has been established that heating of cured meat products accentuates the product's inimical properties in respect to *Clostridium botulinum*. Thus the reduction in the concentration of NO_2^- might conceivably enhance the risk of food poisoning due to this organism. A possible solution would be to diminish a meat's content of NO_2^- and add another preservative that acted synergistically with this anion. Promising results have come from trials with sorbates.

The slicing and packaging of bacon creates ecosystems different from that obtaining on smoked sides of bacon; as the former supports large populations of bacteria (micrococci, staphylococci, yeasts, lactobacilli and vibrios)—mould growth is prevented by vacuum packing in a gas impermeable film—it can be regarded as a semi-preserved product only. It is noteworthy that with this modern method of producing bacon, the salt still selects a flora that is unable to digest the polymers, especially the proteins, of the meat. Thus the product is acceptable to the consumer even though it may be harbouring relatively large numbers of bacteria and it is safe because of the failure of food-poisoning bacteria, especially enterotoxin-producing strains of *Staphylococcus aureus*, to complete with the other components of the flora particularly with storage at c.4°C.

Sugars and Syrups

The high sugar contents of materials such as maple syrup, honey, molasses, etc., mean they are stable unless contaminated with osmophilic yeasts or xerophilic fungi, providing, of course, that the available water is evenly dispersed. If storage conditions result in water sorption or, as may happen in a closed container, chilling causes condensation with water droplets falling on to the surface of the material, then conditions may well permit the growth of yeasts whose requirement for a_w are greater than those of the true osmophiles. This can be a problem when liquid sugar is pumped into storage tanks. After the sugar has had its content of yeasts reduced by filtration through diatomaceous earth filters, it is cooled and pumped into steam-sterilized tanks, the head space of which is flushed with sterile warm air to prevent condensation and localized dilution of the syrup. UV lamps in the headspace may also be used to prevent yeast growth.

The stability of syrups of moderate strength can be improved if invertase is used to increase the concentration of glucose relative to sucrose (Table 13.8). Man has exploited this means of preservation in the storage of candied fruit, the fillings of chocolates and confectionery. Although control of microbial growth can be achieved by adjusting the concentrations of solutes in syrups, manufacturers often pasteurize such products so that they are freed of contamination with osmophilic yeasts. Thus with honey, preservation is assured by pasteurization (71-77°C for a few minutes or 93°C for seconds).

TABLE 13.8.

Foods in Which Sugars aid Preservation

Syrups and candies
Fondant fillings in chocolate*
Honey
Jams and conserves
Candied peel
Dates, sultanas and currants

**Fondant fillings*: >79% of sucrose and invert sugar prevent fermentation by osmophilic yeasts.

In the wine industry an interplay of sucrose and ethanol, which is 4-5 times more inhibitory than the sugar, in yeast fermentation of wine musts is measured in Delle units (Du):

$$Du = x + 4.5y$$

where x = g of reducing sugars 100 ml^{-1} and y = ml of ethanol 100 ml^{-1}.

Drying

The methods used for the desiccation of foods have been cited previously (Table 13.6) and the importance of moisture content on both the chemical and microbial stability of dried products discussed. In some instances, preservation can be achieved by linking a reduced moisture content with a modification of an extrinsic factor. Thus, the successful preservation of cereals having moisture contents of more than the alarm water content of 13% can be achieved by storage in sealed containers where an increase in pCO_2 inhibits mould growth. Likewise, mould growth on cakes can be inhibited by enrichment of the storage atmosphere with CO_2 or the use of a fungistatic agent such as sorbic acid.

Intermediate Moisture Foods

There was little prospect of innovation when the stability of a dried food could be predicted only in terms of its alarm water content. The adoption of the concept of a_w led to: (i) the establishment of the minimum a_w required for the growth of a wide range of spilage and food-poisoning microorganisms

(Table 14.9); (ii) the realization that a_w was only a part of the preservation system in many traditional foods (dried fruits (a_w 0.72-0.80), jams (0.82-0.94), honey (0.75), and (iii) the recognition that by use of appropriate humectants (salt, sucrose, sorbitol, propylene glycol) a food could have its a_w modified without recourse to traditional methods of drying, syruping or salting.

TABLE 13.9

Minimum a_w for the growth of selected organisms

Bacteria	*Min. a_w*	*Yeasts and fungi*	*Min. a_w*
Eubacteria			
Pseudomonas spp.	0.97	*Candida utilis*	0.94
Acinetobacter spp.	0.96	*Rhizopus stolonifer*	0.93
Enterobacter aerogenes	0.95	*Trichosporon pullulants*	0.91
Bacillus subtilis	0.95	*Aspergillus glaucus*	0.70
Escherichia coli	0.96	*Saccharomyces rouxii*	0.62
Staphyloccus aureus	0.86	*Xeromyces bisporus*	0.60
Archaebacteria			
Halobacterium	0.75		

Moreover, the demonstration that the minimum a_w for the growth of an organism is influenced by intrinsic (preservatives, pH) and extrinsic (gaseous environment, temperature) factors led to the possibility of 'modelling' a complex preservative system such that a long storage life at ambient or refrigerated temperatures could be contemplated.

The humectant can be incorporated in the product by: (i) soaking or cooking the food in a solution of the humectant; (ii) soaking a dehydrated food in a humectant; or (iii) regarding the humectant as an ingredient and mixing it with others in the preparation of a food.

The actual effectiveness of the required a_w will be influenced by the pH of a food and its content of preservatives. Thus with the addition of sorbate to control yeast and mould growth, an a_w of c. 0.8 would assure microbial stability. An acid food containing an antibacterial agent can be protected from microbial spoilage by an a_w of 0.85. Although the theory underlying the production of intermediate-moisture foods is relatively simple, practical problems have limited their uses, e.g., special foods required for manned space flights and pet foods. With the latter, a_w is controlled by a mixture of sugar and propylene glycol and shelf-life is aided by pasteurization. Unacceptable organoleptic properties are the major impediments to large-scale production of intermediate-moisture foods for humans. Besides texture and flavour problems due to the humectants available to food manufacturers, the a_w of such food (0.70-0.90) is in the range in which non-enzymic browning, lipid peroxidation and enzyme activity can occur. Fat rancidity can be hindered by anti-oxidants and enzyme activity by heat treatment. As yet, however, there are few reliable remedies for non-enzymic browning.

The microbiological stability of a dried product often calls for critical control of the storage temperature. It is well-known that moisture migration due to wide fluctuations in temperatures can produce 'wet spots' in which micro-organisms grow.

Preservatives

In the Report (1972) on the review of Preservatives and Food Regulations 1962, the following definition of preservative was recommended: 'Any substance which is capable of inhibiting, retarding or arresting the growth of micro-organisms, of any deterioration of food due to micro-organisms, or of masking the evidence of any such deterioration.'

It is obvious that a very wide rage of substances could be used to achieve these ends. In practice, of course, the choice will be limited to those for which there is toxicological evidence that their use does not pose a threat to the health of man. There are other reasons for imposing additional restrictions on the use of preservatives. Thus a preservative should not be used if the microbial stability of a product could be assured by other means, viz. appertization, pasteurization, refrigeration, etc., providing that these do not impose prohibitive costs of give unacceptable organoleptic properties to a food. Moreover, a preservative should not be used merely to remedy faults arising from poor process control or bad hygiene at some or all of the stages in the storage, preparation and distribution of a food. Examples of preservatives are given in Table 13.10; the use of some of these is discussed in the sections concerned with a_w, pH, refrigeration contamination control, etc.

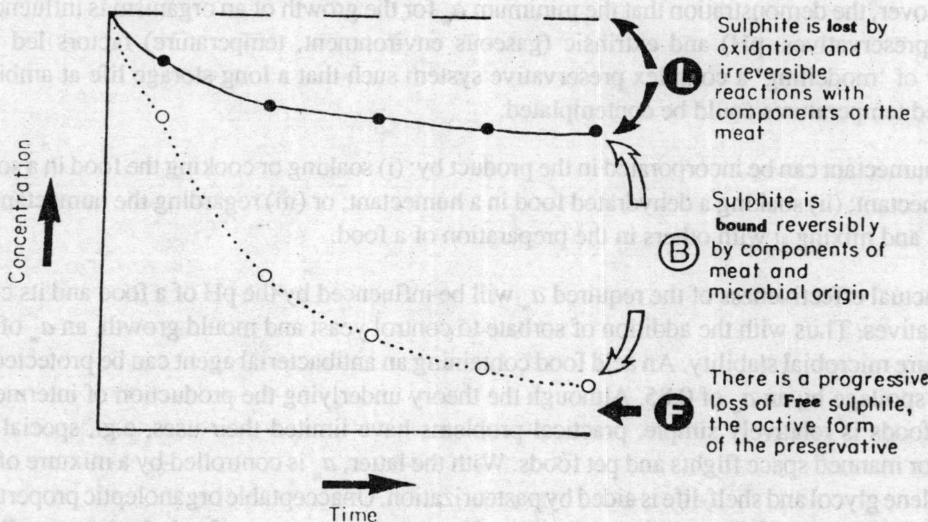

Fig. 13.10. The preservation of British fresh sausage is dependent upon free sulphite, the amount of which diminishes with time. The rate of binding of sulphite is temperature dependent, slow binding occurring at chill temperatures and fast binding at room temperature.

Recent studies of British fresh sausages have shown that the preservative, sulphite—measured as SO_2 and permitted at an initial level of 450 p/10^6—plays many roles. It selects a microbial association in which Grampositive bacteria (*Brochothrix thermosphacta,* yeasts and lactic acid bacteria) predominate and it inhibits the growth of enterobacteria. Comparisons of microbial growth and activity in sulphited and non-sulphited sausages have demonstrated that the preservative curtails by upwards of 1 log cycle the size of the climax populations of members of the association and influences their metabolism so that the pH of sulphited sausages drifts from c. 7.0-6.0, whereas a drift of 2 pH units or more is a feature of unsulphited sausages. In addition the preservative aids colour retention and protects to a limited extent the depolymerization of polysaccharides added to sausage; the combined action of amylase and maltase produces lower concentrations of glucose, maltose and maltotriose than they do in sausages lacking preservative. It is notable, moreover, that sulphite causes these effects even though its occurrence in the free state diminishes rapidly with time (Fig.13.10). The major loss is associated with binding with materials in the sausage, the most likely binder being acetaldehyde of yeast origin.

TABLE 13.10

Food Preservatives

Substance	*Used in*	*To control*
Sulphite	British fresh sausages	Growth of Enterobacteriaceae Acid production by microbial association in sausages
Sulphur dioxide	Fermenting fruit juices	Growth of 'wild' yeasts
	Stored fruits	Growth of fungi, especially those that produce pectinases
	Dried and dehydrated vegetables	Colour loss
Nitrite	Meat products	Growth of spore-forming bacteria
Diethyl carbonate* (diethyl pyrocarbonate)	Fruit juices, wines and carbonated beverages	Growth of fungi
Ethylene or propylene oxide	Spices	Contamination
Acetate	Pickled vegetables	Microbial growth
Benzoate	Cheese and fruit juices	Growth of fungi and yeasts
Esters of p-hydroxybenzoic acids (Parabens) methyl ethyl propyl	Fruit juices	Growth of fungi and yeasts
Propionate Sorbate **	Confectionery, cheese	Growth of fungi
Formaldehyde	Cheese milk	Growth of *Clostridium* spp.
Nisin	Canned foods	Growth of spore-forming bacteria

Pimaricin	Cheese and sausage	Growth of fungi
Nystatin	Bananas	Growth of fungi
Chlortetracycline	Fish	Growth of bacteria

*Of limited use today; dimethyl carbonate may be widely used in future as a 'cold sterilizing' agent of fruit juices, wines, etc.

**See Overview (1982) Food tech. 36 (December), 87-118 for recent studies of the control of growth of *Cl. botulinum*.

Temperature

For practical purposes, 3 storage temperatures for food preservation can be recognized: sub-ambient temperatures (cellar storage), chill storage, —1-4°C, and frozen storage, —18°C or less. If the temperature variation along a 'cold chain' could be minimized, then 'super chill' storage at —5°C would be a useful commercial system for storing perishable foods. The actual choice of temperature will be determined in part by the nature of the food and in part by the length of storage required. Thus with an example discussed previously, long-term storage of butter is achieved at —17 —— —20°C, whereas chill temperatures are used for the storage and distribution of small packs. With fresh eggs having a small air, cell, storage at temperatures below freezing can cause shell fracture due to ice formation in the albumen. Even if the shell is not damaged, disruption of the diffusion gradient at the surface of the yolk will cause Fe^{3+} diffusion into the albumen, the colour of which will become salmon-pink due to Fe^{3+} saturation of ovotransferrin. In other situations low temperatures may cause irreversible damage of a product through breakdown of cell membranes, as with bananas, or unacceptable changes due to enzyme action. Thus chill storage of potatoes causes glucose to be released from starch and, if they were intended for crisp production, the caramelization of glucose would cause discolouration. The blanching of vegetables is intended to destroy enzymes that cause changes in flavour during frozen storage.

Modern slaughter house practices, particularly refrigerated storage, have minimized the risks inherent in cooling meat too slowly. Indeed the practices are such that psychrotrophic micro-organisms grow at the surface of the meat only. With beef, for example, slime production and off-odours are associated with these organisms achieving populations of 10^7-10^8 organisms cm^{-2} of meat surface (Fig. 13.12). Not only will rapid chilling prevent 'bone taint' and delay the manifestation of spoilage at the surface of the meat, it will also minimize evaporation and hence the loss of weight of a carcass. Rapid chilling minimizes also the amount of liquid which drains ('drip loss') from jointed meat. Thus the food microbiologist and the accountant will always hope to achieve rapid chilling of meat, especially of beef and lamb. Too rapid chilling (the meat achieves a temperature of less than 10°C within 10h) can result however in irreversible contraction of the muscles such that the meat is rendered tough ('cold shortening'). This is not an important problem with pork because the onset of *rigor mortis* is fast compared to that of beef and sheep carcasses. With the last two, 'cold shortening' can be avoided by controlling the rate of chilling (no part of the carcass should attain temperatures below 10°C within 10 h of slaughter) or the rate of development of *rigor mortis* through electrical stimulation of the carcass following removal of the hide and viscera.

Broadly speaking, cellar storage is used mainly for ware vegetables and fruits such as apples. Under such conditions, humidity may have to be controlled otherwise plant tissues will wilt or mould growth occur. Some ware vegetables, for example swedes, may be waxed before storage so that water loss is retarded. With some fruits, the natural defence imposed by the integument can be augmented with fungistatic agents. Thus with citrus fruits and bananas, for example, thiabendazole is applied to the skin after harvesting with the object of controlling mould growth. Similarly, the use of diphenyl in wrapping or packaging material is effective in reducing wastage of citrus fruit by mould growth. It has been shown also, that macrolide antibiotics such as nystatin, rimocidin and pimaricin are effective fungistats when sprayed on to soft fruits such as raspberries or strawberries. The fact that some of these antibiotics are medically useful in the treatment of 'thrush' tends to limit their use in the food industry. Current studies of vegetable storage are directing attention at the commercial feasibility of imposing controlled atmospheres, either a reduced atmospheric pressure or an atmosphere of specific chemical composition.

Chill Storage

When considering chill storage it is useful to recognize two situations: chill storage as :

(a) the principal means of preservation;

(b) an adjunct to other means of preservation.

From the discussion of the selective influence of processing of meat, it was noted that a reduction in storage temperature imposes a selective pressure on the heterogeneous population of contaminants acquired by meat during butchering. This results in psychrotrophic organisms being enriched by refrigerated storage. From a consideration of growth rate, it could be anticipated that temperature-induced retardation of growth will extend the shelf-life of a product. Such a view has been supported by studies in which a linear relationship between growth rate and the rate of spoilage has been demonstrated; in other words, shelf-life has been linked directly with an implicit property of micro-organisms, their rate of growth as influenced by an extrinsic property, temperature. With beef, it has been established that spoilage—slime production and off-odours—is associated with populations of 10^7-10^8 micro-organisms cm^{-2} of meat surface. When consideration is given to the initial level of contamination and the shelf-life of a product, it has been noted that these two factors are inversely related (Fig. 13.12). The shelf-life will be determined ultimately by the proportion of organisms in the initial contamination capable of growth during refrigerated storage. Thus the success of such storage will be aided by high levels of hygiene obtaining at all stages of production and the absence of niches within the factory wherein psychrotrophic organisms can grow and spill over on to the product.

It is obvious that reliance on chill storage will impose serious limitations on the shelf-life of a product such as meat. In attempts to overcome this, many attempts have been made to modify the environment offered by packaged meat so that microbial growth is further retarded. In one method extended shelf-life was achieved by storing meat in a modified atmosphere containing 30% (v/v) CO_2 and 70% O_2. This inhibits the growth of members of the *Pseudomonas-Acinetobacter* group and the meat is colonized eventually by the slower growing *Bochothrix thermosphacta*. Likewise, an extension of the storage life of fish in ice can be achieved by adding antimicrobial agents to the ice.

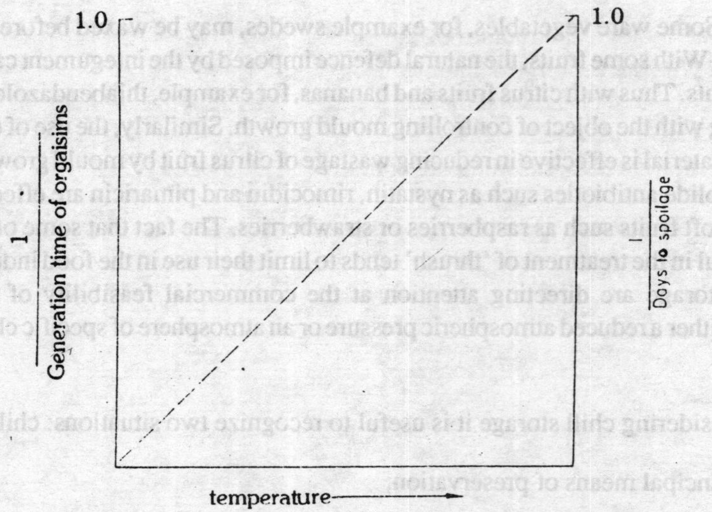

Fig. 13.11. The rate at which meat in a moist atmosphere spoils is determined by an interplay of the generation times of the spoilage organisms and the temperature of storage.

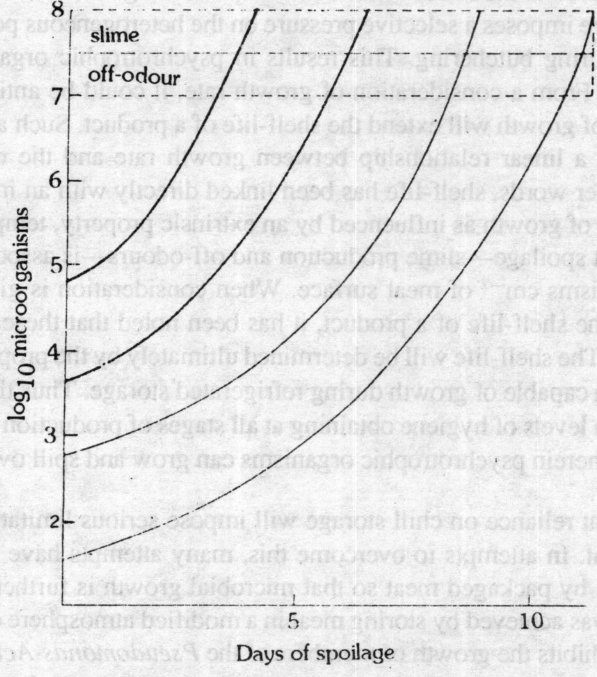

Fig. 13.12. At any temperature within the physiological zone of micro-organisms, the rate at which meat spoils in a moist atmosphere is a function of the initial level of contamination with spoilage organisms.

Deep Freezing

Deep freezing (temperatures below — 10°C) provides a means of long-term storage of food providing that the biological structure of the food is not destroyed and its texture or flavour made unacceptable by enzyme action. Contaminants on the food at the time of preparation are not killed—at least, the majority are not—so that there is probably a need to provide clear instructions so that the consumer may select conditions which allow the product to thaw without encouraging growth of micro-organisms, particularly food-poisoning ones such as *Salmonella*.

The success of food storage at sub-ambient temperatures will depend ultimately on the quality of temperature control, not only during the time that the product is in warehouses but during transit to the shop and home. This is referred to as the 'cold-chain' and, for success, every effort has to be made to ensure that the temperature, be it — 20°C or 1°C, is maintained at all stages of distribution. Perhaps through ignorance or disregard for some of the principles of physics, chilled or frozen foods can be subjected to many abuses during storage and distribution. Indeed the siting of a thermometer in a refrigerated store can be of importance. If it is situated in the air stream coming from the condenser, then the recorded temperature may well be an index of the efficiency of the machine rather than the temperature obtaining within the food in the storage compartment. Likewise, it is possible to entertain the wrong notions about the intended purpose of refrigerated containers. The size of the refrigeration unit and the quality of the lagging of the storage compartment may be adequate to ensure the maintenance of a particular temperature during the distribution of a food providing that the food was at or below this temperature at the time of packing. It will be appreciated that special problems are posed by galleys on commercial aircraft because limitations on the weight of equipment often preclude mechanical refrigeration. Alternative systems—refrigeration based on solid CO_2 (Drikold or dry ice) or freon—are designed so that chill temperatures are maintained rather tan *attained* during flight. With such a system, it is imperative that the food is at the desired temperature when loaded on to the aircraft. This can be assured by storing the food for 3 or more hours in a *large* cold store before putting it in the galleys. Much larger refrigeration units and more efficient lagging would be needed if the intention was to reduce the temperature of the food during transit. Indeed, the effectiveness of the cold chain would

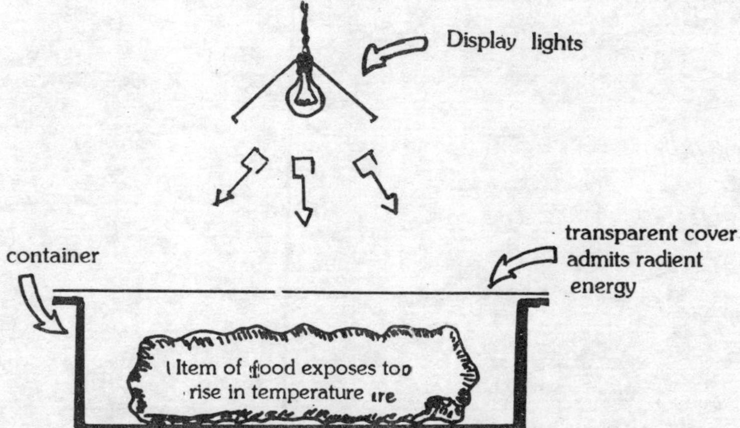

Fig. 13.13. The 'greenhouse' phenomenon. The storage of packs with transparent lids in a chill cabinet situated beneath bright display lights can cause the contents of the packs to heat up.
In future, a very thin coating of a reflective material, such as aluminium, on the lid may be used to reflect heat without impairing perceptibly the transmission of light.

probably be improved if refrigerated containers were rated for their potential to maintain (not achieve) specified temperatures for defined periods in the climatic conditions in which they were operated. Many examples of bad practices in the storage and display of chilled or refrigerated foods may be noted in shops. Thus the display of deep-frozen products in transparent wrapping materials can lead to the greenhouse phenomenon, especially if bright lights are situated immediately above the display cabinet. Likewise a goblect-shaped container (Fig. 13.12) for chilled foods may be an inappropriate shape if the maintenance of a chill temperature depends upon a refrigerant circulating through hollow shelves.

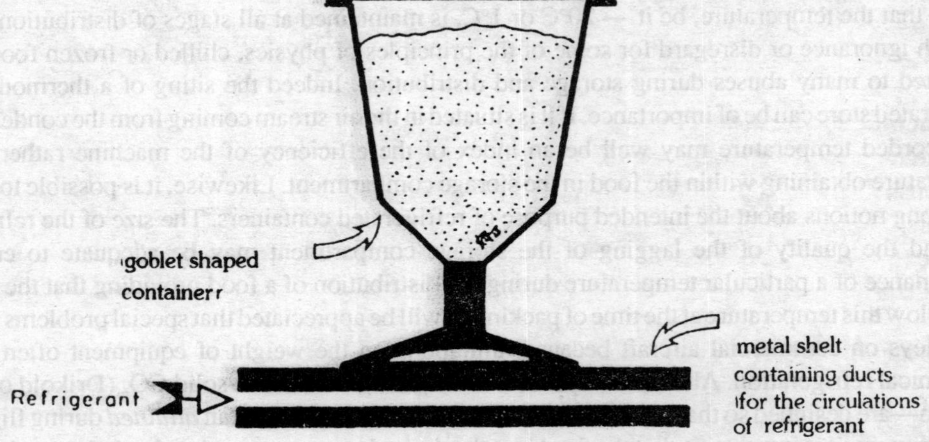

Fig. 13.14. **The design of containers must be such that their contents are maintained at chill temperatures throughout storage. The stalk of a globlet-shaped container may effectively lag the contents when refrigerated shelves are used to display merchandise.**

14

Microbiology of Anaerobic Digestion*

Introduction

The microbiological degradation of organic material in an anaerobic environment can only be accomplished by microorganisms able to use molecules other than oxygen as terminal electron acceptors. This anaerobic decomposition ultimately results in the production of a biogas consisting of methane (50-70%), carbon dioxide (25-45%) and small amounts of hydrogen, nitrogen and hydrogen sulfide. The overall chemical reaction is often simplified to :

organic matter —> $CH_4 + CO_2 + H_2 + N_2 + H_2S$

In fact, the anaerobic degradation of organic material is, chemically, a very complicated process involving hundreds of possible intermediate compounds and reactions, each of which is catalyzed by specific enzymes or catalysts. Many of the transformations may be accomplished by one of several alternate metabolic pathways, and the biochemists continue to attempt to define and more precisely describe the various mechanisms.

The ability of an organism to effect any particular transformation or reaction depends on the availability of the enzyme or catalyst specific to that reaction, and for intercellular reactions, the ability of the substrate to pass through the cytoplasmic membrane. The concentration of substrate material and the possible accumulation of reaction products as well as the presence of enzyme inhibitors may have a profound effect on the reaction rate.

Anaerobic decomposition is generally considered to progress in three stages: hydrolysis, an acid production stage and a methane production stage.

Hydrolysis

The first step is the hydrolysis and liquefaction of the large organic molecules by extracellular enzymes. Carbohydrases catalyze the hydrolysis of glycisidic bonds. Starch and glycogen, for example, are hydrolyzed to a disaccharide by the action of one of a group of enzymes called amylases. These enzymes attack polysaccharides from the non-reducing end of the chain, cleaving alternate glycisidic bonds as shown in Fig. 14.1. The disaccharides are then cleaved to monosaccharides

---

* Excerpts taken from the review published by C. Price, Network, New Jersey.

by a glycosidase. The specific enzyme involved depends on the nature of the glycosidic bond and the monosacharide involved; that is, whether the configuration is alpha or beta, dextro or levo, and the size of the heterocyclic ring. Cellulase and chitinase degrade the structural polysaccharides cellulose and chitin.

Lipases and esterases hydrolyze fats and lipids. When the substrate is a mixed triglyceride of long chain fatty acids similar to those found in natural fats and oils, the hydolysis proceeds in a stepwise manner, with the rapid formation of di- and monoglycerides, followed by slow hydrolysis of the monoglycerides.

Proteolytic enzymes, proteases, catalyze the hydrolytic cleavage of the peptide bonds of proteins. Again, these enzymes are somewhat specific. Exopeptidases are restricted to terminal peptide bonds and endopeptidases must attack peptide bonds that are centrally located in the peptide chains. The enzyme trypsin is specific for bonds involving the amino acids argenine or lysine.[1]

Acid Production Stage

After the hydrolysis and liquefication, the initial degredation products are in a form that can be utilized by the micro-organisms, providing they are able to pass into the cell through the cytoplasmic membrane. This membrane selectively regulates the flux of nutrients, ions and waste products in and out of the cell. It is composed of lipids and proteins. The lipids provide the structural properties, and the proteins provide the membrane with its distinctive functional properties. It is believed that the various proteins embedded in the membrane act as carriers for specific substances or types of molecules. This explains the ability of the membrane to pump against the head of an osmotic pressure gradient, a phenomena called active transport. The lactose transport system of *E. coli*, for example, is capable of producing a lactose concentration 500 times greater inside the cell than, outside. The operation of these protein gates requires a certain amount of membrane fluidity, which is largely determined by the relative amounts of saturated and unsaturated fatty acids in the lipid portion. Bacteria grown at a low temperature have membranes with a greater proportion of unsaturated fatty acids than those grown at higher temperatures.[2]

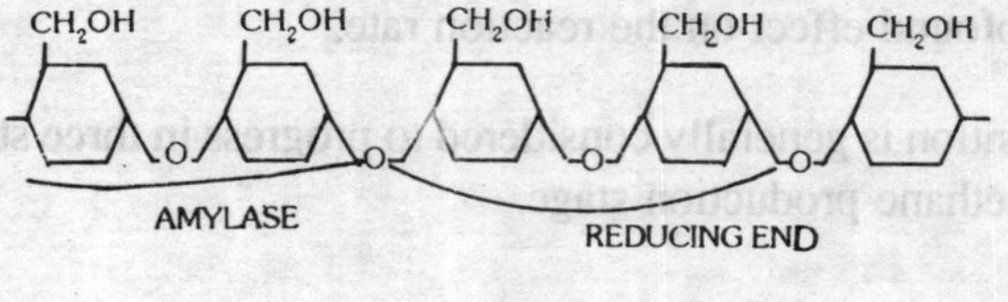

Fig. 14.1. Cleavage of alternate glycosidic bonds of B- amylase.

A typical metabolic pathway for the degradation of monosaccharides is the Embden-Meyerhof-Parnas pathway of glycolysis as shown in Fig. 14.2. Note that the end product of glycolysis in yeast is ethanol, and in bacteria is acetic acid. Carbon dioxide is evolved in either case. ADP and ATP represent adenosine diphosphate and adenosine triphosphate. Energy from exothermic transformation is stored in high energy phosphate bonds of ATP which may be regarded as a charged storage battery. This energy may be released to drive endothermic reactions, at the same time transforming the ATP to ADP. Nicotinamide adenine dinucleotide (NAD) is a coenzyme often involved in hydrogen transfer reactions, which must be regenerated with a coupled reaction.

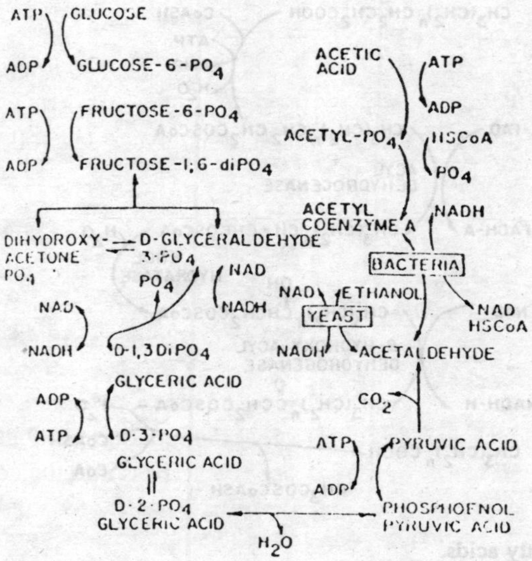

Fig. 14.2. Embden Meyerhol Parnas pathway of glycolysis.

In recent years, carbon 14 tracers have been used in substrates to aid in definition of the metabolic pathways. It is generally agreed that the anaerobic degredation of long-chain fatty acids proceeds primarily via beta oxidation, in which two carbon atoms at a time are split from the chain. The sequence shows the removal of one acetate unit which combines with reduced coenzyme A (CoASH). Acetic acid can then either be liberated from CoA in a subsequent reaction or the acetate can be transferred to other functional compounds. Flavin adenine dinucleotide (FAD) is a prosthetic group for a wide variety of enzymes (E-FAD) associated with hydrogen transfer reactions.

In nature and in anaerobic biological processes of waste treatment, it is suspected that proteolytic species of Clostridium are largely responsible for that anaerobic decomposition of amino acids. They can accomplish this in one of two ways. Some amino acids may be fermented individually by pathways specific to that compound. From their studies using carbon 14 tracers during the anaerobic digestion of glutamic acid, Weng and Jeris have postulated the mechanism shown in Fig. 14.4. The second method is the Stickland reaction whereby pairs of amino acids are fermented. One amono acid acts as the electron donor (is oxidized) while the other is reduced. As shown in Fig. 14.5, the NAD generated by the glycine reduction is used in the oxidation of alanine.[2]

Some other of the species of bacteria which participate in the acid production stage and have been isolated from anaerobic digesters are listed in Table 14.1.[3]

During these hydrolysis and fermentation steps very little C.O.D. is removed. Although some carbon dioxide is formed, the process consists largely of breaking down large molecules into one or two carbon units. The growth yield of fermentative cells is low due to the small net production of A.T.P. by substrate phosphorylation.

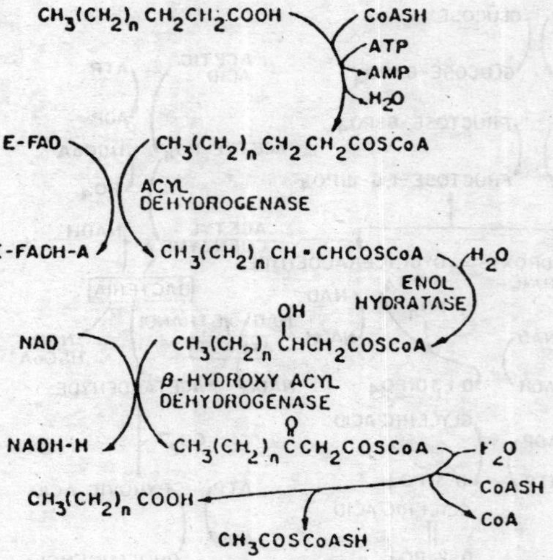

$CH_3(CH_2)_n CH_2CH_2COOH$
CoASH
ATP
AMP
H_2O
E-FAD
$CH_3(CH_2)_n CH_2 CH_2COSCoA$
ACYL DEHYDROGENASE
E-FADH-A
$CH_3(CH_2)_n CH = CHCOSCoA$
H_2O
ENOL HYDRATASE
NAD
$CH_3(CH_2)_n CH(OH)CH_2COSCoA$
β-HYDROXY ACYL DEHYDROGENASE
NADH-H
$CH_3(CH_2)_n C(O)CH_2COSCoA$
H_2O
CoASH
CoA
$CH_3(CH_2)_n COOH$
$CH_3COSCoASH$

Fig. 14.3. Beta oxidation of fatty acids.

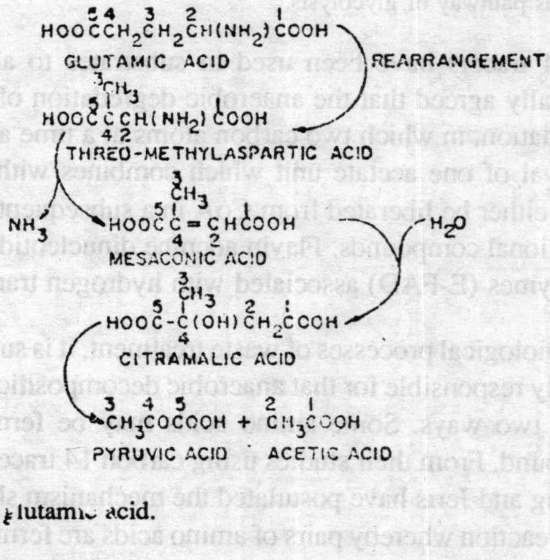

$HOOCCH_2CH_2CH(NH_2)COOH$
GLUTAMIC ACID
REARRANGEMENT
$HOOCCCH(NH_2)COOH$ (CH_3)
THREO-METHYLASPARTIC ACID
NH_3
$HOOCC = CHCOOH$ (CH_3)
MESACONIC ACID
H_2O
$HOOC-C(OH)CH_2COOH$ (CH_3)
CITRAMALIC ACID
$CH_3COCOOH + CH_3COOH$
PYRUVIC ACID ACETIC ACID

Fig. 14.4. Degradation of glutamic acid.

Methane Production Stage

The low-molecular-weight acids produced in the acid production stage are further degraded to methane and carbon dioxide by a highly specialized group of bacteria commonly referred to as the methane producing bacteria. These organisms have the unique ability to couple organic oxidation to reduction of carbon dioxide. In this process carbon dioxide is the terminal hydrogen acceptor and is analogous to oxygen in aerobic respiration.

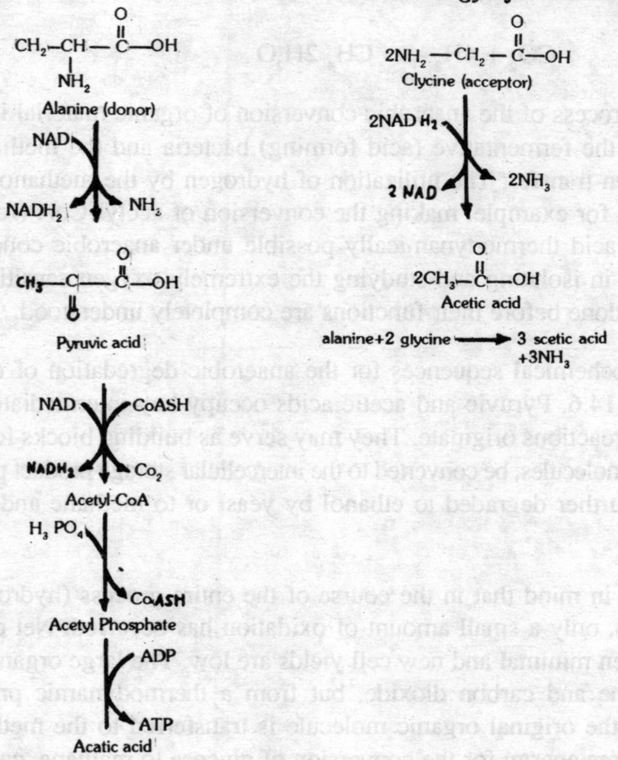

$CH_3-CH(NH_2)-C(=O)-OH$

Alanine (donor)

NAD

$NADH_2$

NH_3

$CH_3-C(=O)-C(=O)-OH$

Pyruvic acid

NAD

CoASH

$NADH_2$

CO_2

Acetyl-CoA

H_3PO_4

CoASH

Acetyl Phosphate

ADP

ATP

Acatic acid

$2NH_2-CH_2-C(=O)-OH$

Clycine (acceptor)

$2NADH_2$

2 NAD

$2NH_3$

$2CH_3-C(=O)-OH$

Acetic acid

alanine+2 glycine ⟶ 3 scetic acid $+3NH_3$

Fig. 14.5. The Strickland reaction.

Four genera of strictly anaerobic bacteria are known to produce methane :

Methanobacterium, a non-spore-forming rod,

Methanobacillus, a spore-forming rod,

Methanococcus, a non-spore-forming coccus; and

Methanosarcina, a non-spore-forming coccus in packets of eight.

We used to believe that a variety of low-molecular-weight alcohols and acids could be used as substrates by particular methanogens. More recent studies suggest that the cultures previously used were not pure and that the methanogens are able to utilize only a few substrates. All are able to form methane from hydrogen and carbon dioxide, and a few are limited to these substrates. Others are able to use formic acid, and at least two species of Methanosarcina are able to form methane from methanol or acetic acid.

In the absence of hydrogen the reaction is :

$$^{*}CH_3COOH \longrightarrow {}^{*}CH_4 + CO_2$$

If hydrogen is available the carbon dioxide is reduced to methane:

$$CO_2 + 4H_2 \longrightarrow CH_4\ 2H_2O$$

In the overall process of the anaerobic conversion of organic material into methane there is an interaction between the fermentative (acid forming) bacteria and the methanogens in the form of interspecies hydrogen transfer. The utilization of hydrogen by the methanogens enhances the acid production stage by, for example, making the conversion of acetyl-CoA (formed in beta oxidation Fig. 14.3) to acetic acid thermodynamically possible under anaerobic conditions. Because of the difficulties involved in isolating and studying the extremely oxygen sensitive methanogens, much work remains to be done before their functions are completely understood.

The general biochemical sequences for the anaerobic degredation of organic compounds are summarized in Fig. 14.6. Pyruvic and acetic acids occupy key intermediate positions from which further biochemical reactions originate. They may serve as building blocks for the synthesis of more complicated organic molecules, be converted to the intercellular storage product poly-beta hydroxybutyric acid (PBH), or be further degraded to ethanol by yeast or to methane and carbon dioxide by the methane bacteria.

It must be kept in mind that in the course of the entire process (hydrolysis, fermentation and methane production), only a small amount of oxidation has occurred. Net energy generated in the form of ATP has been minimal and new cell yields are low. The large organic molecules have been converted to methane and carbon dioxide, but from a thermodynamic prospective much of the chemical energy of the original organic molecule is transferred to the methane. For example, the change in standard free energy for the conversion of glucose to methane, carbon dioxide and water

$$4H_2 + C_6H_{12}O_6 \longrightarrow 2CO_2 + 4CH_4 + 2H_2O$$

is 133 kilocalories per mole, while that for the oxidation of glucose to carbon dioxide and water :

$$6O_2 + C_6H_{12}O_6 \longrightarrow 6CO_2 + 6H_2O$$

is 689 kilocalories per mole.

The methane generated in confined conditions can be collected and used as a fuel, or simply destroyed by torching. In the natural environment (swamps, etc.) most of the methane generated in anaerobic sediments is used as a carbon and energy source by micro-organisms as if diffuses up into the aerobic surface layer.

Denitrification

In the absence of oxygen, some bacteria utilize nitrogen of the nitrate anion as an electron acceptor. The process of the bacteriological conversion of nitrates to nitrogen is called denitrification, and it proceeds in the following steps.

Nitrate (NO_3) —> Nitrite (NO_2) —> Nitrogen (N_2)

TABLE 14.1

Nonmethanogenic Bacteria Isolated from Anaerobic Digesters .[7]

Bacterium / *Isolated on*	*Cellulose*	*Starch*	*Protein*		*Lipid*
			Peptone	*Cesein*	
Aerobacter aerogenes					
Alcaligenes bookerii					
A. faecalis	X				X
***Bacillus* sp**					
B. cereus* var. *mycoides		X		X	
B. cereus	X	X	X	X	
B. circulans			X		
B. firmus			X		
B. knelfelhampi					
B. megateium	X	X		X	X
B. pumilis			X	X	
B. sphaericus			X	X	X
B. subtilis			X	X	X
Clostridium carnofoetidum	X				
Escherichia coli			X	X	
E. intermedia					
Micrococcus candidus		X			
M. luteus					X
M. varians		X	X	X	
M. ureae		X			
Paracolobacterium intermedium				X	
P. coliforme			X		
Proteus vulgaris	X				
Pseudomonas aeruginosa	X				
P. ambigua					
P. oleovorans					X
P. perolens					X
P. pseudomallei					X
P. reptilivora	X				
P. riboflavine	X				
***P.* spp.**	X	X	X	X	X
Sarcina cooksonii					
Streptomyces bikiniensis					

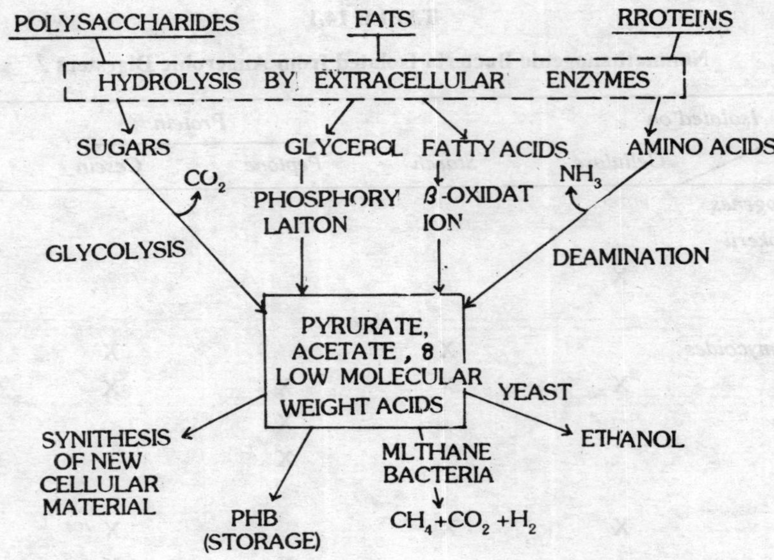

Fig. 14.6. Principal pathways of anaerobic degradation of organic compounds.

Some species form nitrous oxide gas (N_2O) from nitrate. This nitrous oxide may be further reduced to nitrogen[4] or it may escape to the atmosphere. Under acidic conditions nitrite may be decomposed chemically to nitric oxide gas (NO).

In most bacteria the enzyme, nitrate reductase, which is responsible for the transfer of electrons to nitrate is either not formed or is severely inhibited by the presence of oxygen. Denitrification is an important part of the nitrogen cycle, recycling to the atmosphere nitrogen which has been fixed by nitrogen fixing micro organisms. The reduction of nitrate accounts, in part for the lack of fertility of constantly wet soils supporting growth of anaerobic species and, in part, for the function of swampy lands as a nutrient sink.

If organic substrates are available, heterotrophic bacteria such as Pseudomonas denitrificans and Achromobacter carry out the denitrification process. In the absence of organic carbon compounds the autotrophic denitrifiers thiobacillus denitrificans and a facultative autotroph Micrococcus denitrificans predominate. All of the denitrifying bacteria can use oxygen as the electron acceptor if it is available.

Denitrification is also used as a terteriary treatment method, following nitrification, to remove nitrates from sewage treatment plant effluents as well as from high strength nitrate industrial discharges. The deletorious effects of releasing high concentrations of nitrates into surface waters include encouraging eutrophication and two potential health hazards in potable water; i.e., methemoglobinemia and the formation of carcinogenic nitrosomines when nitrate is injested.[2]

Sulfate Reduction

Sulfate reducing bacteria are able to use the sulfate radical as a terminal electron acceptor.

Unlike the denitrifiers who are able to use oxygen if it is available, they are strict anaerobes who can use neither oxygen nor nitrate as an electron acceptor. The sulfate reducers belong to two genera, Desulfovibrio, a Gram-negative non-sporeforming curved rod, and desulfotomaculum, a Gram and negative spore-forming rod. All can use organic compounds as carbon and energy sources, but some can utilize hydrogen as an energy source. Sulfate reducing bacteria may inhibit methane formation by competing with the methanogens for hydrogen.[5] These species are widespread in nature and are responsible for the odors in marine and brackish waters by forming hydrogen sulfide. Not only does hydrogen sulfide have an unpleasant odor, but it is highly toxic to plant and animal life. It has also been implicated in the corrosion of iron pipes and the crown corrosion of sewers.

Factors Affecting the Anaerobic Process

Widespread application of the anaerobic process has been hampered somewhat by its reputation as being easily upset and unreliable. Further development of anaerobic process technology is dependent on a better understanding of the factors associated with the stability of the biological process involved. Process instability is usually indicated by a rapid increase in the concentration of volatile acids, with a concurrent decrease in methane gas production, indicating that the more fastidious metha nogens are the most susceptible to upset.

Optimum conditions and ranges for anaerobic digestion have been studied by many investigators who are not always in agreement. One reason for this may be that their studies are conducted using different feed materials and different methodologies. The nature and composition of the substrate material dictate the microbial regime present, and it appears that a single set of parameters is not valid for all situations.[6]

Many process failures may be the result of insufficient acclimation periods. Acclimation of the micro-oganisms to a substrate has been reported to take more than five weeks, but sufficiently acclimated bacteria have shown considerable stability toward stress-inducing events.

Temperature

Digestion and gas production can occur over a wide range of temperatures as long as the temperature is relatively constant. Once a temperature range is established and the micro-organisms have become adapted, fluctuations can result in process upset. Rates of gas production versus temperature are shown in Fig. 14.7.[15]

Although most sludge digesters are operated in the mesophilic range (30-40°C), methanogenesis can occur as low as 4°C. While decomposition is considerably slower at lower temperatures, most of the degradable material will eventually be destroyed. Complex materials such as lipids and long chain carbohydrates are the most resistant.[7]

Thermophilic digestion (45-60°C) has the advantages of shorter solids retention times, increased digestion efficiency, better sludge dewatering characteristics and increased destruction of pathogenic organisms. It disadvantages are that the thermophilic bacteria are more susceptible to upset and they require more careful buffering. thermophilic operations must also provide superior mixing to ensure better heat distribution and more uniform feeding. The increased energy requirements for heating may be an additional disadvantage.

Denitrification was found to occur rapidly at 25° and 15°C, but at 4°C, it began only after a prolonged lag time.[4]

The optimum temperature for the yeast fermentation process is generally considered to be approximately 35°C.[8]

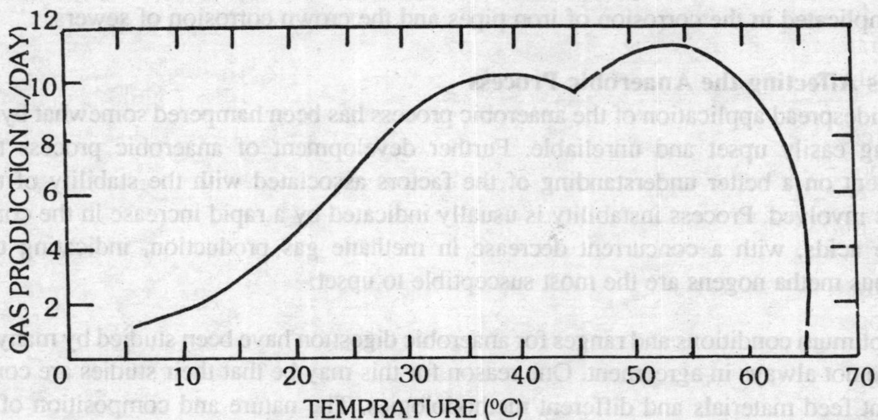

Fig. 14.7. Effect of temperature on gas production.

pH, Acidity and Alkalinity

Optimum pH ranges of 6.9-7.2, 6.4-7.2 and 6.6-7.6 have been reported for the methane bacteria. The non-methanogenic organisms are not nearly as sensitive and are able to function in a range of pH from 5 to 8.5. Process pH results from the interaction of the carbon dioxide-bicarbonate buffering system present with the volatile fatty acids and ammonia formed by the process. It is important that there be sufficient buffering capacity for the acids produced in order that they not lower the pH to a level upsetting to the methane bacteria.

There has been considerable debate concerning the toxicity of the ammonia and volatile acids themselves, independent of pH. Current thinking is that only the unionized volatile acids in the concentration range 30-60 mg/1 are toxic, and process inhibition by ammonia result from excessive concentrations of free ammonia rather than ammonium ions.[9]

Optimum pH for fermentation and ethanol production has been reported as pH 4.0,[8] and maximum denitrification is reported to occur at pH 7.[4]

Moisture

Moisture is required by all bacteria, but they are able to tolerate conditions raging from slight amounts of moisture to dilute solutions of nutrients. This means that very wet wastes can be used without energy consuming drying.

In studying the effect of various moisture concentrations in solid wastes, i.e., in landfills, it was found that maximum methane production occurred at moisture concentrations between 80 and 99%.[10] On the other hand, if the excess moisture (rainfall) contains dissolved oxygen, the methanogens will be inhibited.

Nutrients: Nitrogen, Phosphorus, Sulfur and Carbon

A microbial cell contains a C:N:P:S ratio of approximately 100:10:1:1.[22] C:N and C:P ratios of 25:1 and 20:1 respectively have also been suggested as appropriate for anaerobic digestion[4]. Therefore, in order for growth to occur, these elements must be present and available, and a scarcity of any one can be rate limiting. It is often necessary to add nitrates (or ammonium ions) and phosphates to industrial wastes[11].

At concentrations of 9 millimoles per liter all inorganic sulfur compounds, other than sulfate, inhibit both cellulose degradation and methanogenesis in the order.

thiosulfate > sulfite > sulfide > hydrogen > sulfide

Sulfate is inhibitory only in the respect that the sulfate reducing bacteria compete with the methanogens for hydrogen.

Many organic compounds may inhibit anaerobic digestion. These include organic solvents, pesticides, alcohols and high concentrations of long chain fatty acids[7].

The effects of the chemical structure of petro-chemicals was investigated, and it was found that chloro substitution, aldehydes and double bonds exhibited toxicity, while the addition of hydroxyl groups and increasing carbon chain length decreased toxicity. May compounds found to be toxic to unacclimated cultures were degraded after a period of acclimation.

Cations

All cations are capable of producing toxic effects in any organisms if the concentration are high enough, but the relative toxicity of the cation varies. Generally, toxicity increases with valence and with atomic weight. The results of studies on cation effects in anaerobic processes indicate that the methane bacteria show the same basic responses to the presence of cations as do other organisms; however, they exhibit a greater sensitivity to these effects than do the acid formers. Three basic effects are observed: toxicity, antagonism and stimulation. The toxicity of a cation can vary with the presence of other cations. This ability of one cation to increase or decrease the toxicity of another is called antagonism.

Low concentrations of cations have a stimulating effect on microbial metabolism. For example, although sodium, potassium, ammonium, calcium, and magnesium ions can be toxic, stimulation is achieved at concentrations approximately one-half toxic levels. Cations play a nutritional role in the metabolism of all organisms, serving as metabolic activators for a wide variety of enzymes and trace amounts are required. Popular theories all postulate interaction between cations and enzymes. Such interaction is suggested to produce stimulation when the correct metallic activator unties with the enzyme, but toxicity results when the enzyme unites with an improper cation. Antagonism can be described as a sort of competition between the functional and non-functional cations for the enzyme.[12]

Cation effects can be summarized as follows :

1. Cation effects in anaerobic digestion are a function of all the types and concentrations of cations present.
2. Concentrations lower or higher than optimum result in less than maximum efficiency.
3. Inhibition caused by excessive concentrations of any one ion can be antagonized (minimized) by the addition of optimum concentrations of another ion.
4. Maximum antagonism of an inhibiting cation can be obtained by the addition of optimum concentrations of several other cations.

The heavy metals are more toxic than the light metal ions, although at very low concentrations even extremely toxic cations such as lead and mercury produce stimulating effects. The toxicity of a heavy metal in anaerobic digestion depends on its chemical form, i.e., heavy metals in the form of precipitated sulfides are of little consequences to biological systems, and high concentrations of toxic heavy metals can be tolerated if sufficient sulfide can be provided to act as a precipitant.[4]

Rate-limiting Processes

Aside from limitations imposed by essential nutrient or co-factor deficiencies, there are four potential rate limiting steps in the conversion of cellulose to methane :

1. conversion of cellulose to soluble sugars by extracellular ezymes;
2. formation of volatile acids by acid forming bacteria;
3. conversion of volatile aids to carbon dioxide and methane by methanogens; and
4. transfer of dissolved products from the liquid to the gas phase.

The third step has often been considered to be rate limiting since decreasing the average microbial residence time has increased the volatile acid concentration. It was theorized that these build-ups occurred because of the slow metabolic and consequently slow reproduction rate of the methaogenic bacteria. These bacteria were thought to be largely displaced from the digester at solids retention times of two days or less. The evidence, however, does not distinguish between possibilities 3 and 4, or biological conversion and phase transfer as the rate limiting step.

Finney and Evans[13] postulate that the transfer of the gas phase is actually rate-limiting, and, further, that normal metabolic activity of the methanogenic bacteria is inhibited by product gases. In support of this hypothesis, their experiments using vigorous agitation at elevated temperatures and reduced pressures showed a substantially increased methane production rate. They suggest that microbubbles in the vicinity of the cell wall decrease the bacterial surface area available for the permeation process.

Studies conducted on the bioconversion of cellulose, however, indicate that the hydrolysis step may be rate limiting[10]. Apparently, depending on the conditions of substrate composition, mixing, and concentration of both substrate and micro-organisms, steps 1, 3 and 4 could all be rate limiting.

REFERENCES

1. Frutton, J.S. and Simmonds S., *General Biochemistry*, New York: John Wiley and Sons, Inc. (1959).
2. Gaudy and Gaudy, *Microbiology for Environmental Scientists and Engineers*, New York: McGraw-Hill (1980).
3. Bisseli, C. et. al., *Urban Trash Methanation—Background for a Proof-of-Concept Experiment*, NSF-RA-N-75-002, Mitre Corporation, McLean, VA (1975).
4. Gosh, S., et al., "Anaerobic Process—Literature Review," *J. Water Poll. Control Fed., 50* (10) (1978).
5. Kristjansson, J.K., et al., "Different K. Values for Hydrogen of Methanogenic Bacteria and Sulfate Reducing Bacteria, An Explanation for the Apparent Inhibition of Methanogenesis by Sulfur," *Arch. Microbiol.*, 132-278 (1982).
6. Clárk, C.S. et. al., "Laboratory Scale Composting Studies," *J. of Environ. Eng. Div., ASCE*, 104 (EE1) (1978).
7. Stevens, M.A. and D.D. Schulte, "Low Temperature Anaerobic Digestion of Swine Manure," *J. of Environ. Eng. Div., ASCE*, 105 (EE1) (1979).
8. Cysewaski, G.R. and Wilke, C.R. " Process design and Economic Studies of Alternate Fermentation Methods for the Production of Ethanol," *Biotechnol. Bioeng.*, XX (1977).
9. Hobson, P.N. and Shaw, B.G. "Inhibition of Methane Production by Methanobacterium Formicicium," *Water Res.*, 9:649-652 (1976).
10. De Walle, F.B. and Chian, E.S.K. "Kinetics of Substrate Removal and a Completely Mixed Anaerobic Filter," *Biotechnol. Bioeng.*, XVIII (1976).
11. Payne, J., "Energy Recovery from Refuse: State of the Art," *J. Environ. Eng. Div. ASCE*, 102 (EE2) (1976).
12. Kugelman, I.J. and McCarty, P.L. "Cation Toxicity and Stimulation in Anaerobic Waste Treatment," *J. Water Poll. Control FEd., 37*(1) (1965).
13. Finney, C.D. and Evans, R.S. "Anaerobic Digestion: The Rate-Limiting Process and the Nature of Inhibition," *Science*, 18 (1975).

15

Strategies for Selection of Appropriate Biotechnologies for Industrial Applications*

introduction

The new technologies of cell fusion, genetic manipulation and immobilised cell, enzyme systems, and bioreactors have dominated the headlines and research in recent years, not only because of their novelty and potential for innovative research, but also because of the incredible advances. DNA sequencing possible at a rate of about 350 nucleotides per man year in 1975 can now be elucidated at a rate in excess of 1000 nucleotides per man week. A synthetic gene sequence can be prepared in a biologically active form at a rate around 100 bases per man week compared with 3-5 bases per man year only 10 years ago. Restriction enzymes, upon which the specificity of gene editing and splicing depends, were rare 10 years ago. Perhaps 12-15 were identified, though not available commercially. Today over 150 are described in the literature, about 80 being available commercially.

Such advances continue, and to date have given rise to a plethora of systems which in theory can be used to prepare commercial quantities of scarce materials at realistic cost. To some extent the established techniques of culture enrichment, yield improvement by mutation or simple attention to physiological conditions of growth, have taken second place. In developing a strategic approach to increasing yields of products, or utilising waste materials or tackling other biotechnologically significant problems, all technologies whether established, new or futuristic should be carefully evaluated.

Many of these technologies are complementary, sometimes supplementary or competitive, and details of their specific use are described in other chapters in this handbook. A good example of the alternative use of the technologies is presented by the problem of producing low calorie beer. The requirement is to remove small molecular weight dextrans from beer, such dextrans being refractory to fermentation by normal brewing strains of yeast. This problem has been overcome by the transfer of dextranase enzyme into yeast strains by both the techniques of cell fusion and gene cloning and

---

* Excerpts taken from the review published by W.J. Harris, Musselburgh, Scotland.

both approaches seem to have been equally successful. An alternative process is the immobilisation of dextranase, or dextranase-containing cells upon a column through which standard fermentation broths are finally percolated. The relative commerical value of these alternative approaches requires careful evaluation. In mammalian systems, where the aim is to immortilise cell lines secreting a particular product such as a monoclonal antibody, the current technique of cell fusion is now challenged by direct transformation of cells with viruses, oncogenes or oncogene fragments. Gene cloning, achieved by integration of genes into vectors by the Cohen-Boyer procedure, can now be compared with *in situ* gene amplification or DNA-mediated transformation and integration of foreign genes using transposons.

Each individual case must be considered on its own merits but it is possible to lay down some guidelines and this chapter aims to provide an indication of the necessary analysis. A major factor in such an analysis is the state of art within a particular technology and the third part of the chapter summarises the current state of development of various techniques.

Preliminary Assessment of Feasibility of Project

All biotechnology projects are multidisciplinary in nature and require access both in manpower expertise and resources, to different scientific disciplines. There are few existing organisations with the capability in-house to take a project from inception to market. The processes concerned in the development of a genetically engineered pharmaceutical product are indicated in Table 16.1 and Table 16.2 illustrates the steps required to achieve commercialisation of a monoclonal antibody. Inevitably collaboration with a variety of other laboratories for clinical evaluation, toxicological or efficacy studies is required and such potential collaborators should be identified at an early stage.

TABLE 15.1

Processes Involved in Commercialisation of a Genetically Engineered Pharmaceutical

		Time (man months)
1.	Cloning of gene in laboratory micro-organism	9-18
2.	Sub-cloning of gene in organism to achieve commercial levels of yield	6-12
3.	Process development scale-up	12-24
4.	Large scale fermentation development	12-24
5.	Purification development	6-12
6.	Animal testing	9-48
7.	Clinical testing	24-48
8.	Regulatory approval	6-24
9.	Marketing product	6-12

TABLE 15.2

Process Involved in Commercialisation of Monoclonal Antibodies

		Time (man months)
1.	**Production of hybridomas**	4-9
2.	**Technical evaluation of commercial merit**	6-9
3.	**Robust assay development**	3-6
4.	**Clinical evaluation**	6-12
5.	**Routine monoclonal production**	
6.	**Kit assembly and quality control**	
7.	**Marketing product**	

1. **At least two primary screen assays should be used, one based upon immunoassay (ELISA etc.) and a "functional" assay. The likelihood of ultimate use of a cocktail of monoclonals should be borne in mind as well as the likely use of different monoclonals for separate processes.**
2. **Market survey should ideally be completed before embarking upon a project. Such a survey should take into account the "secondary" commercial potential of monoclonals as well as simple direct end-product use i.e. potential of monoclonals as research reagents for additional inhouse project development. Initial technical survey should establish whether monoclonals have desired specificity/sensitivity for routine use, i.e., lack of cross-reaction with related materials for diagnostic kits, or suitable reversible binding affinities for immunoaffinity purification or strong irreversible inactivation of antigen for detoxification/removal.**
3. **Initial assay systems used in the research laboratory are unlikely to be suitable for routine use in less sophisticated laboratories. In some cases simple modifications of ELISA, IRMA, etc., may be sufficient, but in others it may be necessary to develop simple paper-strip direct visualisation systems.**
4. **Routine monoclonal production either *in vivo* as ascites in mice or *in vitro* requires extensive animal facilities or continually operating well furnished tissue culture facilities and capability to purify the product. In the long term routine production by gene cloning in micro-organisms may be necessary.**
5. **Kit assembly and quality control must conform to appropriate Good Manufacturing Practice Regulations.**

Preliminary evaluation of the viability of a project can be divided into the two broad categories of technical and commercial assessment and both these categories must be assessed in terms of laboratory scale and commercial scale required to meet market requirements.

Table 15.3 summarises the factors which must be considered and it is apposite to expand on these to some extent:

Technical Feasibility

Laboratory Scale

This requires consideration of the state of the art with appropriate techniques and these are discussed separately in detail. Gene cloning or gene transfer studies requires a basic knowledge of genetics, mechanisms of transcription and translation within the proposed host cell, a technique to transfer DNA into these cells, a suitable vector to carry and express the information and a rapid and convenient method to measure the production of the product in question.

Lack of basic scientific knowledge is a severe limitation in many systems, particularly plant and algae. In micro-organisms, mechanisms for DNA transfer include conjugation, transduction, transformation, and transfection with phage particles assembled *in vitro*. The applicability and relative efficiencies of these procedures varies considerably between organisms, and readers should refer to appropriate texts.

The technique of most general applicability across a wide range of prokaryotes is uptake of purified DNA directly into receipient cells rendered sensitive by development of competence, treatment with $CaCl_2$ or conversion to protoplasts.

Competence. A particular physiological state developed by growth under carefully controlled conditions. Both Gram-positive and Gram negative bacterium can develop competence. Transformation frequencies with chromosomal donor DNA are of the order of 10^7 transformants/μg DNA but frequencies with large plasmids or phage DNA can be 10^3- 10^4 lower than this. Frequencies with small plasmids can be 10^7-10^8 transformants/μg DNA.

TABLE 15.3

Factors Concerned with Project Assessment

Technical	Laboratory scale. Commercial scale.
Economic	Product cost v. market value Capital outlay v. existing product(s) Feedstock costs v. existing processes Purification costs v. future processes
Safety Evaluation	
Social Factors	Pollution, fashion.
Political	Government attitudes, geography, multinational power.

$CaCl_2$ Treatment. This is generally the most acceptable method for gram negative bacteria with plasmid transformation frequenceis in *E. coli* of the order of 10^7 transformants/μg of DNA. Conversely to competence, frequencies with chromosomal DNA are only 10^3-10^4 transformants/μg DNA.

Protoplast Formation. This is the technique of choice for plasmid transformation of gram positive bacteria and microbial eukaryotes with transformation frequencies with *B. subtilis* as high as 4×10^7/μg DNA and with *S. cerevisiae* of the order of 10^4- 10^5/μg DNA.

Techniques for uptake of DNA into mammalian cells are compared in another section. A major stumbling block, particularly to agricultural research is the failure to date to introduce foreign DNA into mitochondria or chloroplasts.

Commercial Scale

The problems of large-scale fermentation are considered in other chapters but readers should be aware of these problems prior to the design of laboratory studies. It is clear for example that there are problems in the large-scale fermentation of *E. coli and B. subtilis,* two of the most popular organisms for laboratory gene cloning studies. An important aspect is the control of expression of a foreign gene. While in the laboratory it is reasonable to activate a foreign gene by linking it to an antibiotic marker such that the presence of the antibiotic causes expression, the large-scale use of antibiotics is severely limited by regulatory authorities. The general problems of scale-up include plasmid stability, transcription errors, stability of protein and cellular location of the foreign protein.

The achievement of high expression of a foreign product can cause considerable problems of purification. A product could represent 20 per cent of total bacterial protein of *E. coli* whose final concentration can be as high as 50 g wet weight of cells per litre in fermentation. In such cases, it is common for the foreign protein to exist intracellularly as an insoluble aggregate, resulting in problems of downstream processing. The ultimate solution is development of appropriate technology achieving secretion of the product from the host cell, and progress to this end is summarised later.

Economic Factors

The economics of large-scale production have been the subject of much review[2]. Evaluation of market value and market size requires detailed market analysis beyond the scope of this publication. Suffice it to say that it is essential to generate data regarding the annual requirement of product in terms of weight and purity, evaluate the cost of routine production and compare this with a realistic assessment of the likely market price. Capital cost of fermentation plant or large scale cell culture plant is estimated at \$25-60 per litre of installed capacity with an exponential scale-up factor of about 0.75, while routine running costs for a typical fermentation can be assessed at about 50 per cent for raw materials, 20 per cent for labour and 20 per cent as fixed costs.

The size of production must be sufficient to supply the proposed market and can vary from 1-5 m^3 for vaccine production, 20-100 m^3 for industrial enzyme production, 50-200 m^3 for antibiotics to 1000 m^3 -5000 m^3 for bulk commodities of single cell protein and organic chemicals such as ethanol, citric acid, etc. Hence the scale of a genetically engineered product, may need to be increased 10^5 or more. *Purification* of a product is currently a major factor in determining the success of a process, and requires solids/broth separation, cell disruption, extraction and purification. These have been extensively reviewed with much future emphasis based upon liquid-liquid extraction, 2-phase polymer extraction or chromatography, techniques that can be relatively easily scaled-up.

Safety Evaluation

The role of toxicology in evaluating the safety of a material is to allow establishment of a risk/benefit ratio. This cannot be achieved accurately but a satisfactory margin of safety can be assessed by taking into account the nature of the biological product, its properties, and effect upon the patient or beneficiary, researchers, surrounding personnel and environment. Safety evaluation can be an expensive and time-consuming phase of development, the degree of study required varying considerably depending upon the use of the product. Several reviews of requirements can be consulted. A major role for the new biotechnologies will result in the development of new processes for an existing product. Table 16.4 indicates likely costs and time required to complete adequate safety evaluation.

One example which illustrates the wisdom in a researcher appreciating the ultimate safety requirements for a product can be cited. The immediate application of genetic engineering has been the development of human products such as insulin, growth hormone and interferon. They provide examples of the use of alternative technologies as discussed in the proceeding section. However, these alternative sources of origin are important also from the point of view of safety of extracted product. Traditionally, such materials have been extracted from normal human or animal tissues, and have therefore been regarded as originating from a "safe" source. Recent fermentative production from *E. coli* requires clear demonstration of the lack of bacterial endotoxin associated with extracted material, while purification from mammalian cell lines in culture raises several considerations. Most cell lines are derived from tumors and in many cases, the increased growth of these cell lines is demonstrably the result of expression of viral genes. B-cell lymphoblastoid cell lines for example express Epstein-Barr Virus antigens. The presence of retroviruses and oncogenes must be considered a factor for concern. It has been suggested that as many as one in every 1000 cellular genes is a retrovirus gene. Although many virogenes may be defective, highly regulated and expressed only infrequently, they may be concerned in the aetiology of some human diseases. Treatment of product with 2.5 k rad. of radiation as well as treatment with deoxyribonuclease and ribonuclease may eliminate these possibilities. Nevertheless, the choice of host for a product should consider various microbial organisms and a variety of potential mammalian viral vectors and host systems.

TABLE 15.4

Costs of Safety Evaluation of Products

	New Products *$(million)*	*Existing Product/ New Process* *$(1,000)*
Pharmaceutical:		
(a) terminal/ occasional use	0.5-0.6	10-25
(b) repeated use	1.2-1.5	50-100
Food additives	1.2-1.5	20-40
Agriculture	1.2-1.5	10-25
Cosmetic	0.02-0.15	5-10
Chemical	0.03-0.30	5-10

Case Studies

The enormous breadth of projects which can be tackled in biotechnology precludes any attempt at an exhaustive treatment of case studies. Indeed every problem will carry unique features which make one approach more sensible than another. However, to illustrate that for every problem there are likely to be several approaches possible, it is pertinent to consider three general problems.

Mammalian Protein Products

Examples : hormones, interferons, growth factors, immune modulators.

Potential technologies available are summariesed in Table 16.5. In most cases the fundamental choice is large scale production from either mammalian cell cultures or microorganisms and the section on Current Technologies summarises the capacity of current techniques. A good example is provided by interferon which can be produced from fibroblastic cells grown in monolayers, from lymphobastoid cells in suspension or from *E. coli, B. subiilis* or *Saccharomyces*. Based upon a natural yield of 10^6 units/mg protein from fibroblasts, one could envisage a yield of 4 x 10^6 units/litre with cells immobilised upon beads. Suspension cultures of lymphoblastoid cells with production capacity of 10^5 units/mg could yield 7 x 10^7 units per litre of culture. Recombinant techniques, currently yielding 10^7 units/mg protein could provide 10^{10} units per litre fermentation broth. Application of gene amplification or gene cloning within mammalian cells could possibly increase yield from mammalian cells to 10^7-10^{10} units/litre.

A comprehensive list of cloned eukaryotic genes has been prepared by Davies (1982).[3] Random scission of mammalian DNA (3 x 10^6 kb per haploid genome) will produce 600,000 different fragments of average size 5 kb which is a convenient size for cloning. Strategies for such shot-gun cloning can be consulted. Cloning via C-DNA is the preferred route particularly if the gene contains introns. Various methods of DNA cloning are described by Maniatis including cloning vectors constructed to permit direct expression of cloned DNA segments.

Gene specific mRNA can be partially purified by immunoprecipitation of microsomes with appropriate antisera or using a synthetic gene probe. Providing it is possible to purify a few micrograms of the desired protein by polyacrylamide gel electrophoresis, a 5-15 amino acid sequence from the N-terminal can be determined, sufficient to dessign synthetic probes, the specificity of which will depend on the level of ambiguity due to the degeneracy of the genetic code.

Potential and sensitivity of remaining techniques are discussed in proceeding sections.

End Products of Mocrobial Pathways

Examples : Vitamins, antibiotics, ethanol and organic acids.

Traditionally such products are produced by batch liquid fermentation. Yield improvement or production cost can be reduced either by a new process of continuous culture or immobilised cell systems, or by increasing the yield per cell of product. Table 16.6 summarises the potential techniques. A comparison of batch and continuous culture fermentation is expanded in the section of Mixed Culture Fermentation and is the topic of much discussion. One point which must be borne in mind is that continuous production requires continual processing and purification if the product is unstable. Also, if two products are being produced simultaneously, since it is unlikely that both products will have equivalent continual sales, control of production is more difficult with a continuous system. It is perhaps easier to increase batch size or run more batches than increase line sizes throughout in continuous systems. Generally, therefore, the continuous process would be more attractive for products with a highly predictable market and good stability. The attractiveness of immobilised cell systems is expanded in the section about in situ Expression from Immobilized Systems, and one example is the successful continual production by the Kyowa Kakko Kogyo Co.,

TABLE 15.5 Mammalian Protein Products.

I Extraction from Natural Source	II Mammalian Cell Culture	III Production in Microorganisms	IV Chemical Synthesis
1. Raw Material: Low cost High process cost Limited supply Variable supply	1. Capacity: 8-10,000 litres Continuous production possible Infection potential high Media cost high	1. Unlimited capacity: Continual production possible Inexpensive growth medium	1. Gene synthesis:
2. Purification: Complex Expensive	2. Purification relatively expensive	2. Purification relatively inexpensive	2. Peptide synthesis (up to 50 amino acids)
Technologies for Increasing Yield	Technologies for Increasing Yield	Technologies for Increasing Yield	
None	Gene amplification Self-cloning on mammalian vector Gene transfer to alternate cell source by DNA mediated transformation Immortalisation of cell culture by (a) cell-cell fusion (b) viral/oncogene transformation	Shotgun cloning cDNA cloning	

Japan, of ethanol from live yeast cells (Saccharomyces) entrapped within calcium alginate beads. A pilot plant consisting of two fluidised bed column reactors in series (total volume 4 kl) has been producing 2.4 kl of ethanol a day continually since September 1982. Sucrose conversion yield is 95 per cent with a production rate of ethanol of 20 g. litre^{-1} at a concentration of 9 per cent (v/v). This is 20 times higher than that from conventional batchwise fermentation.

TABLE 15.6

Techniques for Increasing Yield of End Products of Metabolism

(i)	Process Development	
	Liquid culture fermentation	batch
		continuous
	Immobilized cell production	
(ii)	Strain Improvement	Potential Yield Increase
	Random mutation and selection	10-20% per process
	Protoplast fusion	2-5 fold
	Self-cloning	5-20 fold
	Vector-medicated gene transfer	10-100 fold
	Transposon-mediated gene transfer/amplification	5-20 fold

Mutation is a stepwise process with only minor yield increase likely at each stage. Where a wide variety of related strains exist, protoplast fusion is the technology of choice. Both protoplast fusion and mutational studies are possible with little knowledge of the genetics and biochemistry of the metabolic pathways leading to end product. On the other hand, self-cloning and gene transfer require a good knowledge of the metabolic pathway as well as plasmids suitable for conversion into vectors. Careful appraisal must therefore be made whether commitment to understanding reaction sequences is justified.

Utilisation of Complex Feedstocks

Examples : Industrial wastes, agricultural residues, etc.

Potential techniques are listed in Table 15.7. Conceptually the most efficient process would seem the sequestration of a number of genes capable of utilising all materials within the medium into a single microbial organism. In practice, this is rarely possible and indeed arguments against a single microbial organisms can be made (see under the section on Mixed Culture Fermentation). The establishment of a stable mixed culture community where individual organisms utilise different components of a mixture can be a slow process. Because of interactions among the organisms themselves, only a few strains grow rapidly consuming initially the easily assimible components of the media. Once these are exhausted, other strains grow on less digestible components. Substrate utilisation is therefore sequential. However, this system can adjust to fluctuations in the relative mixtures of materials and additional advantages are discussed later. *In vivo* genetic manipulation can be approached where the media components are substrates for genes carried upon self-transmissible plasmids, though such studies generally still result in a community of two or three species. Protoplast fusion between three or more species or sequential fusion across species barriers can in theory result in the gradual accumulation of various activities within a single organism, and examples of diversification of an organism's flexibility in feedstocks are described later. A mixed microbial community in suspension culture or in direct contact requires mutual freedom from potentially deleterious products of each other. A possible approach to allow co-metabolism of complex mixtures with a community of organisms which do interfere with each other is the immobilisation of cell lysates or permeabilised cells such that required biodegradative reactions continue while preventing the generation by one species of products which interfere with the metabolic capacity of a second.

Current Technologies

The following section considers the state of the art of 14 techniques of relevance to biotechnological research. The pace of advances and the intensity of research makes it impossible to present all but a superficial consideration of each topic. No attempt has therefore been made to achieve an extended cover of the literature. References have been selected so as to provide a starting point for readers with emphasis placed upon recent books and reviews where possible.

TABLE 15.7

Techniques for Increasing Utilisation of Complex Feedstocks

Mixed Culture communities
In vivo genetic manipulation
Protoplast fusion
Mixed immobilised cell systems

Strain Enrichment and Mutagenesis

Mutagenesis still represents an important initial activity in improving the yield of a fermentation product. Step-by-step progress is necessary since each mutation step will generally only increase the yield of product 10-20 per cent. Because of this, a prime factor in mutation selection is the design of accurate and sensitive assay procedures capable of screening a population at large, and identifying slight yield improvements. Considerable refinement in mutation and selection techniques have occurred in recent years. Such stepwise improvement of *Penicillium* strains have increased the yield of penicllin over a thousand-fold over the last 30 years.

The essential feature in a successful mutagenesis improvement protocol is the selection of a number of mutagens capable of giving a variety of transition, transversion, missense mutations, as well as high mutation to lethality ratio and the recent technique of Awerbach and Sinskey is valuable in this respect. The frequency of mutation can be increased also either by inhibiting error-free excision-repair, or inducing error-prone repair, though at present the application of this technology to industrial strains is limited. Nitrosoguanidine preferentially alkylates replicating segments of DNA and in this way introduces clusters of mutations at these sites. This results in the phenomenon of co-mutation whereby if MNNG-induced mutants are selected for a particular marker, the mutation frequency in nearby genes can be up to 20 times higher than in unselected clones.

The detection and isolation of mutants is often dependent upon the ingenuity of the investigator in the design of chemical analogues suitable for exerting selection pressure, but where this is not the case, the process becomes more random. In this case, to minimise effort, statistical and automated protocols have been proposed.

With the advent of the sophisticated techniques discussed below, mutagenesis *in vitro* can be applied to achieve site-directed mutagenesis which permits the deliberate insertion of nucleotides of choice at any site within a cloned gene. This is expanded upon in the section about Oligonucleotide and Peptide Synthesis.

Cell/Protoplast Fusion

The ability of fuse cells *in vitro* was for many years limited to mammalian cell systems with Sendai virus as the fusing agent. With the disovery that polyethylene glycol could effect cell fusion, the potential for wide-spread fusion of cells and protoplasts was realised. Today, protoplast fusion is one of the most powerful and promising techniques for genetic manipulation of industrial micro-organisms, and is the principal technique in the generation of monoclonal antibodies.

The aim of cell fusion of micro-organisms is principally the generation of organisms expressing new and novel products, increasing the yield of products, or combining the desirable attributes of several organisms into one. Fusion of mammalian cells has the main purpose of immortalising a cell line synthesising a valuable product. It is pertinent therefore to consider them separately.

Micro-organisms

Protoplast fusion has been applied successfully to fungi, yeast, bacteria, Streptomyces, Micromonospora, and a variety of interspecific, intergeneric crosses have been described : Fusions have also been achieved between yeast protoplasts and hen erythrocytes, Azotobacter and fungal protoplasts, Cyanobacteria and tobacco protoplasts, plant/algal chloroplasts, and a variety of cell lines, mammalian mitochondria and yeast. Cell fusion within fungi have been reviewed by Elander[4] describing successful fusions of industrially important strains of *S. cerevisiae, S. pombe, Aspergillus nidulans, Acremonium chrysogenum, Candida tropicalis, Arthrobacter luteus and Cytophaga.*

Frequency of viable hybrid formation in interspecies hybrids is variable from 5×10^{-4} with yeasts to 1×10^{-6} in Brevibacterium, though clearly such values are highly dependent upon individual handling of techniques for protoplast generation and regeneration of cells from the fused protoplasts. Recombination frequencies within viable hybrids are high (1×10^{-3}) and involve a large proportion of the genome and long stretches of DNA. Also, genetic exchange can occur at high frequency between more than two parents. It is not essential therefore to introduce selectable markers into parental strains, a decided advantage in using industrial strains.

Successful applications include the selection from fusions between Cephalosporium strains, of a breeding strain showing a 40 per cent increase in yield of cephalosporin C as well as increased growth rate and sporulation rate. Other reports have included the transfer of genes for the metabolism of malic acid from *S. pombe* to *S. cerevisiae* to improve flavour during wine fermentation, and the selection of strains which have reduced the fermentation time for Saki production from months to days.

Protoplast fusion is also providing a wealth of information on genetics of a variety of organisms including *B. subtilis, B. megaterium, Brevibacterium flavum, Providencia alcalifaiciens, S. aureus, Lactic streptococci* (for review).[5]

Mammalian Cells

Techniques for the fusion of mammalian cells are well established. The major commercial use of this technique is the fusion of myeloma and spleen or β-lymphocyte cells in the production of monoclonal antibodies.

Mouse/mouse, rat/rat, hybridomas are well-established as the favourite systems for production of monoclonals for diagnostic purposes and immunoaffinity purification purposes, with rat products beginning to gain more favour due to the higher yield of antibody from hybridoma cells and the greater production capacity in the rat. Cell fusion studies using human lymphocytes and myeloma cells have been reported but techniques are poorly reproducible at the present time.

The continued use of cell fusion for the generation of monoclonal antibodies, compared with immortalisation of cells with viruses or oncogenes will greatly depend upon further development of both these technologies and should be carefully evaluated. Recent developments in hybridoma technology include fusion of cells by electrofusion, claimed to be 10^4 more efficient than fusion using polyethylene glycol and the use of APRT-myeloma cells. *In vitro* immunisation of spleen or lymphocyte cell cultures by exposure to antigen for 2-4 days in the presence of less than 10 $\mu g.ml^{-1}$ of antigen offers a more rapid route to hybrid formation and requires ten to twenty-fold less antigen compared with *in vivo* immunisation.

Vector Expression within Micro-organisms

The use of particular micro-organisms as hosts for foreign genes and the production of large quantities of foreign or endogenous products requires either (a) the host organism possesses an endogenous plasmid capable of conversion to a vector, or, (b) the host will maintain stably a vector derived from a separate species either directly or after assimilation of parts of the host genome into the foreign vector. The extent of fulfilment of these requirements in different species is summarised below. Earlier general summaries of microbial vector systems can be found in a number of publications. A variety of vector systems have been lodged with the American Type Culture Collection.

E. coli

The development of *E. coli* is well advanced and innumerable vectors have been developed. Many books and reviews have been published describing the cloning and successful expression of a variety of prokaryotic, microbial eukaryotic, mammalian, viral, plant and human products.

Shuttle vectors, capable of propagation in both *E. coli* and alternative hosts have been described and a number are deposited in the American Type Culture Collection. A variety of operator/promoter systems to achieve low and high expression of foreign products have been described including, *lac*, *trp*, lambda promoters and bacteriophage MS-2 and fd. Of these, only *lac* and *trp* should be considered suitable for scale-up purposes. Systems under *lac* control are induced by the addition of IPTG, while *trp* systems are activated by depletion of tryptophan from the growth medium. The *trp* promoter is ten-fold stronger than *lac* and most high expression systems employ hybrids of the *lac* operator and one or two *trip* promoter regions. The final yield of protein produced in *E. coli* can approach 20 per cent though proteolytic digestion reduces this to 5-10 per cent. Proteolytic cleavage can be reduced by the use of *lon* mutants as host though such cells filament.[6] Nevertheless, final yields of foreign product are usually 100-200 mg per litre.

B. subtilis

Prokaryotic, eukaryotic, mammalian, and viral proteins have all been expressed in *B. subtilis* though the level of expression described for *E. coli* has not yet been described. There are plasmids endogenous to Bacillus species but they do not contain easily identifiable genet:c markers. Hence chimeric plasmids are constructed using tetracycline or chloramphenicol resistance markers from *S. aureus* or *E. coli* vectors or based upon the bacteriophages of *B. subtilis* carrying a *thy*$^+$ into a *thy*$^-$ host cell (thereby restoring the ability of the cell to grow in the absence of thymine.[7] DNA can be introduced by transformation into cells or protoplasts, the latter being considerably more efficient with up to 80 per cent efficiency of transfer into cells having been described. Current progress in *B. subtilis* cloning systems have been recently summarised in a book which adequately reviews the molecular biology of this organism.

A major problem with *B. subtilis* is the instability of vectors due to accumulation of deletion mutations though Barstow et. al.[8] have recently described a series of plasmids which are apparently stable. Relatively little success has been achieved in reducing the proteolytic activity of this organism due to a multiplicity of proteases which are essential for cell growth and sporulation, though direct secretion of product is possible (see later section). Nevertheless, *B. subtilis* remains one of the most promising gene cloning systems potentially suitable for use of the intact living organism in the environment.

Yeast

S. cerevisiae (a budding yeast) and *S. pombe* (a fission yeast) have been extensively studied as hosts for foreign products *S. cerevisiae* in particular has been used to successfully express bacterial, viral, and mammalian products. Yeast cannot process mammalian introns or recognise mammalian promoters nor carry out the correct glycosylation. Hence cloning is through C-DNA, commonly using the yeast PGK-1 (phosphoglycerokinase) gene promoter for expression. Three types of vectors have been developed : Type I, transforms at low frequency but results in stable integration of the vectors ; Type II, transforms at high frequency (20,000 transformants ug^{-1} DNA) with the vector maintained as an autonomous plasmid by inclusion in the vector of all or part of the yeast 2 um plasmid. Stable maintenance of this vector as 5-10 copies per cell requires continual selection pressure; Type III is maintained as an episome by the inclusion of a yeast sequence carrying a chromosomal replicon, e.g., *trp*-1 or *arg*-4 regions.

The fission yeast *S. pombe* can maintain vectors based upon *S. cerevisiae* but specific vectors suitable for cloning foreign and endogenous genes in this organism have been described.[9]

Foreign products under production from *S. cerevisciae* cells in commercial quantities include α-interferon, β-interferon, bovine growth hormone and human hepatitis B antigen, while genetically engineered strains directly exploited include *S. cerevisiae* strains carrying dextranase for the fermentative production of low carbohydrate beer, and strains for the food industry possessing the "killer" characteer — a toxin that "self-cleans" the system.

Pseudomonas and Gram Negative Organisms with Natural Plasmid Transfer

Plasmids occur naturally in many species of micro-organisms and a considerable number are capable of natural transmission by conjugation.

Furthermore, many plasmids have an extensive host range and a single plasmid can replicate in host cells as taxonomically far apart as *Escherichia*, soil bacteria *Azobacter*, and stalked bacterium *Caulobacter*. Such plasmids therefore have considerable potential as vectors covering a wide range of bacterial species. The basic molecular biology of plasmids has been described by Broda,[10] and the development of vectors from wide host range plasmids has been summarised.[11] Some of the most versatile vectors described to date are the pKT series derived from plasmid RSF1010, a plasmid which encodes resistance to streptomycin and sulphonamide and can be co-transferred among different bacterial strains by conjugation within strains carrying additional conjugative plasmids. A model example of the use of these vectors is presented by the work of Timmis.[12] The *Pseudomonas* gene for the aromatic ring cleavage enzyme catechol-2,3-oxygenase was cloned into pKT vectors with resultant increase in the ability of the bacteria to biodegrade aromatics such as toluene and xylene. The use of such vectors to increase the rate of catabolic processes has important consequences in the design of bioreactor systems for efficient degradation of xenobiotics using immobilised cell/enzyme systems.

Additional examples of the use of wide host range vectors from gram-negative bacteria can be found.

In vitro manipulation as described above, gives the investigator control over the experimental system and generally yields predictable results. However, since these plasmids are self-transmissible, *in vivo* genetic rearrangements among plasmids are possible. Such work is considered in the following section concerned with mixed culture fermentation.

Streptomyces species. Streptomyces are the sources of the vast majority of antibiotics in current use and over the last few years techniques for gene cloning have been developed in the laboratory of Hopwood. Three types of vector systems have been developed and recently reviewed and the use of these vectors in increasing the knowledge of antibiotic biosynthesis is discussed.

Lactic acid bacteria. Genetic advances in lactic Streptococci and related organisms have been slow over the last years. Several characteristics essential for milk fermentation are linked to plasmids including lactose fermentation, casein digestion, citrate fermentation, and factors concerned in the production of diacetyl. Though vector systems have not yet been described for lactic streptococci, a number of systems exist for *S. sanguis and S. pneumoniae*.

Other bacteria. Thermophilic organisms are of considerable potential industrial interest and some studies indicating the presence of plasmids in such species are beginning to appear. The thermophile *Thermus thermophilus* possess a plasmid suitable for cloning purposes; *Rhizobium* possess large numbers of plasmids, as do *Listeria*, Corynebacterium and Brevibacterium, and shuttle vectors useful between *E. coli* and *Zymomonas* and *Clostridium* have been described, though they are often unstable and need special media to retain stability.

Gene cloning in *Cyanobacteria* is feasible with the recent insertion of a transposon conferring antibiotic resistance into a natural plasmid of *Anacystis nidulans* and a plasmid has recently been detected in a fresh isolate of anaerobic methanogenic bacteria. Methylotrophs are resistant to many antibiotics and selection of a vector system is difficult. However, a shuttle vector between *E. coli* and methylotrophic bacteria has been constructed and has been used industrially to increase the yield of single cell protein bacteria.

Other fungi. There is as yet no report of plasmid contrained genes which function directly in the production of any industrial product but other plasmid controlled functions have been reported in fungi since 1977, in organisms such as Podospora, Anserina, and Fusarium.[13] Composite plasmids which transform into, and replicate in *N. crassa* have been describe[14, 15]. While a transformation system is described for *Aspergillus nidulans*, suitable plasmids are not available. Plasmids have been described in *Streptosporangium*[16].

Transposon Mediated Gene Transfer

Transposons, designated Tn are lengths of DNA encoding several genes flanked by two copies of a specific DNA sequence in either the same or inverted orientation (IS sequence). They can be plasmid or chromosomally bound and can code for drug or heavy metal resistance. The entire element is translocated intact, independent of the host recombination system to a new site on the same or another genome often causing either gene activation or inactivation by insertion. In *E. coli*, for example, insertion of IS-I sequences reduces gene expression when integrated in either orientation, but integration of IS-2 in orientation II causes constitutive synthesis of genes 3-5 fold higher than the normal rate. In many cases, the element contains genetic determinants for its own translocation. Clearly then transposons may act as *in vivo* genetic manipulation elements and if a gene is introduced in a transposon it becomes as moveable as if it were in the hands of a genetic engineer.

A number of publications can be consulted describing the molecular, genetic, and biochemical properties of transposons. Transposons have been described in gram negative bacteria, *S. aureus*, *S. cerevisiae*, *Rhizobium*, *Drosphila*, *Zea Mayo*.

It it likely that any gene, if placed between two adjacent IS elements will behave as a transposon providing it is introduced into a host cell encoding the transposition apparatus, or the transposon used carries the required enzymes. In this respect transposons Tn7 of *E. coli* is particularly suitable having a high transposition frequency, low specificity of insertion sites, broad host range, and several restriction sites suitable for cloning. It also carries determinants encoding resistance to trimethoprin. Determinants carried by other transposons include resistance against heavy metals, sugar fermentation, and *E. coli* enterotoxin.

Transposon as Vector

Studies describing the use of Tn7 as a vector in gram negative bacteria, *E. coli, M. methylotrophus, and Pseudomonas* have been described recently. A foreign gene can be inserted into Tn7 within a vector and co-transformed into a host cell with a second helper plasmid to promote the transposition functions lost due to insertion of the foreign gene. Integration of the transposon into the chromosome stabilises inheritance of the foreign gene.

Few reports of the application of this approach to yeast and higher organisms have as yet appeared but the potential is indicated by the report that the alcohol dehydrogenase gene of maize is flanked by a repeat sequence, raising the feasibility of genetic engineering in monocotyledenous plants.

Transposon Mutagenesis

Transposons provide a universal technique for identifying and mapping genes and can give information on the size of transcriptional units. Transposon integrate at many sites in the chromosomes. Insertion within a gene causes mutation while insertion beside a gene has either no effect or modifies expression of adjacent genes. The frequency of insertion of a transposon at any single locus resulting in loss of activity occurs at rates of 5×10^{-3} to 1×10^{-4}. Isolation of defective mutants is therefore 10 to 100-fold more efficient than traditional analysis, with the primary selection of mutants achieved by scoring for inheritance of drug resistance carried by the transposon. In a similar fashion, it is possible to screen for over-production of an end-product by selecting all mutants using the drug-resistance marker. In this case, frequencies can approach 1×10^5 or 5×10^4 or better. Representative examples of transposon mutagenesis can be found in the following publication.[17]

Large-Scale Mammalian Cell Culture

Animal cell cultures can be grown an monolayers, upon microcarriers, in suspension culture, or in hollow fibers and have been the subject of recent reviews.[18, 19] The current status of plant cell culture has similarly been summarised. The method of choice at present remains suspension culture when possible, and suspension culture vessels have gradually increased in size until today vessels of 8,000 litre are available for the production of vaccines, growth factors, hormones, enzyme immunomodulators, etc. Anchorage dependent cells can be grown in suspension systems by attachment to glass beads or microcarrier beads, and fragility of cells to stirring can be reduced by micro-encapsulation. Typical cell concentrations achievable in different systems are compared in Table 16.8 and can be used as a basis for estimating potential yields of products.

A major factor in cell culture systems is the cost of growth media particularly animal serum, though with increasing knowledge of serum growth factors and their availability, serum replacement media are becoming available. In addition to hormones, the crucial factor seems to be a requirement to mimic serum lipids needed to preserve cellular membranes.

TABLE 15.8

Maximum Mammalian Cell Yields

	Cells ml
Monolayers	4×10^5
Microcarrier beads	4×10^6
Suspension culture	7×10^6
Immobilised cells (hollow fibres)	1×10^8

The great demand for commercial supplies of monoclonal antibodies has stimulated much research in the last few years for the development of large-scale techniques for the production of these materials. Growth of hybridomas as ascites in host animals can give high yields of monoclonal antibody (10 mg.ml^{-1} or greater, equivalent to approximately 50 mg per mouse). However, immunoglobulins from such preparations are likely to be contaminated with 10 per cent of antibodies from the same species as the monoclonal as well as a high percentage of contaminating proteins. *In vitro* production is therefore preferable but standard yields are likely to be of the order of 100 μg.ml^{-1} in batch cultures. A simple method for continuous production of monoclonals from suspension cultures has recently been described by Fazekas St. Groth[20] and can provide yields of monoclonal up to 40 mg per day from a 500 ml culture vessel (equivalent to approximately 100 μg.ml^{-1}), though yields vary considerably between different hybridoma cell lines. This system has also been adapted to serum-free media, though the preparation of soybean lipid additive is complex and tiresome to perfect. Micro-encapsulation of hybridoma cells in calcium alginate gels (Damon "ENCAPCEL") prior to addition to serum containing growth medium allows growth of cells and accumulation of monoclonal antibody within the capsules thereby effecting a useful preliminary purification of antibody from serum proteins. Future developments based upon improved serum-free media and immobilisation of hybridomas should ultimately provide technologies for unlimited production of large quantities of highly pure monoclonal antibodies.

Recently, hybridomas secreting monoclonals have been immobilised in agarose beads and demonstrated to continually produce antibody for one week, and a lymphoblastoid cell line continuously secreted interleuken-2 for 14 days.[21]

Gene Amplification

Gene amplification is a technique used by several biological systems to produce large amounts of specific gene products at particular phases of life cycles, and cells in culture can be forced to amplify genes for certain proteins by selective techniques. Thus, gene amplification may permit cells overcome deficiencies in essential enzyme activities resulting from exposure to enzyme inhibitors. Amplification of the gene for dihydrofolate reductase in response to exposure of mammalian cells to methotrexate and genes concerned in the biosynthesis of UMP have been described and recently the amplification of glutamine synthetase within Chinese Hamster Ovary cells reported (Wilson, personal communication). In most cases, the size of the amplified unit is quite large (up to 20000 kb) with up to 1000-fold amplification.

This technique can therefore be considered as a means of producing significant amounts of a protein. Yields of 20 per cent of total protein have been reported which, coupled with large scale mammalian cell growth may provide sufficient capacity to satisfy commercial needs of a particular product. Secondly, amplification can be used as a significant means of enriching the genome prior to cloning studies. In this regard, it is usually possible to isolate the amplified part of the genome at a high level of purity.

Third, if it is not possible to devise an appropriate selection technique to directly encourage amplification of the desired gene, one can consider linking the requisite gene to, for example, the DHFR gene *in vitro,* transforming into host cells and subsequently applying amplification pressure upon the DHFR genome, a procedure which also amplifies attached DNA.

Mammalian Cells as Hosts for Foreign Genes

Transfer of foreign genes into mammalian cells can be achieved in a variety of ways, some of which have been reviewed. Methods include :

(a) Enclosure within infective viral particles *in vivo* and transmission into the host cell by viral infection.
(b) Co-precipitation with calcium phosphate of metaphase chromosomes (chromosome-mediated gene transfer, or DNA fragments (co-transformation)).
(c) Enclosure of DNA fragments within liposomes.
(d) Microinjection of DNA.
(e) DNA insertion with electric pulses.
(f) *In vitro* assembly of DNA fragments with polyomalike particles (PLP).

These differing techniques result in the deposition of DNA within recipient cells either naked, vesicularised or precipitated. This affects the efficiency with which DNA can finally be integrated into the host chromosome to produce permanent cell lines expressing the foreign genotype. Table 16.9 compares published data of frequencies of stable inheritance of foreign markers. It is essential that frequencies exceed spontaneous mutation frequency if the technique is to be suitable for cloning unselectable markers. For most purposes, the calcium phosphate technique is adequate and simplest to perform. Microinjection is time consuming and only 1000-2000 cells can be injected at any one time. This technique is therefore best restricted to unique donor DNA samples representing only a few genes.

TABLE 15.9

Efficiency of Gene Transfer Techniques in Mammlian Cells

Transferring Agent	*Frequency of Stable Inheritance (mutants/cells/µg DNA*
Infectious virus	1×1^{-2}
Calcium phosphate	10^{-4} -- 10^{-5}
Liposomes	4×10^{-4} -- 1×10^{-5}
Microinjection	1×10^{-3}
Polyoma-like particles	6×10^{-3}
Electric pulses	7×10^{-5}
Spontaneous mutation frequency	10^{-6} -- 10^{-7}

Selection of cells is time-consuming, typically requiring many weeks for the final establishment of producing cell lines.

At present, the inserted gene is often present only as a few copies and poorly expressed, though this latter property is greatly determined by the host cell line used. However, this technique of co-transformation is important and has considerable potential if used in conjunction with gene amplification and *in vitro* linkage of foreign DNA to regulatory elements such as those discussed in the selection on Gene Regulatory Systems in Vectors. Furthermore, this technique is suitable for systems which lack suitable vectors and can give rise to mammalian cells expressing a product without involvement of tumor virus or oncogenes.

The technique of co-transformation has been described in several publications[21, 22]. When cultured mammalian cells are exposed to DNA the exogenous DNA is integrated in the chromosome of a sub-population of the cells at frequencies of 1×10^{-6} — 1×10^{-5}. As a selection technique, the required DNA is co-transformed with TK^+, $APRT^+$, or $HGPRT^+$ DNA, the host cell line being respectively deficient in these enzyme activities. Co-transformation frequencies is commonly 80-90 per cent and on integration of the DNA into the host cell line being respectively deficient in these enzyme activities. Co-transformation frequencies are commonly 80-90 per cent and on integration of the DNA into the colony/10^6 cells/40 pg of DNA have been reported. With genomic DNA therefore one could effect the transfer of a specific gene at a frequency of 1 in 10^6 per 20 µg of genomic DNA. It is feasible therefore to transfer single copy genes using the total genomic DNA as donor.

With unknown host cell lines as recipient, the best transformation process can be established using DNA from a vector carrying an easily measured enzyme such as chloramphenicol acetyltransferase or galactokinase and such vectors are available.

Mammalian Vectors

Progress in the development of eukaryotic mammalian vectors has been rapid over the last few years. Most vectors are based upon DNA or RNA tumour viruses and monographs describing the detailed properties of these virus groups are available. The commonest vectors are based upon SV-40, adenovirus, bovine papilloma virus, and RNA tumour viruses and a variety of systems using these have been collated recently. The salient features of the systems can be summarised :

SV-40

These vectors have been reviewed [23, 24], and a number of the most useful have been deposited in the American Type Culture Collection. A series of vectors based upon SV-40 genome have been constructed containing part of the *E. coli* vector pBR322 which allows growth and maintenance of the vector in *E. coli* for convenience. Into this basic structure have been inserted the mouse dihydrofolate reductase gene, bacterial kanamycin resistance, or *E. coli* xanthineguanine phosphoribosyl transferase, all of which are suitable selection markers for mammalian cells inheriting the vector. Inheritance involves integration of the vector into the host chromosome as one or a few copies making these vectors unsuitable for expression of large quantities of product. The length of the foreign DNA fragment cannot exceed 2500 base pairs, with replication of the vectors limited to either monkey or mouse cells (dependent upon particular vector construct).

Adenovirus

Adenovirus have good potential as vector systems being able to accommodate up to 30,000 base pairs of foreign DNA. Furthermore, these viruses selectively inhibit host cell protein synthesis late in infection thereby simplifying detection of products which can represent a large percentage of the final protein population. Adenovirus themselves will only grow in human cell lines, commonly Hela cells which can be easily maintained in suspension cultures. Recently, vectors consisting of Adenovirus SV-40 hybrids have been constructed which will grow in both human and monkey cells.

Bovine Papilloma Virus

The unique feature of papilloma virus which distinguishes it from other vector systems is its ability to persist as a stable, unintegrated multi-copy plasmid (= 100 copies) in mammalian cells. Integration of the viral genome is not necessary for the initiation or maintenance of papilloma-induced transformation in mouse cells. The viruses themselves are not successfully propagated in tissue culture, though vectors capable of simple maintenance in *E. coli* are available. Vectors of this type can accept 6-8 kb of foreign DNA, and have good potential for the production of products from continuous cell culture.

Retroviruses

Retroviruses are a diffuse group of unique structure. The general structure has striking similarities with transposable elements and indeed could be used for biotechnological purposes as described in that section. The genomes are found in multiple copies in chromosomes of most vertebrates, and are capable of inserting into many regions of the host chromosome causing either activation or inactivation of adjacent genes. They also have the ability to acquire and transduce foreign genes by incorporation of DNA into virus particles, transduction into host cells, and reverse transcription and recombination into the new host genome. This leads to intronless pseudogenes or "processed genes". Retroviruses have a capacity to accommodate about 7.5 kb of foreign DNA.

Vaccinia

Vaccinia, the virus currently in small-pox vaccines has been used recently as a vector for foreign genes, the ultimate aim being the possible use of a recombinant virus carrying a foreign antigen as a live vaccine for protection against the normal host of the foreign protein. In this respect successful studies have been reported in inserting the gene for hepatitis B surface antigen, and inserting influenza haemaglutinin gene. It seems possible to insert up to 25 kb of foreign DNA into this virus without loss of infectivity.

Non-virally Based Vectors

A large number of eukaryotic vectors have been developed that depend upon integration into host chromosomes. These include PBR322-based plasmids that carry the Herpes simplex virus TK-gene. Recently, to overcome the restriction on use of only TK^- cells as hosts for such vectors, a number of dominant selectable markers have been constructed containing for example the xanthin-guanine phosphoribosyl transferase gene (XGPrt) of *E. coli.* The inheritance of this gene permists cells to survive in mycophenolic acid-xanthine-HAT medium. Similarly, the inclusion of Tn5 phosphotransferase confers resistance to the antibiotic amino-glycoside G418. Lengths of DNA up to 40 kb can be placed in such vectors.

General Conclusions

On the basis of current studies, it is clear that three strategies for exploitation of mammalian vectors can be discerned. The use of infective viruses such as SV-40 and Adenovirus will allow a burst of production of a foreign protein late in infection, though since the infection cycle results in lysis of host cells, commercial production must anticipate batch culture technology. These DNA viruses can be used to establish continual cell lines expressing a foreign product but only one or a few copies of the gene are integrated into the host chromosome. High levels of product are therefore unlikely. Papilloma viruses would seem to offer the opportunity for continual high level of synthesis of production from a continuous cell culture process, while retroviruses are the only vehicles for high-efficiency stable integration of foreign genes into the genome without cell lysis or transforming the host cells.

In Situ Expression from Immobilised Systems

As discussed in the section on Preliminary Assessment of Feasibility of Project, a major factor in the cost of commercial production is purification of product from cell and fermentation broths. The development of immobilised systems in the form of enzymes, organelles, or intact cells could reduce these difficulties. The requirement to grow large quantities of anchorage-dependent mammalian cells has focused research on this problem over many years and this has been discussed in a previous section. The objective of most of the studies however has been the development of system to support rapid multiplication of cell number. Techniques are now being sought to immobilise cells which will undergo little or no cell division while continuing to synthesise large amounts of products. This problem is equally applicable to prokaryotes, eukaryotic microorganisms, animal and plant cells. The latter is described in other chapters in this handbook.

Immobilisation of single enzymes is a well-established technology and is the starting point for understanding the basic problems of immobilisation of biological systems. Many books and reviews have been published. In recent years, mild methods of immobilising prokaryotic cells to carry out sequences of reactions has been achieved, based principally upon the use of Alginates, agarose or acrylamide systems. In selecting appropriate supporting materials, emphasis is placed upon a number of factors described in Table 15.10. Such systems have been used successfully to immobilise a variety of microbial cells. Recent developments include immobilisation of *Cyanobacteria* to produce ammonia and hydrogen and *S. cereviscíae* for the continual production of ethanol discussed in the section about Case Studies.

TABLE 15.10

Properties of Supporting Materials for Immobilisation of Cells or Organelles

1.	High carrier ability
2.	Availability in quantity
3.	Low cost
4.	Easy scale-up
5.	Mechanical strength
6.	Safety of material

Useful industrial examples include the use of *E. coli* cells trapped in polyacrylamide for the stereoselective conversion of fumaric acid to L-aspartic acid developed by Tanabe Seiyaku. This system has a half-life of 120 days and is estimated to cut costs by 40 per cent compared with a process employing free cells in suspension. At present, it is still unclear to what extent immobilised systems will replace traditional fermentation. Immobilisation allows the use of dense microbial populations but accurate control of development to particular growth phases is more difficult, and this is a requirement for efficient production of some secondary metabolites.

Immortalisation of Mammalian Cells

Normal cells can be distinguished from cancer cells by a number of properties (Table 15.11). From a biotechnological point of view, the most valuable properties of tumour cells are their apparently infinite life-span, ability to grow in suspension and in soft agar, and form multilayered colonies on solid substrates. *The in vitro* transformation of normal cells by DNA/RNA viruses or chemical carcinogens to display some or all of these tumorigenic properties has been the subject of much research as a model system for *in vivo* tumorigenesis.[25] For many years such study was limited since (a) irrespective of source tissue, established cell cultures were always fibroblastic in nature, and (b) such cultures had lost expression of specialised or tissue specific products. In recent years however considerable progress has been made in development of procedures to maintain cell cultures which continually produce specialised products for several weeks, and epithelial cell lines can also be established and transformed. Furthermore, human cells, which proved refractory to *in vitro* transformation by procedures successful with rodent cells, can be transformed.

TABLE 15.11

Distinction Between Normal and Transformed Cells in Culture

	Normal	*Transformed*
Growth	As monolayers on solids substrates. Saturation densities 5 x 10 cells/cm^3	In suspension, soft agar colonies, multi-layered complexes on solid substrates. Saturation density 10^7 cells/cm^3
Lifespan	Limited e.g. 50 passages for human cell cultures.	Essentially infinite.
Phenotype	Gradually changes with senescence.	Relatively stable. Changes with variation of growth conditions.
Growth Requirement	Rich medium 10-20% serum.	Possibly 0-5% serum requirements.

While these results are far removed from the industrial aim of immortalising cell lines such that they will continue to synthesise specialised products, they have provided the background to current developments. At present, two techniques of immortalisation show considerable potential for achieving the above target.

The earliest report of immortilisation of cells which continually secrete a product was the *in vitro* transformation of human normal β-lymphocytes by Epstein-Barr virus. They continued to secrete polyclonal immunoglobulins. Progress in the development of this system for the production of human monoclonal antibodies has continued and a variety of cell lines secreting, e.g., IgM antistreptococcal carbohydrate A, IgG anti-tetanus toxoid, IgG and IgM anti-Rhesus antigen D, have been described. This system provides some historical data and pinpoints the current problems with such transformed cell lines, viz. relatively low yield of immunoglobulin (less than 0.5 μg.ml^{-1}) and loss of secretion of product after about six months. Attempts to circumvent this problem involve subsequent fusion of EBV-transformed cells with lymphoblastoid cell lines. The use of EBV is limited to cells which carry appropriate viral receptor sites i.e. β-lymphocytes. To extend the range of susceptible cells, EBV can be reconstituted with Sendai-virus envelopes and introduced by cell fusion. EBV receptors can similarly be co-reconstituted with Sendai virus envelopes and transplanted on to the surface of cells which can then be directly exposed to intact virus.

Recent advances in our understanding of retrovirus biology and oncogenes have provided additional viral transformation techniques. Several recent reviews discuss the use of transforming viruses and retroviral transforming genes. When a retrovirus infects a cell, its RNA genome is copied by viral reverse transciptase into a "minus strand" DNA, which can then be copied to form complementary double-stranded DNA—a molecule recognised as the provirus. This is linear and flanked by long terminal repeats (LTRs), a repeated sequence a few hundred base pairs in length.

The retroviral LTRs contain the promoter region which is the signal for activation of viral genes. Integration of the proviral DNA into the host chromosome can cause transformation in at least two ways: by directly inserting a viral oncogene that is expressed in abnormal quantities, or by inserting the LTR promoter region of the virus near a cellular oncogene and causing its expression. Abelson murine leukemia virus, for example, is a replication-deficient retrovirus that encodes a single protein responsible for its transforming activity, and will transform fibroblastic and haemotopoietic cells. Mouse β-lymphocytes have been immortilised with frequencies of 1×10^{-3} to 2×10^{-4}, a frequency which makes the technique competitive with cell fusion for the production of monoclonal antibodies.

DNA Transfection

Retrovirus carrying oncogenes transform cells with high efficiency, but the number of such viruses is small, and include Abelson murine leukemia virus, Moloney Sarcoma Virus of mice and Rous sarcoma virus, avian myeloblastosis virus, and erythroblastosis virus of chickens. No equivalent human virus has been described. It is now known however, that normal cells carry a variety of such oncogenes in the latent state *(onc genes)*, and these can be activated spontaneously, by viruses or by chemicals. Human cells may contain more than 16 such oncogenes, and during the last three years a variety of such oncogenes have been isolated and characterized. Furthermore, it has been demonstrated that such oncogenes can effect immortalisation of cell lines by transformation of the recipient cells with isolated DNA, by techniques reviewed in the section on Mammalian Cells as Hosts for Foreign Genes. The extent of transformation with oncogenes has been discussed by Cooper.[26] Efficient transformation activity has been reported from human bladder, lung, colon, and mammary carcinoma cell lines, primary tumour isolates and sarcoma cell lines, DNA from these sources efficiently transforming mouse cells lines with frequencies of 0.1 to 1.0 transformants per microgram of donor DNA. The immortalisation of mouse spleen cells by this technique has been reported by Kennet's group. They treated mouse α-lymphocytes with DNA from human leukemic cell line *Rep* and found that 30 per cent of the cell population were transformed, though no data regarding the production of immunoglobulin is available. Though this line of research is at an early stage with respect to industrial application, progress is rapid. The realisation that immortalisation of cellular oncogenes can be achieved through viral infection, gene amplification of chromosomal translocation, lays open the possibility of specific and directed activation of a cellular oncogene to achieve immortalisation without subsequently affecting the expression of synthesis of specialised products.

Secretion of Foreign Products from Cells

Since purification is a substantial part of the costs of manufacture, considerable research is underway to achieve secretion of products from cells.

Micro-organisms

Bacteria export a wide variety of proteins through cytoplasmic membranes using a process which has been highly conserved during evolution. A short signal pepide at the amino terminus directs the protein to the membrane, the peptide being removed during transport. Gram positive organisms with no outer membrane barrier secrete such proteins directly into the culture medium and hence represent the simplest system. For this reason, *B. subtilis* is currently the secreting host of choice and the requirements for secretion of a foreign protein have been described.[27] The secretion of interferon from *B. subtilis* has been described recently.[28] The genus Bacillus accounts for the majority of industrial enzymes present and is likely to dominate for some-time with representative psychropiles, mesophiles, thermophiles, alkalophiles, nitrophiles, and acidophiles all secreting extracellular enzymes. In gram negative organisms, foreign proteins become entrapped in the periplasm and the only way at present of achieviing release into the medium is the use of leaky mutants though research into the secretion of periplasmic proteins is intense.[6]

Secretion of foreign protein from yeast has been reported[29] and studies are in progress to achieve secretion of foreign proteins from streptomyces, which normally secrete proteins and secondary metabolites.

Mammalian Cells

Eukaryotic cells also have involved a common mechanism for transferring secretory and some membrane proteins, again using a short signal peptide.

In the case of secretory proteins, the entire protein is transferred across the cell cytoplasmic membrane while a hydrophobic "anchor" peptide causes membrane protein constituents to be trapped in the membrane structure. The construction of rectors to achieve secretion, or induced deliberate trapping of a foreign protein in the cytoplasmic membrane is at an early stage, but encouraging. Hybrid vectors consisting of SV-40 T-antigen linked to the haemaglutinin (HA) gene of influenza virus have been constructed lacking either the "signal" or "anchor" sequences of haemaglutinin. The "anchor-minus" HA was efficiently synthesised and was almost entirely secreted into the medium. The wild type HA remained cell associated while the "signal-minus" HA was only expressed at low levels.

Progress in these fields should be rapid and contribute to the development of the technology of immobilised cells continually secreting a specialised product.

Mixed Culture Fermentation

In recent years there has been a revitalisation of the interest in mixed culture fermentation systems, and it is becoming clear that the deliberate construction and establishment of stable microbial communities is a viable approach to many biotechnological problems. The value of naturally occurring microbial communities in waste disposal, milk products, and brewing industry is well-known and several excellent texts are available.[30, 31] In recent years, the deliberate construction of mixed cultures in the laboratories to achieve efficient biodegradation of xenibiotics have been achieved, and the principles involved in the design of such systems described. In most cases, the final stable culture consists of a microbial community of several (up to seven have been described micro-organisms. The strategies and techniques involved in such studies have been summarised by Bull.[32]

Mixed culture fermentation techniques are also applied to achieve *in vivo* genetic manipulation or plasmid evolution whereby strains possessing different plasmids encoding required attributes are grown under appropriate selection conditions to encourage the evolution of a recombinant plasmid encoding all the required attributes within the same organism. This technique was successfully applied by Chatterjee.[33] The degree to which the final stable culture consists of one or several organisms is unpredictable but clearly, the more complex the conversion, the lower the potential to generate a single micro-organism.

An example of a stable mixed culture is described by Flickinger[34] consisting of *Clostridium thermocellum* and *Clostridium thermosaccharolytic* for the conversion of cellulose to ethanol. Such a mixed culture can produce 32 g. litre^{-1} or ethanol from corn stover as substrate.

The laboratory establishment of mixed cultures and the application of environmental stress to achieve strain enrichment can be achieved either by batch or continuous culture. Evidence at present mostly favours the use of continuous culture systems. The relative values of these techniques have been reviewed. Continuous culture is a useful technique for the selection of catabolic systems in which there is constant selection for one or more organisms best suited to utilise available carbon and energy sources. The end result however is the selection of a culture system which will function optimally in the enriched environment and may deteriorate if exposed to more natural environments. Some problems with mixed culture systems from the industrial point of view are the difficulties of patenting (one needs to define all the organisms), and problems with freeze-drying the culture. In the latter case, certain species may be preferentially killed. Nevertheless, mixed cultures are advantageous when the feedstock is complex and variable since they yield higher productivity due to increased culture stability and greater resistance to contamination. Furthermore, the general nutrient requirements are likely to be less since individual organisms will feed each other.

Bioaugmentation, the addition of viable microbial formulations in liquid or dry from to start or maintain waste treatment systems, inoculate soil or silage, is becoming increasingly accepted. Genetic engineering and other technologies will permit the deliberate construction of organisms for this purpose, and an important feature of such developments will be the demonstration that such organisms can survive and establish a stable relationship within a natural microbial community.

Oligonucleotide and Peptide Synthesis

Chemical procedures for the solid-state synthesis of oligonucleotides and peptides have advanced dramatically in the last few years and can in some cases, offer a serious challenge to genetic engineering technology, while in other cases, they fulfill a valuable complementary role.

Synthetic Oligonucleotides

These can be used to synthesise a gene completely, produce restriction sites for cloning, hybridisation probes, primers for cloning, gene alteration, editing, mutation, and to construct analogues of genes which are analogues of the parent protein. Three basic chemistries are used and several instruments carrying out automatic synthesis are available commercially. The advantage of such instruments should be carefully evaluated particularly at the current high prices. The fastest machine may reduce the time for synthesis by about 30 per cent. A 20-nucleotide molecule may take two to three days to synthesise manually compared with 10 h in an efficient instrument. However, removal of protecting groups, purification, verification of nucleotide sequence can still take some two weeks. The largest molecules synthesised to date are 584 base pair gene for human growth hormone and 514 base-pair gene for α-interfereon. Both synthesis were achieved by the initial synthesis of oligonucleotides of 14-21 nucleotides in length. Today, one could envisage assembling a 500 base-pair nucleotide with an experienced staff of three to four in two to three months (equivalent to a peptide of 160 amino acids).

Site-directed Mutagenesis

The ultimate in mutagenic studies is to pinpoint a particular nucleotide within a gene and introduce any desired type of mutation at that site without any constraint upon the target. This is realised with site-directed mutagenesis.

In essence, this technology can be used not only to introduce a single base-pair mutation, but also to detect the presence of such mutations *in vivo* since a single base-pair mismatch can cause a difference of 10-15° C in Tm (melting temperature) of an oligonucleotide. The size of synthetic oligonucleotide constructed as a probe for mutant detection depends upon the complexity of the organism. An oligonucleotide will be unique if :

in *E. coli* it consists of 12 oligonucleotides
in *S. cerevisciae* it consists of 13 oligonucleotides
in Man it consists of 17 oligonucleotides

In practice the construction of a synthetic 19-mer is ideal in that a single base-pair change will be sufficient to destabilise the helix. The ideal annealing temperature can be calculated from the formula T- 2nAT + 4n GC °C, the ideal operating temperature being 2-5°C below this.

Synthetic Peptides

Solid state peptide synthesis can be performed both manually and automatically with peptide synthesisers available commercially.

With automation, several hundred miligrams of a peptide less than 20 amino acids in length can be synthesised in 2-3 days. Techniques for synthesis manually have been reviewed by Barany and Merrifield[35] and a list of synthesis to date is available.[36]

For small polypeptides (less than 50 amino acids) chemical synthesis and genetic engineering may be competitive. Despite recent advances, assuming a 95 per cent yield at each stage, then the synthesis of a 20 amino acid polypeptide will give a product of which only 34 per cent is the desired material. Purification is therefore essential, as well as facilities for amino acid sequence confirmation.

In addition to yielding proteins as the desired end product, synthetic peptides can be used to raise anti-bodies which will cross react with the complete protein should the latter prove impossible to obtain in pure form. While antibodies made to a macromolecular protein do not usually react with partial peptide chain, antibodies to the peptide do seem to interact with the parent macromolecule. In the same vane, a protein not previously identified but predicted from a DNA sequence, can be isolated by first preparing a synthetic peptide based upon the DNA sequences and using this peptide to raise antisera subsequently used to detect the intact protein.

Currently, great industrial interest in synthetic peptides lies in their possible use as viral vaccines.

Gene Regulatory System in Vectors

Cloned genes vary greatly in the efficiency with which they are expressed in both prokaryotic and eukaryotic systems. In many cases, the most efficient expression system has been constructed using highly efficient promoter operators derived from the host organism itself. These have been referred to previously as well as several additional reviews.

Two systems however have potential as wide host range expression systems viz., heat shock and metallothionine. *Heat shock* brings about the strong activation of many previously silent genes as well as repression of active genes. This appears to be a universal phenomenon and in most species is a response to stress. Such regulatory elements could therefore be isolated and serve to control expression of foreign genes in innumerable systems. The extent of this phenomenon has been detailed in a recent monograph.[37]

Metalliothionines are low molecular weight cysteine-rich proteins that bind to divalent heavy metal cations such as zinc, cadmium and copper. These proteins are present in all higher eukaryotes examined to date and are induced by the presence of a heavy metal ion. The regulatory systems of these genes have potential value in gene cloning in yeast, animals and plants. The metallothionine promoter/regulator system is similar in a number of species and a 60 base-pair sequence 5' to the structural gene has been identified as responsible for induction of expression. The most advanced study of this system is that described by Palmiter[38] wherein mouse growth hormone gene, within the *E. coli* plasmid pBR322, was placed under control of the metallothionine regulatory sequence. This was injected into the male pronucleus of fertilised mouse embryos and when the progeny mice were fed zinc they grew to nearly twice their normal size. An important observation was that growth hormone synthesis was restricted to liver and kidney, the tissue normally responsible for metallothionine synthesis. Progeny mice also expressed the growth hormone gene.

These developments clearly demonstrate that techniques suitable for precise control of foreign gene expression are near at hand. Systems equivalent to metallothionine but not based upon the use of toxic metals are under development.

Conclusions

An attempt has been made in this review to impart a flavour of the potential variables which impinge upon the successful development and completion of a bio-technological process. Examples selected and references cited have been selected from a large volume of excellent texts. Progress is rapid and undoubtedly during the next few years additional technologies will be discovered to extend further the choice available to the researcher.

REFERENCES

1. Willets, N. and Skurray, R., *Ann. Rev. Genetics*, *14*, 41-76 (1980).
2. Hansen, R.W., in *Issues in Pharmaceutical Economics*, R.I. Chien, ed., Lexington:Health (1979).

3. Davies, K.E., in *Genetic Engineering, 3,* R. Williamson, ed., Academic Press, London, pp. 144-162 (1982).
4. Elander, R.P., *Biotechnol. and Bioengineering, XXII,* 49-61 (1980).
5. Hopwood, D.A., *Ann. Rev. Microbiol., 35,* 237-272 (1981).
6. Mount, D.W., *Ann. Rev. Genetics, 14,* 279-319 (1980).
7. Young, F.E., *J. Gen. Microbiol., 119,* 1-15 (1980).
8. Barstow, D.A., Primrose, S.B. and Atkinson, T.A., *Plasmid, 9,* 273-285 (1983).
9. Beach, D. and Nurse, P. *Nature, 290,* 140-142 (1981).
10. Broda, P. *Plasmids,* W.H. Freeman and Co., Oxford (1982).
11. Timmis, K.N. and Puhler, A., eds. *Plasmids of Medical Environmental and Commercial Importance.* Elsevier/North Holland Biomedical Press, Amsterdam (1979).
12. Timmis, K.N., in *Genetics as a Tool in Microbiology,* D.A. Hopwood and S.W. Glover, eds., Cambridge University Press, Cambridge, pp. 35-47 (1981).
13. Grimaldi, G., Guardiola, J., and Martini, G., *Trends in Biochem. Sci., 3,* 248 (1978).
14. Hughes, K., Case, M.E., Geever, R., Vapnek, D., and Giles, N.H., *Proc. Nat. Acad. Sci., U.S.A., 80,* 1053-1057 (1983).
15. Stohl, L.L. and Lambowitz, A.M., *Proc. Nat. Acad. Sci., U.S.A., 80,* 1058-1062 (1983).
16. Fare, L.R., Taylor, D.P., Toth, M.J., and Nash, C.H. *Plasmid, 9,* 240-246 (1983).
17. Schlessinger, D., ed. *Microbiology, 1981.* American Society of Microbiology, Washington, pp. 107-138 (1981).
18. Brodelius, P. and Mosbach, K., *J. Chem. Tech. Biotechnol., 32,* 330-37 (1982).
19. Fowler, M.W., *J. Chem. Tech. Biotechnol., 32,* 338-346 (1982).
20. Fazekas de St. Groth, S., *J. Immunol. Methods, 57,* 121-136 (1983).
21. Nilsson, K., Scheirer, W., Merten, O.W., Ostberg, L., Liehl, E., Katinger, H.M.D., and Mosbach, K., *Nature, 302,* 629-630 (1983).
21. Wigler, M., Pellicer, A., Silverstein, S., Axel, R., Urlaub, G., and Chasin, L., *Proc. Nat. Acad. Sci., U.S.A., 76,* 1373-1376 (1979).
22. Wigler, M., Sweet, R., Sim, G.K. Wold, B., Pellicer, A., Lacy, E., Maniatis, T., Silverstein, S., and Axel, R., *Cell, 16,* 777-785 (1979).
23. Berg, P., *Science, 213,* 296-303 (1981).
24. Elder, J.T., Spritz, R.A., and Weissman, S.M., *Ann. Rev. Genetics, 15,* 295-340 (1981).
25. Pienta, R.I., in *Carcinogens. Identification and Mechanisms of Action, A. Clark,* Griffin, and C.R. Saw, eds., Raven Press, New York, pp. 121-141 (1979).
26. Cooper, G.M. *Science, 218,* 801-806 (1982).
27. Palva, I., Sarvas, M., Lehtovaara, P., Sibakov, M., and Kaariainen, L., *Proc. Nat. Acad. Sci., U.S.A., 79,* 5582-5586 (1982).
28. Palva, I., Lehtovaara, P., Kaarianen, L, Sibakov, M., Cantell, K., Schein, C.H. Kashiwagi, K., and Weissman, C. *Gene, 22,* 229-235 (1983).
29. Hitzeman, R.A., Leung, D.W., Perry, L.J. Kohr, W.J., Levine, H.L. and Goedel, D.V., *Science, 219,* 620-625 (1983).
30. Bull, A.T., and Slater, J.H., *Microbial Interactions and Communities,* Vol. 1. Academic Press, London (1982).
31. Bushell, M.E., and Slater, J.H. eds. *Mixed Culture Fermentations,* Academic Press, London (1981).
32. Bull, A.T., in *Contemporary Microbiol Ecology,* pp. 107-136 (1980).
33. Chatterjee, D.K., Kellogg., S.T., Watkins, D.R., and Chaukabarty, A.M. in *Molecular Biology Pathogenicity and Ecology of Bacterial Plasmids,* S.B. Levy, R.C. Clowes, and E.C. Koenig, eds., Plenum Press, New York, pp. 519-528 (1981).
34. Flickinger, M.C. *Biotechnol. and Bioengineering, XXII, Suppl. 1,* 27-48 (1980).
35. Barany, G., and Merrifield, R.B. in *The Peptides,* Vol. 2, E. Gross and J. Meienpiper, eds., Academic Press, London, pp. 3-84 (1980).

36. Galpin, I.J., in *Amino Acids, Peptides and Proteins*, Vol. 14, 338-373, Royal Society Chemistry, London (1983).

37. Schlessinger, M.J., ed. *Heat Shock: From Bacteria to Man.* Cold Spring Harbour Symp. New York (1982).

38. Palmiter, R.D., Brinster, R.L., Hammer, R.E., Trumbauer, M.E., Rosenfield, M.G., Birnberg, N.C., and Evans, R.M., *Nature, 300*, 611-615 (1982).

16

Fermentation and Selection

Introduction

While studying the structure, metabolism and genetic information of microbes, it is rarely essential to deal with large quantities of the organisms. A flask filled with about a litre of nutrients and growing microbes is generally sufficient. Biotechnology, however, sets out to produce large amounts of valuable materials — far more than even the most productive microbes can supply when grown in the confines of laboratory vessels. To create an efficient and economic biotechnological process it is generally necessary to scale up operations so that they work in huge metal vats. All the investigations of biologists who have discovered useful microbes, and the genetic engineers who have created new ones, would count for little in biotechnology if it were not for fermentation processes.

'Fermentation' in biotechnology means any process by which microbes are grown in large quantities to produce any type of material, not merely alcohol as in the common, restricted, sense of fermentation.

Success in biotechnology also depends crucially on discovering which type of microbe is best suited to large-scale fermentation processes and which is able to give the highest yields of the particular substance sought.

Even though the term biotechnology had not been coined at the time, the discovery and development of penicillin contains many of the most important elements which characterize biotechnology today. The story of how penicillin was discovered and how it came to be the first antibiotic manufactured in sufficient quantities to save the lives of millions, is, therefore, used to illustrate the general principles of selection and fermentation.

Discovery of Penicillin

In 1928, Alexander Fleming was working at St. Mary's Hospital, London, where he spent much of his time studying bacteria growing on small glass dishes in his rather untidy laboratory. One day in September he noticed that one of his dishes had been contaminated with a mould, *Penicillium notatum*. What really caught his attention was the fact that no bacteria had grown in the area around the mould (Plate 4). He surmised, correctly, that the mould had secreted some substance which inhibits the growth of bacteria, and he named the substance penicillin. Strangely, Fleming did not seize on this observation and follow it right through. He performed some more experiments to confirm his findings, and showed that the material was not toxic to mice. He did not, however, try to purify the active ingredient in the mixture of materials obtained from the mould, nor did he test the material's curative effects on infected mice.

In discovering that a certain type of mould makes some material that harms bacteria, Fleming supplied the first and crucial ingredient of most biotechnologies — he made an important scientific discovery about a property possessed by a living organism. Within a decade of 1928 the climate was ripe for his ideas to be picked up and put to practical use. Changes in scientific attitudes also encouraged the development of penicillin as a therapeutic agent, but two other factors had a greater influence: with the outbreak of the Second World War there came a pressing need to find an effective antibiotic to treat the wounded; and a group of scientists and technologists with diverse skills intent on bringing the project to fruition were brought together.

Three main strands of research and development that were intertwined in the penicillin story were (i) the best type of mould had to be found; the mould used in large-scale production of penicillin had to manufacture as much of the antibiotic as possible and be fairly easy to handle in large quantities. (ii) Methods for separating penicillin from all the other substances produced by the mould were required, (iii) it was necessary to design vessels in which the mould could be grown and from which the valuable harvest of penicillin could be recovered.

The first stages in the development of penicillin were carried out in Oxford by Howard Florey, Ernst Chain and their colleagues. They began work in 1938 and by 1940 American scientists had become heavily involved in the work. In total, several hundred scientists in thirty laboratories took part. This endeavour was pursued with such enthusiasm and skill that by 1944 there was enough penicillin to treat all the serious British and American casualties during the invasion of Europe.

Samples of the mould that happened to fall on Fleming's dish had been kept alive in St. Mary's Hospital. Unfortunately, this particular mould does not produce very much penicillin, so the search was on to find other, slightly different moulds that would yield more. By 1951, it had become possible to obtain 60 mg of penicillin from a 1-litre (0.22-gal) flask full of mould and liquid nutrients. Since then further improvements have been introduced which allow 20g of penicillin to be recovered from the same volume of material — 10,000 times as much as could be produced in Florey's laboratory.

New and sophisticated techniques for separating penicillin from the mass of other materials in the flasks played a vital role in this remarkably improved yield, but equally important was the work of scientists who managed to improve the mould. The word 'improve' in biotechnology is obviously anthropocentric; it means that some characteristic of an organism, in this case penicillin production, has been altered in a direction that suits our purposes. The change is unlikely to be of direct benefit to the organism concerned, for once the mould is able to manufacture enough penicillin to keep bacteria away from it, it gains nothing by escalating the submicroscopic arms race to a level at which it can kill the bacteria ten times over. Biotechnologists must create an artificial environment which encourages organisms to overproduce (from their own point of view) the materials we demand.

The techniques for improving an organism can be divided into two broad categories (i) genetic engineering, which, of course, was unknown until the seventies; and (ii) conventional techniques, which played a central role in the penicillin story and continue to do so in most areas of biotechnology today. Unlike genetic engineering, by which desirable characteristics can be introduced into organisms in a fairly controlled manner, conventional techniques for improving organisms must rely on more random procedures, making use of what nature provides. This does not mean that scientists are reduced to being mere observers of nature's activities, simply picking up the unconsidered trifles she casts out. Man, not nature, writes the rules of this evolutionary game.

The key to finding the organisms which are most suitable for a particular application lies in the existence of variations in nature. Not all individuals of the same species are identical. We can see this by observing the plants and animals around us. Variations also occur in microbes, and to a greater extent. The appearance of two cats may differ considerably, but their basic biochemistry is very similar. To survive, both require the same substances in their food, and both consist of very much the same chemical compounds in about the same amounts. The variations between two microbes of the same species are likely to be greater. If the differences are sufficiently interesting and well defined, the two microbes will be said to be different strains of the species.

The early workers on the penicillin project searched among many individual *Penicillium* moulds to see if a strain that manufactured an unusually large amount of penicillin could be found. This search succeeded after an enormous amount of painstaking and tedious work. Samples of mould were collected from as many sources as possible — one of the best was found growing on a cantaloupe in a New Jersey market. Samples of each mould were grown in the laboratory and tested to see how much antibiotic they produced. The labour involved in this screening process was immense but the rewards, when they came, were even greater.

Once the promising strain of *Penicillium* mould was unearthed, the scientists set about improving it further by inducing mutations in the organism. The differences between strains of *Penicillium,* including their ability to manufacture penicillin, arise from the fact that each has a slightly different set of genetic instructions. A mutation is simply a change in the normal genetic make-up of an organism. Mutations occur quite naturally — indeed, without them evolution would be impossible, for it is on slight variations between individuals in a species that natural selection acts. Some mutations will be immediately lethal for the organism involved; a few will be beneficial; and some will be more or less neutral, conferring no clear advantage or disadvantage on the organism possessing the mutation.

However, natural mutations takes place infrequently. In search of a microbe that is more useful to humans, the rate at which mutations arise is increased, thereby increasing the variety of strains of the species available. It is of no concern to the biotechnologist that most of the individual moulds do not survive so long as one is created that fits the bill — in this case, by producing more penicillin.

By 1951, Felming's original mould, *Penicillium notatum,* had been discarded as a potential large-scale source of penicillin and had been replaced by a similar species, *Penicillium chrysogenum.* Scientists began to search for the most suitable strain of this species, and to increase the natural rate of mutation they treated the mould with X-rays, ultraviolet light and the poison, nitrogen mustard. These harsh treatments induced mutations and, occasionally, threw up an improved strain. This strain would then be subjected to another series of experiments. In over twenty such steps, the strain in commercial use today was created.

From Laboratory to Industry

In their search for the best strains of *Penicillium,* scientists generally grew the mould in small glass or earthenware vessels. With such small-scale, labour-intensive operations there is no hope of producing enough penicillin to treat the vast number of patients who can benefit from the drug. To meet the massive demand for penicillin, the pharmaceutical companies of the eighties have installed gigantic metal vats with capacities of up to 100,000 litres (22,000 gals).

To design the best fermentation processes, biotechnologists must pay careful attention to many factors, including the supply of the best nutrients, preventing contamination and controlling the fermentation conditions, such as temperature and acidity.

Nutrients

If the *Penicillium* mould is to grow and provide penicillin, it must have access to a variety of nutrients which supply it with energy and the raw materials it needs to synthesize the compounds that make up its cells. In terms of bulk, the major components of nutrients are sources of carbon and nitrogen. These provide both energy and building blocks for the cell's compounds and, since almost all of the cell's constitutents contain carbon, the demand for carbon-containing nutrients is high. The biotechnologist's choice of nutrients is influenced by economic and biological factors.

In laboratory experiments, scientists are free to try out a vast range of potential nutrients to see which suits the microbes best. In an industrial operation, however, cost constraints are far more formidable. The prospects for any biotechnology are bleak if it becomes appparent that the organism needs expensive nutrients. Fortunately, this is not often the case and, indeed, some of the most successful biotechnologies utilize waste materials from industrial, agricultural or domestic sources.

Penicillin is still something of an enigma. There is much detailed information about the way it destroys bacteria by affecting their cell walls, but no-one is quite sure just *why the Penicillium* mould manufactures penicillin. The most obvious explanation is that it uses penicillin to keep foreign microbes at bay, preventing them from competing for the precious food resources. However, for most of its life-span *Penicillium* produces almost no penicillin at all; the synthesis of the antibiotic seems to be turned on around the time the mould cells stop growing. Biotechnologists must overcome this and persuade the mould to start making penicillin as quickly as possible. Despite the lack of agreement about the biological reasons for *Penicillium's* apparent reluctance to make penicillin, biotechnologists have been quite successful in achieving their goal.

Penicillium can use many sorts of compounds as sources of energy and carbon, most notably sugars of various kinds. One sugar in particular, glucose, is consumed very eagerly and the mould grows rapidly, but does not produce much penicillin. It was discovered that another sugar, lactose (which is found in milk), is less 'digestible' to *Penicillium* so the mould grows more slowly but produces much more penicillin. The pharmaceutical industry now feeds mixtures of glucose and lactose to the moulds, thus maintaining the optimum balance between cell growth and penicillin production.

Investigations into the most suitable nutrients are key factors in all biotechnological developments and sometimes, as in the case of penicillin, a stroke of luck can have a major impact. When corn (maize) is processed to manufacture starch, a liquid by-product called corn-steep liquor remains. At one time this material was dried and sold as cattle feed, a means of disposing of the waste that was, at best, only barely profitable. For many years the US Department of Agriculture sought a more profitable use for corn-steep liquor, but to no avail. However, during the Second World War some of the Department's scientists were closely involved in the development of penicillin, and they decided to see if the *Penicillium* mould could grow on corn-steep liquor. The results were startling; not only did the mould happily consume this waste material, but yields of penicillin more than doubled. It is now known that corn-steep liquor contains an excellent mixture of carbon compounds, but more importantly, one of these compounds is a precursor of penicillin — a substance that the mould can use directly to build up penicillin molecules. While such serendipity cannot be guaranteed to bless other, newer biotechnologies it certainly serves to illustrate the pleasant surprises nature can spring upon aspiring biotechnologists.

Most of the multitude of compounds inside a cell are composed of a skeleton of carbon atoms with various other types of atoms attached. Clearly biotechnologists must ensure that these, too, are supplied to the growing organisms. In practice, this creates few difficulties. Hydrogen, the most abundant element in the cell, is present in almost every type of carbon compound and, of course, in water. The other elements of major importance in life can easily be provided in the form of simple, cheap compounds — nitrogen (used, for example, to construct proteins, DNA and RNA), phosphorus (DNA, RNA and cellular membranes), sulphur (many proteins), plus smaller quantities of iron, potassium, sodium, zinc and other metals. As the cellular chemistry of micro-organisms is so versatile, most can synthesize everything they need from very simple starting materials

Oxygen

All the most familiar forms of life need oxygen to survive. Inside cells, oxygen is combined with molecules derived from foodstuffs, liberating energy which can be used to maintain all the cell's vital functions. Many of the microbes used in biotechnology, including *Penicillium,* also need oxygen. (Some do not — indeed some microbes are actually killed by oxygen, the most notorious example being *Clostridium botulinum,* the cause of botulinus food poisoning). Controlling the amount of oxygen in a microbe's environment is hence often vital.

When *Penicillium* mould grows in small containers it can obtain all the oxygen it needs directly from the air; oxygen in the atmosphere diffuses into the mass of growing cells. However, oxygen cannot diffuse over large distances and cells near the bottom of the huge fermentation vats used in industry would die for lack of oxygen unless special arrangements were made to keep them well supplied. For this reason, air is pumped into the base of fermentation vessels.

It is relevant to note here another of the significant breakthroughs which enabled penicillin production to reach : , present level of efficiency. *Penicillium notatum,* the first penicillin mould, only grows well when it is near the surface of a vessel. Clearly, there would have been little point in constructing vessels several metres deep if the mould inhabited only the top few centimentres. Fortunately, the strains of *Penicillium chrysogenum* discovered in the forties not only make more penicillin, but also thrive when submerged.

Maintaining an Equable Environment

A key factor that favours biological industries over many chemical methods of producing the same materials is that the former do not need high temperatures. Metallic catalysts often require temperatures of 100° C (212°F) and more to work efficiently. Most biotechnological processes operate best at 30-50°C (86-122°F), about the temperature of the human body.

However, there is a drawback : it is generally necessary that the temperature of the microbes is kept within a narrow range. The vessel's temperature must be monitored constantly, and a little heating or cooling applied as necessary. Living organisms are intolerant of extremes of temperature — say, below minus 10°C (14°F) and above 100°C (212°F) — because of the natue of their chemical compounds. If proteins, DNA or many of the other molecules essential for life are heated, their structure becomes disrupted and they no longer operate correctly. On the other hand, when the temperature drops, the catalytic effect of enzymes diminishes very rapidly. The almost incredible efficiency of enzymes has the penalty that they work well only at or very near a certain temperature. This temperature varies from species to species; polar fish are adapted to life at or below 0°C (32°F) and their enzymes can cope with the cold, whereas bacteria which inhabit hot geothermal springs have enzymes that remain active at temperatures as high as 85°C (185°F).* The ability to maintain a constant body temperature is thought to have been one of the most important factors that allowed the mammals to supplant the reptiles as the dominant animals on land. In most mammals, the body temperature varies by only a degree or so, regardless of the external temperature. This means that they enzymes can work efficiently at all times, allowing them to remain active rather than dropping into a reptilian torpor at night or in cold weather.

Penicillium generates heat as it consumes nutrients. If the mould is grown in a small container, this excess heat is rapidly dissipated through the walls of the vessel. However, the large vessels used in industry must be cooled to prevent the heat building up and the mould being harmed. Fig. 16.1 shows cooling jackets encircling the fermentation vessels and pipes and valves through which acid or alkali may be added to the fermentation medium.

Anyone who has made their own wine will know that it is vital to maintain the correct degree of acidity (pH) if a palatable drink is to result. Just as organisms have a preferred temperature, so a certain degree of aciditiy or alkalinity suits them best. In fact, penicillin production is particularly sensitive to the acidity of the environment — too much or too little causes the yield of antibiotic to fall considerably.

---

* Recently, it has been claimed that some forms of bacteria live in volcanic vents at the bottom of the oceans. The pressure in this environment is very high and the temperature is about 300°C. There are certainly a number of exotic worm-like creatures living near these vents and it may be that they feed on these bacteria. No one knows how bacteria could survive in this very hostile environment.

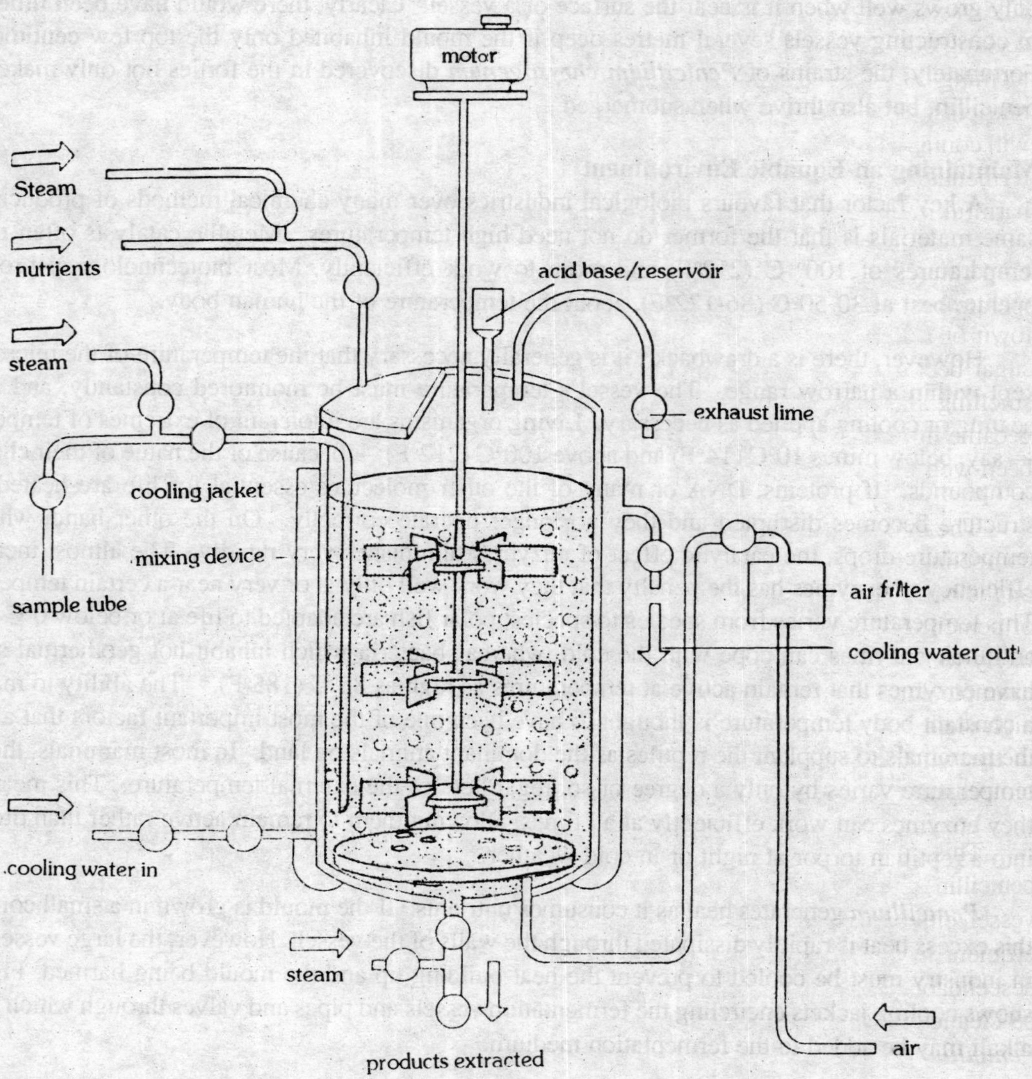

Fig. 16.1. A typical fermentation vessel. The microbes are grown on nutrients placed in the vessel at the start of the fermentation. The vessel is cooled by a water jacket. Air is pumped into the bottom of the liquid, and acid or alkali added as necessary. A stirrer keeps the contents well mixed. Steam lines are provided so that the vessel can be sterilized after each fermentation batch.

Sterilization

From the cold, dark depths of the ocean to the steaming heat of geyser pools, microbes can be found everywhere on Earth. The resilience, diversity and versatility of the microbial world are crucial to the use of microbes in biotechnology, but their very hardiness does bring disadvantages. The vast majority of biotechnological processes use only a single species. There is little point in seeking out the best strain of *Penicillium chrysogenum* and then trying to grow it in a vessel that is teeming with other, quite different microbes. These intruders may compete for the available nutrients, slowing the growth of *Penicillium*. They may even excrete some dangerous toxin which will contaminate the product. Careful sterilization of all equipment and materials introduced into the fermentation vessel is, therefore, essential. This sounds easier than it is in practice. Potentially harmful microbes lurk everywhere and fastidious cleanliness is required to eliminate them.

Louis Pasteur was instrumental in delineating this central principle of biotechnology. In the eighteen-sixties, the production of alcohol from sugar beet was an important industry in the French town of Lille. Occasionally, for reasons quite incomprehensible to the manufacturers, the vats of sugar beet and yeast stubbornly refused to yield alcohol, becoming filled with all manner of foul-smelling substances. Pasteur discovered that the fermentation process went wrong when the vats became infected with other microbes. Ever since then the dangers of microbial contamination have been well-understood, but infections still cause problems from time to time.

The microbes used in most biotechnologies are particularly at risk from being overwhelmed by competing organisms. This is because the strains employed have been artificially selected by humans on the basis of certain characteristics they possess — most obviously, in this case, greater production of penicillin. any organism which overproduces (from its own point of view) any substance is very likely to be at a disadvantage among other organisms. This phenomenon is also encountered, for example, in farming. Wheat has been bred to exhibit characteristics which benefit humans, including high yields of seeds and short stems. These plants would fare badly among the natural competition in the wild, and it is the farmer's job to see that they get special treatment. In the same way, biotechnologist must ensure that their microbes are protected from competition.

This competition may come not only from foreign species (many of which are unaffected by penicillin), but also from related strains of *Penicillium*. A rogue mould may appear in the fermentation vessel, either by infection from outside or by a mutation in the production strain. If this rogue is more efficient in utilizing the available resources—perhaps by ceasing to overproduce penicillin — its descendants will soon outnumber those of the original strain. Once this has happened the vessel must be cleaned out, resterilized and the process started again with a fresh batch of the good strain of *Penicillium*. It is for this reason, among others, that most biotechnologies operate using batch processes. The whole system is set up and the microbes allowed to grow and produce the desired material (this takes about nine days for penicillin fermentation). The vessel is then emptied, the product purified (about fourteen hours for penicillin) and the vessel cleaned before a new batch is started.

Clearly, there are advantages to be gained from a shift towards continuous processes, which could operate for weeks or months with nutrients being gradually fed into the system and products being drawn off in a steady stream. Much effort is being put into the development of continuous systems in several areas of biotechnology.

Purifying the Product

When the fermentation process is complete, the vessel is full of a thick broth of microbial cells, some unconsumed nutrients and dissolved penicillin. The extraction and purification of fermentation products varies widely according to the particular process. In the case of penicillin, a fairly small molecule which the cells excrete into the surrounding liquid, the problems are not too difficult to overcome. First, the cells are filtered from the penicillin-laden liquid and discarded. When certain potassium compounds are mixed into the liquid, the penicillin forms crystals which settle to the bottom of a container and can be collected. This basic form of penicillin, named penicillin G, may then be chemically modified to form a wide range of semi-synthetic penicillins with names such as ampicillin and methicillin. The chemical modification of antibiotics produced by microbes is a common feature in the pharmaceutical industry; biotechnologists and chemists cooperate to produce many different antibiotics, each with a particular application.

In other biotechnological processes it may be necessary to break open the microbial cells to release the desired product. Most enzymes, for example, remain locked inside the cell and are not excreted naturally. If it is necessary to disrupt the cells, the process of purification of the desired product is made more complicated by the presence of large amounts of cell debris. This may add substantially to the costs of the whole operation. It is likely, however, that before long genetic engineers will help reduce costs by inducing microbes to start secreting the required products into the fermentation liquids.

17

Fermentation Reactions of Anaerobic Digestion*

GENERAL

Metabolic Groups and Pathways

Two-and Three-phase Models

Anaerobic digestion consists of a complex series of reactions, which are catalyzed by a consortium of bacteria and accomplish the conversion of organic compounds to the terminal products methane and carbon dioxide. The sum of these reactions is a fermentation which converts a wide array of substrate materials having carbons at various oxidation/reduction states to one-carbon molecules in the most oxidized (CO_2) and most reduced (CH_4) states. The latter arrangement is thermodynamically the most stable state (at atmospheric pressure), and these molecules are not further degraded in ecosystems where light and inorganic electron acceptors such as nitrate or oxygen are absent.

The traditional model of anaerobic digestion divides the reactions into two phases, or stages. The first, acidogenesis, rearranges (ferments) the organic matter to carbon dioxide, hydrogen, and volatile organic acids, including acetate, propionate and butyrate. The methanogenic phase converts these extracellular intermediates to the terminal products, methane and carbon dioxide.

This division into two phases has for many years been a valuable model for anaerobic digestion, because it emphasizes important distinctions between organisms of each phase. For instance, the acidogenic bacteria are generally fast growing and resistant to inhibition, in contrast to the slow growing and fastidious organisms of the methanogenic phase. However, the model obscures important interactions between the bacteria of the acidogenic phase and those of the methanogenic phase. Recent advances in the understanding of these interactions has led to a more complete model, diagrammed in Fig. 17.1 having three stages instead of two. The traditional model groups together in the methanogenic phase all bacteria responsible for the conversion of volatile organic acids to

---

* Excerpts taken from the review published by Prof. David R. Boone, Los Angeles, USA.

methane and carbon dioxide, even though some of these bacteria cannot themselves produce methane. The extended model recognizes as a group (separate from the methanogens) the recently discovered obligate proton-reducing acetogenic bacteria, whose substrates include propionate, butyrate, and higher fatty acids. These compounds are converted to accetate, molecular hydrogen, and carbon dioxide, which in turn are substrates for the true methanogens.

This three-phase model includes in the methanogenic category only those organisms which can themselves produce methane. In digestors they produce methane almost entirely by decarboxylating acetate or by reducing carbon dioxide with molecular hydrogen. Methanogens can produce methane from a limited range of other compounds including formate, methyl-substituted amines, and methanol, but these are quantitatively of limited importance in anaerobic digestion. Thus acetate and hydrogen are important extracellular intermediates in anaerobic digestors, intermediates through which essentially all of the reducing equivalents (COD) present in the original substrate must pass. The non-methanogens in anaerobic digestors must convert this assemblage of organic matter to these simple compounds which the methanogens can degrade; the methanogens co-operate by maintaining these compounds (acetate and hydrogen) at low concentration which facilitates their continued production by non-methanogens.

Acetate is quantitatively the most important methane precursor, the source of about two thirds of the methane produced in anaerobic digestors. It is degraded by a decarboxylation reaction, in which the carboxyl group is oxidized to carbon dioxide, and the methyl group is reduced to methane, with its substituent hydrogens intact. *Methanosarcinae* and *Methanothrix soehngenii* are responsible for this reaction.

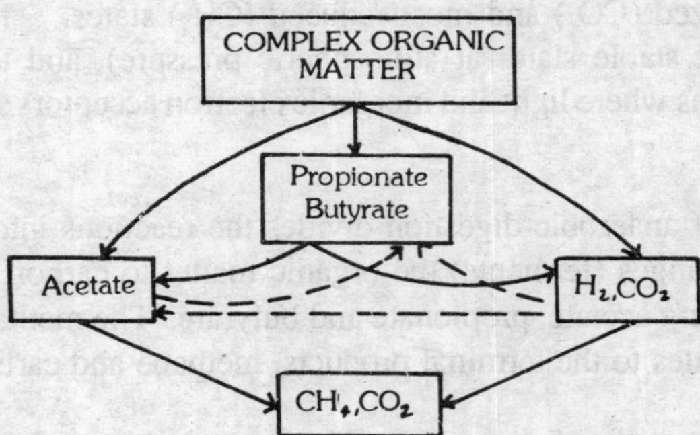

Fig. 17.1. Three-stage model for complete anaerobic digestion of organic matter of methane and carbon dioxide. Dotted lines represent "black reactions" in which hydrogen is used in making acetate and higher volatile organic acids.

Nearly all the methane produced in anaerobic digestors from sources other than acetate is the result of carbon dioxide reduction with molecular hydrogen; this is the preferred energy-yielding reaction of nearly all the other methanogenic species, and concentrations are kept very low in anaerobic digestors by the continuous removal of hydrogen by these methanogens.

Some acetate-degrading methanogens, including *Methanobacterium soehngenii and some Methanosarcinae,* are unable to utilize hydrogen for methanogenesis. For others (e.g., most *Methanosarcinae),* hydrogen is the preferred substrate. However, those methanogens which cannot use acetate show a great affinity for hydrogen, and they may maintain hydrogen at levels so low that acetate-degraders cannot dissimilate it, such that the two methanogenic precursors may each be degraded by separate groups of methanogens. Sometimes hydrogen and acetate are used by the same species : in the studies by McInerney et al., of butyrate enrichments from the rumen, a *Methanosarcina* species was the predominant hydrogen-utilizing species. It was also apparently responsible for acetate degradation, but it was not determined whether the hydrogen and the acetate produced from butyrate were degraded simultaneously or sequentially.

Interspecies Hydrogen-Transfer

Recent advances in understanding anaerobic digestion can be traced to the resolution of "Methanobacterium omelianskii" into constituent organisms. "Methanobacterium omelianskii" had been reported to oxidize ethanol to acetate, using the generated electrons to reduce CO_2 to methane; however Bryant and co-workers found that the culture is not axenic (pure), but rather contains two distinct strains of bacteria, neither of which alone can completely degrade ethanol.

When the organisms are separated, one, a methanogen later named *Methanobacterium bryantii,* grows on hydrogen and CO_2, but no other substrates. The other species, S. organism, oxidizes ethanol to acetate. The electrons generated from this oxidation are transferred to a pyridine nucleotide carrier (NAD), and ultimately used to reduce protons to molecular hydrogen.[1] Because hydrogen production from ethanol (or from NADH) is thermodynamically possible only at low hydrogen concentrations, S organism requires the removal of hydrogen for complete ethanol conversion. In the case of "Methanobacterium omelianskii" this is accomplished by the action of *Methanobacterium bryantii.*

S organism can grow alone on other substrates such as pyruvate, without the removal of hydrogen by a methanogen. However hydrogen removal is always beneficial, allowing more growth and substrate degradation, and possibly a greater growth yield via the phosphoroclastic reaction. Without hydrogen removal, acetogenic intermediate are diverted as hydrogen sinks to ethanol production.

"Methanobacterium omelianskii" was the first demonstration of hydrogen removal by methanogens affecting the fermentation of non-methanogens, a phenomenon now known as "interspecies hydrogen-transfer." It is now clear that there are many reactions which can occur in anaerobic digestors, but which are exergonic only when hydrogen concentration is low. Some compounds degraded by these reactions (e.g., ethanol and lactate) are not important intermediates in the anaerobic digestion process, but they may be present in the influent substrate. A number of examples of interspecies hydrogen-transfer reactions were subsequently studied those which degrade small, relatively high-energy compounds; fermentative reactions; and those which degrade very low-energy compounds.

The first, catabolizing small organic molecules such as lactate and alcohols, produces hydrogen and requires its removal by methanogens or other hydrogen-utilizing bacteria. S organism belongs in this group. The bacteria catalyzing these reactions can grow axenically on other substrates (e.g., S organism grows on pyruvate), but catabolize the alcohols or lactate only when hydrogen is removed. In the absence of hydrogen removal, traces of substrate are oxidized and traces of hydrogen accumulate, inhibiting further conversion. Some *Desulfovibrio* species grown in the absence of sulfate fall into this category. With sulfate present as an electron acceptor, compounds such as lactate can be oxidized. When it is absent, protons can replace it, but only when molecular hydrogen is kept at low concentration by interspecies hydrogen-transfer.

The second category of interspecies hydrogen-transfer occurs with fermentative bacteria catabolizing carbohydrates. These bacteria grow well in the absence of hydrogen removal, producing little or no hydrogen. However when hydrogen-utilizing species such as methanogens are present, the fermentative bacteria greatly increase their hydrogen production, often increasing their growth yields as well. When fermentative bacteria catabolize carbohydrate, electrons are generated and stored temporarily as reduced pyridine nucleotides (Fig. 17.2). These nucleotides must be recycled, by conversion back to the oxidized form, in order for the fermentation to continue. One possible target of the electrons carried on reduced pyridine nucleotides is use by hydrogenase enzymes for proton reduction, producing hydrogen gas. In this case the products are two moles of acetate, two moles of CO_2, and four moles of molecular hydrogen per mole of hexose. The thermodynamics of hydrogen production from reduced pyridine nucleotides allow it to proceed only at low hydrogen concentrations. If hydrogen accumulates, this reaction is blocked, and the fermentative bacteria must find other targets for the electrons in order to reoxidize the pyridine nucleotides. This is accomplished by transferring the electrons to glycolytic intermediates, forming reduced products such as lactate, ethanol, propionate or butyrate, at the expense of acetate production. This represents a loss of potential growth-energy for the fermentative bacteria, because with the phosphoroclastic reaction, acetate production is accompanied by the production of ATP. The bacteria catalyzing these reactions and the specific effects of interspecies hydrogen-transfer on their fermentation have been recently reviewed.

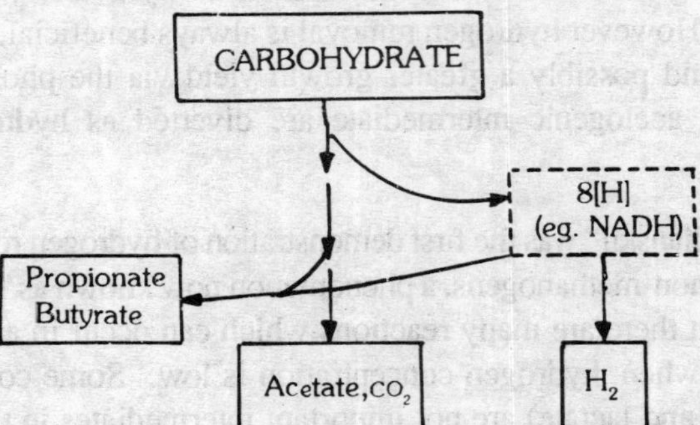

Fig. 17.2. Fate of electrons on reduced pyridine nucleotides in many fermentative bacteria. When hydrogen concentration is low, hydrogen is produced, but when hydrogen concentration is high, electrons are used to reduce organic intermediates of the fermentation, diverting them from acetate production.

In the third category of interspecies hydrogen-transfer reactions, hydrogen production is an obligate process. Unlike S organism or *Desulfovibrio* species, bacteria in this group have no alternate substrates with which they can grow in axenic culture, and so are obligate syntrophs, requiring partners which can remove molecular hydrogen. They have been named the "obligate proton-reducing acetogens," and accomplish the conversion of propionic and longer monocarboxylic acids to the methanogenic substrates acetate and hydrogen. The other unique characteristic of this category of reactions is that the concentration of hydrogen required to make them exergonic is must lower than that required for reactions of the other two categories.

The first discovery of bacteria belonging to this group was reported by McInerney, *et al.*[2, 3]. *Syntrophobacter wolfei* oxidizes butyric and longer (up to octanoic) normal monocarboxylic acids, producing acetate and molecular *hydrogen* (and propionate in the case of odd-chain legnth acids). It can also oxidize isoheptanoic acid, from which it produces acetate, hydrogen and isobutyrate. These findings are consistent with a B-oxidation mechanism, although this has not been established. Thermodynamic considerations require hydrogen concentrations below 10^{-4} atmospheres for these reactions to occur, and so the organism can function only when co-cultured with hydrogen-utilizing bacteria. It was isolated using lawns of hydrogen-utilizing bacteria (desulfovibrio or methanogens). Colonies in roll tubes take many weeks to develop, and, when picked, contain *Syntrophomonas wolfei* and only one other species, the lawn organism. Monoxenic co-cultures of *Syntrophomonas wolfei* have also been prepared using *Methanospirillum hungatei* or other methanogens as the hydrogen-utilizing species. These co-cultures accomplish the degradation of butyrate in a manner analogous to ethanol oxidation by "Methanobacterium omelianskii." The oxidations of the fatty acids do not occur when hydrogen is present at significant levels or when hydrogen removal by the hydrogen-oxidizing bacterium is inhibited.

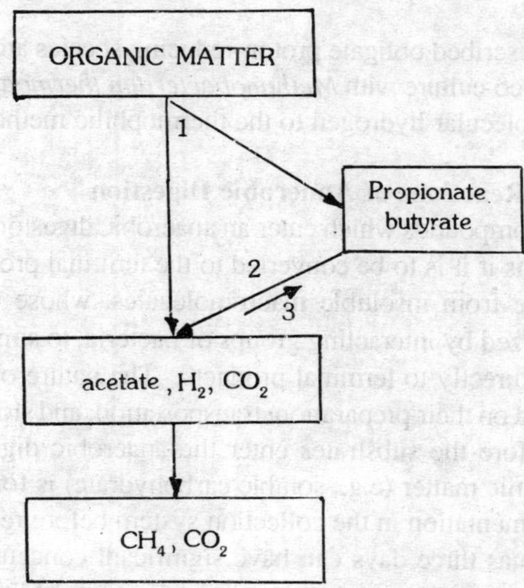

Fig. 17.3. Hydrogen control-points in anaerobic digestion. The concentration of hydrogen controls the flow of carbon in anaerobic digestors in at least three places: (1) the fraction of organic matter converted to higher fatty acids; (2) the rate at which those higher fatty acids were degraded; and (3) the rate of the back reaction in which higher fatty acids are produced from acetate and hydrogen (and sometimes CO_2).

Syntrophobacter wolinii oxidizes propionate to acetate and carbon dioxide, like *Syntrophomonas wolfei,* produces molecular hydrogen only at low concentrations. It was also isolated in monoxenic co-culture with *Desulfovibrio strain* G11 as the hydrogen-utilizing species, and grows even more slowly than *Syntrophomonas wolfei* (minimum generation time 3.6 days). The generation times of the bacteria which oxidize propionic and longer volatile organic acids are similar to the minimum solids retention times necessary to maintain populations of propionate- and butyrate-degrading bacteria in anaerobic digestors.

A benzoate-degrading bacterium was also isolated which requires hydrogen removal for growth. It can degrade no other of the tested aromatic compounds, and grows even more slowly than the other obligate proton-reducing acetogens, with a minimum doubling time of 5.5 days. Obligate proton-reducing acetogens are not the only species which degrade aromatic compounds, as some bacteria accomplish this fermentatively.

One interesting class of obligate proton-reducing bacteria has been found to be not acetogenic, but rather acetate degrading. Zinder[4] reported the isolation in co-culture of a bacterium which oxidizes acetate to CO_2, producing molecular hydrogen; this also requires hydrogen removal, and occurs only at low hydrogen concentrations. The organism is particularly interesting in that it degrades acetate by a reaction which conforms to the van Niel's CO_2, reduction theory of methanogenesis. Acetate degradation is the main exception to the theory, which predicted that all organic matter is oxidized to CO_2, and generated electrons are used to reduce part of that CO_2 to methane, so that all biogenic methane is the result of CO_2, reduction. Most acetate in anaerobic digestors is degraded by decarboxylation, but these acetate-oxidizers may be of limited importance there.

All but one of the described obligate proton-reducing species are mesophilic. One thermophilic bacterium, in monoxenic co-culture with *Methanobacterium thermoautotrophicum,* oxidizes butyrate to acetate, transfering molecular hydrogen to the thermophilic methanogen.

Hydrolysis and Initial Reactions of Anaerobic Digestion

The wide range of compounds which enter an anaerobic digestor must be converted to substrates utilizable by methanogens if it is to be converted to the terminal products methane and CO_2. These substrate materials range from insoluble macromolecules, whose degradation may require many different reactions catalyzed by interacting groups of bacteria, to simple compounds such as acetate, which can be degraded directly to terminal products. The nature of the substrates of an anaerobic digestor of course depend on their preparation, transportation, and storage history. Initial degradative reactions may occur before the substrates enter the anaerobic digestor. For instance very little readily fermentable organic matter (e.g., soluble carbohydrate) is found in domestic sewage sludge, because it undergoes fermentation in the collection system before reaching treatment plant. Animal waste stored for as little as three days can have significant concentrations of lactic acid, although lactic acid is not present in fresh wastes, nor is it produced in healthy anaerobic digestors.

The compounds which actually enter an anaerobic digestor may enter into the generalized pathway for degradation at a number of different places. Polymers, such as cellulose, starch, or proteins, are first hydrolyzed, mainly to oligomers. These oligomers, as well as monomeric carbohydrates, amino acids, and some organic acids (such as lactic) are degraded in digestors by fermentative reactions. Fats are hydrolyzed to glycerol and long-chain fatty acids; glycerol is subsequently degraded by fermentative reactions, but the fatty acids are apparently metabolized by obligate proton-reducing acetogens, entering the metabolic scheme where propionate and butyrate are degraded. Common to the pathway for degradation of these or any other organic compound degraded completely in anaerobic digestors is the conversion to substrates of methanogens.

Hydrogen Concentration and Effects on Metabolic Pathways

Molecular hydrogen is an important extracellular intermediate in anaerobic digestors, with about 30 per cent of the methane produced by the reduction of CO_2 using molecular hydrogen. In addition, hydrogen concentration influences other reactions at several control points. By its action at the three control points, high hydrogen concentration causes increased propionate and butyrate production at the same time their degradation rate is decreased, resulting in their accumulation. Increased hydrogen concentration can occur in digestors due either to inhibition of methanogenesis or to an increase in hydrogen production.

At low hydrogen concentrations, protons are the preferred electron acceptor for excess electrons, resulting in the production of molecular hydrogen as the electron sink, plus acetate and CO_2. At higher hydrogen concentrations, the production of hydrogen from reduced pyridine nucleotides is endergonic, and alternate electron acceptors are required. In digestors this results in increased production of propionate and butyrate. The influence of hydrogen concentration on the production rates of propionate and butyrate has been termed hydrogen control point number 1 by McInerney and Bryant.[5]

The second control point of hydrogen concentration is the inhibition of propionate and butyrate degradation, which is complete and rapid except when hydrogen concentration is very low.

The third location at which hydrogen concentration can influence fermentation patterns in anaerobic digestors affects the production of volatile organic acids (acetate and longer) from hydrogen, acetate to CO_2. There are a number of bacteria which have been isolated from digestors and other sources which can grow by using H_2 to CO_2 or acetate, producing higher volatile organic acids. These reactions do occur in digestors, and bacteria which can catalyze them to obtain energy can be found there, suggesting that conditions for which the reactions are exergonic occur there. These reactions can be thought of as "back reactions," since they are the reverse of those which lead to the terminal products methane and CO_2; thus under identical conditions (i.e., concentrations of reactants and products) they will have a free-energy change equal but of opposite sign to the respective forward reactions. Both kinds of reactions can be simultaneously exergonic only if there are distinct microenvironments in the digestor having different concentrations of compounds involved in those reactions. Of the possible compounds, hydrogen has a turnover rate much more rapid than the others, and so is the most likely to have significant concentration gradients in digestors. Hydrogen production from a particulate organic matter may be a point source of hydrogen which may generate these gradients.

Apart from the three control points described above, high levels of hydrogen can influence digestors by inhibition of acetate degradation by acetoclastic methanogens; this effect has been demonstrated in enrichments and pure cultures and in digestors. When substrates from which hydrogen is produced and degraded in mixed-culture systems also containing *Methanosarcinae*, acetate degradation usually starts only after that hydrogen—generating substrate is depleted. (However, if hydrogen remains at very low concentrations during its production, such as when it is produced from butyrate, acetate degradation can accompany hydrogen production. Formate also inhibits acetate degradation in enrichments, probably because hydrogen is produced from it. If hydrogen is bubbled into digestors, inhibition of acetate degradation occurs. Hydrogen-grown *Methanosarcina barkeri* strain MS has a long lag phase after hydrogen is removed before it can rapidly degrade acetate. Although some strains of methanogens in culture can convert acetate to methane only in the presence of hydrogen, the conversion rate is very small, and is probably not quantitatively important in digestors. Hydrogen does not inhibit cultures which are almost exclusively *Methanothrix soehngenii;* unlike most *Methanosurcinae*, they cannot catabolize hydrogen.

FACTORS AFFECTING RATE AND EXTENT OF METHANOGENESIS

Rate-limiting Reactions

For anaerobic digestion of most waste materials, it is hydrolysis of polymers which limits the extent of methanogenesis from organic wastes. With a properly designed and operated digestor, the theoretical potential of methane production represented by the soluble organic matter in the effluent is usually very small compared with the total amount of methane produced. When viewed in this way, the potential for improvement of conventional digestion lies in increasing the extent of hydrolysis of polymers. However, the fermentative and methanogenic reactions can limit methanogenesis in two ways: (1) instabilities in the fermentation can cause digestors to go sour, and (2) the slow growth of bacteria accomplishing the conversion of volatile organic acids to methane and CO_2 require long solids retention times in order to maintain adequate populations in digestors.

When digestors are operated at high loading rates there is a rapid turnover of volatile organic acids, so that they accumulate rapidly if their degradation is inhibited. When their levels begin to increase, halting substrate addition sometimes allows the acetogenic and methanogenic microflora to recover. However, the continued fermentation of slowly-degrading substrates may cause the increase to continue. The ultimate result of volatile acid accumulation is a drop in pH, which further inhibits methanogenesis. This situation is known as "sour digestion," and it can be expensive and inconvenient.

When very dilute or soluble compounds form the substrate, they are rapidly fermented, and it is often the growth rate of the slow-growing methanogens and acetogens which limits methanogenesis. Various methods have been used to increase the hydraulic loading rate of digestors while maintaining solids (i.e., bacteria) retention time at a level high enough to maintain adequate populations of slow-growing bacteria. The traditional technique is to recycle solids, but more novel techniques are now receiving increased attention. The ability of some bacteria to flocculate or adhere to inert surfaces can be exploited to keep them within the fermentor. The anaerobic filter and related techniques rely on this ability, allowing liquid to flow through the system rapidly while organisms remain within the reactor. This can give hydraulic retention times as short as a few hours while maintaining solids retention times long enough for methanogens and acetogens to florish. The principle of the downflow system is different, but its operating characteristics are similar. Recently immobilization with alginate of the methanogenic and acetogenic bacteria was shown to be an effective method for maintenance of populations in systems with short hydraulic retention times.

Temperature

There is some controversy as to whether there exist two temperature optima (i.e., mesophilic and thermophilic) for anaerobic digestion. Pfeffer[6] examined anaerobic digestion of domestic refuse at temperatures between 35° and 60°C at 5°C intervals, and found 40°C consistently more favourable than 35° or 45° when the retention time was between 4 and 30 days. With the same retention times he found a second optimum at 60°C even more favourable than at 40°C. Digestors perform more satisfactorily at 60°C than either 55° or 65°C. There are only small differences in gas production of completely-mixed anaerobic digestors at temperatures between 30° and 60°C when the loading rate is not high (less than 7 g/L) and the retention time is at least six days. However, when loading rates are high or retention times short, thermophilic digestors usually outperform digestors operated at lower temperatures.

Buhr and Andrews[7] designed a dynamic process model to describe effects of temperatures. This model predicts that the temperature which gives minimum volatile organic acid concentration increases with decreasing retention time so that at very low retention times (3.5 days) the optimal temperature is high (about 50°C). It also predicts greater maximum methanogenic rates at increasing temperatures up to 60°C. These predictions are generally consistent with later studies which tested shorter retention times than were previously investigated (a thermophilic [55°C] cattle waste digestor operated at a 3-day retention time produced more methane than digestors operated at longer [6, 9, or 12 day] retention times, although volatile organic acid levels were higher. Steady-state conditions can be achieved with a retention time as short as 2.5 days, the shortest reported for a completely mixed digestor. Stable conditions at this short retention time were achieved only at 60°C when tested as temperatures in 5°C increments between 30°C and 65°C.

In these studies, the digestors were always allowed a period of acclimation to the temperature at which tests were done. To obtain thermophilic digestion, digestors can be started up at the desired temperature with a low initial loading rate, or temperatures of mesophilic digestors can be shifted slowly (about 1°C per week). When the temperature is raised gradually there is a reproducible decrease in gas production at 43°C, but maximum gas production is again achieved at 45°C. More rapid changes in temperature may not cause ill effects if the digestor is not already operated near maximum loading rate or minimum retention time; animal waste digestors can normally be shifted abruptly from 55° to 60°C, giving increased methane production. Of the culturable bacterial population in a mesophilic digestor, 9 per cent can grow at 50° and 1 per cent can grow at 60°;[8] temperature acclimation of digestors may then be accomplished by a selection of thermophilic subpopulations at high temperatures.

Little is known about effects of sudden decreases in temperature of digestion (as would result from failure of heating systems) but the model of Buhr and Andrews[7] predicts that a drop from 50°C to 40°C would cause digestor failure within two to three days.

Advantages of thermophilic digestion aside from its ability to operate at shorter retention times and higher loading rates include better destruction of bacterial and viral pathogens and better dewaterability of the sludge. Its chief disadvantage is the heat requirement to attain and maintain the higher digestor temperature.

Temperatures higher than 60°C consistently been found to give less favourable fermentations. Volatile acids concentrations were higher at 65° than 60°C, but stable fermentations can sometimes be achieved.[9] Pohland and Bloodgood[10] were not able to achieve a stable methanogenic fermentation at 60°C. There have been no detailed investigations of digestion temperatures higher than 65° although one report found no acclimation of the digestor to 65° even after 50 days. It is possible that because the complex nature of anaerobic digestion requires many interacting groups, it is unable to function normally at very high temperatures. In extreme environments such as high temperatures, microbial diversity is known to be limited.

pH

In general the optimal pH for mesophilic digestors is between 6.7 and 7.4, and they do not function well if the pH is below 6 or above 8.[3] The optimal pH range for thermophilic digestors has not been investigated, but the normal operating pHs for thermophilic digestors tend to be higher. The higher vapor pressure of water dilutes gaseous CO_2, giving a lower partial pressure, and CO_2 itself is less soluble,[11] giving a lower aqueoous concentration of that acidic gas.

The pH in anaerobic digestors is controlled mainly by the bicarbonate buffer system. Other compounds either do not change their extent of ionization greatly over the normal pH range (e.g., volatile organic acids and ammonia are essentially completely ionized at near neutral pHs), or they occur at concentrations too low to function as effective buffers (the pKa's of NaH_2PO_4 and H_2Saq are close to operating pHs of anaerobic digestors, but they do not function significantly as buffers because the bicarbonate concentration is so much higher). Therefore the pH of digestors is controlled by the partial pressure of CO_2 in the produced gas and the concentrations of acidic and basic substances in soluble intermediates or the added substrate.

Acidity in anaerobic digestors comes predominantly from aqueous CO_2 dissociation and from volatile organic acids, which dissociate. One source of basic compounds in digestors is ammonia, produced from proteinaceous material in the digestor or present in the substrate added to it. This can be an important source, especially when the C:N ratio is low. Other basic substances can occur in the substrate, in the form of inorganic salts or salts of organic acids. The total of these basic substances must balance the concentration of biocarbonate (from CO_2 dissociation) and volatile organic acid salts in the digestor to achieve pH stability.

If volatile organic acids are produced by the fermentative bacteria more rapidly than they are removed by the obligate proton-reducing acetogens and methanogens, the acids will accumulate. If the problem is not resolved, a drastic drop in pH occur, inhibiting the methanogens and exacerbating the situation. This can be a significant problem in commercial digestors, and generally addition of substrate is halted until recovery. There has been some debate about whether the low pHs or the acids *per se* are inhibitory to methanogens, and whether adding alkaline materials to raise the pH is effective in enhancing recovery. McCarty and co-workers showed that the acids are not inhibitory, but that the cation of the alkaline compound added could be inhibitory (e.g., NaCl is as inhibitory as sodium actate). They also showed that a proper balance of monovalent (Na and K) and divalent cations gives minimal inhibition of methanogenesis.

Stoichiometry

When molecules containing only the atoms, C, H and O are fermented to the gasses CH_4 and CO_2, stoichiometric calculations of the theoretical products can be made using the following formula :

$$C_n H_a O_b + \left(n - \frac{a}{4} - \frac{b}{2} \right) H2O \longrightarrow \left(\frac{n}{2} - \frac{a}{8} + \frac{b}{8} \right) CO_2 + \left(\frac{n}{2} + \frac{a}{8} - \frac{b}{4} \right) CH_4 \qquad (1)$$

The equation can be modified for other atoms in fermentations if the reaction path is known. For instance, sulfur present in the organic matter is converted to hydrogen sulfide, and to alter the equation to include this, the equation would become :

$$C_n H_a O_b S_c + n \left(- \frac{a}{4} - \frac{b}{2} + \frac{c}{2} \right) H_2O \longrightarrow \left(\frac{n}{2} - \frac{a}{8} + \frac{b}{4} + \frac{C}{4} \right) CO_2 + \left(\frac{n}{2} + \frac{a}{8} - \frac{b}{4} - \frac{c}{4} \right) CH_4 + cH_2S \qquad (2)$$

These stoichiometric equations, however, do not take into account the dissociation of carbon dioxide to bicarbonate. The "CO_2" in the equation refers to all forms of carbon dioxide, including gaseous, aqueous, bicarbonate, carbonate, and insoluble salts. The amount of this "CO_2" which leaves the digestor with the effuent will depend on two factors, the rate at which effluent leaves the digestor and the concentrations of dissolved and precipitated species. So digestors having a short hydraulic retention time would tend to have more CO_2 leaving in the liquid phase, and a higher proportion of methane in the gas. The concentrations of dissolved species depend on the partial pressure of CO_2, its solubility, and its extent of dissociation. The solubility and extent of dissociation are determined by pH, temperature, and ionic strength.

The equations above can be used with complex wastes if the elemental composition is known; the empirical formula is then used in the calculation. Also in this way the effects of inorganic electron acceptors present in the substrate can be seen on the products. When significant quantities of inorganic electron acceptors such as sulfate, nitrate, or oxygen enter the digestor, they can change the stoichiometry of the reaction. If their quantities and their products in the fermentation are known, they can be figured into the stoichiometric equation by modifying the first term of the equation above to include them in an overall empirical formula for the substrates of the digestor. For instance if glucose is the substrate, but oxygen concentration is 10 per cent of the glucose concentration, the empirical formula of the substrate would then be $C_6H_{12}O_6 + 0.1O_2$, or $C_6H_{12}O_{6.2}$. These equations assume that the substrate is completely converted to CH_4 and CO_2, and so should be applied only to the biodegradable portion of the organic matter.

Also a portion of the biodegradable organic matter will be converted to microbial cells rather than converted to methane and CO_2. Although this portion varies depending on the energy content of the degraded organic matter, it is roughly 10 per cent of the total.

Nutrition and Inhibition

Nutritional requirements for anaerobic fermentative bacteria usually include a fermentable carbohydrate or amino acid, minerals, and often some B vitamins.[5] As in many other aspects of anaerobic digestion, much information about bacterial nutrition is inferred from isolates of the rumen ecosystem, which has been more thoroughly studied. The differences between those ecosystems should be considered when making these inferences, however. The organic loading rate of the rumen is much higher than most anaerobic digestors, the retention time of liquid and solids is shorter, and the concentrations of volatile organic acids and other compounds which may function as growth factors are higher. An environment, such as the rumen, where these potential growth factors are present in high concentrations tends to have more bacteria which require them than does an environment which does not have as high levels of these growth factors (such as digestors). The "minimum-requirement mechanism" described by van Niel indicates that, for instance, bacteria having a nutritional requirement for a compound available in the rumen would tend to predominate there over similar organisms which did not have that requirement. If digestors did not have the required concentration of that growth factor, the rumen species would be outcompeted there by bacteria which did not have that requirement, and rumen species may tend to have more nutritional requirements than digestor species.

Phosphate and iron are normally essential for anaerobic digestion, while nickel, cobalt, and molybdenum are required for some methanogenic bacteria. Nickel is required for factor F430, a coenzyme of methanogenic bacteria, although very high levels of nickel have been shown to be more inhibitory to methanogenic bacteria than to fermentative bacteria. The very high levels of iron addition required to give maximal levels of methanogenesis in biodegradability assays was probably not due to requirements for high concentrations of iron, but rather due to traces of nickel in the iron solution which satisfied the nickel requirement only when large amounts of iron were added. Iron could also be stimulatory by precipitating phosphate, which otherwise could precipitate other required trace minerals.

Other metals, especially calcium, magnesium, sodium, and potassium, are often required as well. The concentrations of metals relative to each other and to certain other metals, especially heavy metals, are as important as the absolute concentrations, in this respect they can be considered as inhibitors as well as nutrients. The levels of phosphate and sulfide can also effect the concentrations of these metals by precipitation. Sulfide is the normal sulfur source for bacteria in digestors, although some species require or are stimulated by cysteine or methionine, and a few can use sulfate.[5] Ammonia is required as a nitrogen source by many species, but amino acids or short polypeptides are stimulatory to some, and required by a few. In addition, some species can use urea or nitrate as a nitrogen source. Some rumen species are stimulated by high levels of acetate, although for many the carbon sources for growth can be the intermediates of the fermentation. Some require other volatile organic acids, especially isobutyric, isovaleric, and 2-methylbutyric acids. These, as well as butyrate and valerate, may be required for fatty acid or amino acid synthesis.[12] In some rumen strains CO_2 is required as a carbon source, and in some it is also necessary in dissimilatory reactions which produce succinate from pyruvate.[13]

When nutrients are present at concentrations much higher than required they can become inhibitory. High concentrations of ammonia are inhibitory, although digestors can become acclimated to them.[14] Sulfide can also be inhibitory[15]. The importance of mineral nutrients in anaerobic digestors can be studied with batch or continuous culture experiments to determine concentrations which give optimal rates of methanogenesis.

For other nutrients, individual bacteria grown in pure culture may have requirements which are normally met in the digestor by cross-feeding from other species, and so are not required as additions to the digestor. These nutrients, such as several B- and other vitamins should be included when axenically culturing bacteria from digestors, although their addition to digestors may not be stimulatory.

When animal wastes are digested there may be antibiotics present at inhibitory/concentrations. These can cause digestors to fail, although acclimation of the microflora to these antibiotics can occur.[16] Both mesophilic and thermophilic digestors have been demonstrated to adapt to monensin or chlortetracycline containing animal wastes if they are given time to acclimate.[16]

REFERENCES

1. Reddy, C.A., Bryant, M.P. and Wolin, M.J. "Ferredoxin- and Nicotinamide Adenosine Dinucleotide-dependent H_2 Production from Ethanol and Formate in Extracts of S Organism from 'Methanobacillus Omelianskii'," *J. Bacteriol., 110*:126-132 (1972).
2. McInerney, M.J., Bryant, M.P. and Pfennig, N. "Anaerobic Bacterium that Degrades Fatty Acids in Syntrophic Association with Methanogens," *Arch., Microbiol. 122*:129-135.

3. McInerney, M.J., Bryant, M.P. Hespell, R.B. and Costerton, J.W. *"Syntrophomonas wolfei* Gen. Nov. Sp. Nov., An Anaerobic, Syntrophic, Fatty Acid-degrading Bacterium," *Appl. Environ. Microbiol., 41*:1029-1039 (1981).
4. Zinder, S.H. and Koch, M. "Acetate Oxidation by a Thermophilic Methanogenic Syntrophic Coculture," *Abst. Annu. Meeting, Amer. Soc. Microbiol.*, 146 (1983).
5. McInerney, M.J. and Bryant, M.P. "Review of Methane Fermentation Fundamentals," in:D.L. Wise, ed. *Fuel and Gas Production From Biomass,* Vol.1. CRC Press, Inc., Boca Raton, Fl., pp. 19-46 (1981).
6. Pfeffer, J.T., "Temperature Effects on Anaerobic Fermentation of Domestic Refuse," *Biotechnol, Bioeng., 16*:771-787 (1974).
7. Buhr, H.O. and Andrews, J.F. "The Thermophilic Anaerobic Digestion Process," *Water Res., 11*:129-143 (1977).
8. Chen, M., "Adaptation of Mesophilic Anaerobic Fermentor Populations to Thermophilic Temperatures," *Appl. Environ. Microbiol., 45*:1271-1276 (1983).
9. Bryant, M.P., Varel, V.H. Frobish, R.A. and Isaacson, H.R. "Biological Potential of Thermophilic Methanogenesis from Cattle Wastes," in: H.G. Schlegel and Barnea, J. eds. *Microbial Energy Conversions.* Erich Goltze KG, Gottingen, pp. 347-358 (1976).
10. Pohland, F.G. and Bloodgood, D.E. "Laboratory Studies on Mesophilic and Thermophilic Anaerobic Sludge Digestion," *J. Water Pollut. Control Fed., 35:*11-42 (1963).
11. Harned, H.S. and Davis, R. Jr., "The Ionization of Carbonic Acid in Water and the Solubility of Carbon Dioxide in Water and Aqueous Salt Solutions from 0° to 50° C," *J. Amer. Chem. Soc., 65*:2030-2037 (1943).
12. Allison, M.J., Bryant, M.P. and Doetsch, R.N. "Studies on the Metabolic Functions of Branched-chain Volatile Fatty Acids, Growth Factors for Ruminococci, I. Incorportion of Isovalerate into Leucine," *J. Bacteriol., 83:*523-532 (1962).
13. Dehority, B.A., "Carbon Dioxide Requirement of Various Species of Rumen Bacteria," *J. Bacteriol., 105:*70-76 (1971).
14. van Velsen, A.F.M., "Adaptation of Methanogenic Sludge to High Ammonia-nitrogen Concentrations," *Water Res., 13:* 995-999 (1979).
15. Anderson, G.K., Donnelly, T. and McKeown, K.J. "Identification and Control of Inhibition in Anaerobic Treatment of Industrial Wastewaters," *Process Biochem., 17(4):*28-32, 41 (1982).
16. Varel, V.H. and Hashimoto, A.G. "Effect of Dietary Monesin or Chlortetracycline on Methane Production from Cattle Waste," *Appl. Environ. Microbiol., 41:*29-34 (1981).

18

Biosensors in Fermentation and Environmental Control*

Introduction

On-line measurement of raw materials and products are important for the control of fermentation processes. Monitoring of organic pollutants is also needed for environmental control. Spectrophotometry and chromatography can be used for the determination of these compounds but they are unsuitable for on-line measurements. Electrochemical determination of these compounds has definite advantages. For example, wide concentration ranges are measurable without pretreatment and there is no need for the test sample to be optically clear. Recently many biosensors have been developed and provided methods of rapid and continuous measurement of various compounds. Enzyme and microbial sensors can be used for fermentation and environmental control. However, microbial sensors are suitable for the industrial process because they are stable for a long time. Two different types of microbial sensors were developed for measurement of organic compounds.

First, microbial sensors made up of immobilized whole cells and an oxygen probe were used for determination of substrates and products. The concentration of compounds was determined from microbial respiration activity which could be directly measured by an oxygen probe.

Second, a microbial sensor made up of immobilized microorganisms and an electrode was used for determination of organic compounds. The concentration of compounds was indirectly found from electroactive metabolites like proton, carbon dioxide, hydrogen, formic acid, and reduced cofactors which could be measured by the electrode.

In this chapter, microbial sensors for fermentation and environmental control are described.

BIOSENSORS FOR NUTRIENTS

Glucose Sensor

The determination of glucose is quite important for the process control. Assimilation of glucose by micro-organisms can be found by an oxygen electrode because respiration activity increase after

* Excerpts taken from the review published by Prof. Isao Karube, Nagatsuta, Japan.

assimilation of organic compounds. Therefore, it is possible to construct a microbial electrode sensor for glucose using immobilized whole cells that utilize mainly glucose and an oxygen electrode.

A microbial electrode made up of immobilized whole cells of *Pseudomonas fluorescens* and an oxygen electrode was developed for the determination of glucose. Furthermore, the microbial sensor was applied to the continuous determination of glucose in molasses.

A schematic diagram of the microbial sensor is illustrated in Fig. 18.1. The electrode consisted of bacteria membranes and an oxygen electrode. The bacteria membrane is in direct contact with the oxygen electrode and is tightly secured with rubber rings.

The microbial sensor was inserted into a sample solution and the sample solution was saturated with dissolved oxygen and stireed magnetically while measurements were taken. The temperature of a sample solution was maintained at 30.0 ± 0.1 °C. The current was measured by a milliammeter and the signal was displayed on a recorder.

Fig. 18.2 shows the typical response curves of the sensor. The current at time zero was that obtained in a sample solution saturated with dissolved oxygen. The bacteria began to utilize glucose in a sample solution when the sensor was placed in it. Then consumption of oxygen by the bacteria in collagen membrane began. Consumption of oxygen by bacteria caused a decrease in dissolved oxygen on the membrane. Because of this the current of the sensor decreased with time until a steady state was reached. The steady state indicated that the consumption of oxygen by the bacteria and the diffusion of oxygen from the solution to the membrane were in equilibrium. The steady state current was attained within 10 min at 30° C. The steady state depended on the concentration of glucose.

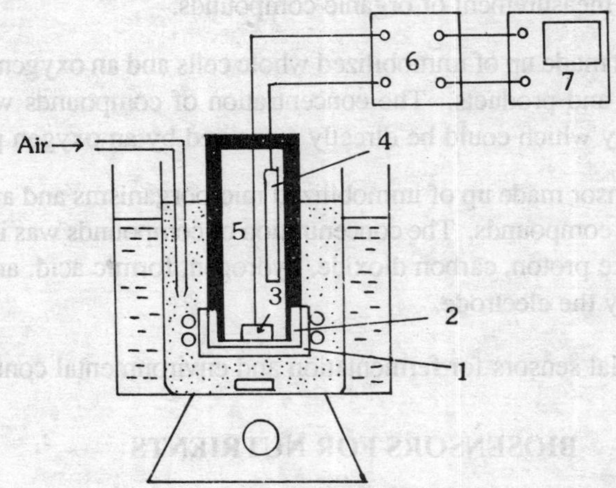

Fig. 18.1. Scheme of the microbial electrode sensor for glucose. 1, Bacterial collagen membrane; 2, Teflon membrane; 3, Platinum cathode; 4, Lead anode; 5, Electrolyte (KOH); 6, Ammeter; 7, Recorder.

When the sensor was removed from the sample and placed in a solution free of glucose, the current of the microbial sensor gradually increased and returned to the initial level within 15 min at 30° C.

The sensor responded slightly to fructose, galactose, mannose and saccharose. However, no response was observed in the case of amino acids. Therefore, the selectivity of the microbial sensor for glucose was satisfactory.

A linear relationship was observed between the current and the concentration of glucose below 20 mg l^{-1} by the steady state method. The minimum concentration for determination was 2 mg glucose l^{-1}. The current was reproducible within $\pm$ 6% of relative error when a sample solution containing 10 mg l^{-1} of glucose was employed. The standard deviation was 0.6 mg l^{-1} in 20 experiments.

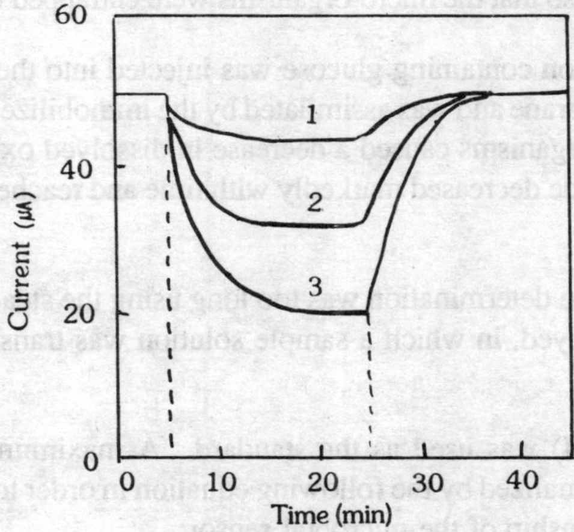

Fig. 18.2. Response curves of the microbial electrode sensor. Glucose concentrations were 1, 3.6 mg l^{-1}; 2, 10 mg l^{-1}; 3, 16 mg l^{-1}.

The microbial sensor for glucose was applied to molasses. The concentrations of glucose was determined both by the microbial sensor and by the enzymatic method.[1] Satisfactory comparative results were obtained. Glucose in molasses was determined with an average relative error of 10% by the microbial sensor.

The reusability of the microbial sensor was examined. The operation was repeated more than ten times per day. No decrease in current output was observed over a two week period and 150 assays.

Assimilable Sugar Sensor

On-line measurements of substrate concentrations in culture broths are required in the fermentation industry. In the cultivation of micro-organisms in cane molasses, which contains various sugars, determination of the total assimilable sugars in a broth is important for the control of the fermentation process. Reduced sugars and sucrose in culture broth are determined by the ferricyanide method.[2] However, the method is not completely reliable because unassimilable substances are also determined.

As described above, assimilation of glucose by micro-organisms can be determined from the respiratory activity of the microorganisms, which can be directly measured by an oxygen electrode. It is therefore possible to devise a microbial sensor for total assimilable sugars using immobilized micro-organisms. The micro-organisms used for the sensor was for preference, the same species as that being cultivated in the fermentation. The sensor was applied to the determination of total assimilable sugars in a fermentation broth for glutamic acid production. *Brevibacterium lactofermenium* AJ 1511 was used for the sensor.

A strip of nylon net was coated with the cells (0.015 g). The nylon net retaining micro-organisms was placed on the surface of the Teflon membrane of an oxygen electrode and covered with a cellophane membrane so that the micro-organisms were entrapped between the two membranes.

When the sample solution containing glucose was injected into the system, glucose permeated through the cellophane membrane and was assimilated by the immobilized micro-organisms. Oxygen consumption by the micro-organisms caused a decrease in dissolved oxygen around the membrane, and the current of the electrode decreased markedly with time and reached steady state within 10 min (the response time).

The time required for the determination was too long using the steady state method. Therefore, the pulse method was employed, in which a sample solution was transferred to the flow cell for a certain time.

Glucose solution (l mM) was used as the standard. A maximum decrease in current (peak height) for a sample was normalized by the following equation in order to correct for variations in the current-concentration relationship of the microbial sensor:

Normalized peak height =

$$\frac{\text{observed peak height}}{\text{peak height of the standard solution}} \qquad ...(1)$$

Fig. 18.3 shows a calibration curve of the microbial sensor for assimilable sugars such as glucose and fructose. Peak heights were normalized against the standard solution (l mM glucose solution) by equation (1). The calibration curves for glucose, fructose and sucrose respectively, can be represented by the following equations derived by the method of least squares:

$$I_1 = 0.07 + 0.95\ C_1 \qquad ...(2)$$

$$I_2 = 0.05 + 0.77\ C_2 \qquad ...(3)$$

$$I_3 = 0.03 + 0.91\ C_3 \qquad ...(4)$$

where C is the concentration of sugar (mM) in the flow cell, I is the normalized peak height and subscripts 1, 2 and 3 represent glucose, fructose and sucrose, respectively. The ratio of the sensitivity to glucose (180 mg l^{-1}), fructose (180 mg l^{-1}) and sucrose (360 mg l^{-1}) was 1.00 : 80 : 0.92. The sensitivity of the microbial sensor for sugars corresponds to the amount of oxygen consumed by the immobilized micro-organisms assimilating them.

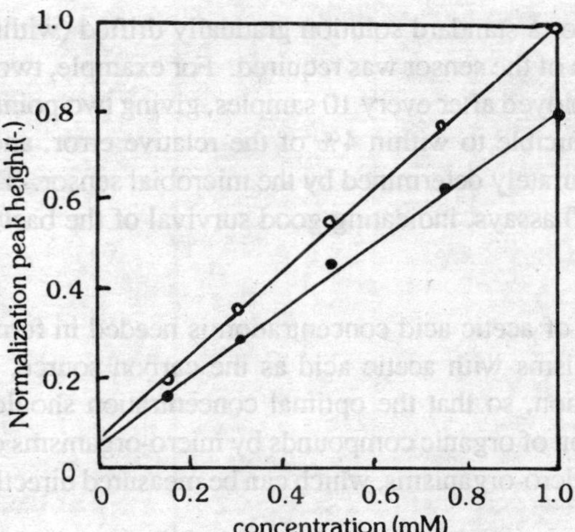

Fig. 18.3. Current concentration relationship of glucose (O) and fructose (●). A sample solution (0.21 ml) containing various amounts of glucose and fructose was injected into the system for 0.5 min.

The sensor scarcely responded to the sugars, which might be slowly utilized by the micro-organisms; on the other hand, the sensor gave little response to glutamic acid. Therefore, the selectivity of the microbial sensor for assimilable sugars was satisfactory for fermentation control.

When a mixture of assimilable sugars was applied to the system, the observed normalized peak height was compared with that calculated by the following equation :

$$I = 0.15 + 0.95\, C_1 + 0.77\, C_2 + 0.91\, C_3 \qquad ...(5)$$

where I is the normalized peak height and C_1, C_2 and C_3 are the concentrations of the assimilable/ sugars, as defined above. Equation (5) means that the response of the microbial sensor to a mixture of the assimilable sugars is equal to the algebraic sum of the responses to the individual assimilable sugars. The difference between the observed and the calculated values was within 8%. These results suggest that total assimilable sugars can be measured by the microbial sensor.

The microbial sensor for total assimilable sugars was applied to a fermentation broth (cane molasses) for glutamic acid production. In this method, the concentration of assimilable sugars in the broth was determined as glucose, because the sensor was calibrated with a standard solution containing glucose. On the other hand, the concentration of reducing sugars in the broth was determined by the ferricyanide method.[2]

Results obtained by the conventional method were higher than those obtained by the microbial sensor because the conventional method was influenced by reducing substances other than the assimilable sugars.

The reusability of the sensor was examined by using a standard solution containing glucose and the broths. When the microbial sensor was used continuously for a long time, the maximum current decrease of the electrode for a standard solution gradually drifted (within 15% during 10 days use) thus, occasional calibration of the sensor was required. For example, two standard solutions (0.4 and 1.0 mM glucose) were employed after every 10 samples, giving two points of calibration. The values obtained were then reproducible to within 4% of the relative error, and the concentration of total assimilable sugars was accurately determined by the microbial sensor. This sensor could be used for more than 10 days and 960 assays, indicating good survival of the bacteria in the electrode.

Acetic Acid Sensor

On-line measurement of acetic acid concentration is needed in fermentation processes. In the cultivation of micro-organisms with acetic acid as the carbon source, acetic acid inhibits growth above a certain concentration, so that the optimal concentration should be maintained by on-line measurements. Assimilation of organic compounds by micro-organisms can be dertermined from the respiration activity of the micro-organisms, which can be measured directly with an oxygen electrode.

A microbial sensor consisting of immobilized yeasts, *Trichosporon brassicae,* a gas-permeable Teflon membrane and an oxygen electrode is described for the determination of acetic acid.[3] The sensor was applied to the continuous determination of acetic acid in fermentation broth.

The system Fig. 18.4 made up of a jacketed flow cell, a magnetic stirrer, a peristaltic pump, an automatic sampler and a current recorder.

The temperature of the flow cell was maintained at 30.0 ± 0.1° C by warm water passed through the jacket. Tap water was adjusted to pH 3 with 0.05 M sulfuric acid and saturated with air. Then, it was transferred to the flow cell at a rate of 1.4 ml min^{-1} together with air at a flow rate of 200 ml min^{-1}. When the output current of the sensor became constant, a sample was passed into the system. When the sample solution containing acetic acid entered the system, acetic acid permeated through the gas-permeable membrane and was assimilated by the micro-organisms. Oxygen was then consumed by the micro-organisms so that the concentration of dissolved oxygen around the membranes decreased. The current decreased until it reached a steady state.

The time needed for the determination of acetic acid is long by the steady state method, therefore samples were passed into the flow cell for only 3 min.

In this case, the measurement could be done within 4 min. The total time required for an assay of acetic acid was 30 min by the steady state method and 15 min by the shorter method.

The calibration graphs obtained showed linear relationships between the current decrease and the concentration of acetic acid up to 54 mg l^{-1} by the steady state method and up to 72 mg l^{-1} by the shorter method. The minimum concentration for determination was 5 mg of acetic acid per liter. The current difference was reproducible within ± 6% for an acetic acid sample containing 54 mg l^{-1}. The standard deviation was 1.6 mg l^{-1} in 20 experiments.

The microbial sensor for acetic acid was applied to a fermentation broth of glutamic acid. The concentration of acetic acid was determined by the microbial sensor and by a gas chromatographic method. Good agreement was obtained; the regression coefficient was 1.04 for 26 experiments. Whole cells in the broth did not affect the electrochemical determination of acetic acid.

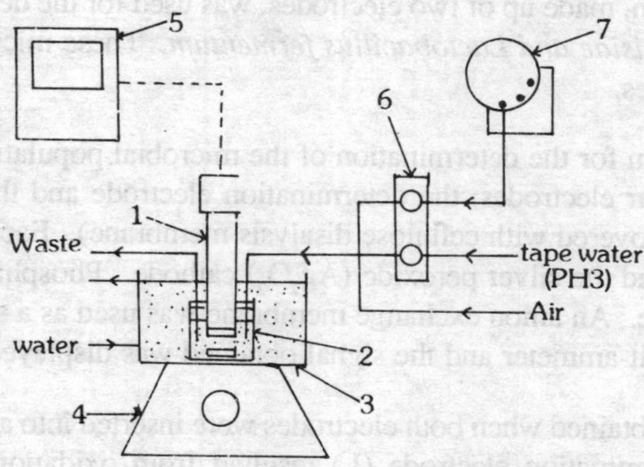

Fig. 18.4. The sensor system for acetic acid. 1. microbial electrode; 2, flow cell; 3, jacket; 4, magnetic stirrer; 5, recorder; 6, peristatic pump; 7, sampler.

The long-term stability of the microbial sensor was examined for acetic acid solutions (72 mg l^{-1}). The current output (0.29-0.25μ A) of the sensor was constant (within ± 10% of the original values) for more than 3 weeks and 1500 assays.

SENSOR FOR CELL POPULATION

Fuel Cell Type Electrode

The determination of microbial populations in fermentors is important for the control of fermentations processes. Several methods, such as hemacytometer counts, electronic particle counts, and colony counts, have been used for the determination of cell numbers. However, most of these methods are time consuming and are not suitable for the continuous determination of cell numbers in fermentors. Turbidimetry is a simple method for monitoring the number of cells in a fermentor. But sampling is needed, and a coloured or suspended broth is not suitable for this method.

Recently, the impedance measurement of culture media has been proposed as a method for cell number determination.[4] Nutrients are converted to various charged metabolites, such as organic acids and other compounds, by bacteria. Consequently, the impedance of media increases with increasing cultivation time. This method is suitable for the determination of small numbers of bacteria. An electrochemical method based on the potentiometric determination of hydrogen molecules produced by bacteria also has been developed for the estimation of cell numbers of hydrogen-producing bacteria, such as Enterobacteriaceae, etc. However, both electrochemical methods determine cell numbers indirectly from bacterial metabolites. The results obtained sometimes are not correlated with true cell numbers. Simple and continuous methods for the direct determination of cell populations are still required for fermentation control. It was found that bacteria were oxidized directly on the surface of the anode and a current was generated. This electrochemical system may be applicable to the determination of microbial populations.

The electrode system, made up of two electrodes, was used for the determination of populations of *Saccharomyces cerevisiae and Lactobacillus fermentum.* These micro-organisms are currently used in the food industries.

The electrode system for the determination of the microbial populations is shown in Fig. 18.5. In consists of two similar electrodes; the determination electrode and the reference electrode (the surface of the anode is covered with cellulose disalysis membrane). Each electrode was composed of the platinum anode and the silver peroxide (Ag_2O_2) cathode. Phosphate buffer (0.1 M, pH 7.0) was used as the catholyte. An anion exchange membrane was used as a separator. The current was measured with a millivolt-ammeter and the signal obtained was displayed on a recorder.

The currents were obtained when both electrodes were inserted into a culture broth. The current obtained from the determination electrode (I_1) resulted from oxidation of micro-organisms and electroactive substances. The current from the reference electrode (I_1) was attributed to oxidation of electroactive substances, because microorganisms could not penetrate through cellulose dialysis membrane. Consequently the current difference between two electrodes ($\Delta I = I_1 - I_2$) was proportional to the number of microbial cells in a culture broth : $\Delta I = I_1 - I_2 = kn$, where n is the number of micro-organisms and k is a constant.

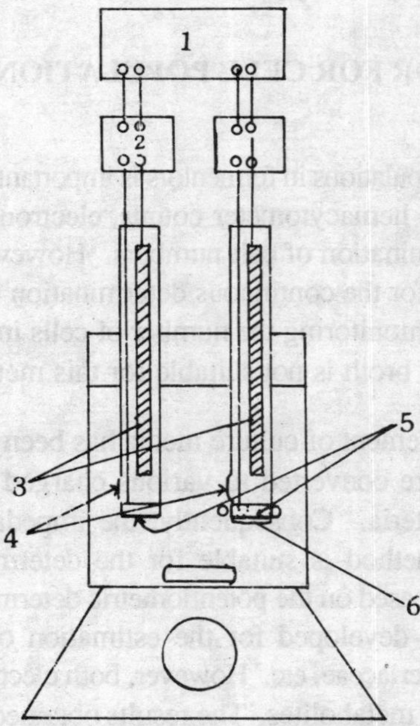

Fig. 18.5. Scheme of electrode system for determination of microbial population, 1. Recorder; 2, ammeter; 3, cathode (Ag_2O_2); 4, anion exchange membrane; 5, anode (Pt); 6, cellulose dialysis membrane.

The open circuit anode potential of the determination electrode was –370 mV versus the saturated calomel electrode, and the cathode potential was 220 mV versus the saturated calomel electrode. The anode potential of the reference electrode was –150 mV versus the saturated calomel electrode.

As the anodes polarized, high currents were obtained initially from both electrodes. Then, the anode potentials became constant. The diffusion of micro-organisms and electroactive substances to the determination electrode became the rate-determining factor, and a steady state current was obtained. The response time (the time required for the current to reach a steady state) of the determination electrode was 10 min. On the other hand, the response time of the reference electrode was 15 min. Because the reference electrode was covered with cellulose dialysis membrane, the response time was longer than that of the determination electrode.

S. cerevisiae in a culture broth was excluded by centrifugation. Then the electrodes were immersed in the culture broth. The open-circuit anode potentials of both electrodes were –150 mV versus the saturated calomel electrode, and the same current was obtained after 10 min. of incubation.

A linear relationship was obtained between the current difference and the number of cells below 4×10^{-8} cells per ml. The current difference was reproducible, with an average relative error of 5% when a medium containing 2×10^8 *S. cerevisiae* cells per ml was used for experiments.

Furthermore, the cell population of *Lactobacillus* was determined by the electrode system. A linear relationship was observed between the current difference and the number of cells measured by colony counts. This system could also be applied to anaerobic bacteria.

The cell population of *S. cerevisiae* in a fermentor was determined continuously by using the electrochemical method.

The cell number plots measured by visual counting are located near the line which was determined by the electrochemical method. A little deviation was observed in exponential growth of *S. cerevisize*. This may be caused by budding of the yeast. Many buds were observed at this time with a light microscope.

The current differences between the electrodes decreased markedly when the micro-organisms were inactivated by heat, ultrasonic and nystain (an antibiotic which destroys the cell membrane) treatments. These facts suggest that only living micro-organisms contribute to the current generation. Current is scarcely obtained from dead cells and ordinary particulate materials. Consequently, the number of viable cells is determined by this electrochemical method. Measurement is not interfered with by non-viable cells and non-microbial particles.

Potentiostatic System

Bacteria were oxidized directly on the surface of the anode and a current was generated when an electrode composed of a platinum anode and a silver peroxide cathode was inserted in a microbial medium. The system composed of two electrodes has been applied to the determination of cell populations as described above. The cell populations of yeast and anaerobic bacteria could be determined by using a new electrode system. However, this electrode system was not suitable for continuous determination of cell populations because the cathode (Ag_2O_2) was decomposed by sterilization.

A new potentiostatic method was developed for the determination of cell numbers, the new system has a determination and a reference part.[5] Each component consists of two platinum electrodes and a saturated calomel electrode. This system was used for continuous monitoring of cell populations of *Bacillus subtilis* which are currently used for the production of antibiotics and enzymes in fermentation industries.

The apparatus contains a similar system for determination and for reference (the surface of the anode is covered with cellulose dialysis membrane). Each system has a working electrode (platinum), the counter electrode (platinum) and the saturated calomel electrode (S.C.E.). Saturated KCl was used as an electrolyte. Three electrodes were connected through a potentiostat. The current was measured by a millivolt-ammeter and the signal obtained was displayed on a recorder.

The effect of potential on the current different between both systems was examined. The currents of both systems increased with increasing potential. The current difference also increased, and became constant above + 0.2 V. Therefore, the experiments were performed at the anode potential of + 0.2 V relative to the S.C.E.

Two systems were immersed in a culture broth containing 1.4×10^9 cells ml^{-1} of *B. subtilis*. The potentials of both systems were at + 0.2 V relative to the S.C.E. High currents are obtained initially and then the steady state current was obtained. Therefore, the diffusion of micro-organisms and electroactive substances to the electrode seems to become the rate determining factor. The response time of the both systems was 3-5 min. The steady state current was measured 5 min. after the insertion of the system in a medium. A culture broth containing 1.4×10^9 cells ml^{-1} of *B. subtilis* was sterilized for 10 min. at 110° C. Then, both systems were immersed in the broth. The current difference of both systems was very small. Therefore, the current difference between both systems might result from reaction of the living bacteria with the anode. A linear relationship was obtained between the current difference and the cell population of *B. subtilis* below 2.0×10^9 cells ml^{-1}. The current difference was reproducible with an average relative error of 4% when 30 samples of culture broths containing 1.4×10^9 cells ml^{-1} were employed.

Fig. 18.6 shows the continuous determination of microbial population in a fermentor. The time course of cell population determined by the electrochemical system was represented by the solid line. The point shows the cell population determined by the colony method. The cell population plots measured by the conventional method are located near the solid line obtained by the electrochemical methods. The continuous monitoring of cell population in the actual cultivation was repeated twenty times. The same growth curves as presented in Fig. 18.6 were obtained. The current difference was reproductible with an average relative error of 8%. This shows that the systems can be used for more than 400 h.

The sterilization of electrodes is needed for the practical application of the system for continuous determination of cell population. After sterilization for 10 min., in an autoclave at 110° C, the systems were immersed in the broth containing 1.4 x 10^9 cells of *B. subtilis* and the currents were measured. As a result, there was no difference between currents before and after sterilization. The system was sterilized 30 times, however, the output currents did not change. Therefore, the system can be sterilized in an autoclave.

Piezoelectric Membrane System

The measurement of sludge concentration in wastewaters was performed with an ultrasonic method. Ultrasonic waves can transmit into sewages, and the velocity of waves changes as the solution components change. A new membrane called a piezoelectric membrane, which consists of a polyacetal resin, chlorinated polyethylene, and Pb (Zr Ti) O_3, has been developed as an oscillator. One advantage of this membrane is its flexibility, and it can also be cut into any shape easily.

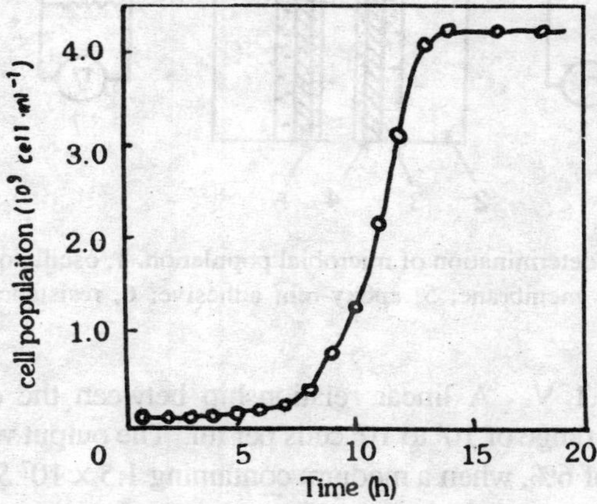

Fig. 18.6. Continuous determination of cell populations — population determined by the electrochemical method, O, population determined by the colony count method.

We have developed an apparatus for the determination of cell populations with the piezoelectric membrane. It was applied to the continuous monitoring of microbial population in a fermentor. Furthermore, the apparatus was applied to the continuous determination of the *Saccharomyces cerevisiae* population in media. The apparatus used for the measurement of cell populations is shown in Figure 9.7 (the distance between the two piezoelectric membranes was 2.5 mm). The surface of the piezoelectric membrane (1.5 by 2.0 cm; thickness, 0.2 mm) was coated by an epoxyresin adhesive for electrical insulation. The apparatus was put in a medium (50 ml), and an arbitrary voltage was charged to the piezoelectric membrane by an oscillator. The ultrasonic waves generated were transmitted in the medium, and they vibrated the other piezoelectric membrane. The output voltage generated was measured with an AC voltmeter and the signal was displayed on a recorder. When the piezoelectric membrane system (input voltage = 5.1 V) was immersed in the suspension of *S.cerevisiae*, 20 to 100 mV of alternating current potential was measured as an output voltage. The output voltage gradually increased with increasing cell population in preliminary experiments. The standard deviation of the determination was about 10% when a reaction medium of the same cell population (5×10^7 cells per ml) was used for experiments.

At first, the effect of the ultrasonic frequency of the output voltage was investigated with an *S. cerevisiae* suspension. Wet cells were suspended in 50 mM phosphate buffer. The experiment was performed at 26.0 ± 0.5° C. The output voltage of the system increased linearly with increasing ultrasonic frequency. It shows that the piezoelectric membrane does not have an intrinsic resonance frequency (the specific frequency at which a piezoelectric material resonates) in this frequency range. The frequency applied to the system was fixed at 40 kHz in subsequent experiments.

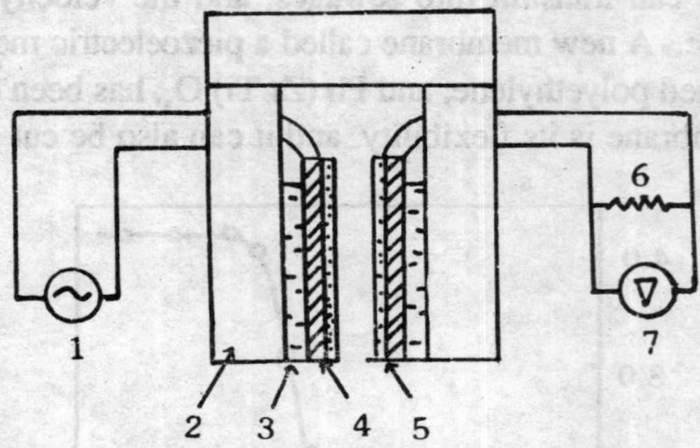

Fig. 18.7. Apparatus used for the determination of microbial population. 1, oscillator; 2, plastic plate; 3, silicon rubber; 4, piezoelectric membrane; 5, epoxy-rein adhesive; 6, resistance (100 kΩ); 7, alternating current voltmeter.

The input voltage was 5.1 V. A linear relationship between the output voltage and cell population was observed in the range of 10^6 to 10^8 cells per ml. The output voltage was reproducible, with an average relative error of 6%, when a medium containing 1.5×10^7 *S. cerevisiae* cells per ml was used for experiments. The system was also available for the determination of *S. cerevisiae* cell populations in the range of 10^8 to 10^{10} cells per ml, although the slope of the calibration curve was different from that for 10^6 to 10^8 cells per ml. This system could be applied not only to yeast cells, but also to gram-positive and gram-negative bacteria.

The continuous monitoring of *S. cerevisiae* cell population during cultivation in the medium was performed by the piezoelectric membrane system. The experimental conditions were as follows : temperature, 31.0 ± 0.5° C; rotation with magnetic stirrer, *ca.* 500 rpm; aeration with minipump, *ca.* 1 vol vol^{-1} min^{-1}; frequency of input, 40 kHz; input voltage, 5.1 V; and medium volume, 50 ml. The values determined by the system were slightly higher than those measured by hemacytometric and turbidimetric methods. The deviations were especially large in the logarithmic phase of bacterial growth. This may be caused by the budding of *S. cerevisiae*. However, desirable agreement was obtained between this method and the conventional ones.[10] Additionally, the conductivity of the medium increased during cell growth. This was caused by electrolytes produced by bacteria and by charges on cells.

Colored media, such as molasses, are used for fermentation by micro-organisms. Therefore, the system was appliable to the continuous determination of *S. cerevisiae* cell populations in cane molasses. The medium used contained 5% molasses (pH 7.0). A similar good agreement with hemacytometric and turbidimetric method was obtained with a relative error of 7%. This system could be used for the determination of cell populations in coloured media. Additionally, little effect of aeration, rotation, and foaming of the medium on the output voltage was observed during fermentation.

Dye-coupled Electrode System

As described above, a new electrode system for the continuous determination of cell numbers in fermentation media. A new electrode system using a fuel cell-type electrode, a redox dye, and a porous acetylcellulose membrane filter for trapping micro-organisms was developed.[6] This system was used to determine small numbers of cell populations in polluted waters.

The electrode composed of the platinum anode and the silver peroxide cathode. Phosphate buffer (0.1 M, pH 7.0) was used as the catholyte. An anion-exchange membrane was used as a separator. The electric current was measured with a millivolt-ammeter, and the signal obtained was displayed on a recorder. When the electrode was immersed in water containing 1.4×10^7 cells of *E. coli* per ml, scarcely any current was generated. To amplify the current, a redox dye was added to the sample solution containing *E. coli,* and the steady state current generated was measured. The currents were low when methylene blue, methyl viologen, tetrazolium red, and phenazine methosulfate were used. On the other hand, 2,6-dicholorphenolindophenol (DCIP) gave a higher current.

Without adding redox dye, a slight current (0.02 μA) was obtained from the electrode. However, a higher current was obtained from all of the bacteria tested in the presence of DCIP.

Micro-organisms in water (100 ml) were concentrated on a membrane filter, and the membrane was fixed on the anode of the electrode. One hundred milliliters of water containing 5.4×10^5 cells of *E. coli* per ml was used. When the electrode was immersed into the phosphate buffer solution, a small current was obtained. Then DCIP (40 μM) was added to the solution, and the current was continuously measured. The current of the electrode increased markedly with increasing reaction time. The response time was 15 min. The current, however, decreased markedly when the micro-organisms were inactivated by heat treatment (110° C, 10 min). Various numbers of micro-organisms were concentrated on the membrane and fixed on the anode of the electrode. Although the current generated varied from organism to organism when equivalent cell concentrations of different species were tested, a linear relationship was obtained between the current of the electrode and the number of cells measured by the colony count method above 10^4 cells ml^{-1}. The electrode was used to determine cell populations in industrial waters. Micro-organisms in 2 to 100 ml (depending on the micro-organism content) of industrial waters from a paper mill company and a petrochemical company were concentrated on the membrane and attached on the anode of the electrode. The cell numbers were calculated from the calibration curve for *E. coli.* The cell population was also determined by the conventional colony count method. Although considerable deviation from a standard curve was observed, each test system showed a good correlation above cell numbers of 10^4 cells per ml.

Combinations of micro-organisms concentrated on a membrane filter, a redox dye, and an electrochemical device enabled rapid and simple measurement of micro-organisms in the range of 10^4 to 10^6 cells ml^{-1}. The effects of redox dyes as electron carriers are well-known. DCIP in particular is often used as an artificial electron mediator in some enzyme reaction systems. Therefore, the dye was also expected to act as an electron mediator between living micro-organisms and an electrode. Co-factors such as reduced flavin mononucleotide and $NADH_2$, are the most probable electron donors on the surface of micro-organisms, though the details about redox compounds on the cell membrane are not clear. Our results show that DCIP is effective as an electron mediator from living micro-organisms to a platinum anode, as schematically.

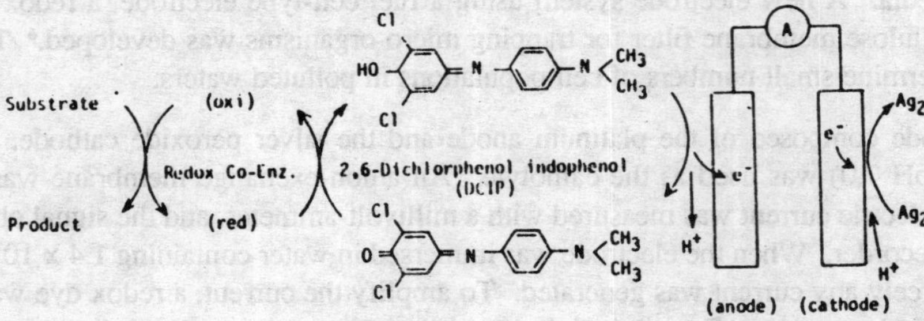

Fig. 18.8. Principle of current generation.

Membrane Filter-oxygen Electrode System

The system for the determination of microbial populations is sufficient for continuous monitoring of micro-organisms in microbial reaction mixtures. However, it is not suitable for the determination of small numbers of viable cells.

An electrode system using an oxygen electrode and a porous acetylcellulose membrane filter for trapping micro-organisms was developed, and used to determine small numbers of viable cells. The system consisted of an oxygen electrode and a membrane filter (Millipore type HA, 0.45 µm pore size, 25 mm diameter, 150 µm thick). The current of the electrode was converted to the voltage by 2 kΩ resistance and displayed on a recorder.

The pyrex Filter Holder was used for retaining micro-organisms on the acetylcellulose membrane. The apparatus was trapped and autoclaved in the conventional manner before use. The buffer solution (pH 7.0, 50 ml^{-1}) containing glucose (500 mg l^{-1}) was saturated with oxygen before measurement. When the membrane filter-electrode containing various amount of *E. coli* was inserted into the solution, glucose was assimilated by the bacteria on the membrane. Oxygen was then consumed by the micro-organisms so that the oxygen concentration around the membrane decreased. The current decreased until it reached a seady state. It required 10 min to reach the steady state and the steady state current depended on the number of cells on the membrane, when the glucose concentration of the sample solution was constant. The current decreased with increasing cell members. Linear relationships were obtained below 3.2 x 10^8 cells for *Bacillus subtilis and*

Flavobacterium arabrescence. Therefore, this method can be used to determine variable cell numbers of various aerobic bacteria. The current was reproducible, with an average relative error of 5% when the electrode contained 2×10^8 of *E. coli.*

Linear relationships were also obtained between the current and the yeasts, *Saccharomyces cerevisiae and Trichosporon cutanium* cell numbers below 3.7×10^7. This method is also applicable, therefore, to the determination of the number of yeast cells.

The electrochemical method using a membrane filter was applied to the determination of cell populations in water and fermentation broth. 10 ml of sample solution containing an appropriate amount of bacteria was immobilized on the membrane filter. The cell population of *E. coli* was determined by the electrochemical method and by conventional colony counting. A good agreement was obtained; the correlation coefficient was 0.98 for 10 experiments.

Lactate Sensor

A lactate sensor composed of lactate oxidase and a voltammetric device to determine H_2O_2 is described and used to determine the numbers of lactic acid producing bacteria.

The system consisted of the lactate sensor and a membrane filter for trapping lactic acid producing bacteria. The lactate sensor was prepard as follows. The enzyme solution (100 μl containing 7.2 I.U. of lactate oxidase) was dropped onto a porous acetylcellulose membrane and immobilized on the membrane with slight suction. The enzyme membrane was attached to the platinum electrode (0.5 cm^2) of the voltammetric device, being held in position by a cellulose dialysis membrane. The membrane contained 0.35 I.U. of lactate oxidase. The platinum electrode was positioned at 0.6 V vs saturated calomel electrode by the potentiostat and the production of hydrogen peroxide by lactate oxidase was monitored with the voltammetric device.

The principle of the system for the determination of cell numbers is as follows. Lactic acid producing bacteria such as *Lactobacillus, Streptococcus and Leuconostoc* produce mainly lactate as a metabolite from glucose. The rate of lactate production depends on the cell numbers retained on the membrane filter, when the glucose concentration of the sample solution is constant. Therefore, the current obtained from the lactate sensor is also affected by the cell numbers on the membrane filter.

The response of the sensor was studied when the membrane filter containing various amounts of *Leuconostoc mesenteroides* was attached to the electrode. When the glucose solution was transferred to the cell, lactate was produced from glucose by the bacteria on the membrane. Therefore, the current gradually increased until it reached a steady state. It took 10 min. to reach the steady state and the steady state current depended on the cell numbers on the membrane. The current increased with increasing cell numbers.

The medium was diluted as required and micro-organisms were immobilized on the membrane before measurement. A linear relationship was obtained below 1.2×10^8 cells for *Leuconostoc mesenteroides*. The current was reproducible with an average relative error of 6% when the sensor contained 5×10^7 cells.

The system is suitable for the specific determination of lactic acid producing bacteria.

BIOSENSOR FOR PRODUCTS

Alcohol Sensor

On-line measurements of ethyl alcohol concentration in culture broth are required in fermentation industries. Furthermore, in the cultivation of yeasts using sugar as a carbon source, it is well-known that ethyl alcohol as a by-product, decreases the sugar basis yield of whole cells. In the cultivation of micro-organisms using methyl alcohol as a carbon source, the concentration of methyl alcohol must be maintained at the optimal level to avoid substrate inhibition. Many reports on applications of enzyme electrodes have been published for the determination of alcohols. However, enzymes are generally expensive and unstable.

It is well-known that many micro-organisms utilize alcohols as carbon. Assimilation of alcohols by micro-organisms can be determined from the respiration activity of micro-organisms. The respiration activity is directly measured with an oxygen electrode. Therefore, it is possible to construct a microbial sensor for alcohols using immobilized micro-organisms and an oxygen electrode.

A microbial electrode consisted of immobilized yeasts or bacteria, a gas permeable Teflon membrane, and an oxygen electrode was prepared for the determination of methyl and ethyl alcohols. Furthermore, the microbial sensor was applied to continuous determination of alcohols in a fermentation broth.

Unidentified bacterium AJ 3993 and *Trichosporon brassicae* CBS 6382 were used for the methyl and ethyl alcohol sensors, respectively. The porous membrane retaining micro-organisms was fixed on the surface of the Teflon membrane of the electrode. Therefore, the micro-organisms were trapped between the two membranes. Furthermore, a gas permeable membrane was placed on the surface of the electrode and covered with a nylon net. These membranes were fastened with rubber rings. The steady state currents obtained depended on the concentration of ethyl alcohol. The response time was within 10 min at 30° C.

However, the time required for the determination of ethyl alcohol is long by the steady state method. Therefore, the pulse method was employed for the determination. The assay can be done within 6 min by the pulse method. The total time required for the assay of ethyl alcohol was 30 min by the steady state and 15 min pulse methods. The minimum concentration for the determination was 2 mg ethyl alcohol l^{-1}. The current difference was reproducible within $\pm$ 6% of the relative error when a sample solution containing 16.5 mg l^{-1} of ethyl alcohol was employed. The standard deviation was 0.5 mg l^{-1} in 40 experiments. The sensor did not respond to volatile compounds such as methyl alcohol, formic acid, acetic acid, propionic acid, and other nutrients for micro-organisms such as carbohydrates, amino acids, and ions. As the microbial sensor was covered with a gas permeable membrane, only volatile compounds can penetrate through the membrane. However, an ethyl alcohol-utilizing yeast, *T. brassicae,* did not utilize methyl alcohol. The selectivity of the microbial sensor for ethyl alcohol was satisfactory.

The microbial sensor for ethyl alcohol was applied to fermentation broths of yeasts. The concentration of ethyl alcohol was determined by the microbial sensor and by gas chromatography. Satisfactory comparative results were obtained between them. The correlation coefficient was 0.98 with 20 experiments. The reusability of the microbial sensor for ethyl alcohol was examined. Ethyl alcohol solutions (from 5.5 to 22.3 mg l^{-1}) were used for long-term stability testing of the sensor. The current output of the sensor was almost constant for more than three weeks and 2100 assay.

A microbial sensor consisted of immobilized bacteria, a gas permeable membrane, and an oxygen electrode was applied to the determination of methyl alcohol. A methyl alcohol-utilizing bacterium (AJ 3993) was employed for the electrode. A linear relationship was also observed between the current decrease and the concentration of methyl alcohol. Therefore, the sensor can also be applied to the determination of methyl alcohol.

Formic Acid Sensor

Formic acid is now attracting attention as an intermediate of biomass conversion which is easily converted to hydrogen. It has been determined by gas and ion-exchange chromatography and spectrophotometry. However, coloured sample cannot be determined by spectrophotometric method. Moreover, these conventional methods are not suitable for on-line measurement. Some anaerobic bacteria such as *Escherichia coli, Clostridium butyricum,* and *Rhodospirillum rubrum* produce hydrogen from formic acid. The reactions involved are :

$$\text{Formic acid} \longrightarrow \text{Ferredoxin}_{\text{reduced}} + CO_2$$

$$\text{Ferredoxin}_{\text{reduced}} \xrightarrow{\text{hydrogenase}} \text{Ferredoxin}_{\text{oxidized}} + H_2$$

$$\text{Formic acid} \xrightarrow{\text{formate dehydrogenase}} \text{cytochrome C}_{\text{reduced}} + CO_2$$

$$\text{Cytochrome C}_{\text{reduced}} \xrightarrow{\text{hydrogenenase}} \text{cytochrome C}_{\text{reduced}} + H$$

Therefore, determination of formic acid is possible by using *Clostridium butyricum* and a fuel cell type electrode.

A specific microbial sensor composed of immobilized *C. butyricum,* two gas permeable Teflon membranes and a fuel cell type electrode was developed for the determination of formic acid.[7] The sensor was applied to the determination of formic acid in fermentation broth.

The fuel cell type electrode consisted of a Teflon membrane, a platinum anode, Ag_2O_2 cathode and an electrolyte. The bacteria membrane was attached on the surface of the Teflon membrane. This was covered with a porous Teflon membrane.

When the sensor was inserted into a sample solution containing formic acid, formic acid permeated through the porous Teflon membrane. Hydrogen, produced from formic acid by *C.*

butyricum, penetrated through the Teflon membrane, and was oxidized on the platinum anode. As a result, the current increased until it reached a steady state, which indicated that the production of hydrogen by microorganisms and the diffusion of formic acid from the sample solution to the immobilized cells were in equilibrium. The steady state was attained in all cases within 20 min. The steady state current depended on the concentration of formic acid.

A linear relationship was obtained between the steady state current and the formic acid concentration below 1,000 mg l^{-1}. The minimum concentration for determination was 10 mg l^{-1}. The currents were reproducible with an average relative error of 5% when a medium containing 200 mg l^{-1} of formic acid was used. The standard deviation was 3.4 mg l^{-1} in 30 experiments.

The sensor did not respond to non-volatile nutrients such as glucose, pyruvic acid and phosphate ions. As the microbial sensor was covered with a porous Teflon membrane, non-volatile nutrients could not penetrate through the Teflon membrane. Volatile compounds such as acetic acid, propionic acid, n-butyric acid, methyl alcohol and ethyl alcohol can permeate through the porous Teflon membrane. However, no current was obtained from these compounds, because *C. butyricum* did not produce hydrogen from them.

The microbial sensor was applied to the determination of formic acid in the cultivation medium of *Aeromonas formicans*. The formic acid concentration was measured by the gas chromatography and by the microbial sensor. Good agreement was obtained between both methods; the regression coefficient was 0.98 for 10 experiments. The cultivation medium did not affect the current of the sensor. To study the stability of the immobilized *Clostridium butyricum* in the sensor, it was stored in 0.1 M phosphate buffer at 5° C. Formic acid (200 mg l^{-1}) was determined at 5-day intervals with these stored micro-organisms. The current output obtained from each experiment was constant for 20 days.

Methane Sensor

Methane is an attractive energy source and a main component of natural gases. It is used as gaseous furel in many fields, and it is well-known that methane forms explosive mixtures with air (5-14%). Therefore, rapid methods for the determination of methane in air are required in various fields such as coal mining and gasification processes. Methane is also produced by methanogenic bacteria. The biological formation of methane is the result of a specific type of bacteria energy-yielding metabolism. World-wide interest has arisen in the production of methane by fermentation of biomass, renewable ressources.

In general, methane-oxidizing bacteria utilize methane, and oxygen is consumed by the respiration as follows.[7]

$$CH_4 + 2O_2 \xrightarrow{\text{methane-oxidizing bacteria}} CO_2 + 2H_2O$$

The characteristics of a methane sensor system were examined and the sensor system was applied to the determination of methane in air.

Fig. 18.9 shows a schematic diagram of the system. The system consisted of two oxygen electrodes, two reactors, an electrometer and a recorder.

The reactors contained the culture medium, one with and one without bacterial cells. The oxygen electrodes consisted of a Teflon membrane, a platinum cathode, a lead anode and sodium hydroxide electrolyte (30% w/v). These electrodes were fixed to custom-made Teflon flow-through cells. The system was connected up with glass and Teflon tubing.

The system contained two vacuum pumps, one to evacuate the gas sample tube and the other to transport the sample gas through the system. The flow rate of the sample gas through the reactors was controlled with the glass valves at 80 ml min^{-1}. The cotton filter served to remove other micro-organisms in the sample gas and to prevent contamination of the two reactions and gas lines. The sample gas existing from the system was passed into the laboratory extraction hoods via glass tubes.

All the lines were made gas-tight with Teflon tubes and glass tubes, and careful checks were made for leakage. The lines were designed so as to maintain symmetry between the measuring and reference flow lines. The flow rates in each line were balanced by adjusting valves 15 and 16. The reactors were maintained at 30.0° C ± 0.1° C by a thermostated bath.

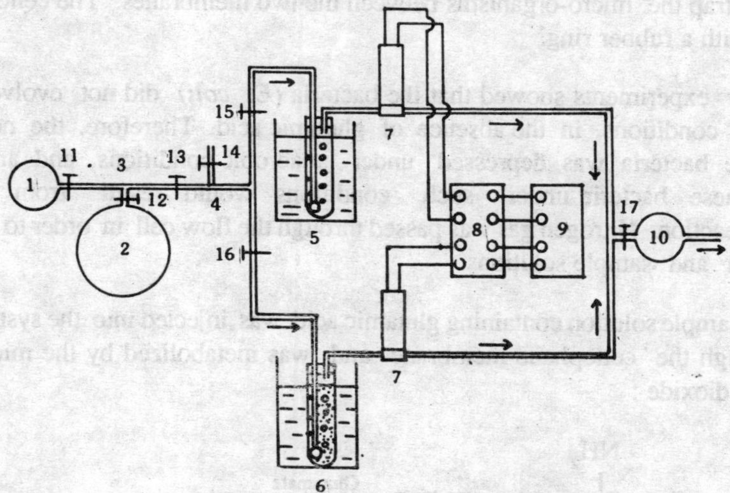

Fig. 18.9. Scheme of microbial sensor system for methane. 1, vacuum pump; 2, sample gas bag; 3, gas sample line; 4, cotton filter; 5, control reactor; 6, methane-oxidizing bacteria reactor; 7, oxygen electrode; 8, amplifier; 9, recorder; 10, vacuum pump; 11-17 glass stopcocks.

When the sample gas containing methane entered the reactor, methane was assimilated by the micro-organsims with consumption of oxygen. The current decreased to a minimum, which indicated that the consumption of oxygen by the micro-organisms, and the supply of oxygen from the air was in equilibrium. As the system contained two oxygen electrodes, the maximum difference of current depended on the concentration of methane in the sample gas. When pure air was again passed through the reactors, the current of the sensor returned to its initial level within 1 min. The response time required for the determination of methane was within 1 min, and the total time required for an assay of methane was 2 min.

The calibration graphs for the system were strictly linear for concentration of methane in the range 0-6.6 mM, the current difference ranging from 0 to 0.35 μA. The minimum concentration for determination was 13 μM. The current difference measured for the same sample (0.66 mM) was reproducible within 5%; the standard deviation was 9.40 nA in 25 experiments.

The microbial sensor system was applied to the determination of methane in air samples that were also analyzed by conventional gas chromatography. Over the range 0.2-3.5 mM methane in air the correlation coefficient between the results of the two methods was 0.97.

Glutamic Acid Sensor

Glutamic acid was produced by fermentation process and used as a food additive so that rapid automatic measurements of glutamic acid in fermentation broth are required. The sensor for glutamic acid was constructed as follows. The freeze-dried cells of *Escherichia coli* was mixed with one drop of water and coated on both sides of a nylon mesh which was placed on the surface of the silicone rubber membrane of the carbon dioxide electrode and covered with a cellophane membrane to entrap the micro-organisms between the two membranes. The cellophane membrane was fastened with a rubber ring.

Preliminary experiments showed that the bacteria *(E. coli)* did not evolve carbon dioxide under anaerobic conditions, in the absence of glutamic acid. Therefore, the normal respiration activity of the bacteria was depressed under anaerobic conditions, and any carbon dioxide produced by these bacteria under such conditions would result from the glutamate decarboxylase reaction. Nitrogen gas was passed through the flow cell in order to remove dissolved oxygen in buffer and sample solutions.

When the sample solution containing glutamic acid was injected into the system, glutamic acid permeated through the cellophane membrane and was metabolized by the micro-organisms to produce carbon dioxide :

$$HOOC—CH_2—\underset{}{\overset{NH_2}{\overset{|}{CH_2}}}—CH—COOH \xrightarrow[\text{decarboxylase}]{\text{Glutamate}} HOOC—CH_2—CH_2—CH_2—NH_2 + CO_2$$

The enzyme reaction was carried out at pH 4.4, which was sufficiently below the pK_a value (6.34 at 25° C) of carbon dioxide to allow the carbon dioxide around the membranes to increase. Due to this the potential of the carbon dioxide gas-sensing electrode increased with time. The assay can be done by using a 1-3 min injection period and measuring the maximum potential with little loss of senstivity, and consequent increase in the sample throughout rate.

Fig. 18.10 shows the response of the microbial sensor to various concentrations of glutamic acid. The plot of the maximum potential vs. the logarithm of the glutamic acid concentration was linear over the range; the slope over this range was approximately Nernstian. When a glutamic acid solution (400 mg l^{-1}) was measured in replicate, the standard deviation was 1.2 mg l^{-1}(20 experiments). The sensor responded to glutamic acid and glutamine and very slightly to some other amino acids. The response to glutamine can be decreased, if necessary, using acetone-treated *E. coli*. The microbial sensor did not respond under the anaerobic conditions to organic substances such as glucose and acetic acid. The influence of inorganic ions on the response was negligible.

The microbial sensor was applied to the determination of glutamic acid in fermentation broth. Known amounts of glutamic acid were added to a broth and the concentration of glutamic acid was determined. Satisfactory recovery data (99-10.3%) were obtained. The concentrations of glutamic acid in some fermentation broths were determined by the microbial sensor and by the Auto Analyzer method. The results were in good agreement.

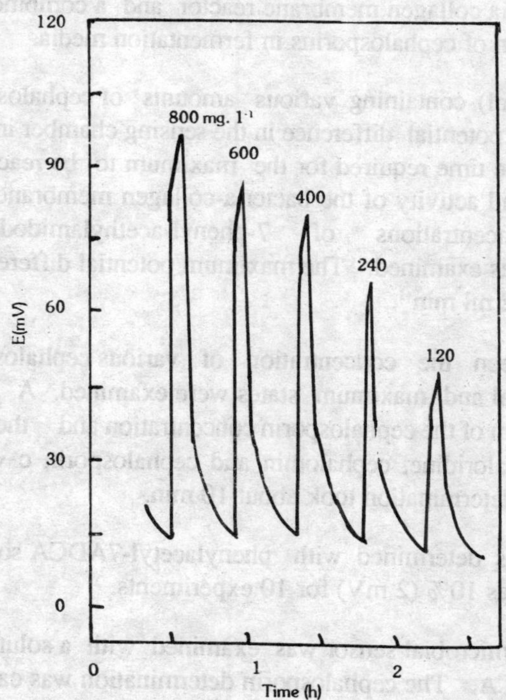

Fig. 18.10. Responses given by the electrode for glutamic acid solutions of the concentrations stated. Sample solution (3 ml) was injected for 3 min. The determination was carried out under the recommended conditions.

Glutamic acid solutions (240-800 mg l^{-1}) and fermentation broths of glutamic acid were used to test the long-term stability of the sensor. The response of the sensor was constant for more than 3 weeks and 1500 assays.

Cephalosporin Sensor

For control of an antibiotic fermentation, antibiotics are usually determined by microbioassay based on turbidimetric or titrimetric methods. However, these methods require a long time for cultivation of bacteria.

It was found that *Citrobacter freundii* produced cephalosporinase, which catalyzes the following reaction of cephalosporin, which liberates hydrogen ions :

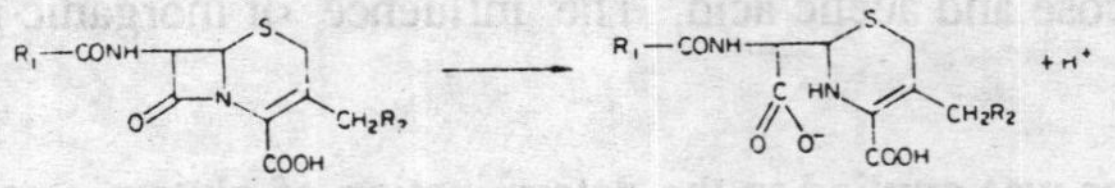

Cephalosporin may therefore be determined from the proton concentration generated in a medium by using immobilized cephalosporinase. Immobilization of cephalosporinase was difficult because the molecular weight of the enzyme is only 30,000 and the enzyme is unstable. Whole cells of *Citrobacter freundii* were immobilized in a collagen membrane. A microbial sensor composed of a bacteria collagen membrane reactor and a combined glass electrode could be applied to the determination of cephalosporins in fermentation media.

Sample solutions (10 ml) containing various amounts of cephalosporins were transferred to the reactor. The electrode potential difference in the sensing chamber increased with time until a maximum was reached. The time required for the maximum to be reached (the response time) depends on the flow rate and activity of the bacteria-collagen membrane. The response of the sensor for various concentrations of 7-phenyl-acethylamidodesacetoxysporanic acid (phenylacethyl-7 ADCA) was examined. The maximum potential difference was attained in 10 min at a sample flow rate of 2 ml min^{-1}.

The relationship between the concentration of various cephalosporins and the potential differences between the initial and maximum states were examined. A linear relationship was obtained between the logarithm of the cephalosporin concentration and the potential difference. Phenylacethyl-7ADCA, cephaloridine, cephalothin and cephalosporin c were determined by the cephalosporin sensor. Each determination took about 10 min.

The reproducibility was determined with phenylacetyl-7ADCA solution (125 μg ml^{-1}); the relative standard deviation was 10% (2 mV) for 10 experiments.

The reusability of the microbial sensor was examined with a solution containing 125 μg ml^{-1} of phenylacethyl-7ADCA. The cephalosporin determination was carried out several times a day, and no change in the potential difference response was observed for a week.

The system was applied to the determination of cephalosporin c in a broth of *Cephalospolium acremonium,* and was compared with a method based on high-pressure liquid chromatography (h.p.l.c.). The relative error of the determination by the microbial system was 8%. Accordingly, the method is suitable for continuous analysis of fermentation broths.

Electrochemical Microbioassay of Nystatin

Many polyene antibiotics have been isolated from the culture broth of *Streptomyces* species. These antibiotics are active against fungi, yeast and other organisms. However, most are too toxic for clinical use. Only nystatin and some other polyene antibiotics are useful for clinical applications.

The polyenes are believed to bind with the sterol present in the membranes of sensitive cells leading to the formation of pores. The subsequent death of the micro-organisms is preceded by the leakage of cellular materials. The death of the micro-organisms can be detected with an oxygen electrode. Therefore, an electrode consisting of the microorganisms and an oxygen electrode should be applicable to the microbioassay of polyene antifungal antibiotics. A yeast electrode composed of a yeast membrane and an oxygen electrode was developed for the microbioassay of nystatin.[8]

The electrode contains a collagen membrane and an immobilized yeast, *Saccharomyces cerevisiae*, membrane attached to an oxygen electrode. The collagen membrane was used to prevent leakage of yeast cells.

If sufficient nutrients are present in the solution, the electrode current is not affected by changes in nutrient concentration. The steady state current decreases linearly with increasing glucose concentration, and becomes constant at concentrations exceeding 300 mg l^{-1}. The rate of metabolism in the yeast cells may be rate-determining in this range. The current depends on the total respiration activity of the yeast cells.

The total respiration activity of yeast and therefore the current is affected by the number of yeast cells in the membrane. The current decreased with increasing number of cells but a linear relationship was obtained for 1×10^7 — 4.6×10^7 yeast cells. Consequently, a membrane containing 4.6×10^7 cells was used thereafter for the electrode.

When the yeast electrode was inserted into the glucose-buffer solution containing nystatin, a steady state current was obtained and then the current began to increase giving a sigmoidal curve. When the electrode was inserted into the glucose-buffer solution not containing nystatin, no current increase was observed. In the presence of nystatin, the electrode current ultimately reached the level of the electrode in the absence of yeast. The rate of current increase is a measure of the nystatin concentration, and is most easily measured as the linear slope at the mid-point of the sigmoidal curve.

Fig. 18.11 shows that the relationship between the rate of current increase and the nystatin concentration is linear below 54 units ml^{-1}. The rate of current increase had a standard deviation of 1.2 units ml^{-1}, when a sample solution containing 27 units ml^{-1} of nystatin was analysed (30 experiments).

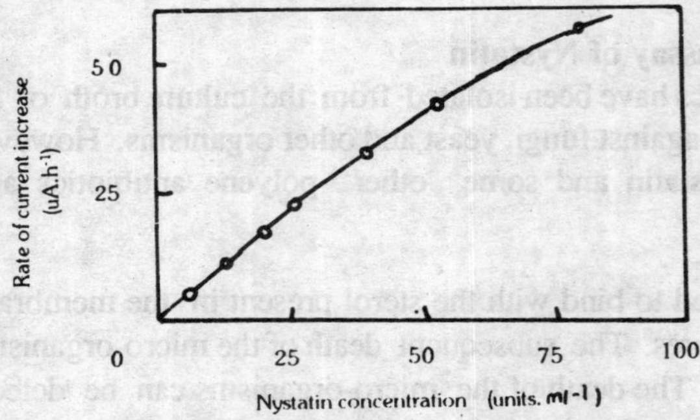

Fig. 18.11. Calibration curve for nystatin obtained under the recommended conditions.

Electrochemical Microbioassay of Vitamin B_1

As bacteria require specific nutrients for their growth, bioassay of amino acids, vitamins and antibiotics is possible with the use of special bacteria such as *Lactobacilli, Streptococci,* etc.

A new electrode system is described for the microbioassay of vitamin B_1. *Lactobacillus fermenti* (ATCC 9338) was employed for the assay of vitamin B_1. An electrochemical sensor for microbioassay consisted of a silver peroxide (Ag_2O_2) cathode (1 cm x 4 cm) and a platinum anode (diamter, 2.2 cm). Phosphate buffer (0.1 M, pH 7.0) was used as the catholyte. An anion exchange membrane was used to separate the electrochemical cell from the culture broth. The current was measured by a milliammeter and the signal was displayed on a recorder.

The time-courses of bacteria growth and the current generated were examined. *L. fermenti* grew normally in the culture medium containing 50×10^{-9} g ml^{-1} of vitamin B_1. After injection of the *L. fermenti* suspension, the electrode was inserted in the culture medium and the current was measured. The current increased with increasing incubation time. The maximum current was obtained at the middle point of the exponential bacterial growth. There was no increase in current from the medium in the absence of vitamin B_1.

The time required to obtain the maximum current decreased with increasing amount of the injected bacterial suspension. The minimum time for microbioassay was 6 h when 5×10^{-6} g wet cell ml^{-1} was injected into the incubation medium. Further increase of the amount injected did not shorten the incubation time. An amount of 5×10^{-6} g ml^{-1} was therefore employed for further work.

The culture medium for microbioassay was incubated for 6 h at 37° C before the electrode was inserted. The opten-circuit anode potential was –420 mV vs. SCE and the cathode potential was 220 mV vs. SCE. As the anode was polarized, a high current was obtained initially. Then the anode potential became constant. Diffusion of the electroactive substances produced became the rate determining factor and a steady state current was obtained. The steady state current was attained within 15 min. at all cases.

A linear relationship was obtained between the steady state current and the vitamin B_1

concentration below 5×10^{-9} g ml^{-1}. The steady state current was reproducible within ± 8% of the relative error when a medium containing 25×10^{-9} g ml^{-1} of vitamin B_1 was used. The standard deviation was 10^{-9} g, when various concentrations of vitamin B_1 were measured by this method.

Electrochemical Microbioassay of Nicotinic Acid

Some bacteria require specific nutrients for growth; these bacteria produce mainly lactic acid as a metabolite. A rapid method for determination of micotinic acid, by using immobilized bacteria and a combined glass electrode to measure the lactic acid produced, is described here, *Lactobacillus arabinosus* ATCC 8014 was employed for the assay of nicotinic acid and immobilized in agar gel.

Preliminary experiments showed that the immobilized *L. arabinosus* produced mainly lactic acid. As the membrane potential of the glass electrode is proportional to the logarithm of proton activity in the solution, the lactic acid produced by the immobilized whole cells can be determined by the electrode. The potential increased with increasing concentration of nicotinic acid in the medium. The potential difference (ΔE_1) between the initial medium and the medium incubated for 1 h with immobilized bacteria was calculated. The potential difference (ΔE_2) between the incubated blank and sample media was also calculated. The rate of production of lactic acid by immobilized bacteria in the media containing nicotinic acid was higher than that in the blank medium.

The potential difference (E) between E_1 and E_2 was found to be proportional to the logarithm of the nicotinic acid concentration :

$$\Delta E = E_1 - E_2 = k \log [nic]$$

where E is a potential difference, [nic] is the nicotinic acid concentration and k is a constant. The potential difference first increased linearly with increasing incubation time, but reached a plateau after incubation for 1 h, because both ΔE_1 and ΔE_2 then changed at the same rate. An incubation time of 1 h was selected for the assay of nicotinic acid.

The relationship between the potential difference and the logarithm of the nicotinic acid concentration was examined. A linear relationship was obtained for 5×10^{-8} — 5×10^{-6} g of nicotinic acid in the 1 ml aliquot added.

The potential difference was reproducible with an average relative deviation of 5% when a medium containing 5×10^{-7} g of nicotinic acid per ml was employed; the standard deviation was 2×10^{-8} g ml^{-1} in 30 experiments.

To study the stability of the immobilized bacteria, they were stored in the physiological saline at 5° C. Nicotinic acid was determined at 10-day intervals with this stored immobilized bacteria. The potential difference obtained from each experiment was constant for 30 days. The bacteria immobilized in agar gel matrix are therefore active for a month.

The number of bacteria in the gel increased from 4.1×10^{10} to 5.6×10^{10} g^{-1} of gel during incubation in the medium containing nicotinic acid. Without nicotinic acid, growth of *L. arabinosus* in the gel was not observed.

OTHER SENSOR FOR FERMENTATION

Electrochemical Assimilation Test of Micro-organisms

Assimilation characteristics of micro-organisms are very important for classification and utilization of microorganisms. A number of methods for the assimilation test have been developed, but they are based on detection of growth in a basic medium containing a specific carbon compound which serves as the sole source of carbon. These methods require a long cultivation time (24-72 h) and special techniques such as aspetic precautions. Recently, many devices have been developed for the electrochemical monitoring of biochemical reactions. Microbial electrodes consisting of immobilized whole cells and electrochemical devices such as oxygen electrode, a fuel cell electrode or a combined glass electrode have been developed. and applied to the determination of organic compounds as described above. Assimilation of a substrate by microorganisms can be detected from their respiration activity. The respiration activity is directly measured by an oxygen electrode as described above. Therefore, it was possible to detect assimilation characteristics of micro-organisms by using a microbial sensor consisting of the immobilized whole cells and an oxygen electrode. The microbial sensor was applied for assimilation tests of several species of molds, yeasts, bacteria, actinomycetes and activated sludges.

Micro-organisms (*ca.* 0.015 g) were coated on the surface of an acetylcellulose membrane. The membrane retaining the micro-organisms was placed on the Teflon membrane of the oxygen electrode so that the micro-organisms were trapped between the two membranes. The membranes were covered with a nylon net and fastened with rubber rings. When a sample solution containing an assimilable substrate was injected into the system, the substrate permeated through the porous acetylcellulose membrane and was assimilated by the micro-organisms. This respiration of the micro-organisms was activated and dissolved oxygen around the membranes was consumed. As a result, the current of the electrode decreased markedly with time until a steady state was reached. The steady state current depends on many factors among others such as the concentration of the substrate, the assimilation activity of the immobilized micro-organisms and the rate of metabolism (oxidation) of the substrate. When the injection of the sample solution was stopped, the current of the microbial sensor returned to the initial level. It is, therefore, possible that assimilation of a substrate by the micro-organisms can be tested by the current of the electrode.

Fig. 18.12 A, B shows typical response curves of the microbial sensor to various substrates (29 substrates shown in Table 19.1). The pulse method was employed for the assimilation test. A linear relationship was observed when the concentration of substrate was below 100 mg l^{-1}. In this case, the rate of metabolism (oxidation of a substrate) of the immobilized microorganisms does not seem to be a rate determining step and the current decrease obtained from the microbial sensor for a definite concentration of a substrate is, therefore, proportional to the apparent rate of metabolism (oxidation) of the immobilized micro-organisms. Therefore a concentration of 50 mg l^{-1} or 100 mg l^{-1} for each substrate tested was employed in this study. Various microbial sensors were prepared and substrates were applied to the sensor. In order to generalize the results, glucose was used as a standard substrate and maximum current decrease for a substrate was normalized in the following way :

$$\frac{\text{maximum current decrease for a substrate } [\mu A] \times \text{concentration of glucose } [mg\ l^{-1}]}{\text{maximum current decrease for glucose } [\mu A] \times \text{concentration of a substrate } [mg\ l^{-1}]} \times 100(\%)$$

The normalized maximum current decrease for a substrate gives the apparent rate of metabolism (oxidation) of the substrate by the micro-organisms being examined. As shown in Fig. 18.12 (Number 1 in Fig. 18.12 A, B), the maximum current decrease for glucose increased gradually with time, since the activity of the microbial gradually increased. Therefore, the normalization of maximum current decrease for a substrate was carried out by injecting a glucose standard solution for every 10 samples. Table 19.1 shows the results of assimilation tests for molds *(Aspergillus sojae and A. niger)*. As shown in Table 19.1 molds can assimilate a wider range of substrates as well-known. In the case of molds, the test must be carried out as soon as possible after cultivation. Otherwise, molds form spores which show little or no respiration activity.

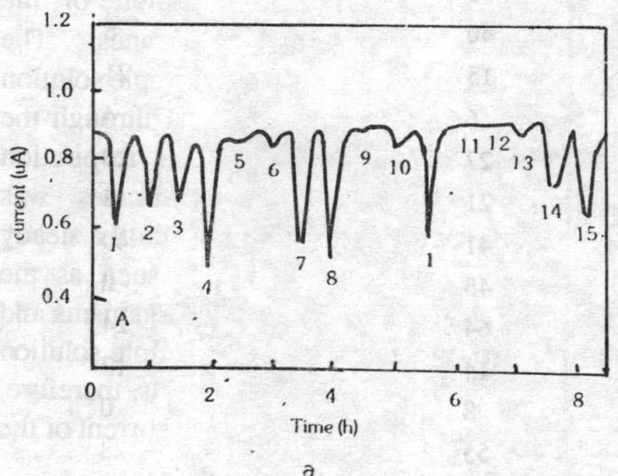

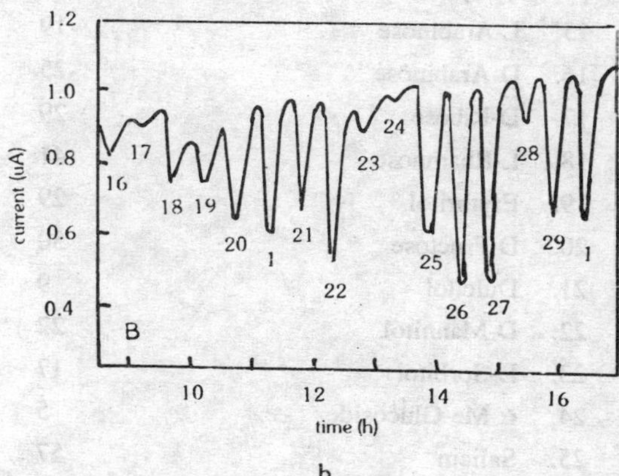

Fig. 18.12. Typical response curves of the microbial electrode for various substrates. Immobilized *Aspergillus sojae* was employed for the electrode. A sample solution (0.5 ml) containing 10 mg l^{-1} mg glucose and 100 mg l^{-1} of various substrates (shown in Table 18.1) were injected during a period of 5 min into the system. The test was carried out under the described standard conditions.

TABLE 18.1
Results of the Assimilation Test for Molds

		Aspergillus sojae IFO 4200		*Aspergillus niger ATCC 6275*
		Normalized maximum current decrease[a] *(%)*		*Normalized maximum current decrease* (%)*
No.	*Substrate*	*Run 1*	*Run 2*	*Run 1*
1.	D-Glucose	100	100	100
2.	D-Galactose	31	41	36
3.	Sucrose	55	37	41
4.	D-Maltose	61	67	60
5.	Lactose	6	2	0
6.	L-Sorbose	9	10	4
7.	D-Cellobiose	64	57	58
8.	Trehalose	68	66	44
9.	Melibiose	5	3	18
10.	Raffinose	2	11	9
11.	Melezitose	0	0	8
12.	Inulin	0	0	0
13.	Soluble starch	3	8	0
14.	D-Xylose	21	31	20
15.	L-Arabinose	19	40	8
16.	D-Arabinose	25	15	21
17.	D-Ribose	29	6	4
18.	L-Rhamnose	41	27	20
19.	Erythritol	29	21	8
20.	D-Fructose	50	41	49
21.	Dulcitol	9	43	0
22.	D-Mannitol	22	64	24
23.	D-Sorbitol	17	14	12
24.	α Me-Glucoside	5	8	0
25.	Salicin	57	55	28
26.	DL-Lactic acid	60	71	8
27.	Succinic acid	11	71	28
28.	Citric acid	11	16	0
29.	Arbutin	43	48	40

* Calculated from Eq. (1).

The assimilation characteristics of various micro-organisms could be tested with the microbial sensor system. There is a fundamental difference between this method and conventional growth tests. In the case of the electrochemical method, substrate assimilation by micro-organisms is determined by detecting an increase in oxygen consumption of the micro-organisms. Therefore, a substrate which is gradually utilized by micro-organisms such as soluble starch is difficult to be tested since the increase in oxygen consumption by immobilized micro-organisms is too small to be detected. Similarly, a substrate which requires the induction of an enzyme is also difficult to test because the time for measurement is too short to permit enzyme induction. In spite of these problems, the results of this method showed fairly good agreement with those of the conventional growth tests. This study provides a rapid, simple and automatic method for an assimilation test. This system is also applicable to the following tests :

- Simple identification of micro-organisms.
- Selection of culture medium for micro-organisms.
- Measurement of the enzymatic activity of micro-organisms.
- Estimation of the biodegradability of substances in waste waters.
- Selection of micro-organisms for waste water treatment.
- Assimilation test of activated sludges.
- Estimation of biological degradation of materials.

BIOSENSOR FOR ENVIRONMENTAL CONTROL

BOD Sensor

The Biochemical Oxygen Demand (BOD) is one of the most widely used and important tests in the measurement of organic pollution. The conventional BOD test requires a five day incubation period and values of the test results depend on the skill of an operator. Therefore, rapid and reproducible methods are desirable for the BOD test. Yeasts, *Trichosporon cutaneum,* which was utilized for waste water treatment was employed for the sensor. The micro-organisms immobilized on a porous acetylcellulose membrane were sandwiched between an oxygen permeable Teflon membrane and a porous membrane. Then, the membrane was directly fixed on the surface of the platinum cathode of an oxygen probe. A continuous flow system using a new microbial sensor was developed for automatic estimation of 5-day BOD.

When the sample solution containing glucose and glutamic acid was injected into the system, organic compounds permeated through the porous membrane and were assimilated by the immobilized micro-organisms. Consumption of oxygen by the immobilized micro-organisms began and caused a decrease in dissolved oxygen around the membranes. As a result, the current of the electrode decreased markedly with time until a steady state was reached within 18 min. The steady state current depended on BOD of the sample solution. Then the current of the microbial electrode sensor finally returned to the initial level.

A linear relationship was observed between the current decrease and the 5-day BOD of the standard solution below 60 mg l^{-1}. The minimum measurable BOD was 3 mg l^{-1}. The current was reproducible within ± 6% of the relative error when BOD 40 mg l^{-1} of the standard solution was employed. The standard deviation was BOD 1.2 mg l^{-1} in 10 experiments. The current means the current differences hereinafter.

The microbial sensor was applied to estimation of 5-day BOD for untreated waste waters from a fermentation factory. The 5-day BOD of the waste waters was determined by JIS method.[9] As shown in Fig. 18.13 good comparative results were obtained between BOD estimated by the microbial sensor and those determined by JIS method. Regression coefficient was 1.2 in 17 experiments and the ratios (BOD estimated by the microbial electrode/5-day BOD determined by JIS method) were in the range from 0.85 to 1.36. This variation might have been caused from change in composition of organic waste water compounds.

Furthermore, the BOD of various kinds of untreated industrial waste waters were estimated by the sensor. The BOD values estimated by the sensor depended on compounds in waste waters. Stable responses to the standard solution (BOD 20 mg l^{-1}) were observed for more than 17 days (400 tests). Fluctuations of the current and the base line (endogeneous level) were within ± 20% and 15% respectively for 17 days. The microbial sensor could be used for a long time for the estimation of BOD.

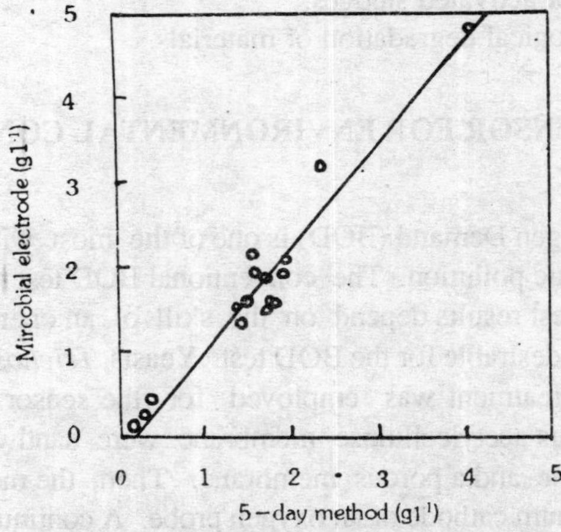

Fig. 18.13. Comparison between 5-day BOD obtained by JIS method and BOD estimated by the microbial electrode. Waste waters from a fermentation factory were employed for experiments. The line was fitted by the least square method.

Ammonia Sensor

The determination of ammonia in a sample solution is important in various fields such as medical, environmental and industrial process analyses. A nitrifying bacterium *Nitorosomonas* sp., utilizes ammonia as a sole source of energy and oxygen is consumed by the respiration as follows :

$$2NH_3 + 3O_2 \xrightarrow{\textit{Nitrosomonas}\ \text{sp.}} 2HNO_2 + 2H_2O$$

Therefore, ammonia may be determined by the microbial sensor using immobilized nitrifying bacteria such as *Nitrosomonas* sp. and an oxygen probe. Nitrifying bacteria were immobilized on a porous acetylcellulose membrane. The microbial sensor was applied to the determination of ammonia in waste waters. The porous membrane retaining the nitrifying bacteria isolated from activated sludges was carefully attached on the Teflon membrane of the oxygen probe so that the micro-organisms were trapped between the two membranes and covered with Nylon net and fastened with rubber rings.

When the sample solution containing ammonia was injected into the system for 12 min, it permeated through the porous acetylcellulose membrane and was assimilated by the immobilized nitrifying bacteria. The steady state current is obtained within 8 min.

The time required for the determination of ammonia was long by the steady state method. Therefore, the pulse method was employed for the determination. Fig. 18.14 shows response curves of the microbial sensor by the pulse method. In this case, the current reached 80% of that obtained by the steady state method. The assay can be done within 4 min and the electrode recovery time is about 8 min by the pulse method. The total time required for the assay of ammonia was 30 min by the steady state method and 12 min by the pulse method.

A linear relationship was observed between the current difference and concentration of ammonia below 1.3 mg l^{-1} by both the steady state and the pulse method. The minimum concentration for determination was 0.05 mg of ammonia l^{-1}. The reproducibility of the current difference was examined using the same sample. The current difference was reproducible within 4% of the relative error when a standard solution containing 1.32 mg l^{-1} of ammonia was employed. The standard deviation was 0.025 mg l^{-1} in 10 experiments by the pulse method.

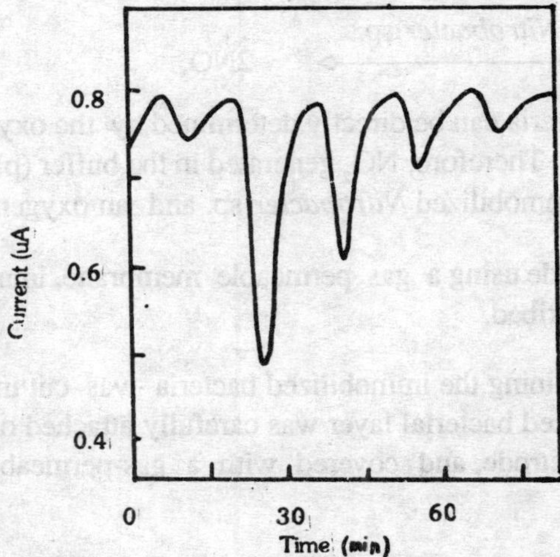

Fig. 18.14. **Response curve of the microbial electrode by the pulse method. Standard solutions (2.4 ml) were pumped into the system for 3 min with 15-min intervals.**

The sensor scarcely responded to compounds such as glucose, ethyl alcohol, glutamic acid, acetic acid and ethylamines, but slightly responded to nitrite, urea and monomethylamine. The microbial sensor was applied to the determination of ammonia in waste waters of a fermentation factory. The concentration of ammonia and the BOD of the sample solutions were determined by the distillation acidimetry and JIS method[10] respectively. A good agreement was obtained between the ammonia concentration determined by the conventional method and that obtained by the microbial sensor. The relative difference between the two methods was less than 6%. The microbial sensor could be used for more than two weeks and 1400 assays. The sensor current decreased gradually; for example, the current decrease diminished to 70% of its initial value after 10-day use. Therefore, occasional calibration of the sensor was required. For example, two kinds of standard solutions (0.44 and 1.32 mg l^{-1} of ammonia) were injected every 10 samples (2 points calibration). In this case, the variation of the successive calibration curves was negligibly small (within 0.3%).

Nitrite Sensor

The principal gaseous oxides of nitrogen of interest in air pollution sampling and analysis are nitric oxide (NO), and nitrogen dioxide (NO_2). During the combustion of all types of fossilfuels at flame temperature, less than 0.1% up to about 0.5% nitric oxide is formed, along with much smaller amounts of nitrogen dioxide. When discharged to the atmosphere, nitric oxide oxidizes at a measurable rate to nitrogen dioxide. Nitrogen dioxide is the most reactive of the gaseous oxides of nitrogen and is a primary absorber of sunlight in photochemical atomospheric reactions that produce photochemical smog. Therefore, the determination of nitrogen dioxide is important in environmental and industrial process and analyses.

Nitrobacter sp. utilize nitrite as the sole source of energy as described above and oxygen is consumed by the respiration as follows :

$$2NO_2^- + O_2 \xrightarrow{\textit{Nitrobacter}\ \text{sp.}} 2NO_3^-$$

Oxygen uptake by the bacteria can be directly determined by the oxygen electrode attached to the immobilized bacteria. Therefore, NO_2 generated in the buffer (pH 2.0) can be determined by the microbial sensor using immobilized *Nitrobacter* sp. and an oxygen electrode.[11]

A nitrogen dioxide electrode using a gas permeable membrane, immobilized *Nitrobacter* sp. and an oxygen electrode is described.

The porous membrane retaining the immobilized bacteria was cut into a circle and soaked in the buffer. This immobilized bacterial layer was carefully attached on the surface of a Teflon membrane of the oxygen electrode, and covered with a gas-permeable Teflon membrane and fastened with rubber rings.

Fig. 18.15 shows schematic diagram of the system which consisted of a jacketed flow cell with microbial sensor, a peristaltic pump, an amplifier, and a recorder.

When the sample solution (sodium nitrite solution) is transferred into the flow cell, nitrous ions change to nitrogen dioxide gas at pH 2.0.

Then nitrogen dioxide passes through the gas permeable membrane. Nitrogen dioxide gas changes to nitrous ions in the bacterial layer and are utilized by *Nitrobacter* sp. as the sole source of energy. The consumption of oxygen around the membrane is determined by the oxygen electrode. Therefore, the concentration of sodium nitrite can be indirectly determined from the current decrease of the oxygen electrode.

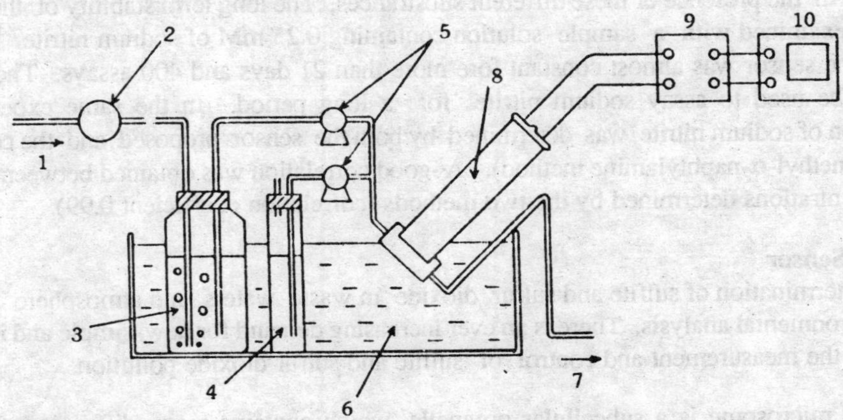

Fig. 18.15. Schematic diagram of the sensor system. 1, air (280 ml min^{-1}); 2, pump; 3, O_2 saturated buffer (pH 2.0); 4, sample solution; 5, peristatic pumps; 6, incubator (30° C); 7, waste; 8, microbial electrode; 9, amplifier; 10, recorder.

When a sufficient quantity of the bacteria is immobilized in the sensor, the current of the sensor for a sodium nitrite solution depends mainly on the rate of diffusion of nitrite from the sample solution to the immobilized bacteria. The O_2 content in sample solutions were checked using differential sensor system (the sensor without immobilized bacterial layer and with immobilized bacterial layer) to prevent an influence of oxygen in the sample solution. Therefore, the steady state current depended on the concentration of sodium nitrite. The differences between the initial and steady state currents were directly proportional to the concentration of sodium nitrite. When only the buffer solution was transferred to the flow cell, the current of the microbial sensor returned to its initial level within about 12 min. The next sample could then be determined by continuously using the same sensor system.

A linear relationship was observed between the current decrease (the current difference between the initial and the steady state) and the sodium nitrite concentration below 0.59 mM (current decrease 0.63 μA). At greater than 0.65 mM sodium nitrite, a linear relationship was not observed between the currents and concentrations. The minimum concentration for the determination of sodium nitrite was 0.01 mM (signal to noise, 20: reproducibility, ± 5%). The reproducibility of the current decrease was examined by using the same sample. The current decrease was reproducible within ± 4% of the relative error and the standard deviation was 0.01 mM in 25 experiments when a sample solution containing 0.25 mM of sodium nitrite was employed.

Thus the amperometric determination of sodium nitrite became possible using the microbial sensor. The selectivity of the microbial sensor for sodium nitrite was examined. The sensor did not respond to volatile compounds such as acetic acid, ethyl alcohol, and amines (diethylamine, propylamine and butylamine) or to involatile nutrients such as glucose, amino acids, and metal ions (potassium and sodium ions). Therefore, the selectivity of this microbial sensor was satisfactory in the presence of these different substrances. The long term stability of the microbial sensor was examined with a sample solution containing 0.25 mM of sodium nitrite. The current output of the sensor was almost constant fore more than 21 days and 400 assays. The microbial sensor can be used to assay sodium nitrite for a long period. In the same experiments the concentration of sodium nitrite was determined by both the sensor proposed and the conventional method (dimethyl-α-naphtylamine method). A good correlation was obtained between the sodium nitrite concentrations determined by the two methods (correlation coefficient 0.99).

Sulfite Ion Sensor

The determination of sulfite and sulfur dioxide in waste waters and atmosphere is important in any environmental analysis. There is an ever increasing demand for new, simple and inexpensive methods for the measurement and control of sulfite and sulfur dioxide pollution.

Hepatic microsome is a subcellular organelle, which contains many different oxidases, and enzymatically oxidizes sulfite to sulfate with consumption of molecular oxygen. An amperometric organelle sensor for the determination of sulfite, composed of immobilized microsome particles, a gas-permeable Teflon membrane and an oxygen electrode. Rat liver S9 fraction (100 μl) containing microsome was filtered through the porous acetylcellulose membrane. The quantity of organelle immobilized was equivalent to 2.7 mg protein. The microsome was retained on the acetylcellulose membrane. The organelle membrane was attached to the surface of the Teflon membrane of the oxygen electrode and covered with another Teflon membrane

When the sample solution of sulfite was injected, sulfur dioxide permeated through the Teflon membrane and was oxidized by the microsome containing sulfite oxidase. Consumption of oxygen by the microsome began simultaneously and caused a decrease in dissolved oxygen around the membrane. As a result, the current of the sensor decreased markedly with time until a steady state was reached after 10 min. When a sufficient quantity of the microsome was employed in the sensor, the current of the sensor depended mainly on the rate of diffusion of sulfur dioxide from the sample solution to the immobilized organelle. The steady state current thus depended on the concentration of sulfite ions.

The response of the organelle sensor increased with increasing amount of organelle, and became constant at an amount of organelle exceeding 2.7 mg protein. An organelle amount equivalent to 7.7 mg protein was therefore immobilized on the acetylcellulose membrane. In this case, diffusion of substrate (sulfite) may also be rate determining.

A linear relationship was obtained between the steady state current and the sulfite ion concentration below 3.4 x 10^{-4} M. The minimum concentration for determination was 0.6 x 10^{-4} M. The currents were reproducible with an average relative error of 7% when a sample solution containing 2.8 x 10^{-4} M of sulfite ion was used. The standard deviation was 0.3 x 10^{-4} M in 30 experiments.

The sensor did not respond to non-volatile compounds such as sucrose, glucose, pyruvic acid, sulfate, phosphate ions, and ammonium ions, since the organelle sensor was covered with the gas-permeable Teflon membrane, and non-volatile nutrients could not penetrate through this membrane. Volatile compounds such as formic acid, acetic acid, propionic acid, n-butyric acid, methyl alcohol, and ethyl alcohol can permeate through the Teflon membrane, but no current was obtained from these compounds, because they are not oxidized by the enzyme system of the microsome. When the sensor was inserted into a sample solution containing 2.8 x 10^{-4} M of nitrite, a current decrease was observed (2.0 μA). This shows that nitrite could interfere with the determination of sulfite, but the response of the sensor to nitrite ion was only 23.5% of that to sulfite ion.

REFERENCES

1. Washka, M.E. and Rice, E.W. *Clin. Chim. Acta, 7:* 542 (1961).
2. Reducing Sugar and Sucrose in Food Products, Industrial method no. 142-71A, Technicon Industrial System (1972).
3. Hikuma, M. Kubo, T. Yasuda, T. Karube, I. and Suzuki, S. *Anal. Chim. Acta, 109:*33 (1979).
4. Hadley, W.K. and Senyk, G. *Microbiol,* American Society for Microbiology, Washington, D.C., p. 12 (1975).
5. Matsunaga, T. Karube, I. and Suzuki, S. *European J. Appl. Microb. Biotechnol., 10:* 125 (1980).
6. Nishikawa, S. Sakai, S. Karube, I. Matsunaga, T. and S. Suzuki, *Appl. Environ. Microbial., 43:* 814 (1982).
7. Matsunaga, T. Karube, I. and Suzuki, S. *European J. Appl. Microbiol. Biotechnol., 10:* 235 (1980).
8. Karube, I. Matsunaga, T. and Suzuki, S. *Anal. Chim. Acta, 109:*39 (1979).
9. Japanese Industrial Standard Committee, *Testing Methods for Industrial Waste Water,* JIS K 0102, p. 33 (1974).
10. Japanese Industrial Standard Committee, *Testing Methods for Industrial Waste Water,* JIS K 0102, p. 36 (1974).
11. Karube, I. Okada, T. Suzuki, S. Suzuki, H. Hikuma, M. and Yasuda, T. *European J. Appl. Microbiol. Biotechnol. 15:* 127 (1982).

19

The New Green Revolution

Introduction

Between 1930 and 1975, the average yield of corn per hectare in US more than tripled. This remarkable increase in productivity is just one of the many achievements of what came to be known as the green revolution. The development of higher yielding varieties of corn, wheat, rice and many other crops, allied to the greater use of irrigation, fertilizers, pesticides and herbicides, have played a large part in ensuring that the world's food shortages are not even more acute. While the green revolution has not been an unmitigated success — for example, the adverse environmental effects of pesticides cannot be denied — there is no doubt that it has made a major contributioin to human welfare. However, the last few years have seen a considerable slowing down of yield increases as the green revolution runs out of steam. Biotechnologists, particularly genetic engineers, are hopeful that they can revitalize many sections of agriculture by enabling us to grow more food at a lower cost and, perhaps, to utilize land that lies idle at present.

Nitrogen—Its Vital Role in Plant Productivity

In building up a complicated and co-ordinated structure there is generally one component that is limiting. For example, it is pointless in a motor-car factory use more people or increase the speed of the assembly line if there are only enough spark plugs to complete the present output of vehicles. In the case of crop plants, the limit to growth is generally set by the availability of nitrogen in the soil. So, there is little value in increasing the irrigation in a field of wheat if the soil lacks enough nitrogen to support extra plant growth.

Nitrogen is needed to construct most of the vital compounds inside cells, including proteins, DNA and RNA. Animals obtain the nitrogen they need by eating plants or other animals, whereas plants must extract it from the soil. A massive industry has grown up to meet the vast and rapidly increasing demand for fertilizers which supplement the land's natural supplies of nitrogen, but the costs are high and likely to increase substantially in the next few years. American farmers are spending nearly $1 billion a year on nitrogen fertilizers for the corn crop alone, and the 40 million tonnes now spread over that nation's fields might require to quadruple by the end of the century if present trends continue. The price rises can only be guessed at, but it is highly significant that the fertilizer industry depends heavily on oil, consuming about 10 per cent of the US total. The urgent need for alternatives to nitrogenous fertilizers is obvious, and 'green gene' engineering could provide an answer.

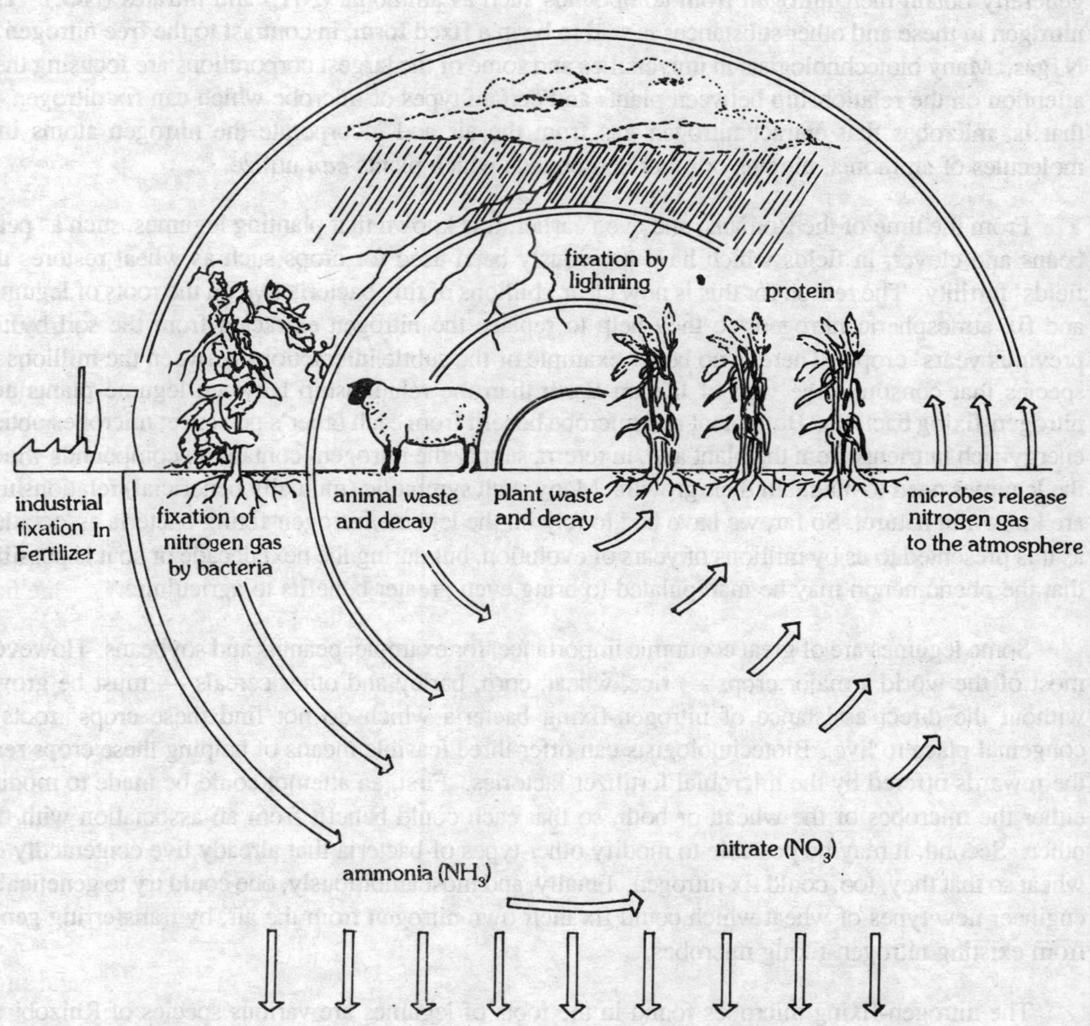

Fig. 19.1. The nitrogen cycle is the movement of nitrogen between plants, animals, microbes, the land and the atmosphere. Nitrogen exists either in the form of nitrogen gas (N_2) or fixed into a wide variety of chemical compounds, including proteins, ammonia and nitrate.

Fig. 19.1 shows the nitrogen cycle, the global processes whereby nitrogen continuously moves between plants, animals, microbes, the land and the atmosphere. It may seem strange that there can be a shortage of nitrogen when 80 per cent of the air that surrounds us is made up of nitrogen gas, but for most forms of life nitrogen gas (N_2) is totally useless. Plants, animals and most microbes have no way of incorporating gaseous nitrogen into their celluar compounds. Plants and most microbes generally obtain their nitrogen from compounds such as ammonia (NH_3) and nitrates (NO_3). The nitrogen in these and other substances is said to be in a fixed form, in contrast to the free nitrogen in N_2 gas. Many biotechnologists in universities and some of the largest corporations are focusing their attention on the relationship between plants and certain types of microbe which can fix nitrogen — that is, microbes that extract nitrogen gas from the air and incorporate the nitrogen atoms into molecules of ammonia, nitrates or other compounds which plants *can* utilize.

From the time of the Romans, and even earlier, it is known that planting legumes, such as peas, beans and clover, in fields which have previously been used for crops such as wheat restores the fields' fertility. The reason for this is now clear : billions of tiny bacteria live on the roots of legumes and fix atmospheric nitrogen, so they help to replace the nitrogen extracted from the soil by the previous years' crops. There is no better example of the subtle interactions between the millions of species that constitute the web of life on Earth than the relationship between legume plants and nitrogen-fixing bacteria. Both plant and microbe benefit from each other's presence; microbes obtain energy-rich nutrients from the plant and, in return, supply the nitrogen- containing compounds which the legumes need to maintain their growth. Many such symbiotic (mutually beneficial) relationships are known in nature. So far, we have had to rely on the legume/nitrogen-fixing bacteria association as it is presented to us by millions of years of evolution, but during the next decade or so it is possible that the phenomenon may be manipulated to bring even greater benefits to agriculture.

Some legumes are of great economic importance, for example, peanuts and soybeans. However, most of the world's major crops — rice, wheat, corn, barley and other cereals — must be grown without the direct assistance of nitrogen-fixing bacteria which do not find these crops' roots a congenial place to live. Biotechnologists can offer three feasible means of helping these crops reap the rewards offered by the microbial fertilizer factories. First, an attempt could be made to modify either the microbes or the wheat, or both, so that each could benefit from an association with the other. Second, it may be possible to modify other types of bacteria that already live contentedly on wheat so that they, too, could fix nitrogen. Finally, and most ambitiously, one could try to genetically engineer new types of wheat which could fix their own nitrogen from the air, by transferring genes from existing nitrogen-fixing microbes.

The nitrogen-fixing microbes found in the roots of legumes are various species of Rhizobium bacteria. These rod-shaped organisms infiltrate the root hairs and take up residence in the root itself, producing nodules which are characteristic of legumes. The subtle interactions between the bacteria and their plant hosts are exceedingly difficult to unravel. At present there is not a full understanding of the way that *Rhizobium* enter the root hairs, penetrate into the body of the root and form nodules, receive the plants' supply of nutrients and make their 'payment' of nitrogen-containing compounds. Before persuading *Rhizobium* to take up new reisdences, say in wheat roots, more needs to be learned about their natural living conditions so that these can be matched as closely as possible. This is a formidable challenge and, at the moment, the other two possible routes to the goal seem more straightforward.

In the last decade or two, an immense amount has been learned about the molecular machinery used by bacteria to fix nitrogen. The central feature is an enzyme called nitrogenase. This takes nitrogen gas and, using energy largely derived from the plant host's photosynthetic activities, converts the gas into ammonia. More than a dozen genes, termed the *nif* genes, are involved in the assembly of the nitrogen-fixation apparatus. At first sight, it would appear to be a herculean task to transfer all these genes into another type of microbe. Fortunately, as is often the case with genes that are involved in performing a single function within cells, these *nif* genes are linked — that is, they are not scattered throughout the vast amount of DNA that makes up the bacterial chromosome, but are all clustered together in one region. This makes it much easier to cut out the relevant stretch of DNA in a *Rhizobium* chromosome and insert the whole batch into another organism.

Genetic engineers have already managed to transfer the *nif* genes from a nitrogen-fixing bacterium into *E. coli* and, equally importantly, the *E. coli* was then able to fix nitrogen. This experiment did not use *Rhizobium* genes, but *nif* genes taken from *Klelolella pneumoniae,* a soil bacterium which lives independently of any plant host. This bacterium has no less than seventeen *nif* genes, and this fact that it was possible to transer all of these into a new home augurs well for future work on the bacteria which now colonize the roots of wheat and other cereals but cannot fix nitorgen.

More enticing still is the prospect of inserting *nif* genes directly into crop plants, dispensing with the need for any sort of nitrogen-fixing microbe. In this approach, a number of various problems are encountered, not least in duping a plant cell into treating bacterial genes as its own. As mentioned earlier, the most significant division in all forms of life is between the eukaryotes (in which DNA is packaged inside a nucleus) and the prokaryotes (which have no nucleus). At the molecular level this distinction is far more important than the more obvious differences between two eukaryotic organisms, such as a rat and a tomato plant. Since all bacteria are prokaryotes and all plants are eukaryotes, the bacterial *nif* genes will not be 'understood' by the crop plants. In particular, ways must be found of ensuring that the plants manufacture the right amounts of the proteins specified by the material *nif* genes.

This presupposes that genes can actually be transferred from bacteria to plants. Many examples of genetic engineering have already been described which involve the transfer of genes from eukaryotes to prokaryotes (for example, human insulin genses into *E. coli*) and from eukaryotes into other eukaryotes (for example, interferon genes into yeasts). These techniques are well developed in comparison with the transfer of prokaryotic gene into eukaryotes, for example, from *Rhizobium* into wheat. In the last few years, however, great advances have been made in the understanding and application of eukaryotic vectors — pieces of DNA that can ferry foreign DNA into eukaryotic cells. As with the vectors employed to carry genes into bacteria, there are two main types, viruses and plasmids, both of which are becoming increasingly employed in plant genetic engineering.

The key to many recent developments is to be found in a strange excrescence called a crown galls, a type of tumour which afficts many flowering plants. The pumpy tumour itself consists of a mass of plant cells which rapidly proliferate because they have escaped the plant's normal growth control micro-organisms and, in this sense, they are analogous to animal tumours. However, in the case of crown galls the causative agent is a bacterium, *Agrobacterium tumefaciens,* inside which are small circular pieces of DNA known and (tumour-inducing) plasmids. A crown gall is created when Ti plasmids are transferred from the bacteria into the chromosomes of the infected plant, the resultant change in the genetic constitution of the plant cells inducing them to grow and divide very rapidly. Here, then, is a potential vector for introducing bacterial genes into plants, and the Ti plasmid is being studied in many laboratories, particularly in Ghent and Cologne where its discoverers, Marc van Montagu and Josef Schell, work.

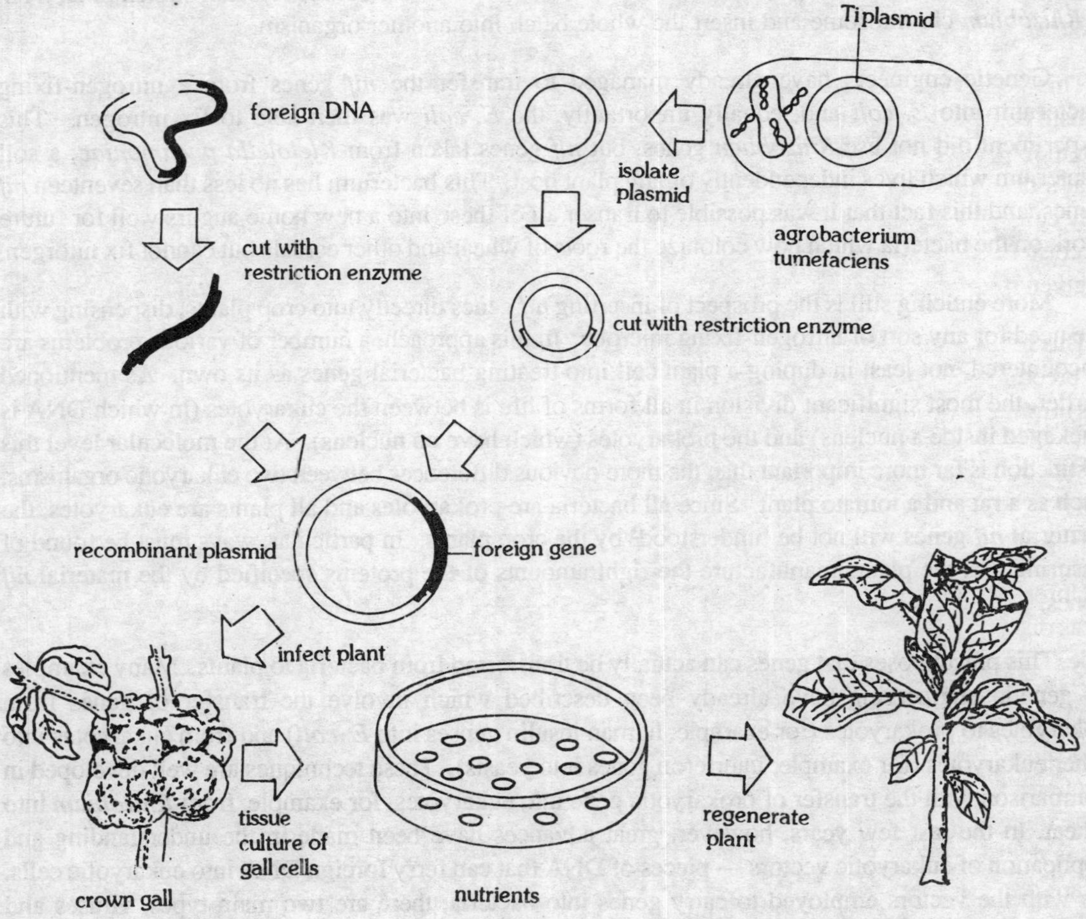

Fig. 19.2. Plant genetic engineering with Ti plasmids from *Agrobacterium tumefaciens*. A foreign gene is inserted into the Ti plasmid and this is used to infect plants, producing a crown gall. The plant cells in the gall all contain the Ti plasmid with its piece of foreign DNA. Gall cells are then grown in culture to produce plantlets, which can be transferred to soil where they grow into mature plants. Since each plant is derived from a single cell carrying the foreign gene, all the cells in the fully grown plant contain that gene.

Once a suitable vector has been found, the panoply of genetic engineering tricks become feasible, and one can also attempt to insert genes from one plant into another plant species. A notable feature of this kind of research is the possibility of regenerating a whole plant from a single cell, at least in the case of some plant species. Gardeners are familiar with a very similar process called vegetative propagation, more commonly 'taking a cutting.' There is something very remarkable about the fact that, with care and attention, a small piece snipped off an adult plant can grow into a fully fledged replica of its parent. The equivalent operation with animal cells produces quite different results. If the isolated animal cells manage to survive at all (which they often do not), one only obtains a mass of cells which are of the same type as the original cell — for example, hamster kidney cells grow to produce more hamster kidney cells, not whole hamsters.

This phenomenon bears directly on one of the greatest puzzles in biology. Every cell in an organism contains all the DNA needed to specify everything required to build up the whole organism, but in individual animal cells certain genes become irreversibly switched off at an early stage in the animal's development. The genes needed to make paws and whiskers are still present in hamster kidney cells, but they seem to be permanently blocked. In many types of plants, this is not the case; tobacco stem cells can give rise to whole tobacco plants because all their genes are potentially active and, given the right conditions, will produce a complete tobacco plant from just one cell.

This has many important practical consequences, both for plant genetic engineering and conventional plant breeding. In the latter, it means that it is often possible to grow thousands of clones (identical plants) from one parent plant.

Fig. 19.2 illustrates how this property may be used by genetic engineers. This has been attempted and, while the results have been mixed, in some cases the trick has worked, and the offspring of the genetically engineered plants also carried the new gene.

Unfortunately, this does not bring the introduction of nitrogen-fixation genes into cereal crops as close as one might think. In fact, few people expect to see nitrogen-fixing cereal crops within the next decade. There are two obstacles, neither of which should prove insuperable.

Although it has been known for several years that *Agrobacterium tumefaciens* (with its Ti plasmid) could infect an enormous range of plant species called dicotlyedons,* it was only in 1984 that scientists realized that it could also infect cereals and other monocotyledons. Hence research into the interaction between the bacterium and the many economically important monocotyledonous plants is not well advanced. A great deal of work is now being invested in this research and, indeed, there are other options available.

One possibility makes use of two tools of the biotechnologist's art encountered in previous chapters — liposomes and protoplasts. It is possible to package pieces of DNA inside liposomes and

* The terms monocotyledon and dicotyledon refer to the one or two seed leaves, respectively, a species possesses. Many forest trees, such as oak, elm and beech, as well as fruit trees, potatoes, tomatoes, peas, beans and cabbages, are dicotyledons. Monocotyledons include bananas, grasses (including the cereal crops), palms, tulips and orchids.

it is not too difficult to remove the sturdy cell walls of plant cells to create protoplasts. The general plan would be to wrap *nif* genes into liposomes and mix these with wheat cell protoplasts. With the protective cell wall removed from the plant cells, there is a good chance that the liposomes could deliver the foreign genes into their new home. This has not yet been achieved for *nif* genes and wheat cells, but similar experiments with other gene/plant cell combinations have proved successful. This technique is clearly likely to be of commercial importance if it is possible to grow whole plants from individual genetically engineered protoplasts, and this brings us to the second major challenge in working with cereal plants.

While many sorts of plant can be propagated from small cuttings or single cells, others, including wheat and corn, cannot at present. This problem may be overcome by fusing, for example, a wheat cell carrying the *nif* genes and a cell from a plant which can easily be regenerated from a single cell. The resulting hybrid cell might be very similar to a wheat cell, but would also contain the *nif* genes and be able to grow into a complete plant. This proposal is analogous to the methods employed to create hybridomas.

Obviously there are many imponderables on the way to cereals which make their own fertilizers, but a fair degree of optimism seems warranted when one considers the recent advances in genetic engineering and cell fusion, nearly all of which were science fiction only a decade ago. Indeed, plant genetic engineering offers almost boundless possibilities for farming in the nineties. The following are just a few of the other projects which plant genetic engineers are undertaking.

Despite its apparent simplicity, nitrogen fixation involves some complicated chemical reactions; not only is gaseous nitrogen converted into ammonia, but hydrogen gas is also produced. This is a wasteful process, for there are great quantities of energy locked up in hydrogen which, if tapped, could be put to good use, particularly to fuel the fixation of yet more nitrogen. This would benefit the bacteria's plant hosts since they provide most of the energy to keep the whole machinery in operation.

A survey of *Rhizobium* associated with soybean plants in the US revealed that many of the bacteria, especially in the north and east of the country, contain hydrogen-uptake (or *hup*) genes. These genes seem to confer the ability to recycle hydrogen gas back into the nitrogenase system which fixes the nitrogen, thus harvesting the energy in hydrogen which would otherwise have been lost to the plant.

The most straightforward application of this discovery is to introduce the *hup* genes into those strains of *Rhizobium* that now lack them. The *hup* genes in certain other types of bacteria are found on plasmids, and if the same is true of *Rhizobium* the *hup*-carrying plasmids can be transferred from one strain of these bacteria to another. The addition of the ability to utilize the energy in hydrogen gas would not guarantee significantly increased crop yields, since any change in an organism's capabilities is likely to have many ramifications, some beneficial and others possibily not. The sheer complexity of life on the molecular level, with its maze of intricate relationships between one function and all the others (in this case, hydrogen uptake and growth rates), makes it almost impossible to predict with certainty the entire consequences of a single alteration. Here, the problem is compounded by the symbiotic relationship between *Rhizobium* and soybean plants, making it even more difficult to estimate the effects on the two organisms. As usual, experiments will have to be performed to test the effects. If the plants really do grow better, one would aim to introduce *hup* genes directly into crops which have also been given the capacity to fix nitrogen.

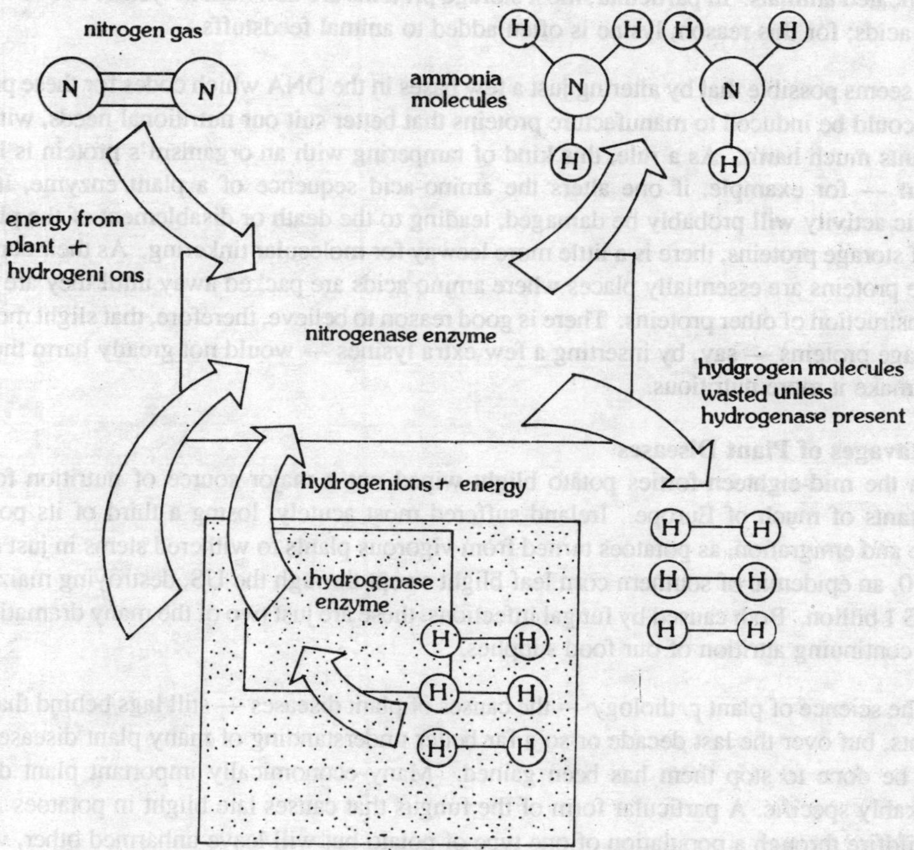

Fig. 19.3. The nitrogenease enzymes of bacteria fix nitrogen by converting nitrogen gas (N_2) into ammonia (NH_3) using energy and hydrogen ions supplied by their plant hosts. This process also produces energy-rich hydrogen gas. Some bacteria also have hydrogenase enzymes which can capture hydrogen gas, converting it into hydrogen ions and releasing energy which can be fed back into the nitrogenase system.

Other genes which are being viewed with interest by genetic engineers are the *osm* genes, which are in some way related to a plant's ability to withstand certain stresses, such as lack of water, heat, cold and salty soils. All these adverse conditions have the effect of forcing water into or out of plant cells by the process of osmosis (hence the name of the genes involved in counteracting it). Millions of hectares of land throughout the world are useless for agriculture because of cold, insufficient water or high salinity. A target for the future (at least a decade ahead) is to introduce *osm* genes into crop plants with the aim of opening up vast tracts of barren land to agriculture.

Perhaps nearer at hand is a comparatively simple modification to plant genes which could be of great value to humans. The balance of amino acids in many of the proteins contained in the seeds of cereals—the storage proteins — is not ideal from the point of view of nutrition for humans and domesticated animals. In particular, most storage proteins are deficient in lysine, one of the essential amino acids; for this reason, lysine is often added to animal feedstuffs.

It seems possible that by altering just a few bases in the DNA which codes for these proteins, the plants could be induced to manufacture proteins that better suit our nutritional needs, without doing the plants much harm. As a rule, this kind of tampering with an organism's protein is likely to be harmful — for example, if one alters the amino-acid sequence of a plant enzyme, its essential catalytic activity will probably be damaged, leading to the death or disablement of the plant. In the case of storage proteins, there is a little more leeway for molecular tinkering. As their name implies, storage proteins are essentially places where amino acids are packed away until they are needed for the construction of other proteins. There is good reason to believe, therefore, that slight modifications to storage proteins — say, by inserting a few extra lysines — would not greatly harm the plant, but might make it more nutritious.

The Ravages of Plant Diseases

In the mid-eighteen-forties potato blight wiped out a major source of nutrition for the poor inhabitants of much of Europe. Ireland suffered most acutely, losing a third of its population to famine and emigration, as potatoes turned from vigorous plants to withered stems in just a few days. In 1970, an epidemic of southern corn leaf blight swept through the US, destroying maize valued at about $ 1 billion. Both caused by fungal infections, these are just two of the many dramatic examples of the continuing attrition of our food supplies.

The science of plant pathology — the causes of plant diseases — still lags behind that of human ailments, but over the last decade or so a far better understanding of many plant diseases and what might be done to stop them has been gained. Many economically important plant diseases are remarkably specific. A particular form of the fungus that causes late blight in potatoes may spread like wildfire through a population of one type of potato but will leave unharmed other, very similar varieties of the plant. An obvious solution may appear to be to plant only the resistant varieties of potato but, needless to say, nature is rarely that simple. A potato that is resistant to one type of fungal infection may be susceptible to another and, furthermore, that potato may not possess the right taste, shape, colour, and all the other factors which make up a marketable product.

Over the centuries, and particularly in the last few decades, plant breeders have had immense success in interbreeding different plant varieties to produce hybrids which possess the good features of both parents. It is clearly important for suppliers involved in the multi-million dollar hybrid seed business to be sure that all batches of seeds are hybrids of the correct sort. At present, this is often checked by growing a sample of the seeds and then examining the plants which result. This takes time and, of course, means that the bulk of the seeds cannot be sold with confidence until the test is completed. New tests are now being introduced which are based on techniques developed by biotechnologists. One method is to analyse the type of enzymes contained in the seeds. Hybrid seeds will have enzymes which are slightly different from those found in either parent. In the future, even more rapid and accurate methods of identification based on DNA probes may well come into common use in the seed industry. DNA probes can also be used to check if seeds or plants are infected with viruses which may later become rampant and reduce crop yields.

Virtually all the world's major crop plants have been developed by conventional techniques, such as cross pollination. However, this type of plant breeding has an inherent limitation — only quite closely related plant varieties can be interbred to give fertile hybrids. The rapidly developing techniques of protoplast fusion can help overcome this natural obstacle to the creation of new plant varieties. The tough walls that surround plant cells can be removed to produce these naked cells. Protoplasts from quite dissimilar plant species can be induced to fuse together forming a single hybrid cell. In some cases, a complete hybrid plant can be grown from this single cell — for example, 'pomato' plants have been produced which are hybrids between tomatoes and potatoes.

Much effort is now being invested in the search for new crop varieties which are resistant to as many diseases as possible. Tobacco plants are particularly easy to manipulate in the laboratory, and protoplast fusion techniques have already yielded a new variety of tobacco plant which seems to be resistant to hornworm, a prevalent disease in tobacco farming. The first successful attempts to introduce pest-resistance into plants by genetic engineering were reported in 1985 by a group of Belgian scientists. They inserted genes from a bacterium into tobacco plants and gave them the ability to make a protein which attacks the intestines of insects. While research into these methods of developing new crops has only just begun, it has already stimulated considerable investment from food-manufacturing firms — for instance, Campbell's Soups have put more than $ 10 million into the search for a better variety of tomato.

The regeneration of plants from single cells taken directly from plant tissues rather than produced by protoplast fusion is, by contrast, of great economic importance today. Asparagus, pineapples, strawberries and oil palms can all be propagated now by separating cells from an especially suitable plant and growing complete replicas of that plant from each individual cell. The same technique is also in general use for the breeding stock of many other species including sprouts, cauliflowers, bananas, carnations and ferns, and even redwood trees are now being tested.

The great advantage of this method of propagating plants is that many thousand identical copies or clones of a *single* plant can be produced. If one simply grew plants from the seeds of a particularly suitable plant, there would be a great degree of variation among the seedlings since each would have a different genetic constitution depending on the way the genes from its *two* parents were distributed. Furthermore, many viral diseases of plants are passed from generation to generation through the seeds. By cloning virus-free plants the spread of these diseases can be reduced.

In economic terms, one of the most significant developments in plant cloning is the introduction of over 1000 cloned oil palms in Malaysia. Palm oil sales are about $ 2.5 billion a year, including its use in margarine. Once an individual palm was discovered to be unusually resistant to disease and to yield 20-30 per cent more palm oil than the average obviously it made sense to produce multiple copies of this plant. Unilever achieved this in the late seventies, and their test plantation is showing good results.

More recently, researches in Australia have been able to clone river redgum trees. This is significant because these trees can grow in very salty soils which are unfit for agriculture, and such land could be reclaimed by planting redgums. The hope is that as they grow, they will pump water from the soil, lowering the water table and allowing rain to wash the salt out of the surface layers of the soil. Since about 400,000 sq. km. (154,000 sq. miles) of land around the world are severely affected by excess salt, this project has immense potential. Even if the redgums or other salt-tolerant trees prove unable to open the way for conventional agriculture, the wood obtained from otherwise unproductive land would be valuable, particularly in countries, such as India, which have large areas of salt-laden earth and a great need for fuel.

20

Biotechnology from Farm to Supermarket

Introduction

Good health without good food is next to impossible. While the finer points about what constitutes 'good' food are much debated, the plain fact is that many millions of people are chronically malnourished by any reasonable standards.

Historically, biotechnology found its first applications in the production of food—in the making of bread and alcoholic drinks, and in the harvesting of algae as food, practised by the Mexican Indians centuries ago. The present and future applications offer food in more abundance, at a lower cost, higher in nutritional value, greater in variety and more appealing in taste than those available today.

The greatest impact of biotechnology so far has been in the food-processing industry, but the coming years will see important advances in the production of food derived from microbes, plants and animals, which should benefit both the malnourished and the well-fed sections of the world's population.

Beefing up the Meat Industry

An outbreak of foot-and-mouth disease is a devastating blow to cattle farmers, entailing the destruction of all affected cattle, which may be worth many thousands of pounds, and all movements of animals in the area must be halted. This disease also affects pigs, sheep and goats, and is endemic in Asia, Africa and South America. Because it spreads so easily, no area of the world is safe from its ravages, and it is the most economically important infectious disease of farm animals.

Foot-and-mouth disease is caused by a small virus related to those that produce polio and the common cold. At present, countries which have only occasional outbreaks (such as Britain and the US) control the disease by slaughtering infected cattle. This is costly, as is the current vaccination process employed in places where the disease is common. Clearly, both areas would benefit greatly from cheaper and more effective vaccines against the disease, and these should become available very soon.

The importance of the problem has attracted many scientists in industry and government establishments, notably the Animal Virus Research Institute at Pirbright, UK and the Plum Island Animal Disease Laboratory, New York State. The gene which codes for one of the virus's coal proteins, VP1, has already been cloned and inserted into *E. coli*. Once purified, this protein is very likely to act as an effective vaccine against the foot-and-mouth disease virus.

Over one billion doses of the present vaccine are injected each year but this vaccine, which is obtained by infecting cultures of hamster cells, is both expensive and must be given annually to maintain a high level of immunity. The VP1 vaccines produced by genetic engineering could offer substantial advantages in terms of cost and effectiveness.

The vital economic importance of eradicating, or at least controlling, viral diseases in animals and the role genetic engineering can play in such endeavours has been recognized by the United Nations Industrial Development Organization. This body has assigned a high priority to the development of genetically engineered vaccines against several diseases, including foot-and-mouth, rabies, which causes economic losses of tens of millions of dollars, especially in South America; African horse sickness, which erupts periodically and, for example, killed 300,000 horses, mules and donkeys in the Middle East in the late nineteen-fifties; and a sheep disease, bluetongue, which is concentrated in Africa but also appears in southern Europe and the US.

Veterinary vaccines are important, not only for their intrinsic value but also because they can be used as proving grounds for technologies that later may produce vaccines for human diseases. The first product of genetic engineering ever to go on sale was a vaccine against a bacterial disease of pigs. Scours kills about 10 per cent of piglets grown in intensive rearing houses in The Netherlands, and in 1982 the Dutch firm Akzo began to market a genetically engineered vaccine against this disease. Similarly, poultry bred in factory farms are susceptible to Newcastle disease virus, and there is a large and valuable market for a vaccine against this illness. Even greater benefits would accrue from a vaccine against animal forms of sleeping sickness; subduing these diseases, which are transmitted by tsetse flies, could open up vast tracts of Africa for the breeding of cattle.

Animal interferons, the natural anti-viral agents, also have the potential to cut farming costs. In 1984 the US Food and Drug Administration licenced the use of genetically engineered interferon to treat 'cattle shipping fever' in Texas. This complex disease is quite common in cattle that have suffered the stress of being transported from farms to auctions. Estimates of the effects of this fever have placed the cost to US beef producers as high as $700 million a year.

As well as keeping farm animals healthy, biotechnology may increase yields very significantly. Injections of extra bovine growth hormone can increase a cow's milk production by up to 40 per cent, and cause cattle to put on 10-15 per cent more weight than normal. Given the enormous importance of the dairy and beef industries, it is not surprising that at least four major genetic engineering companies are aiming for a share of the vast potential market for bovine growth hormone-$250-500 million a year in the US alone. The gene has been cloned and it is likely that the product will become available during the mid-eighties, probably to be followed by its sheep and pig counterparts.

The self-shearing sheep sounds a most unlikely beast, but in a sense it already exists. Sheep wool, like human hair, grows out of cells in the skin and its rate of growth is controlled, in part, by a hormone called epidermal growth factor (EGF). If extra doses of EGF are given to sheep, the wool grows more rapidly and each hair becomes narrower. A dose of EGF, therefore, creates a weakness in the hairs as they are being constructed inside the skin. After a few days, this weak point has emerged from the skin into the visible woolly coat, and at that stage simple brushing snaps off the hairs, yielding a harvest of wool without the need for conventional shearing.

As one might expect, most of the research into this strange phenomenon is taking place in Australia, but whether it can be turned to commercial use depends on at least two problems being solved. First, biotechnologists must clone the EGF gene and so produce large quantities of cheap EGF protein; this should present no insuperable difficulties. The second problem is more mundane, but equally important. As the break-point in the wool takes a few days to emerge from the skin, the wool cannot be plucked off until a few days after the EGF injection. Keeping sheep penned for this time is expensive, but releasing them into their grazing lands would result in the wool being shed on their first encounter with a brush, fence or even each other. The task of collecting wool scattered over hundreds of hectares is probably greater than shearing the sheep in the time-honoured fashion. Obviously the economic realities — cost of EGF plus cost of penning sheep versus cost of conventional shearing — will decide whether this intriguing offshoot of biotechnology becomes a commercial proposition.

Microbes Maketh a Meal

The term 'food processing' certainly has an unnatural ring to it and today, when packaging can take precedence over content, its negative connotations are not entirely undeserved. However, processing is the rule rather than the exception, for mere cooking is a type of processing. Of more concern in this section are the very common forms of food processing which require the assistance of microbes—for example, the processing of wheat to make bread, grapes to make wine, barley to make beer, milk to make cheese and yogurt, and vegetables to make pickles. Including substances which are added to food to preserve it, render it more palatable or improve its nutritional value — many of which are derived from microbes — makes the influence of biotechnology on our diets even more evident.

Humans have known how to make use of microbes, particularly yeasts, since Neolithic times, long before people were even aware that microbes existed. The discovery that intoxicating drinks could be manufactured from grapes, rice, barley and other cereals must surely have been accidental — a chance contamination of the juice by a suitable species of yeast, such as those which form the bloom seen on grapes. Over the centuries brewers and wine makers learned how to control this capricious process. Some types of microbes produce a stronger and more appealing drink than others, and these would have been selected for future use. Eventually a taste for particular types of drinks became established, and one aspect of biotechnological development — the selection of the most appropriate microbes—declined in importance.

The use of different strains of yeast will produce changes in, for example, a beer, and any changes must usually be kept to a minimum once the product has become accepted by the consumers. The brewing industry would like a yeast which produced a beer higher in alcoholic content since this would reduce processing costs. However, the flavour and appearance of beer depend on many factors, some of which are only poorly understood. It is difficult, therefore, to find a new strain of yeast which possesses desirable qualities, such as making a stronger product, while retaining the correct blend of flavours and other characteristics. For these and other reasons, two types of yeast dominate in the brewing industry. *Saccharomyces cerevisiae* is employed to manufacture traditional British-style beers, termed top fermentation beers because this type of yeast tends to float on the surface of the brewing vessel. Lager beers, by contrast, are made with the aid of a bottom fermentation yeast, *Saccharomyces carlsbergensis*, so-named because it was first isolated in the Carlsberg Institute, Copenhagen, by Emil Christian Hansen.

The brewing industry has laid the foundations of much of the new biotechnological revolution and is the largest biotechnology, with an annual turnover of billions of pounds. It is ironic, therefore, that it stands to benefit rather little from the most recent developments. There are however some possibilities, and one of particular interest concerns light beers. These are beers which have a low concentration of carbohydrates, not to be confused with the more familiar light ales which have been available in the UK for many years. In the US, and to lesser extent in the UK, light beers have gained much popularity. The brewers' usual yeasts cannot ferment substances called dextrins, which make up about a quarter of the carbohydrates in the original fermentation liquid. A related yeast (which is not used in the industry), *Saccharomyces diasticus*, can turn dextrins into alcohol, thus reducing the carbohydrate content of the beer. The genes responsible for this have been introduced into normal brewing yeasts. At present this work is still at the research stage, and so far the taste of the beer has been far from satisfactory, but further investigations are beginning to show how this may be improved.

The role of genetic engineering in the brewing industry of the future may be uncertain but another well-established aspect of biotechnology has already become firmly entrenched — the use of various enzymes obtained from microbes. One of the major costs in beer production is the malting process during which barely seeds are allowed to germinate. The main food store inside barely consists of starch molecules, which are composed of many sugar subunits linked to form a branching chain (see Fig. 20.1). The sugar units by themselves or linked in pairs make an excellent source of energy for yeasts, and when consumed give rise to alcohol. Brewing yeasts cannot make use of the complex starch molecules but, during the germination of barley seeds, enzymes are released which snip up the starch chains into pieces that yeasts can consume.

Over recent years, a means of avoiding some of the expense of malting has been developed. Certain species of bacteria also manufacture some of the enzymes which attack starch, notably amylases. These amylases are now extracted on a commercial scale from bacteria, and this market is worth several million pounds a year. When added to the fermentation vessel, these enzymes can perform some of the tasks traditionally undertaken by the malted barley enzymes, allowing a proportion of the cheaper, unmalted barley to be employed without affecting the final product.

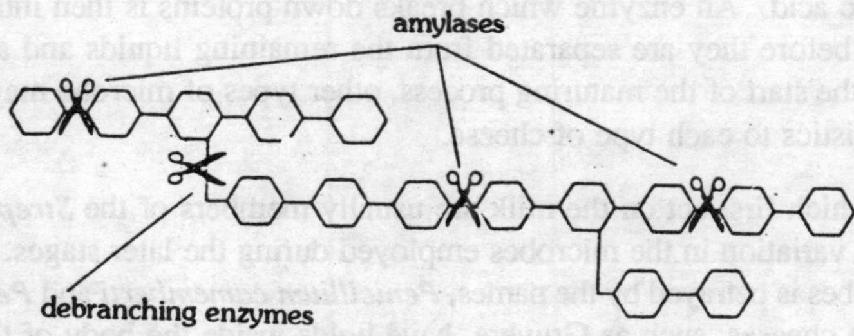

Fig. 20.1. Starch consists of branched chains of sugar subunits. Amylase enzymes produced by germinating barley seeds or bacteria are able to snip the links to yield shorter chains of sugar which yeasts can consume. 'Debranching' enzymes cut the branching points of the chains.

Clarity, as all beer advertisers know, is a strong selling point. There are several reasons why a beer may become hazy, one of which—excess protein in the liquid — can be eliminated with the aid of enzymes. Proteases are, as their name suggests, enzymes which destroy proteins. Unlike most enzymes, proteases are generally quite undiscriminating — present them with any type of protein and they will quickly chop it into small pieces.

The protease most widely used is papain, an enzyme extracted from the papaya fruit (and also employed to tenderize meat by attacking the tough fibres). The Japanese, who are the world leaders in the industrial applications of enzymes, have been investigating microbes which produce proteases. The Kirin Brewing Company has discovered that a mixture of papain and a protease obtained from the bacterium *Serratia marcesens* is more effective in eliminating beer haze than either enzyme alone. Bacterial proteases for these and other applications are sold in vast quantities, over 500 tonnes a year with a value of 60 million worldwide.

Brewers and associated researchers, particularly since the time of Pasteur, have laid a solid foundation for much of biotechnology's future. They have gained unparalleled practical experience over many years in putting microbes to work in an industrial environment—making sure that they are given the right conditions to perform their assigned task, excluding contaminating organisms, dealing with vast quantities of raw materials, refining the products of a biotechnological process to present them in a saleable form, and keeping costs to a minimum. All this knowledge is proving invaluable in transforming newer, and perhaps more glamorous biotechnologies into economic realities.

Second only to the alcoholic beverages industry in terms of sales are the industries which employ microbes to manufacture foods from dairy products, chiefly cheeses, buttermilks, sourmilks and yogurts.

The principles of cheese-making vary comparatively little from one cheese to another, but the specific techniques, particularly the enormous range of microbes employed, differ greatly. The first step is to add bacterial to the fresh milk, allowing them to turn it sour as the milk's lactose sugar is converted into lactic acid. An enzyme which breaks down proteins is then introduced, causing the solids to coagulate before they are separated from the remaining liquids and allowed to mature to form a cheese. At the start of the maturing process, other types of microbe may be added to impart particular characteristics to each type of cheese.

The bacteria which first act on the milk are usually members of the *Streptococcus* group, but there is much more variation in the microbes employed during the later stages. The applications of some of these microbes is betrayed by the names, *Penicillium camemberti* and *Penicillium roqueforti*, for example. Some cheeses, such as Gruyere, have holds inside the body of the cheese which are created by adding to the milk bacteria that give off carbon dioxide gas. Much the same principle applies in the use of carbon-dioxide-generating yeasts to leaven bread.

One of the cheese industry's biggest problems is the occasional failure of the bacterial starter cultures to initiate the fermentation of the milk. In Australia, for example, thousands of dollars worth of milk is wasted by this difficulty, but scientists in Melbourne foresee a solution. Certain types of viruses (bacteriophages) infect and kill bacterial cells. Although some species of *Streptococcus* bacteria are resistant to bacteriophage attack, the strains which are best for cheese-making are susceptible to the viruses. By employing genetic engineering techniques or cell fusion, these scientists hope to create hybrid bacteria which are both efficient milk fermenters and able to fight off bacteriophage infections.

Traditionally, the protein-degrading enzyme employed in cheese-making has been rennin, which is obtained from the stomachs of calves. Both the cost and potential shortage of calf rennin stimulated investigations into alternative enzymes obtained from microbes. Over the past fifteen years or so, microbial rennin has become increasingly widespread in the cheese-making industry. Three species of fungi, *Mucor pusillus* (developed in Japan), *Mucor miehei* (US and Denmark) and *Endothia parasitica* (US) make rennins which have been approved for use in most countries. This is another traditional area of biotechnology that has tempted the genetic engineers, who hope to reduce the cost of these enzymes. At least two groups of researchers, in the UK and Japan, have managed to clone the calf rennin gene in bacteria and yeasts.

Other dairy products (buttermilk, sourmilk, yogurt and so on) are made by mixing specific types of microbes with fresh or pre-treated milk. Vegetables may be preserved by fermenting them with bacteria that produce acids to yield foods such as sauerkraut. Even the kind of pickling done in the home depends indirectly on the action of microbes; vinegar is made when certain microbes feed on grape juice or infusions of cereal grains. Civilizations in the Far East have been even more adventurous in the use of microbes. In Japan, China and Indonesia, delicacies such as natto, sufu and tempeh kedele are made by growing moulds on soybean preparations, and the ubiquitous soy sauce is manufactured with the aid of several fungi, bacteria and yeasts.

Bug Grub

It is not only in preserving and altering the tastes of foods that microbes have a role to play — microbes themselves can be eaten. The prospect of trying to sell a food consisting of little more than squashed, dried microbes or 'bugs' is far from appealing to marketing consultants. To avoid public revulsion — unjustified though it may be—the more wholesome-sounding, but less accurate, term single-cell protein (SCP) has been adopted.

The origins of human consumption of microbes can be traced at least as far back as the early sixteenth century. In 1521, Bernal Diaz del Castillo described how the inhabitants of Mexico ate small cakes with a cheese-like flavour, which they prepared from an 'ooze' extracted from lakes. This 'ooze' was almost certainly *Spirulina maxima*, a species of alga that still thrives in the highly alkaline waters of Lake Texcoco. It is not known quite when the Aztecs discovered the nutritional value of this alga, nor, indeed, how long the Kanembu people of Chad, thousands of kilometres away in Africa, have been eating a related microbe, *Spirulina platensis*. It is clear, however, that human interest in microbial food far antedated the intensive research it has received in the last two decades.

The modern story of SCP, which may contain up to 70 per cent protein by weight, began in Germany during the First World War. Scientists and technologists in Berlin devised a process by which brewers' yeasts were grown in large quantities and added to sausages and soups to supplement the nation's diminished food resources. Eventually the yeasts made up more than half of the country's restricted food imports. Again in the Second World War, Germany turned to yeasts, this time employing *Candida* species. In Britain, brewers' yeasts have made a contribution to the diet for many years, the excess microbes being sold as animal food or processed to form products such as Bovril and Marmite.

The next major events in the SCP story — which have since been called grandiose follies — took place in the nineteen-sixties. Swept along by a tide of enthusiasm, many of the world's largest firms, especially those in the oil industry, invested heavily in SCP production. The results were little short of disastrous — millions of pounds disappeared into ambitious projects, sometimes without a penny in return.

Of all the foods which are in short supply in the world the most acute deficiency is in high-quality protein foods. Protein in animal meat is generally more nutritious than plant proteins, but feeding plants to animals is a very inefficient way of enhancing protein quality.

Obviously SCP production makes good sense, but despite this the recent SCP ventures were dismal failures. The reason for this was that, inconceivable though it may seem to us today, accustomed as we are to high petrol costs, some firms decided that there was an economic future in feeding *oil-based* chemicals to microbes which would then be harvested and sold as animal, or even human, food.

British Petroleum, for example, spent tens of millions of pounds developing a factory in Sardinia, which was to manufacture Toprina, an SCP product based on yeasts grown on oil-derived nutrients. Today the plant lies idle and is up for sale with a price tag of only £10 millions. The price of oil rocketed, there were political difficulties within Italy, and the company could not convince the authorities that Toprina was safe to use as animal food. A whole chain of objections were put up, many of which impartial observers considered spurious. One of the main concerns over safety centred on the relatively high concentrations of nucleic acids (DNA and RNA) in Toprina. Although every cell in our bodies contains nucleic acids—and human nucleic acids are chemically virtually identical to those found in microbes — a question mark hangs over the safe concentration of these substances in food. Yeast SCP has about 10 per cent by weight of nucleic acids, compared to only 1-2 per cent in plants and 4 per cent in Spirulina.

ICI has preserved with SCP production. The company's plant at Billingham grows the bacterium *Methylophilus methylotrophus* on methanol derived from natural gas. Current production of Pruteen is over 50,000 tonnes a year and it is sold as animal feed. The process may face competition from Hoechst in West Germany, who are developing a similar product, Probion, from which much of the nucleic acids and other materials have been removed, and it might eventually be approved as a human food. ICI aims to fight back by reducing the cost of Pruteen with the use of a genetically engineered modification of their current bacterium, one which will make better use of the ammonia that must be supplied to the fermentation vessel to provide the nitrogen essential for microbial growth. The genetically engineered bacterium uses about 7 per cent less methanol; but it has yet to be introduced into commercial operations.

Also in Britain, Rank Hovis McDougall has spent about £30 million in manufacturing an SCP product, mycoprotein, for *human* consumption. This company is investigating a species of *Fusarium* mould which contains about 45 per cent protein and 13 per cent fat, a combination quite as nutritious as many meats. Moreover, the mycoprotein is high in fibrous content, a feature which is viewed with ever-increasing approval by most nutritionists. *Fusarium* grows on a wide variety of carbohydrate sources, so the precise carbohydrate nutrient supplied to the mould could vary according to local conditions—potatoes or wheat starch in Britain, and cassava or other tropical plants where these are plentiful. This product has great potential, particularly since it contains less than the acceptable maximum amount of nucleic acids and has a well-balanced amino acid composition. Several hundred people, both inside and outside the company, have eaten this new food and their reactions have been favourable. Extensive tests with animals have revealed no harmful effects of a diet containing mycoprotein. The use of raw materials which are themselves perfectly edible, helps promote confidence that the product will be safe.

It might be imagnied that the cheapest feed stock would be some material that comes free of charge, but even this can be improved upon. Waste and pollution disposal is big business, and biotechnology can convert industries which are paid to eliminate some noxious substance into industries which also utilize these as raw materials to produce valuable products such as fuels or foodstuffs.

The paper-making industry faces a serious problem in disposing of one of its waste products, sulphite liquor. If sulphite liquor is simply pumped into rivers and lakes, it quickly depletes the water's oxygen reserves with dire consequences to the environment, the death of thousands of fish being only the most obvious effect. In Finland, sulphite liquor is fed to *Paecilomyces* moulds in what is known as the Pekilo process. This not only purifies the waste liquids from paper factories but also yields a rich harvest of microbes which are sold as animal feed. Similar techniques are being developed throughout the world to utilize waste materials from forestry, cheese-making, unwanted fruit pulps and many other materials for which there are few, if any, present uses.

As the oil-based processes proved all too vividly, the economics of SCP production depend critically on the costs of raw materials and the price of conventional alternatives, such as soybean and fish-meal. The USSR, one of the world's major producers of SCP, already has nearly 100 installations, and expects to have enough yeast SCP to fulfil all its needs for protein in animal feeds by the end of the next decade.

The acceptability of SCP as human food is still largely untested. After hesitantly tasting a sample of bacterial SCP, a nutritionist is reported to have remarked, 'Yes, it has all the characteristics. I look for in a new human food; it is odourless, colourless, textureless and tasteless.' This unappetizing description means, of course, that it could be employed as the basis of a wide range of acceptable styles and tastes of food. The idea will not please food purists, but the ever-worsening food crisis may force us to abandon such niceties. Once each form of SCP has passed the relevant safety tests and been constituted into an appealing and preferably familiar form, there is little reason why 'bug grub' (under a suitably attractive brand name) should not play an even greater part in improving the world's diet.

The SCP projects developed in the West, including those already mentioned and the Swedish Symba system which utilizes potato wastes, all require heavy capital investment and fairly sophisticated installations. However, the major protein shortages are found in countries where these resources are scarce. If SCP production takes off in the West, the developing world may well benefit through a knock-on effect, in which more of the well-nourished countries' food can be diverted to those who need it. Indigenous SCP programmes would free protein-starved nations from the vagaries of international politics and economics.

Spirulina algae offer excellent prospects for low-technology SCP production which could be employed in these parts of the world. The yields of *Spirulina* per hectare may be ten times higher than wheat in terms of total weight and many more times greater in protein content. Harvesting the algae from natural or artificial lakes is very simple, and it can be dried in the sun and then flavoured according to local tastes. This corkscrew-shaped organism is photosynthetic and so gains its energy from sunlight, using it to build up compounds from carbon dioxide in the air. Since both light and air are free, and only a few cheap chemicals need be added to the ponds to keep productivity high, *Spirulina* has many enticing characteristics.

Vitamins and Amino Acids

Biotechnology is also contributing substantially to the improvement of the nutritional value of existing foods. The value of vitamins scarcely needs emphasizing, for constant injunctions from advertisers have ensured that very few readers will be unaware of the severe consequences of too little of them in the diet. Indeed, a more important task is to counter the idea that twice the recommended dose of a particular vitamin must be twice as health-giving. Few healthy people in developed countries, eating a normal diet, need extra vitamins. This happy state of affairs is largely due to their ready access to many vitamin-rich foods, and biotechnology has also had an influence.

A lack of vitamin B_{12} (cobalbumin), which may be caused by intestinal disorders or, occasionally, an inadequate diet, results in pernicious anaemia. Too little vitamin B_2 (riboflavin) can produce lip sores, mouth ulcers, skin rashes and eye problems. Today, much of the supply of these two vitamins, used in medicine and as food additives, comes from microbes. The interest in these two substances here lies in the great success of biotechnologists in persuading certain species of microbes to increase their production of these vitamins to a remarkable degree. Wild strains of the mould *Ashbya gossypii* manufacture only minute amounts of vitamin B_2, but by successive selection of high-yielding strains and manipulation of the fermentation conditions, the quantity of vitamin produced has been increased 20,000 times. Similarly, the industrial strains of two types of bacteria, *Propionibacterium shermanii* and *Pseudomonas denitrificans*, make over 50,000 times more vitamin B_{12} than their natural cousins. This achievement of biotechnologists has enabled their microbes to form the basis of very efficient industrial processes, supplying most of the $150 million market for these vitamins.

Cereal grains constitute a major part of the feed for many economically important animals, especially during the winter. Unfortunately, many cereals are deficient in two of the amino acids, lysine and methionine, which all animals need to build up their proteins. These are, therefore, usually added to animal feed to ensure that an adequate diet is provided.

At present methionine is manufactured by chemical processes, but 80 per cent of the $ 200 million sales of lysine is fulfilled by fermentation processes which use bacteria. The ever-growing demand for animal feed has encouraged several firms to invest in larger and more efficient facilities for producing lysine, and biotechnological methods of manufacturing methionine are being sought.

About 40,000 tonnes of lysine are produced each year, chiefly by over-producing strains of *Corynebacterium glutamicum*. These bacteria lack the enzyme homoserine dehydrogenase, and this defect causes them to continue to churn out vast quantities of lysine-far in excess of their needs.

While several other amino acids are currently being manufactured by microbial fermentation processes, one dwarfs all the others in economic importance. A quick examination of the labels on food packages in the average kitchen will reveal the ubiquity of monosodium glutamate (MSG), which is used as a flavour enhancer. The MSG industry now produces over half a million tonnes a year, and its growth has been based on two bacteria — another strain of *Corynebacterium glutamicum;* and *Brevibacterium flavum*, a short rod-shaped organism.

The cost of MSG is relatively low because biotechnologists have discovered a particularly neat way of extracting it from bacteria. *Corynebacterium glutamicum*, like many other microbes employed in fermentation processes, is provided with the sugar glucose as its main source of carbon and energy. It also needs small quantities of the vitamin biotin, which is used to help build up the membrane that surrounds the cell. It the bacteria are supplied with just too little biotin, their membranes become 'leaky', and some substances, including MSG, escape from the cells into the surrounding liquids from which they can easily be isolated. This cuts down the cost of purifying the product, since it is not necessary to break open the bacterial cells to collect the MSG. Furthermore, the amount of MSG manufactured by each bacterium increases because it then replaces the MSG it has lost.

Sweeter than Sweet — Sugary Substitutes

The commonest disease Britain is caries, or tooth decay. Virtually every adult has one or more fillings, and millions have lost all their natural teeth. Sugar, which is added to thousands of our foods, is thought to be the major culprit, nourishing billions of bacteria in our mouths which promptly produce acids to corrode teeth and gums. The enormous scale of the sugar industry has made it a prime target for biotechnologists. Already several alternative sweeteners are available through the skills of microbes, and some of these alternatives offer the prospect of reduced tooth decay.

Chemists use the term sugar to describe a great number of compounds which share certain physical and chemical characteristics, but only a small number of sugars taste sweet. The most familiar of the sweet sugars is sucrose, which is obtained from sugar cane and sugar beet. A similar compound, fructose, is found in fruits and honey. The biotechnological production of fructose is a major and rapidly growing industry, which is perceived as such a threat to sugar beet farmers that the EEC has blocked its use in the Community. While the use of fructose does not promise better dental health since plenty of acid-producing bacteria can feed on it just as well as on sucrose it has attracted great attention for three main reasons: it is sweeter than sucrose; it can prove cheaper than sucrose; and fructose is more suitable for diabetics than ordinary household sugar. In the years since 1970 the use of fructose sweeteners in the US has risen from almost zero to nearly 20 per cent of the sweetener consumption of about 50kg (110lb.) per person per year, and before long half of the sweetening in Coca Cola may be provided by this sugar. Fructose tablets were briefly in vogue in Britain when the alcohol breath test was introduced since there were ill-founded claims that they reduced blood alcohol concentrations.

The remarkable impact of fructose has been based largely on advances in biotechnology, in particular the availability of amylase enzymes which convert cheap starch raw materials into glucose, and another enzyme, glucose isomerase, which turns glucose into fructose. The principles are familiar: the impressive catalytic powers of enzymes are harnessed to provide an economic and efficient means of changing a raw material of low value (starch) into one of higher value (fructose). There is one aspect of these processes that deserves special attention since it has considerable importance in many diverse areas of biotechnology—the immobilization of enzymes.

Since 1928 when James Sumner became the first person to obtain pure crystals of an enzyme — urease from jack-beans-scientists' ability to isolate specific enzymes has increased beyond measure. However, they remain, in general rather expensive materials and so it is vital that any commercial process which employs enzymes should utilize them as efficiently as possible. One of the characteristics of enzymes is that they can perform the same function over and over again; barring accidental damage, a single molecule of the enzyme glucose isomerase will continue to work tirelessly converting glucose molecule after glucose molecule into fructose. Once all the glucose has been turned into fructose the enzyme has nothing left to work with and its undiminished vigour is wasted. Since the enzyme and the fructose are both dissolved in the final mixture, it is very difficult and costly to extract the enzyme and use it again with a fresh bath of glucose.

This waste of expensive enzyme can now be avoided by the simple expedient of fixing the enzyme molecules on to a solid surface; once they have done their job, the fructose is drained away leaving the enzymes stuck on their supports ready for the next infusion of glucose. There has been much research into the theory and practical applications of immobilized enzymes over the last decade, and many different types of enzymes have now been successfully immobilized on a wide variety of solid supports including glass beads, plastics and natural fibres, such as cellulose illustrates. Fig. 20.2 three of the main methods employed to fix enzymes.

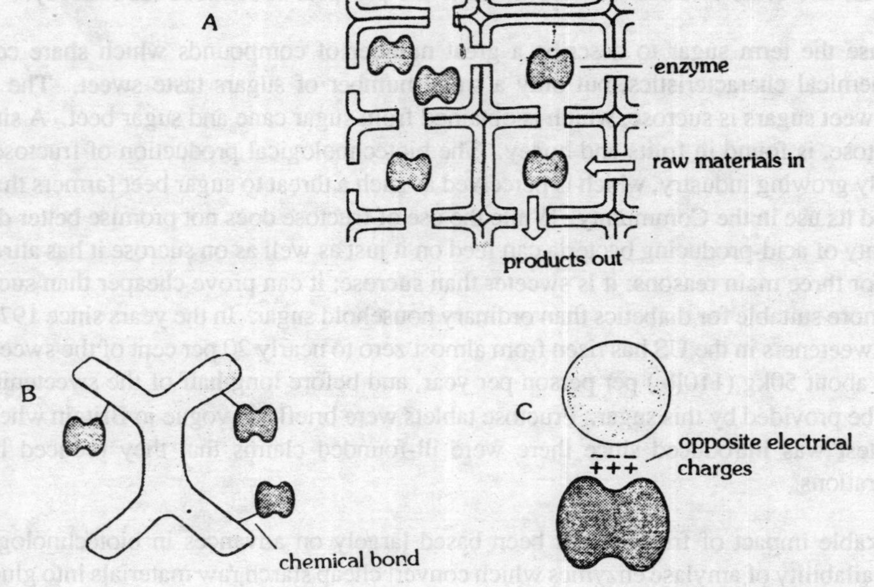

Fig. 20.2. Some methods of immobilizing enzymes. (A) Entrapment in a solid lattice constructed of plastic or other materials. Small gaps in the lattice allow the raw materials to approach the enzymes and the products to leave the compartments. (B) Chemical bonding of enzyme to a support, such as cellulose fibres. (C) Electrostatic attachment to supports such as glass beads. The opposite electrical charges possessed by the enzyme and the support hold the two together.

In early experiments, the activity of enzymes was frequently destroyed during the immobilization process. Now the techniques have been refined to such a degree that nearly all the enzyme molecules will be unharmed when they are linked to the supports. These advances, coupled with the expectation that it will be possible, eventually, to immobilize any desired enzyme, guarantees an increasingly important place for immobilized enzymes in biotechnology's future. At present about fifty microbial enzymes are of industrial importance, but patents have been filed for commercial uses of over 1000 enzymes. There are many and varied reasons why the majority of this large number of potential applications have not been brought to fruition. This high cost of enzymes is certainly a major factor in many cases but immobilized enzyme technology is transforming many industrial processes. The re-utilization of a single bath of enzymes can turn a previously uneconomic biotechnological process into one with great potential. This has happened in the production of fructose, of the food additive malic acid, and of several amino acids.

Although fructose production is the most significant contribution of biotechnology to the sweetener industry at present, at least two other types of sweeteners will probably become important in the next few years. One of these, a synthetic compound called aspartame, has recently been approved for use in the US and the UK as a low-calorie sweetener. The other, thaumatin is being investigated intensively.

Aspartame is quite a simple material compound of two linked amino acids, aspartic acid and phenylalanine. Both the constituents can be manufactured by microbes and aspartame itself will probably soon be produced by a microbial fermentation process. This market is worth over $4 million in the US and rising rapidly, so there is a large incentive for biotechnologists to break into it.

Thaumatin is a much larger compound, consisting of 207 amino acids. It is remarkably sweet — about 2,000 times as sweet as cane or beet sugar. Obviously this means that tiny quantities would satisfy the tastes of even the most sweet-toothed consumers. Thaumatin is found in West African bushes, but the cost of extracting it from plants is high and is likely to limit its use in the food industry. Recently the gene for thaumatin has been cloned and there are now excellent prospects that genetically engineered microbes will manufacture it at a significantly lower price.

21

Composting of Agricultural Wastes

Introduction

The Shorter Oxford Dictionary defines composting as the process of making into a mixed manure of organic origin. Biddlestone and Gray define composting as the decomposition of heterogeneous organic matter by a mixed microbial population in a moist warm aerobic environment. Bell and Pos define it as an aerobic thermophilic decomposition process while Muller states that in its broadest sense composting may be defined as incomplete microbial degradation of organic waste, where the microbial processes may vary from aerobic to anaerobic and may lead to more or less extensive degradation.

Composting differs from other decomposting systems in that the temperature and rate of decomposition are usually altered by positive human intervention. Materials are collected into a pile or box, and mixed or placed in layers. If water is then added the indigenous organisms present will begin the biodegradation process. This will be accompanied, to a greater or lesser extent by a rise in the temperature of the compost. Most composts are deliberately composed of a mixture of materials. Even when an apparently homogeneous substrate such as hay or straw is composted it will contain a number of distinct chemical fractions each with its own decomposition characteristics. Usually the process is aerobic. Anaerobic processes, such as silage making, although similar in some respects, are probably better considered separately as fermentation processes. Although in some cases the intention may simply be bulk reduction, composting is usually intended to produce a product, compost, which can be used as a fertiliser and soil conditioner. Decomposition is therefore incomplete in almost all compost systems.

In this paper composting will be taken to mean the incomplete, artificially accelerated, decomposition of heterogeneous organic matter by a mixed microbial population in a warm, moist, aerobic environment.

Composting is not a simple unitary process. Rather it is the sum total of a series of complex metabolic processes and transformations which are brought about as a result of the activities of a large mixed population of micro-organisms. The precise chemical changes, and the microbial species involved in them, will vary according to the composition of the composting materials. This inherent variation in the types of materials that are composted complicates any attempt to compare reports on composting. Several authors have drawn attention to this problem. Variation will arise where the materials to be composted arrive from different sources, where they have been stored for different periods of time and due to seasonal factors. It is however possible to make some general observations about the composting process.

Composting is one of the oldest solid waste treatment methods known to man. Compost making has been practised since biblical times but modern interest probably stems from an extended visit to China, Japan and Korea in 1909 by Professor F. H. King of the U.S. Department of Agriculture. His observations were stuided by Sir Albert Howard, a botanist employed by the Indian Government, who developed the Indore composting process. Initially the process involved the construction of pits 30 feet long by 14 feet wide. These were dug two feet deep to conserve heat and moisture, except during the monsoon season when they were built above ground. The heap was constructed by building up a two inch thick layer to plant material, a thin layer of animal manure and a dusting of soil and wood ashes. This was watered and the process repeated until a height of two feet was reached. The heap was turned after 16, 30 and 60 days, intermittently watered, and carted to the fields after 90 days. Subsequently the process was improved and developed. The pits were dug three feet deep, the depth of the plant layer was increased to six inches, the thickness of the animal manure layer increased to two inches, and the layering continued up to a height of five feet. Vertical aeration vents were made, using a crowbar every three to four feet across and along the heap of wastes. Turning was only carried out twice, after 2-3 weeks and after five weeks, and the composting was completed after 90 days. During this period Howard's field work was complemented by the laboratory studies of Waksman. He and his co-workers studied the breakdown of composting plant materials with particular reference to the hemicellulose, cellulose and lignin content of the composts. They also demonstrated the significant role of thermophilic actinomycetes and fungi in these processes. They established that the successful preparation of composts of plant residues depended upon the nature of the organic materials, the proportion of nitrogenous compounds to carbohydrates, the temperature of decomposition, and the microbial population of the compost. They showed that composting was more rapid when the compost was composed of a mixture of plant materials rather than a single material such as wheat straw. They also concluded that no single pure microbial culture could give as extensive or rapid a decomposition as the total population of a thermophilic compost. Waksman also showed that the "humus" remaining at the end of the breakdown process was composed not only of lignin which had not been utilised by the micro-organisms but also materials synthesised by the decomposer organisms.

As a waste disposal method composting has a number of advantages:

1. it reduces the bulk volume of the waste
2. it yields a stable product
3. it leads to the destruction of pathogenic bacteria and of the eggs of parasites
4. it allows organic wastes to be recycled without the destruction of their innate high energy value.

For agricultural wastes composting offers two additional advantages. It reduces the risk of polluting water-courses and it helps to conserve increasingly expensive fertilisers. Composting reduces the soluble nitrogen content of agricultural wastes as the nitrogen is assimilated into microbial biomass. If these wastes are spread directly on to land the highly soluble nitrogen compounds can easily be washed into waterways or lost as ammonia. Composting also results in phosphorous compounds becoming bound up in new microbial cells so that run-off can again be avoided. An increase in crop yields of 25-30% has also been claimed for composted versus directly applied animal manures. An additional problem in the disposal of agricultural wastes is that large volumes of animal wastes may be generated on a relatively few sites, often with no local means of utilisation although these wastes might well be of value if they could be transported elsewhere. Land utilisation patterns vary with climate with arable production often concentrated in drier regions while intensive dairy and beef production have developed in areas with higher rainfall. The manure produced in the dairy regions is used to feed cattle but in some cases an excess may be produced. The wet climate itself may produce problems as access to land for manure spreading may be restricted and the application of high-moisture content slurries to already wet ground can lead to accelerated run-off into watercourses.

Composted farm and garden wastes are of value as fertiliser. On a percentage dry weight basis the compost will contain 0.4-3.5% nitrogen (as N), 0.3-3.5% phosphorous (as P_2O_5) and 0.5-2.0% potassium (as K_2O). It has been shown that, when compared to inorganic fertilisers of similar nutrient value, compost has increased crop yields by as much as 10%. It may also be better economics to use a low cost compost rather than a high cost artificial fertiliser to greater nutrient value, and to accept a lower crop yield. Composts have the additional advantages that they add humus to the soil and improve the aeration and water-holding capacity of the soil.

The overall process of composting is summarised in the flow-sheed diagram. Once the moisture content of the heterogeneous organic material is brought to a suitable level (50-60%) the activity of the indigenous micro-organisms accelerates. The organic matter is partially broken down and heat is generated.

Temperature

When dead organic materials are decomposed by mesophilic micro-organisms they first utilise the most readily decomposable carbohydrates and proteins. If this dead organic matter has been gathered into piles or heaps the heat generated by its rapid decomposition is conserved within the heap by the inherent insulating properties of the organic materials. The precise shape and magnitude of the temperature *v* time curve is dependent upon the nature of the organic matter which is being composted, the availability of nutrients within that organic matter, moisture content, size of the heap, insulation, particle size of the compost components, and the degree of agitation and aeration. The typical shape of the temperature *v* time curve for the centre of a compost heap is shown in Fig. 21.2. It is to be expected that the higher the the proportion of easily degradable components in compost the greater will be the amplitude of the temperature *v* time curve and the faster will be the rate of decomposition. This is confirmed by the work of Ardidi, summarised by Gray *et al.* which shows that the decomposition rate is faster in materials with a higher proportion of readily available or degradable carbon substances.

A rapid warming occurs as micro-organisms multiply in the composting mass. When the temperature moves past 40° C the mesophilic stage is succeeded by the thermophilic stage. This is usually after 2-7 days of composting. After the temperature in the middle of the pile peaks at around 70° C it gradually cools to ambient temperatures. The temperature gradient across the compost heap depends on its size, the insulating properties of the compost, and on the insulating properties of the container, if one is in use. Variation is greatest in small heaps which have the largest surface: volume ratio. Gray *et al* reported that to achieve satisfactory temperatures at least 0.7 m³ (1 yd³) of wastes is needed in some sort of protected enclosure, and that in such heaps the heat extends through the compost to 76-152 mm from the edges. Dawson records similar heat distribution in compost prepared in 0.5 m³ boxes. A typical profile as found in a compost box.

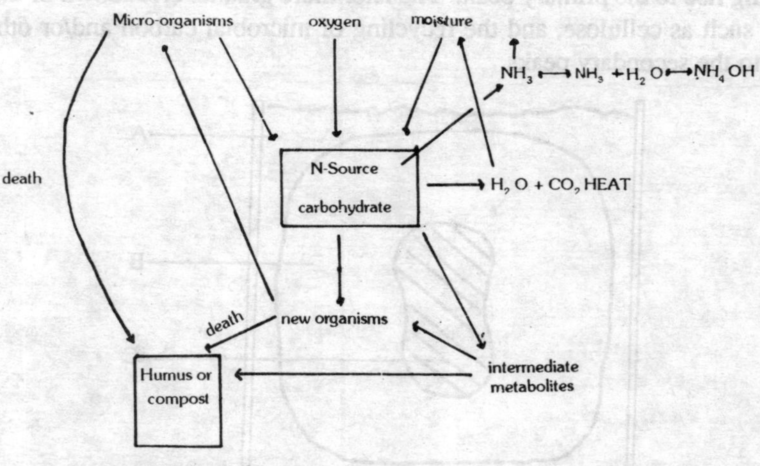

Fig. 21.1. The composting process.

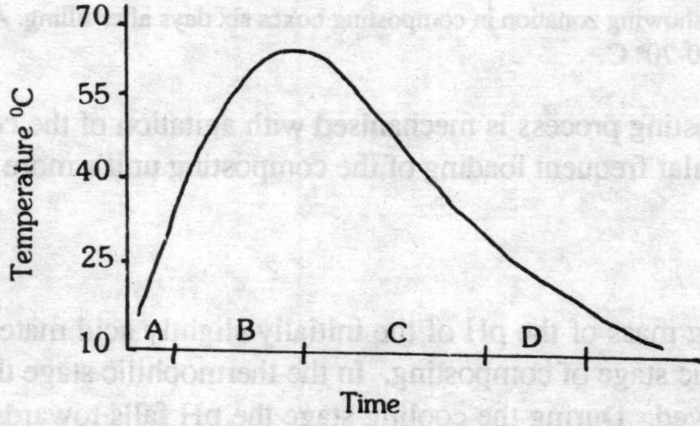

Fig. 21.2. Temperature variation with time for a typical compost indicating the phases of microbial activity. A, mesophilic; B. thermophilic; C. cooling; D. maturing.

In the Indore process and when preparing mushroom compost the compost is turned shortly after it has passed peak heat. Repetition of this procedure gives rise to a series of secondary or subsidiary peaks as the relatively less decomposed outer material is brought into the centre of the heap of composting material. A temperature profile of this shown in Fig. 21.4Secondary peaks have however also been recorded in undisturbed composts. When wheat straw and grass composts which had been made in 1 m^3 boxes were studied, the temperature profile for the grass composts followed the pattern, but that for wheat straw compost produced a secondary peak about 10 days after the initial main peak. Similarly an investigation of self-heating of old stored hay in Dewar flasks exhibited a multipeaked heating pattern and a multiple peaked temperature profile has also been recorded for moulding baled hay. Multiple peaked temperature profiles have also been shown for a variety of wood based composts and in composting broiler litter. The common factor in these apparently anomalous systems seems to be the presence of relatively large amounts of less readily degradable fibrous material. It seems that the readily available materials are first utilised by the micro-organisms thus giving rise to the primary peak. The later more gradual breakdown of the less easily degraded materials such as cellulose, and the recycling of microbial carbon and/or other minerals may then give rise to the secondary peaks.

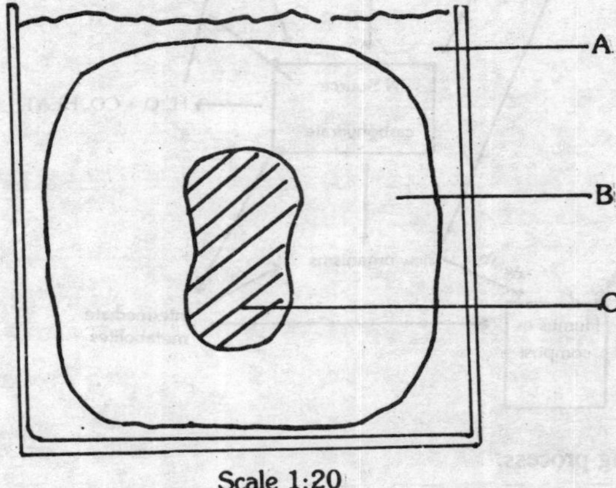

Fig. 21.3. **Temperature profile showing zonation in composting boxes six days after filling. A. ambient temperature—50°; B. 50-60°; C. 60-70° C.**

If, however, the composting process is mechanised with agitation of the compost and/or forced aeration coupled with a regular frequent loading of the composting unit a more constant temperature will be recorded.

pH

In a typical composting mass of the pH of the initially slightly acid material falls to the range 5.0-5.5 during the mesophilic stage of composting. In the thermophilic stage the pH rises rapidly to 8.0-9.0 as ammonia is evolved. During the cooling stage the pH falls towards neutrality.

Modification of compost pH away from this pattern is difficult as the evolution of ammonia exerts a powerful influence on the compost environment.

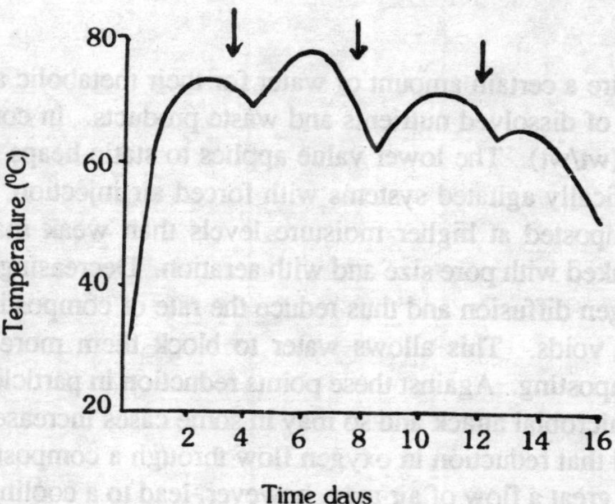

Fig. 21.4. Temperature variation with time in a compost heat which has been turned (turning of the heap is indicated by arrows).

Aeration

If composting is to prroceed rapidly, an adequate oxygen supply to the middle of the mass of composting material must be maintained. The diffusion of oxygen into the heap from the atmosphere will be reduced where the heap is large, where the air spaces within the compost are very small, or if the moisture content of the compost is so high that many of the voids are filled with water. The supply of oxygen can be increased by blowing air into the compost, agitation, the provision of air vents into the base of the composting mass, or by turning. If, however, the rate of air flow through the compost is too great, heat losses and desiccation will occur, and the rate of composting will be reduced. An advantage of an agitated system may be that abrasion of particles will occur exposing new surfaces to microbial attack thus speeding the composting process. Inermittent agitation has been found to give better results than continuous agitation. Oxygen consumption by composts is directly related to compost temperature is shown in Fig. 21.5. At temperatures higher than 70° C the oxygen consumption rate would be expected to decline at higher temperatures.

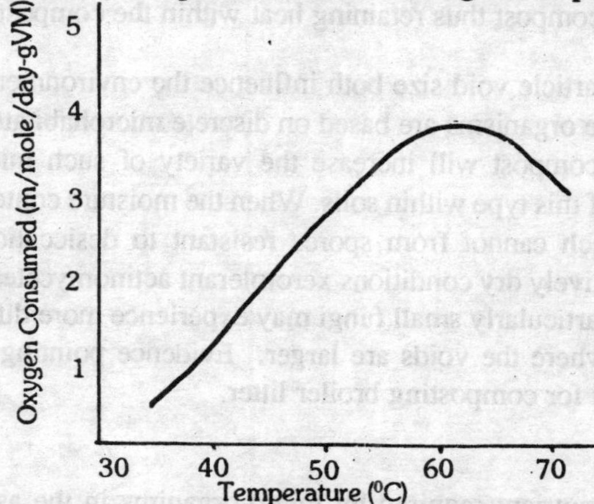

Fig. 21.5. Effect of temperature on oxygen consumption rate for mixed refuse.

Moisture Content

Micro-organisms require a certain amount of water for their metabolic activities as it provides a medium for the transport of dissolved nutrients and waste products. In composting the optimum moisture range is 50-70% (wt/wt). The lower value applies to static heaps while the upper one is more applicable to mechanically agitated systems with forced air injection. More rigid materials, such as straw, can be composted at higher moisture levels than weak materials such as paper. Moisture level is closely linked with pore size and with aeration. Decreasing the particle size of the compost will decrease oxygen diffusion and thus reduce the rate of composting. It also reduces the size of the interparticulate voids. This allows water to block them more readily also reducing aeration and the rate of composting. Against these points reduction in particle size does increase the surface area available for microbial attack and so may in some cases increase the rate of activity. It has also been demonstrated that reduction in oxygen flow through a compost leads to a lowering of compost temperature. Too great a flow of air may, however, lead to a cooling down of the compost and some degree of drying. At certain times this cooling effect may be desirable to prevent the compost heating beyond the point where microbial activity is possible.

Particle Size

Particle size is a critical factor in composting. This is not only because of factors directly relating to the particles themselves but also to the spaces between them, the interparticulate voids. Particle size has a significant influence upon the moisture content and the aeration of composts. If, for example, particle size is too small, the interparticulate voids are reduced in size and the rate of oxygen and carbon dioxide diffusion will be reduced, especially during the thermophilic stage of composting when the oxygen demand is greatest.

When particle size is small, the rate of microbial decomposition is enhanced because there is a greater surface area available for microbial attach. Composting grass/straw mixtures have been shown to break down more rapidly when macerated than when intact. Decomposition of relatively small compost particles is also facilitated because the intraparticulate depth for oxygen and carbon dioxide diffusion and for microbial penetration is reduced. Small particle size will also improve the insulating properties of the compost thus retaining heat within the compost.

Particle size and interparticle void size both influence the environment of the compost micro-organims. The lives of these organisms are based on discrete microhabitats within the compost and the heterogeneity of most compost will increase the variety of such micro-habitats. Hattori has investigated microhabitats of this type within soils. When the moisture content of the compost varies, gram-negative bacteria which cannot from spores resistant to desiccation will be placed under particular stress. Under relatively dry conditions xerotolerant actinomycetes will be favoured. If the interparticulate spaces are particularly small fungi may experience more difficulty in penetrating the compost than they would where the voids are larger. Evidence pointing to a similar situation in composts has been obtained for composting broiler litter.

Nitrogen

Nitrogen is the major nutrient required by micro-organims in the assimilation of the carbon substrate in organic wastes. Alexander suggests that 20-40% of the carbon substrate is assimilated into new microbial cells, the rest of the carbon being oxidised. These cells contain approximately 50% carbon and 5% nitrogen. Therefore, the microbial nitrogen requirement is 2-4% of the initial carbon, that is, a C/N ratio between 25:1 and 50:1. Microbial activity will tend to reduce this towards 10 : 1. Thus if the initial ratio is greater than 25-50:1, several cell generations are required to reduce the carbon. If the ratio is lower than this, then nitrogen will be lost from the system, usually as ammonia. A complicating factor in such a calculation is the availability of the carbon and nitrogen to the micro-organisms. Carbon present as cellulose or lignin is insoluble and thus nitrogen may be lost although the C/N ratio would seem to be too high for this to occur. Nitrogen present as keratin will similarly be unavailable and thus complicate the calculations. Composts have been made successfully where the initial C/N ratio was as large as 200:1, while others have been reported where the initial ratio was less than 10:1. Composting with an initially high level of nitrogen will induce the production of free ammonia because all the nitrogen cannot be converted into microbial cell components or incorporated into nitrogen-rich "lignin-humus complexes." The nitrogen present in the "lignin-humus complexes" will not be available unless the lignin is broken down. The overall loss of nitrogen from nitrogen-rich systems may be accompanied by an increase in the protein content of the compost. This increase may be two-three times the initial values.

Fibre

Once the more readily degradable components have been utilised the major carbon sources remaining in the compost are hemicellulose, cellulose and lignin. A common feature of these three less easily degradable compost components is that they are all high molecular weight sugar polymers. These large molecules cannot pass through the cell walls of the micro-organisms responsible for decomposition. The micro-organisms secrete extracellular enzymes which hydrolyse the polymers into short lengths which are usually the basic sugar units of which the polymers are composed. Only a proportion of the micro-organisms which inhabit composts can carry out these hydrolyses but most are capable of utilising the products of the extracellular breakdown. Of these high molecular weight polymers hemicellulose and cellulose are more readily degraded than lignin.

During composting the decomposition rates of both hemicellulose and cellulose can be represented by exponential curves. There is a consensus of opinion in the literature that the decomposition rates of hemicellulose and cellulose are roughly parallel for at least the first 20 days of composting. In the latter stages of composting there is a tendency for the rate of cellulose decomposition to exceed the rate of hemicellulose decomposition so that finally the percentage of the initial cellulose lost is greater than the percentage of the initial hemicellulose lost. This change in the degradability of hemicellulose during the course of composting reflects the heterogeneity of hemicellulose relative to cellulose. Whereas cellulose is composed only of glucose molecules hemicelluloses are composed of a variety of pentose and hexose sugars. Hemicelluloses are classified on the basis of their main sugar component : xylan and araban for polymers of the two principal pentose sugars and mannan and galactan for the two hexose sugars. Some of these hemicelluloses will be degraded more readily than others. It has, for example, been shown that mannan and xylan are more easily degraded than galactan. An additional complicating factor is the presence of hemicellulose in the hyphae of some fungi. It is therefore possible that some of the hemicellulose detected in the latter stages of composting is newly synthesised fungal hemicellulose rather than undegraded hemicellulose which has persisted throughout the composting process.

Compost temperature has also been shown to affect the decomposition of hemicellulose and cellulose. Work on the changes in the hemicellulose and cellulose of decomposing wheat straw after 16, 48, 105 and 273 days showed that the decomposition rate of hemicellulose was slower than that for cellulose particularly at lower temperatures. The difference was greatest at 7° C and least at 37° C, the latter being the highest temperature used. This may be explained by studies carried out on grass composts. This showed that in grass composts during the first four days 62% of the hemicellulose was lost while the cellulose was "more or less unattacked." During the next eight days of composting the cellulose was attacked. It was concluded that the initial thermophilic attack appeared to be on the proteins and hemicelluloses of the grass. It seems therefore that a thermophilic environment selectively favours the decomposition of hemicellulose.

There is evidence that cellulose is utilised by thermophilic fungi and that the rate of loss of cellulose is greatest during periods of active thermogenesis. It has also been shown that the rate of decomposition of both hemicellulose and cellulose was greatest at 50° C and slowest at 75° C. Cellulose decomposition was faster at 28° C than at 65° while hemicellulose decomposition was faster at 65° C than at 28° C.

It must therefore be concluded that hemicellulose is selectively degraded in the initial thermophilic phase of composting while the temperature builds up, and perhaps at peak temperature. Cellulose decomposition begins around the peak temperature and continues in the post peak phase while the temperature is still relatively high. The slow cooling phase after peak heat would seem to be better suited to cellulose decomposition than the initial heating and peak heat periods. There is general agreement that "Among the various ingredients of natural organic material, such as straw, the lignins are the most resistant to the action of fungi and bacteria". In the bulk of published work lignin shows little evidence of being decomposed in the first two-three months of composting, although one worker asserts that it is attacked or modified as soon as composting starts. The nitrogen content of the substance increases as methoxyl-groups are split off and nitrogenous substances are built in. This results in the formation of ligno-proteins which are of importance in the nutrition of *Agaricus bisporus*. Only the higher basidiomycetes possess the phenoloxidase enzymes which are able to break down the altered lignin. During mushroom composting the lignin content of the compost remains practically constant until spawning and only then does it begin to decrease. In other composts lignin disappears only very slowly. One piece of research showed that only after 480 days at ambient temperatures had 50% of the lignin been degraded. Temperature affects the rate of lignin decomposition, the half-life being reduced to 105 days at 37° C while at 27° C only 17% of the lignin had been broken down after 105 days.

Various attempts have been made to describe decomposition processes mathematically. Exponential functions have been found to be most useful for describing the breakdown of easily degraded materials. In such a function the fractional loss rate (k) is the slope of the regression of $\log_e W_{to}/W_{tn}$ where W_{to} is the initial dry weight of the material and W_{tn} is the dry weight after n times units.

Where, however, more complex, less easily degradable, substrates were present these substrates accumulated to a greater extent than was predicted by the exponential functions. To deal with such less readily degradable materials a double exponential model was proposed with separate rates for quickly and slowly decomposing constituents. A greater degree of refinement was achieved by assigning a separate exponential function to all the major chemical constituents in decomposing material. Howard and Howard adopted a different approach to the mathematical expression of decomposition. They tested a number of mathematical functions against their observed data from a study of tree and shrub leaf litters. The best fit was obtained using the asymptotic regression $W = A + B \times r^t$ where W is the percentage of the original weight remaining, "t" is the time in days, W = A at t and W = A + B at t_o. This model assumes that even after an infinite period of time a proportion of the litter will remain undecomposed. Asymptotic curves have been fitted to data for compositing broiler litter. They were found to give a better fit than exponential curves both for the total carbon content of the compost and for five individual compost components. A useful feature of the model is that the lower asymptote, A, can be calculated and thus an estimate of the compost residue predicted. This is particularly useful where the purpose of compost is to achieve bulk reduction.

Micro-organisms

Composting is a dynamic process is which the physical and chemical changes are caused by a rapid succession of mixed microbial populations. The precise nature of the succession and the number of organisms present at each stages depends on thè nature of the composting material and upon the preceding organisms in the succession. The importance of a mixture of organisms has been demonstrated and it has been shown that no single organism, no matter how active, can compare, in the rapidity of decomposition of manure with a mixed population. Individual organisms in pure culture can, however, often decompose a high percentage of a single component of a compost. The range of numbers of bacteria, fungi, and actinomycetes which have been recorded in composts is shown in Table 22.1.

At the start of the composting process the material is at ambient temperature. As the indigenous mesophilic (autochthonous) organisms increase in numbers the temperature rises. These mesophilic organisms are then succeeded by thermophilic bacteria and fungi as the temperature rises above 49° C. These in turn are replaced by thermophilic actinomycetes and thermophilic spore-forming bacteria as the compost reaches peak heat. As the compost cools the middle of the compost is re-invaded.

Bacteria are involved in the initiation of heating. The acid producing bacteria are numerically important at this stage. In self-heating hay the increase in bacterial numbers from 10^6-10^9 bacteria per gram dry weight in the first day of heating was shown to be due mainly to an increase in Gram-positive cocci. Chang and Hudson state that during the initial heating phase of composting the role played by the mesophilic bacteria resembles that of the thermophilic fungi rather than that of the mesophilic fungi. They ascribe this to the relatively high maximum growth temperatures of some mesophilic bacteria. At higher temperatures thermophilic bacteria become active and at high peak heats only spore-forming bacteria will survive in the compost core. As temperatures fall below 40° C mesophilic bacteria recover in number. In the initial heating phase the bacteria utilise the simple, easily degradable, organic substances present in the compost. The bacteria occurring later in the process may attack the more complex materials, moribund organisms in the compost, fungal mycelium, or they may exploit substances released from the less degradable substrates due to the action of the extracellular enzymes of other organisms.

TABLE 21.1
Composting Microflora

	Numbers/g of compost
Bacteria	10^8-10^9
Fungi	10^4-10^6
Actinomycetes	10^5-10^8

Fungi also play a part in brining about the initial rise in temperature. The fungi present initially represent the natural flora of the composted material. Those which show an initial increase in numbers are usually the rapid growing species which have the ability to utilise the more readily available substrates. At temperatures higher than 40° C the activity of thermophilic fungi becomes important. Thermophilic fungi have been defined as those which have their maximum temperature for growth at or above 50° C and their minimum temperature at or above 20° C. Thus their activity will have commenced and their numbers will have built up during the initial mesophilic phase of composting. Thermotolerant fungi will also be active through the mesophilic phase and at least the beginning of the thermophilic phase of composting. These fungi can grow at high temperatures but their lower limit for growth is below 20° C. Above 60-65° C the activity of the thermophilic fungi ceases since these temperatures represent the upper limit for the growth of eukaryotic organisms. As peak heats are attained in composts fungi tend to disappear from the central zone of the compost. In grass and straw composts where a peak heat at 70° C was recorded the thermophilic fungi disappeared from the compost core for a period of 3 days. As the compost temperatures fall below 60° C thermophilic fungi reappear in the middle of the compost. It is usually considered that organisms which have survived in the cooler outer parts of the compost, or airborne spores are responsible for this recolonisation. Extreme peak heats seldom persist for more than a few days and it is possible that in some cases fungi survive the peak heats in and around the compost core. Harrington *et al.* recorded four species of fungi which were able to survive a temperature of 70° C for three days in composted conifer bark. Fungi thus left would be ideally placed to exploit the central region of the compost as the temperature fell. Mesophilic fungi also disappear from the compost core at peak heat. Chang and Hudson noted that they were absent from the middle of the compost for 12 days. Their reappearance coincided with the temperature dipping below 40° C and they only became numerous around 30° C. Later in the composting process lignin often remains as the major energy source and it may be attacked by basidiomycetes. The introduction of the spawn in mushroom composting corresponds with this phase. An example of typical changes within the fungi flora of a compost is shown in Fig. 22.6 which is taken from Chang and Hudson.

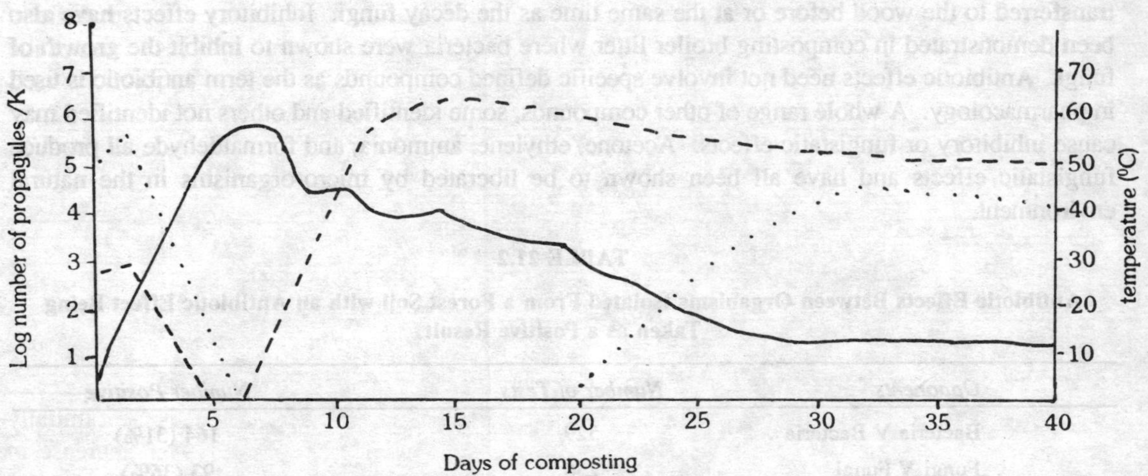

Fig. 21.6. Changes in the numbers of mesophilic (.....) and thermophilic (. . .) fungi in a wheat straw compost. The temperature is superimposed.

The attaining of a peak heat in the region of 60-70º C represents the typical compost situation, but cooler composts have been investigated and in such composts fungi remain in the middle of the pile. Gregory *et al*. recorded that in self-heating hay where the temperature did not rise above 50º C mesophilic fungi remained numerous. Many of these fungi must have been surviving rather than actively growing because the numbers isolated on plates incubated at 25º C were only half of those recorded from plates incubated at 40º C. Similar results were obtained by Festenstein *et al*. who recorded a peak heat of 58º C in old self-heating hay.

The disappearance of fungi from a compost may in some cases be attributable to factors other than temperature. Ammonia is known to have fungicidal properties and has been used to preserve stored grain from fungal attack. When nitrogen rich materials such as animal wastes are composted excess nitrogen is lost to the environment as ammonia. This ammonia may in some circumstances have a negative effect on fungal numbers. Fungi are also vulnerable to antagonistic effects within composts. Antagonism is probably the commonest type of interaction between micro-organisms. Three aspects of antagonism have been identified. These are antibiosis, where one species produces a substance which is harmful to a second species, exploitation in which one species directly uses another for its own benefit, and competition which occurs when two species are rivals for some feature of the environment which is of limited availability.

Antibiotic effects can readily be demonstrated in the laboratory. The results of one study are summarised in Table 21.2. Henningsson examined the effect of bacteria on fungi, both groups of fungi having been isolated from the same decaying wood. He showed that bacteria could produce fungistatic substances which, if favourable conditions existed, could diffuse and inhibit the growth of fungi at considerable distances from the bacterial colony. When bacteria and fungi were grown together on sterilised wood, the activity of the decay fungi was reduced if the bacteria were transferred to the wood before or at the same time as the decay fungi. Inhibitory effects have also been demonstrated in composting broiler litter where bacteria were shown to inhibit the growth of fungi. Antibiotic effects need not involve specific defined compounds as the term antibiotic is used in pharmacology. A whole range of other compounds, some identified and others not identified may cause inhibitory or fungistatic effects. Acetone, ethylene, ammonia, and formaldehyde all produce fungistatic effects and have all been shown to be liberated by micro-organisms in the natural environment.

TABLE 21.2

Antibiotic Effects Between Organisms Isolated From a Forest Soil with an Antibiotic Effect Being Taken as a Positive Result

Opponents	*Number of Tests*	*Number Positive*
Bacteria V Bacteria	529	164 (31%)
Fungi V Fungi	256	93 (36%)
Bacteria V Fungi	368	161 (44%)

Actinomycetes are associated with the thermophilic stage of composting. Thermophilic actinomycetes are responsible for producing the peak heats in the hotter composts. The multiplication rate of actinomycetes within composts has been found to be at its greatest between 45 and 60° C, and at these temperatures actinomycetes are still capable of causing extensive decomposition but at 75° C their ability to break down compost components is much reduced. Having increased in numbers during the latter part of the heating phase of composting they are capable of maintaining a high level of viable propagules throughout the peak heat phase. These organisms are thus well placed to exploit the compost environment as it cools in the immediate post peak heat phase. In the cooling stage actinomycetes can become so numerous that their hyphal strands cause the compost surface to take on a grey or white appearance. During this stage of the composting, actinomycetes have been shown to attack hemicellulose while lignin and cellulose are degraded to only a limited extent.

Composting can have the effect of reducing the burden of pathogenic micro-organisms within the composting material. These organisms include viruses, patheogenic bacteria and toxigenic fungi.

There is considerable variation in the survival times reported for viruses in composts. When an RNA virus similar to several animal viruses, was inoculated into sewage sludge composts to give a concentration in excess of 10^6 plaque-forming-units per gram of compost, the time required to eliminate the virus varied from 50-70 days according to the temperature of the compost. At the other end of the range poliovirus was inactivated in composts after the temperature had been above 50° C for 30 minutes.

There have been a number of studies on the survival of pathogenic bacteria during composting. The destruction of bacteria by heat depends on a time-temperature interaction and survival times will also vary according to species. In one study *Pseudomonas aeruginosa* was inactivated after one hour but various strains of *Salmonella and E. coli* survived for 3-4 hours. Unlike viruses, bacteria can multiply in a compost. It has been shown that while *Salmonella* numbers decrease in the middle of composts they can multiply in the outer 40 cm. and particularly in the outer 20 cm. This increase in numbers peaked at the seventh day of composting and thereafter fell off rapidly although small numbers persisted in the outer 20 cm. for about 40 days. Factors other than temperature are also important in the elimination of pathogens from composts. This was demonstrated in a series of experiments carried out by Platz and his co-workers. In order to separate the temperature effects from other compost effects bacteria were sealed into glass vials which were then buried in the compost (indirect contamination). The survival of these organisms was then compared with bacteria directly inoculated into the compost (direct contamination). It was found that bacteria in the indirect contamination group survived for longer than those in the direct contamination group, and that bacteria in the direct contamination group were eliminated at lower temperatures than those in the indirect contamination group. These experiments demonstrate a biological (antagonism) effect in addition to the temperature effect.

The production of toxic metabolites and competition for nutrients or sites within the compost may be significant factors in such a process. It can be concluded that while composting will not produce a sterile residue it will yield a product which is microbiologically safer as the initial flora is replaced by a compost flora thus reducing or eliminating both plant and animal patnogens.

Conclusion

It is clear that the process of composting is ecologically complex and that it is influenced by a wide range of environmental variables. The relationship between the physical structure of the wastes and the moisture content of the compost appears to be of particular significance. Composting offers a relatively convenient method for the disposal of a variety of agricultural wastes, and it may in some circumstances yield a useful product at the end of the process.

22

Biotransformations

Introduction

Shell, Du Pont, ICI, Exxon, International Nickel and Standard Oil are amongst a few of the multinational conglomerates that have been tempted to take a stake in the future of biotechnology. Many have interests in the pharmaceutical, agricultural and energy industries and, therefore, stand to benefit from the biotechnological developments, but they must be particularly attracted by the potential of biotechnology in their other business activities — the chemical, mining and oil extraction industries which lie at the core of the developed world's economy.

At present much of the world's chemical industry relies on raw materials obtained from oil, gas and coal. Plastics, paints, adhesives and synthetic rubbers are all derived entirely or in part from these diminishing resources. In 1978, in the US alone, over $35 billion worth of petrochemicals were made from relatively few raw materials. Several of these can already be manufactured by biotechnological processes, including acetone, glycerol, butanol and ethanol, and the rapid progress in this field may soon yield others.

Acetone and Butanol — A Biotechnological Renaissance?

War though unfortunate but is one of the most potent catalysts for industrial innovation. Soon after the outbreak of the First World War, Britain and Germany faced a similar crisis: chemicals needed for the manufacture of munitions were in short supply. Britain needed more acetone, a liquid solvent used during the production of cordite, while Germany lacked glycerol, from which dynamite is made. Both countries turned to microbes to solve their problems and met with considerable success. Germany was soon making 1000 tones of glycerol a month. After the war, the rise of the petrochemical industry supplanted biotechnology as the major source of acetone and butanol. In recent years, a better understanding of the microbes that make these chemicals has been gained and the cost of oil has increased substantially. These factors may bring us full circle, reinstating these and other biotechnologies in a central position in the chemical industry. This time they can be put to much more constructive uses, including the manufacture of plastics, fibres and resins.

In 1912, the chemist Chaim Weizmann (who was later to become Israel's first president) was working in Manchester. He developed a process whereby the bacterium *Clostridium acetobutylicum* fermented starch to form the liquids acetone and butanol, and it was this process that was introduced on a large scale during the war. While Weizmann's process has largely fallen into disuse, the demand for acetone and butanol has grown unabated. Over a million tonnes of acetone are consumed by the US chemical industry each year, much of it in the manufacture of plastics. Butanol, a kind of alcohol, is even more versatile; it is employed in the manufacture of resins, protective coatings, paints, synthetic rubber and brake fluids.

In many respects the Weizmann process is similar to the production of fuel alcohol and biotechnologists face the same kind of problems in developing economically attractive systems. There is one particularly significant difference: when yeasts are supplied with sugars in the absence of air, they will always make alcohol, but *Clostridium acetobutylicum* does not always make acetone and butanol. Microbiologists are searching for strains of the bacterium that can be relied upon to produce the desired products. However, this bacterium is unusually sensitive to the conditions under which it is grown and unless the temperature, acidity and other factors are precisely right it will not yield acetone and butanol in significant quantities. This means that the fermentation must be very carefully controlled.

Another difficulty is that the bacteria are harmed by the products they make. Just as yeasts cannot tolerate high concentrations of alcohol in their surroundings, so *Clostridium acetobutylicum* is damaged by acetone and butanol. Unfortunately, here the problem is even more acute, and only 2 or 3 per cent of these materials in the fermentation liquids will inhibit the bacteria from making more of them. This means that large fermentation vessels must be built to obtain the desired quantities of products, and the cost of purifying such a small proportion of acetone and butanol from all the other materials in the vessel is high.

Despite these obstacles, there are good grounds for optimism about the future of the Weizmann process. In particular, recent experiments indicate that immobilizing the bacteria may increase the efficiency of the fermentation by 65 to 200 per cent. If this increase is achieved in practice, the economics of the process will improve radically.

Until the First World War, German industry had manufactured glycerol from imported vegetable oil, but the British naval blockade soon began to stem the flow of the raw materials. This provided the impetus to develop the work of the biochemist Carl Neuberg who, a few years before, had discovered that the small amount of glycerol produced by yeasts when fermenting sugars could be greatly increased by the addition of a simple, cheap chemical—sodium bisulphite. Glycerol has since grown to become an almost indispensable part of the chemical industry, being employed, for example, as a lubricant and softener, a plasticizer for cellophane and a raw material for resin manufacture. Today, Neuberg's process has sunk into commercial obscurity and glycerol is made from petrochemicals and vegetable oils. In a few years, his process may be revived, again due to the rising costs of alternative manufacturing methods. Rather nearer at hand is an entirely novel biotechnological source of glycerol—algae.

Despite its forbidding name, the Dead Sea has been colonized by a few unusually hardy microbes. The high concentrations of salt (up to eight times that of the oceans) in the Dead Sea and the Great Salt Lake of Utah soon kill most organisms. They die as the water inside their cells is pulled out through the membranes into the highly saline environment. Certain microbes, called halophiles (salt-lovers), have evolved ways of preventing this dehydration. One of these, the alga *Dunaliella bardawil*, manufacturers large quantities of glycerol so that the cell contains a high concentration of dissolved material. This counteracts the force — known as osmotic pressure — which tends to draw water out of the cells and into the surroundings.

In Israel, near the Red Sea coast, this alga is grown in specially constructed ponds covering 2 hectare (5 acres). Since *Dunaliella bardawil* is a photosynthetic organism, it obtains most of its energy from the sun, and only a few simple nutrients need to be supplied. Once the algae have been harvested and dried, the glycerol, which accounts for something approaching 40 per cent of the weight of the cells without their water is extracted.

This alga is one of the most promising organisms for biotechnology. It also contains about 8 per cent of beta-carotene, the compound which gives carrots their characteristic colour and finds a ready market as a food colourant. Once the glycerol and beta-carotene have been taken out of the cells, the residue forms an excellent, protein-rich animal feed. In addition, this alga thrives in brackish water, and so it can be grown in semi-arid regions where fresh water is at a premium. The fact that this type of biotechnology does not compete with agriculture for good-quality water is a particularly important factor for many less-developed countries. *Dunaliella bardawil* lends itself to low-technology industries which need little capital investment. Valuable products can be obtained from the cells fairly easily and the growth ponds do not require much attention. In particular, contamination by unwanted organisms — the bane of most biotechnologists — is largely eliminated. Most other organisms are destroyed by the very high concentrations of salt found in the ponds.

Biotechnology and the Plastics Industry

The plastics industry is currently worth over $ 50 billion a year—a tempting market for biotechnological enterprises. As we have seen these could supply acetone, glycerol and butanol to the chemical industry, and the ethanol produced by fuel alcohol factories could also be diverted for use in it. Ethanol serves as a vital starting material for the manufacture of materials as diverse as detergents, dyes, adhesives and resins for synthetic fibres.

Among several novel and ambitious projects, one of the most promising is the synthesis of compounds known as alkene oxides which are widely used in the manufacture of plastics and polyurethane foams commonly used in furnishings. Development of these processes began in the late seventies and may be in commercial operation before the end of this decade, with potential sales of $2-3 billion.

Alkenes are a group of compounds which contain only carbon and hydrogen. Most importantly for the plastics industry, alkenes can be polymerized—that is, the individual molecules can be linked in chains—to form materials such as polypropylene (used, for example, to make containers) and polyethylene (better known as Polythene). Before alkenes are polymerized to form plastics, they must be converted into alkene oxides by the addition of oxygen to the molecules. At present, this is accomplished by purely chemical techniques, but two biotechnological processes are on the horizon. The first, largely developed by the Californian firm Cetus, employs three enzymes — two from fungi and one from bacteria—to perform the same function as the current chemical methods of making alkene oxide.

More promising still is a process patented in 1981 by scientists from the University of Warwick, England. In the famous baths of the city Bath, they discovered a microbe which can add oxygen to alkenes. When supplied with propylene or ethylene gases, the bacterium *Methylococcus capsulatus* can insert an oxygen atom into the molecules to produce propylene oxide or ethylene oxide. This bacterium is particularly enticing because it lives very happily at about 45° C (113° F), and at this temperature the alkene oxide are gaseous. It is much simple to collect the product as a gas than as a liquid, mixed in with all the other materials in a fermentation vessel.

Both the enzyme and the bacterial systems have several advantages over conventional processes. The current chemical methods require expensive chlorine gas, whereas the enzyme technique employs cheap common salt and the bacteria do not even need that. Both biotechnological processes work at lower temperatures, saving on energy costs, and they are flexible in that different types of alkene oxides can be made with the same basic installation. Finally, the pollution created by chemical treatments can be avoided by employing enzymes or bacteria.

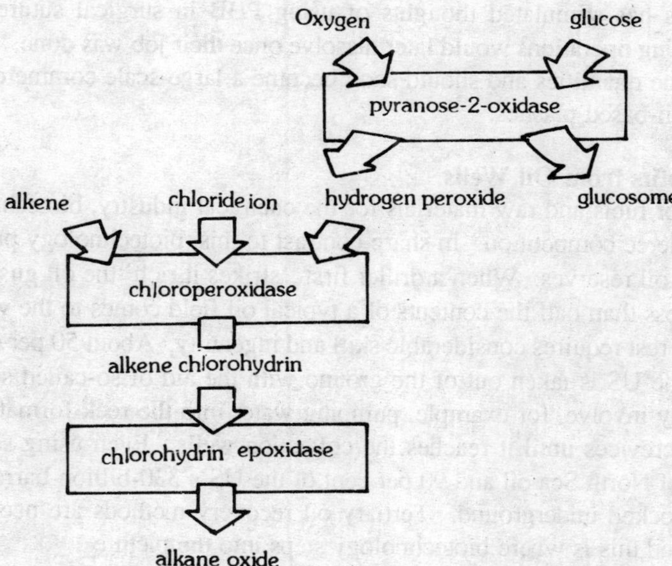

Fig. 22.1. Alkenes, such as ethylene and propylene, are converted into alkene oxides and then polymerized to form plastics, such as polyethylene and polypropylene. A series of three enzymes can be employed to make alkene oxides. First an enzyme, pyranose-2-oxidase (obtained from fungi) catalyses a reaction between oxygen and glucose to produce hydrogen peroxide. Another fungal enzyme, chloroperoxidase, joins hydrogen peroxide and a chloride ion from sodium chloride (common salt) to an alkene chlorohydrin molecule. The chloride and a hydrogen ion are then removed with the aid of an epoxidase enzyme from bacteria to yield an alkene oxide.

A notable aspect of these future biotechnologies is the way that they would be integrated into conventional industries. The alkenes would still be derived from oil or similar materials, and the polymerization of alkene oxides to produce plastics would still be achieved by chemical techniques, with biotechnology making its contribution at the intermediate stage. The reliance of the process on oil as the source of alkenes is an obvious drawback. Looking much further into the future one can envisage biotechnology supplying the alkenes. No known microbe makes significant amounts of alkenes, but, as has happened so often before, a diligent search for organisms that contain the required substances may prove successful. If such a search failed, the problem could be turned over to genetic engineers, for whom designing a microbe to supply alkenes would be a major challenge. There is very little chance that the introduction of a single gene into a microbe could induce it to start synthesizing alkenes, since genes code for enzymes and other proteins, compounds that are quite different from alkenes. It would be necessary to put several genes into a microbe, each of which codes for one of a series of enzymes. These enzymes might then act in concert to convert some substance normally found in the cell into alkenes.

A rather more immediate biotechnological process for the plastics industry involves a compound called polyhydroxybutyrate (PHB). This material, which is similar to synthetic polyesters used in the textile industry, is found in many types of bacteria. Bacterial PHB is rather fragile, but it is biodegradable. This has stimulated thoughts of using PHB in surgical sutures; threads of this material inserted during operations would later dissolve once their job was done. PHB is now being manufactured in tonne quantities and should soon become a large-scale commercial product which may compete with oil-based plastics.

Sweetening the Profits from Oil Wells

In the market for fuels and raw materials for the chemical industry, biotechnology and the oil industry will be in fierce competition. In sharp contrast to this, biotechnology promises to help oil drillers recover vast oil reserves. When a driller first 'strikes it rich' the oil gushes to the surface, but, unfortunately, less than half the contents of a typical oil field comes to the wellhead with such ease. Extracting the rest requires considerable skill and ingenuity. About 50 per cent of the oil now being produced in the US is taken out of the ground with the aid of so-called secondary recovery methods. These may involve, for example, pumping water into the rock formation forcing the oil along subterranean crevices until it reaches the collection wells. Even using secondary recovery £300 billion worth of North Sea oil and 90 per cent of the US's 330-billion-barrel oil reserves may remain stubbornly locked underground. Tertiary oil recovery methods are needed to extract this precious resource, and this is where biotechnology steps into the picture.

Most of the oil in a reservoir is found not as vast pools of liquid, but as a coating on grains of rock. The oil sticks tenaciously to these grains and must be dislodged before it can be brought to the surface. Ordinary water is too thin a liquid to budge most of this oil; it simply flows past the oil-coated grains. In tertiary oil recovery (also known as enchanced oil recovery) materials are mixed with the water to make it more viscous, and one such material is xanthan gum.

Xanthan gum, a polysaccharide produced by the bacterium *Xanthomonas campestris*, is composed of many glucose units linked into a chain with other types of sugars branching off at regular intervals. This knotty structure makes xanthan gum a very efficient thickening agent, allowing the water/gum mixture to act like a piston, pushing oil towards the wells (see Fig. 22.2). Price is the major obstacle at present to its general use in extracting recalcitrant oil deposits. The kind of biotechnological strategies that have been applied so successfully with other organisms could easily reduce the cost of xanthan gum quite dramatically in the near future.

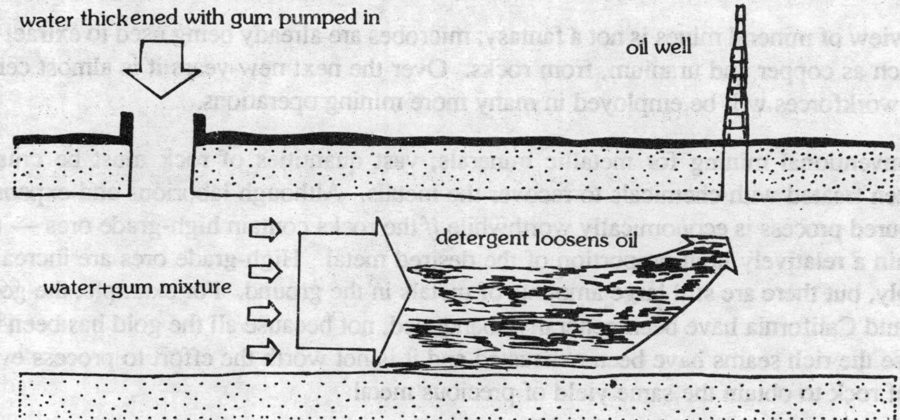

Fig. 22.2. Using xanthan gum to recover oil. Water containing a detergent-like material is first pumped into the ground to loosen the oil clinging to rock particles. Water thickened with xanthan gum is then pumped in to act as a piston, pushing the oil-bearing mixture towards the oil well.

Xanthan gum and other similar materials excreted by microbes have already found a use in the petroleum industry. The huge drills which penetrate rock must be lubricated, and various types of mud perform this task. These drilling muds—mixtures of water, clays and other materials — also counterbalance the upward pressure of the oil. Microbial polysaccharides have been proved to be eminently suitable thickening agents.

If xanthan gum is likely to be so effective in tertiary oil recovery, why go to the trouble of growing the bacteria in factories, extracting the gum and then pumping it into the oil field ? Why not send the microbes themselves underground ? This is exactly the line of thought being pursued by some oil companies. Field tests employing Bacillus and *Clostridium* bacteria rather than *Xanthomonas* are already under way, and the preliminary results are encouraging. Sugars and other nutrients are fed to these bacteria while they are deep below ground. Here they grow, produce chemicals that help wash the oil free, and yield carbon dioxide and other gases which also assist in forcing the oil to the surface.

It is remarkable that these bacteria can survive in conditions of high pressure and temperature, lack of water and oxygen, and large quantities of salt and sulphur, all of which are inimical to these organisms. This shows again that the resilience and adaptability of microbes should never be underestimated, and these qualities, enhanced as necessary by biotechnologists, may soon be put to commercial use in the oil industry.

Microbes in Mineral Mines

The ore extraction plant of the future could have the appearance of a present-day water-treatment plant....free from the dirt and spoil heaps normally associated with mining operations, while far below ground millions of microbes are carrying out the tasks which today are characterized by the roar of machinery and the ring of pick and shovelon rock (Dr. Richard Manchee, Microbiological Research Establishment, UK, 1979).

This view of mineral mines is not a fantasy; microbes are already being used to extract valuable metals, such as copper and uranium, from rocks. Over the next new years it is almost certain that microbial workforces will be employed in many more mining operations.

In conventional mining for metallic minerals, vast quantities of rock must be crushed and ground, then treated with chemicals to recover the metals. Although laborious and expensive, this time-honoured process is economically worthwhile *if* the rocks contain high-grade ores — that is, if they contain a relatively high proportion of the desired metal. High-grade ores are increasingly in short supply, but there are still large amounts of metals in the ground. For example, the gold mines of Wales and California have been generally abandoned, not because all the gold has been removed but because the rich seams have been exhausted and it is not worth the effort to process ever larger amounts of rock to obtain the same yield of precious metal.

Thus, the problem that faces the mining industry is to make use of low-grade ores which cannot be exploited economically by conventional techniques. Mining engineers in Spain, Canada, the US and elsewhere have found that the answer is to rely on microbes to extract the valuable metals and concentrate them into a form that is cheaper and more convenient to handle.

Thiobacillus ferro-oxidans is probably one of the oldest forms of life on Earth, but it was not until 1947 that it was discovered in an abandoned coal mine in West Virginia. This bacterium is now known to be present in many types of rock throughout the world, and there can be many millions in just a handful of material. Many microbes have bizarre dietary preferences, but none so odd as this rod-shaped organism. It does not obtain energy from sunlight (it usually lives in total darkness) nor from organic materials in its surroundings. Instead, it unlocks energy from inorganic compounds, such as iron sulphide, and uses that energy to construct the materials it needs to live from the carbon dioxide and nitrogen in its environment. In the process it also manufactures sulphuric acid and iron sulphate, which explains why *Thiobacillus ferro-oxidans* can be used in mining operations.

The sulphuric acid and iron sulphate it produces attack the surrounding rocks and leach (dissolve) many metallic minerals. For instance, the activities of these microbes will convert insoluble copper sulphide into soluble copper sulphate. As water percolates through the rocks the copper sulphate is carried along and eventually collects as bright blue pools. In this way the copper scattered throughout thousands of tonnes of low-grade ore is concentrated in metal-rich lagoons. The metal is recovered by passing copper sulphate solution over pieces of iron. Eventually a layer of copper is deposited on the iron, and this can be scraped off. Uranium is leached from its ores by the same type of process.

Already about 14 per cent of the copper produced in the US depends on this biotechnology. At present, microbial leaching is mainly employed with the waste materials from conventional mining and extraction processes which leave substantial residues of metal in the discarded rocks. Dumps, up to 370m (1200 ft.) high and weighing 4 billion tonnes, are constructed from these 'waste' materials. Water is sprayed on to the top of the mounds and, as it filters down, it picks up the soluble metal compounds created by the action of the bacteria. The ubiquity of *Thiobacillus ferro-oxidans* means that it is rarely necessary to introduce it into the dumps. The area on which the heaps are built is usually covered with clay or asphalt so that the metal-rich liquids collect in pools at the foot of the heaps instead of seeping right into the ground.

When using microbes in this type of secondary recovery process, miners are still faced with the substantial cost of hauling the ores to the surface. The experience gained at the Stanrock Uranium Mine in Canada shows that even this cost is not always necessary. This mine was opened in 1958, operating on conventional principles. By 1962 it was found that the pools of liquid that had accumulated underground contained about 13,000 kg (29,000lb) of uranium oxide which had been leached out of the rocks. Before long, conventional mining was halted and bacteria were left to do most of the work, with water being hosed on to the rocks to assist the natural leaching process. This underground solution mining has cut costs by a quarter at this mine. Similar techniques are almost certain to be introduced into other mines, especially those with low-grade ores. Semi-industrial processes have already shown the promise of microbial leaching in the recovery of cobalt, lead and nickel, and other valuable metals, such as cadmium, gallium, mercury and antimony, are targets for the future. The potential is truly enormous — for example, the US has several billion tonnes of rock which contain small concentrations of nickel. Since there is, on average, only about 1 kg (2.2lb) of nickel in each tonne of rock, it is uneconomic to extract this valuable resource with conventional mining techniques. Advances in microbial mining could make it possible to tap this source of nickel which is worth about $60 billion and recover a further $10 billion hoard of cobalt in the same rocks.

Microbial mining has the additional advantage for less-developed countries that it eliminates the need for some of the costly (and usually imported) heavy mining equipment. Although a substantial capital investment is required for any mining development, microbes could help these countries preserve precious foreign currency.

Role of Microbes to Clean up the Mess

The threat that pollution poses to ourselves and the environment has become all too apparent in recent decades. Oil spills, pesticides, herbicides, chemical effluents and heavy metals, such as lead and mercury, are just some of the hazards. Biotechnology can be used to tackle these problems in two ways. Firstly, the root causes can be attacked by the introduction of more biotechnological production methods, which are intrinsically less polluting. For example, in manufacturing chemicals for the plastics industry with the aid of biotechnology, microbes are fed on innocuous raw materials such as sugar, whereas conventional processes use oil-based raw materials, some of which inevitably escape to pollute the environment. Secondly, microbes can be developed as voracious scavengers, removing all manner of pollutants.

The principal aim of all the biotechnologies discussed so far is to manufacture specific products; the fact that certain raw materials are incidentally consumed during the process is an often expensive necessity. In biotechnological processes that are targeted towards the control of pollution, the emphasis is reversed; their primary purpose is the destruction of specific raw materials — the pollutants. Despite this fundamental distinction, it should be noted that there is no precise dividing line between product-orientated and raw-material-orientated biotechnologies. For instance, the economic feasibility of the methane generators depends on two factors: the value of the methane fuel produced, and the fact that they perform a useful function in disposing of domestic and agricultural wastes.

Pouring Microbes on Troubled oil

The *Pseudomonas* are a group of bacteria noted for their ability to break down esoteric compounds that most microbes shun. In particular, various strains of *Pseudomonas* can consume hydrocarbons, which constitute the bulk of oil and petrol. However, each individual strain can utilize only one or a few of the many different types of hydrocarbon. The genes that code for the enzymes which attack hydrocarbons are not found on the main bacterial chromosome, but on plasmids, the small, semi-autonomous rings of DNA.

In an attempt to create a 'superbug' which would be able to mop up *all* the types of hydrocarbon in spills, Ananda Chakrabarty of General Electric introduced plasmids from several different strains of *Pseudomonas* into a single cell (see Fig. 22.3). One idea was to grow these recombinant bacteria in the laboratory, mix them with straw and dry them. The bacteria-laden straw could be stored until needed, when it would be scattered over oils licks; the straw would first soak up the oil, and the bacteria would break it down into harmless, non-polluting materials.

Chakrabarty's microbe has not yet been used commercially, and indeed many scientists doubt that it has much a future; they argue that a mixture of wild strains of *Pseudomonas* may degrade oil just as well or even better. Whatever the merits of this particular bacterium, the general approach should prove valuable for controlling other types of pollution. Late in 1981, Chakrabarty and his colleagues announced that they had developed a microbe that attacked 2, 4, 5, T, a very persistent herbicide and the main ingredient of the Agent Orange, which was used to destroy vast areas of jungle in Vietnam.

Although the suitability of genetically engineered microbes has yet to be proved, more conventional types of biotechnology have already been put to good use, as two examples from the US illustrate. When the *Queen Mary* was moved to Long Beach in California about 3,600,000 litres (800,000 gal) of oily water lay in its bilges. Obviously, if this had been discharged into the harbour it would have harmed marine life and disfigured nearby beaches. Therefore, a mixture of several different sorts of bacteria was introduced into the bilges. In six weeks they decomposed the oil, leaving a combination of water, bacteria and innocuous chemicals that could be released safely in to the harbour.

An oil company in Pennsylvania faced a similar problem when a leakage of 27,000 litres (6,000 gal) of petrol threatened to contaminate underground water supplies. Bacteria already living in the vicinity would doubtless have destroyed the petrol eventually, but without human intervention the process might have taken decades. The bacteria could only grow slowly because there were insufficient nutrients in their surroundings to give them the oxygen, nitrogen and phosphorus required for rapid growth. By pumping the missing nutrients into the ground, the bacteria were spurred into action and the petrol was degraded in only a year.

Bacteria are also being used to clean up other types of mess. A particularly impressive example is the addition of bacteria to the contents of ship's tanks very effectively prevented the build up of grease on the tank's interior. This kind of use of bacteria may become important in factories which process meats and poultry where pipes and vessels can become plugged with grease.

The living world has developed complex and highly efficient means of consuming natural wastes and utilizing them. If it were not the constant recycling of materials—in which microbes play the major role—we would all be knee-deep in dinosaur remains! It is the new types of pollution, created by industry, that present the most intractable problems. New chemicals, such as pesticides and substances which previously appeared on the Earth's surface only in small amounts, such as oil and many metals, tend to persist since few of the common microbes in soil or water can use them as food. This has created a growing demand for tailor-made packages of microbes which can remove specific forms of pollution. Several major companies, especially in the US, offer mixtures of microbes and enzymes designed to clean up chemical wastes, including oil, detergents, waste waters from paper mills, and highly toxic materials such as dioxin, the chemical that wreaked havoc on the town of Seveso, Italy. By 1978, the US market for biological pollution control products had reached about $3 million. As public disquiet about the effects of pollution begins to be translated into more effective legislation, the demand for these products seems sure to grow — perhaps to fifty times its present level.

How to Control Pollution at Source ?

The deployment of biotechnology to minimize the effects of oil spills and other pollution emergencies is clearly valuable, but a far sounder approach is to attack the sources of pollution. Just about every city and town in the developed world has already made a substantial investment in one form of biotechnology — sewage processing plants. The majority of these depend on the action of microbes to purify waste water by consuming a wide variety of solid materials in domestic and agricultural effluents. Perhaps surprisingly, there is still no clear understanding of precisely how sewage is broken down. This is because the composition of sewage varies considerably and there are many different species of microbe at work in sewage ponds. However, as the existing techniques are usually very successful there is little incentive to investigate new methods of disposing of general sewage. The situation is quite different in the case of certain industrial pollutants, most notably the heavy metals.

Heavy metals, a group of elements which includes mercury, cadmium and lead, are among the most insidious pollutants produced by modern industry. Mercury, for example, was responsible for the most notorious outbreak of metal poisoning, in which dozens of people in the Japanese fishing village of Minamata died or suffered severe damage to their nervous systems. Mercury discharged from a nearby factory was taken up by fish, which were later eaten by the local people. The dangers of lead pollution, particularly mental retardation of children, are beyond dispute. Many countries, Britain being a notable exception, have already introduced strict controls on the emission of lead from petrol engineers, but there is also a need to remove lead from certain factory effluents.

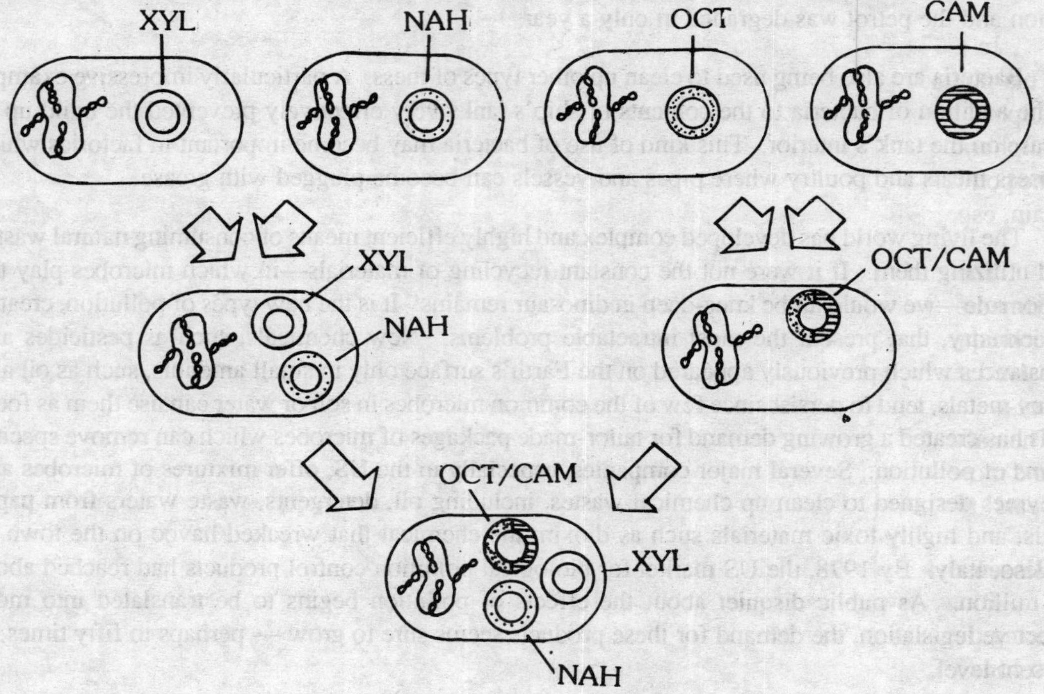

Fig. 23.3. Oil contains several different types of hydrocarbon compounds, the main groups being xylenes, naphthalenes, octanes and camphors. Certain strains of *Pseudomonas putida* bacteria can consume each of these hydrocarbons, but no single strain found in nature can consume all four types. The genes which enable these bacteria to feed on hydrocarbons are found on four types of plasmids, termed *xyl*, *nah*, *oct* and *cam*. By introducing all four sets of genes into a single cell a 'superbug' was created which could digest all four major components of oil. The *cam* and *oct* plasmids cannot coexist inside the same cell, so the relevant genes from each plasmid are first joined into a single plasmid.

Heavy metals are as toxic to most types of microbe as they are to animals including humans, but some species of algae and bacteria avidly extract metals from their surroundings. This strange behaviour has prompted researchers in universities and industry to fashion new biological methods of purifying metal-laden effluent to replace the existing, and costly, chemical techniques. The idea is to grow bacteria or algae in ponds filled with factory effluents and suitable nutrients, allowing the microbes to scavenge metals from their environment and sequester them inside their cell membranes. A strain of the very common bacterium *Pseudomonas aeruginosa*, for example, is known to accumulate great quantities of uranium. As much as half of the total weight of each cell (excluding its water content) may be made up of uranium. These, and microbes which extract other heavy metals, could be filtered from their liquid environment and deposited in special dumps.

The same kind of approach could be used to recover valuable metals for re-use in industry. For example, *Thiobacillus* bacteria, similar to those that leach metals out of rock ores, can accumulate silver. A certain amount of silver inevitably appears in the waste waters from film factories and other industrial sites, and these bacteria could reduce the losses of this expensive resource.

In the last decade, acid rain has come to be recognized as a major threat to the environment. This pollution is caused when materials which contain sulphur, especially coal, are burned. The sulphur is oxidized and emitted into the atmosphere, where it dissolves in water droplets to form sulphuric acid. The acid rain accumulates in lakes, often hundreds of kilometres from the source of the pollution, killing fish and plants in vast numbers. Hundreds of lakes have been devastated by acid rain, especially in Canada and Scandinavia.

The spread of acid rain can be prevented either by removing sulphur from the coal before it is burned or by catching the sulphur oxides before they reach the atmosphere. Conventional chemical techniques for cutting sulphur pollution from coal add about £10 a tonne to its cost, but this might be halved if bacteria were called in to do the job. Several types of bacteria, especially those that inhabit hot-water springs, have a great appetite for sulphur compounds, from which they derive energy. Their activities would separate much of the sulphur from high-sulphur coal, producing a much cleaner and more valuable fuel. A small-scale plant in Ohio has shown that this process works, and in a few years' time the first industrial-scale operations may commence.

Biosensors and Biochips

Biotechnology and the microchip industry share a number of features, one of the most significant being the alacrity with which both have latched onto the very latest scientific research and turned it to commercial advantage. It is not surprising therefore that these two very distinct applied sciences — one based on biology, the other on physics — are now being married to produce some novel and highly promising offspring.

Much of an organism's life depends on its ability to sense and measure the presence of certain materials within its cells or in its environment. Enzymes often form part of complex regulatory networks which ensure that organisms make best use of the available resources. Biotechnologists are now succeeding in employing enzymes and other biological materials to measure the quantities of many diverse substances in a range of environments. They do this, not by copying the subtle and interconnected processes employed by living cells, but rather by utilizing the special properties of biological materials linked to microchip devices. In this way it has been possible to construct 'biosensors' which can measure sugars, proteins and hormones in body fluids, pollutants in water, gases in the air and much else besides.

In principle the operation of a biosensor is quite straightforward (see Fig. 22.4). The biological component, usually an enzyme or an antibody, is chosen because it reacts in a specific way with the substance to be measured. For instance, when the enzyme glucose oxidase comes into contact with the sugar glucose a specific chemical reaction takes place, and this is accompanied by the release of electrons. By coupling immobilized glucose oxidase to a microchip-based device which collects and amplifies the electrical signal it is possible to obtain a direct read out of the amount of glucose in the sample. The possibility of measuring glucose in a diabetic's bloodstream in a precise, rapid and continuous manner opens up some very important possibilities. In the first instance, it should be possible for diabetics to keep a more accurate check up on their need for insulin injections, but the prospects go much beyond this. A normal pancreas continually monitors blood sugar concentrations and reacts promptly to changes by increasing or decreasing its production of insulin. The notion of an artificial pancreas is based on the ability to mimic this process. Biosensors, permanently implanted in the skin, would check the glucose concentration in the blood. The signals from the biosensor would be passed directly to another small implanted device — a minipump filled with insulin. In response to the information provided by the biosensor, the pump would then release just the correct amount of insulin required by the patient at that time.

Glucose sensors and minipumps are already available and it is probable that the combination of the two will be tested on volunteers within the next two or three years. If successful, this application of biotechnology could have major medical benefits since it would even out the peaks and troughs in the amount of insulin in the body produced by occasional injections, giving instead a steady and optimal insulin level and quite probably reducing the eye and kidney damage caused in many diabetics.

New applications of biosensors are being developed almost every month and many of the most interesting will help doctors to make rapid diagnoses. The hormone chorionic gonadotropin is produced during pregnancy; high concentrations of creatinine are often produced after a heart attack; hepatitis antigens are found in the blood during infections by this virus; and abnormal amounts of urea in the blood or urine can indicate kidney disease—all of these substances can now be measured by biosensors which utilize either enzymes or various sorts of antibodies.

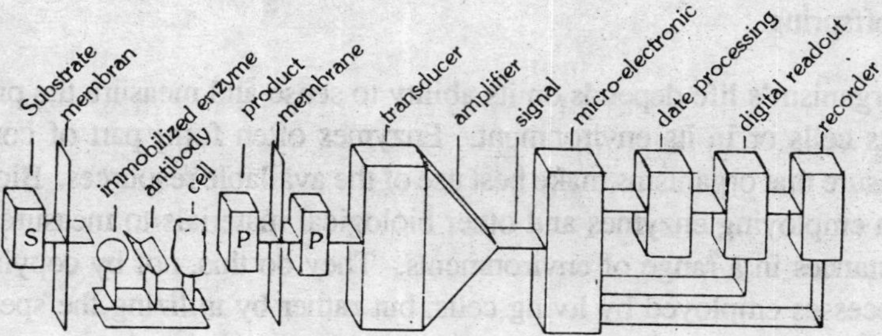

Fig. 22.4. Schematic outline of a biosensor. The substance to be measured (substrate) passes through a thin membrane and then encounters the biological sensing agent — usually an enzyme, an antibody or a whole microbial cell. The substrate and sensor interact to give a product which may be an electrical current, heat, a gas, or a soluble chemical. The product then passes through another membrane to the transducer which detects and measures the product, producing an electrical signal which is amplified and used to give an immediate read-out.

Biosensors are also beginning to find a place in industry where they can be used to measure materials such as acids, alcohols and phenols involved in manufacturing or in industrial effluents. In these cases it is often whole microbes that are built into the biosensor rather than just one or two of their enzymes. The further refinement of these types of biosensors should produce considerable benefits in altering workers to the presence of hazardous substances in the environment and in providing the precise and immediate information needed for computer controlled operations in both chemical and biotechnological industries. The darker side of biosensor research is indicated by the support being given by the military for the development of biosensors which can detect various chemical warfare agents.

It is likely that the next five to ten years will see the introduction of biosensors into a great many areas of medicine and industry. The time-scale appropriate for biochips, the second major product of the interaction between biotechnology and the microchip business, is much longer. Only in the last three or four years have serious scientists been confident enough (or rash enough) to speculate on the possibilities of what are, in effect, biological computers.

Today you can walk into a store and buy a powerful computer which fits comfortably on a desk-top; only a few years ago the equivalent computing power would have been way beyond the budget for ordinary people and the machine would have filled a large room. This incredible reduction in price and size has largely resulted from the development of the silicon microchip containing a multitude of complex circuits. Developments in silicon chips have been so rapid that computer manufacturers will soon be reaching the physical limits of the present technology. The number of electrical circuits that can be squeezed onto a chip is limited by several factors. Most notably the width of the circuit cannot be reduced beyond the wavelength of light used to mark out its path during manufacture. In addition, if the circuits are placed too close together a peculiar phenomenon known as 'electron tunnelling' will create short circuits, ruining the chip's operation. Thirdly, if a very large number of circuits are crammed together, the heat produced by the electric current builds up and the system fails. Biochips, in which silicon is replaced by various biological materials, might avoid all three restrictions on the further miniaturization of computers, opening the way to even more minute and powerful computers.

The basic idea behind biochip research is to insert semiconducting molecules into a protein framework and fix the whole onto a protein support. The circuit would, in essence, be not much more than one molecule wide. A key feature of biological material which is directly relevant to biochip design is that proteins, for example, can assemble themselves into complex and predetermined three-dimensional structures. This property is becoming increasingly well-understood, and indeed Aaron Klug of the MRC Laboratory of Molecular Biology in Cambridge, England received the Nobel Prize in 1982 for his work in finding out how the protein coat of tobacco mosaic virus assembles itself from simpler protein subunits. This phenomenon may mean that proteins employed in biochips would, in a sense, grow to take up the shape required for the electrical circuits.

The problem of electron tunnelling should be less acute if one is using semiconducting organic molecules than contemporary silicon technology and also, because these molecules have a very low electrical resistance, much less heat will be produced and the circuits can be packed together much more closely.

If miniscule computers based on biochips can be manufactured they could be used in a number of ways that are unlikely for their silicon chip counterparts. Implants of several sorts are among the possibilities. Devices implanted in artificial limbs might allow them to respond to natural nerve impulses; heart beats could be regulated and perhaps even blindness and deafness could be overcome by devices that sense external stimuli and convert them into electrical signals which, when passed to the brain, would simulate sight and sound. It should however, be emphasized that of all the possibilities discussed in this book, these are probably the most speculative and lie furthest in the future. Equally, they are among the most dramatic and beneficial for some disturbing military implications.

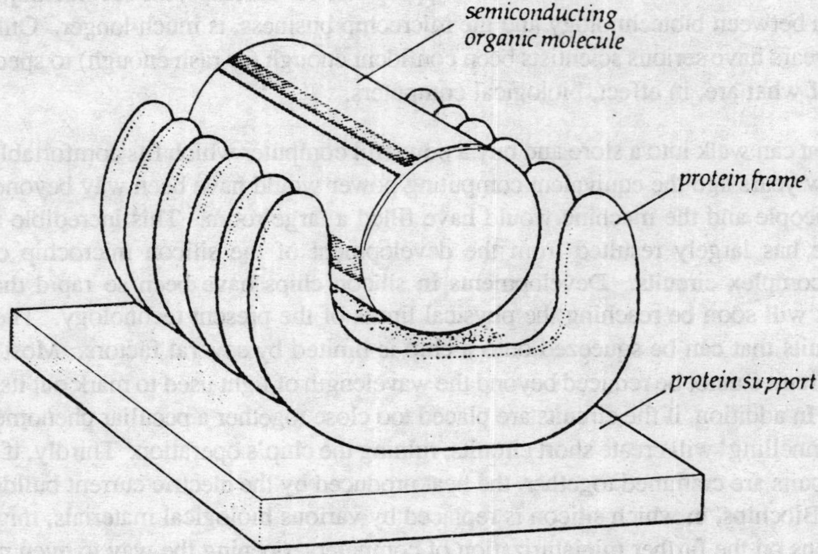

Fig. 22.5. **The basic unit of a biochip is a semiconducting organic molecule which is cradled in a protein frame. Electrical signals can pass along the semiconductor, much as conventional silicon microchips allow signals to pass through certain parts of the chip.**

A couple of amusing problems may also face biochip computer designers. Knowing the opportunism of microbes, it wouldn't come as a great surprise to find that biochips might become infected with hungry bacteria quietly eating their way through the data stores. Clearly the term debugging a computer would take on a more literal connotation. Finally, the opportunity to wipe out information simply by pouring in a protein-digesting enzyme might appeal to many who resent the spread of computer intrusions. The biodegradable computer could be a big step forward !

23

Likely Beneficiaries

Introduction

The ramifications of the bioindustrial revolution will extend far beyond the industries directly affected by it. The potential benefits — better health, more food, renewable energy sources, cheaper and more efficient industrial processes and reduced pollution—are immense, but what of the potential disadvantages ? Any major new technology has profound social, economic and political effects. Biotechnology is no exception, and the possible consequences of the growth of biological industries on the health of workers and the public, on national and international trade, on economic power and on the position of science in society need to be examined.

Genetic Engineering

In the seventies, there erupted a controversy the vehemence of which has only been matched in science by the debate about nuclear power. The realization that humans had the power to transfer genetic material between completely unrelated organisms led to speculation about the possible creation of 'killer bugs'. The emotional atmosphere of the genetic-engineering debate was further heightened by an ignorance of biology among some who professed to represent the 'public interest' and a disturbing arrogance of some 'scientific experts'. In the mid-seventies, rational public debate was the exception rather than the rule.

In time, tempers cooled and, most importantly, more facts were accumulated, making it possible to produce reasonable assessments of the potential risks of specific types of genetic engineering—and how to reduce them. The central concern of those who wanted to regulate, or ban, genetic engineering research was that newly created organisms might produce uncontainable diseases. Two aspects of genetic engineering were claimed to lend special plausibility to this fear: the widespread use of the bacterium *E. coli,* whose normal habitat is the human intestine, and the fact that plasmids which confer resistance against certain antibiotics are often employed to ferry foreign pieces of DNA into microbes. One conjecture was that the foreign DNA might turn a normally benign *E. coli* bacterium into a form which produces a dangerous disease; this bacterium could then infect someone, probably a laboratory researcher, and be spread throughout the population; finally, since the human body is quite accustomed to the presence of *E. coli,* it might not fight back against this dangerous variant, and the fact that the bacterium is resistant to one or more antibiotics would make medical treatment that much more difficult.

Most of the set pieces of the genetic engineering debate took place in the US where this research was generally furthest advanced. Two landmarks were the international conferences held in New Hampshire in June 1973 and the Asilomar Center, California, in February 1975. These discussions led eventually to the drafting of guidelines for genetic engineering research in countries which had the scientific expertise to carry it out, and produced a ban on certain types of experiment. It was not permissible, for example, to try to insert the gene which codes for cholera toxin into other bacterial that can live in the human gut.

The first set of guidelines laid down by bodies such as the US National Institutes of Health and Britain's Genetic Manipulation Advisory Group were quite stringent. They defined the laboratory conditions under which particular types of experiments could be attempted. Most importantly, they specified the precautions that must be taken to prevent microbes escaping, and what kind of microbes could be employed. By 1983 many of the restraints on genetic engineers had been loosened, although a considerable number of important safeguards remain. To understand the scientific basis for this shift towards a liberalization of the rules governing genetic engineering, it is necessary to be specific about what putative dangers should be considered, and the steps that can be taken to prevent them.

Two crucial points must be made at the outset. First, no-one can argue with complete honesty that there is not, and never will be, some conceivable risk in genetic engineering. The concept of zero-risk is virtually meaningless in any area of human activity. Even the simple act of breathing entails the risk of inhaling germs and contracting a disease. The only responsible attitude to any new technology is to find out as much as possible about its conceivable disadvantages and place them in the context of everyday risks. Any extra risk can then be weighed against known or supposed benefits.

Second, discussions concerning the safety of genetic engineering centre on the chances of accidents occurring. There is little doubt that someone sufficiently skilled and deranged could *deliberately* construct a new microbe that would pose a threat to human health. This is clearly a separate problem from that of the regulation of genetic engineering in universities and industry, and it is difficult to see how a ban on future genetic engineering in universities and industry would help stop scientific knowledge being perverted for destructive ends.

Thus, the questions that biotechnologists must answer concern the possibility of genetic engineering accidents and their effects. The vast maj ity of biologists now concur that the risks are neglibible for several reasons. These stem from the fact that a pathogenic (disease-producing) organism must possess several distinct characteristics, and any creation by genetic engineers will be safe unless all are present. Most notably the organism must be able to survive in the special environment of the human body (assuming the particular case of human diseases); it must be capable of being transferred from one person to another; and it must somehow produce symptoms of a disease, perhaps by manufacturing some toxic substance.

The genetic engineers' first line of precaution is to prevent microbes escaping from the laboratory. This can include the containment of organisms in air-tight chambers and the thorough sterilization of all equipment before and after use. Microbiologists have had years of experience in handling very dangerous natural organisms — smallpox viruses, cholera bacteria and so on. There have been few accidents and no known cases in which these have spread disease to the general public.

Despite this good safety record, genetic engineering does not rely on these measuress alone. The most important safeguards are provided by what is known as biological containment. Much of the early concern centered on the fact that most genetic engineering experiments involved *E. coli*. This implied that if a bacterium escaped from the laboratory it would have a good chance of becoming establilshed inside the human body. However, the strain most often employed, *E. coli* K-12, is a feeble creature. This is not surprising since these bacteria have been grown for many generations in the luxuriant surroundings provided for them in laboratories. The result of this pampering has been to produce bacterial which would find it very difficult to survive if they happened to find their way out of this comfortable environment and into the much harsher conditions prevalent in the human body. *E. coli* K-12 is thus biologically contained, as without an artificial environment it cannot thrive.

Even this degree of enfeeblement in microbes for genetic engineering was considered insufficient to provide a good margin of safety. Microbiologists then set about breeding strains of *E. coli* which need to be provided with unusual chemicals if they are to live—chemicals that are not found in the bodies of human beings, other animals or plants. One such strain, developed by Roy Curtiss III in Albama, is *E. coli* X1776, named in honour of the American bicentennial. This requires several unusual chemicals to live and is sensitive to bile salts (found in the intestine), antibiotics and detergents. Curtiss's bacterium has been approved for use in those genetic engineering experiments which specify that the microbe employed must be at least 100 million times less able to survive in nature than normal *E. coli* K-12.

A critical factor in initiating the recombinant DNA debate was a proposal put forward in the early 1970s to insert, into *E.coli,* genetic material from SV 40, a monkey virus. SV 40 is known to cause cancer in mice and the idea of introducing its genes into *E. coli* clearly required the most careful thought. This prompted a series of experiments with another virus, the results of which had a profound influence on discussions about risks. Polyoma virus is the most infective tumour virus known for hamsters, and it was decided to insert this virus into *E. coli* and inject it into animals. The fact that no tumours were caused indicated that once inside the bacterium this virus was much less dangerous than normal.

The 1982 guidelines which apply to all work carried out in connection with the US National Institutes of Health—the major source of funding in that country—prohibits six main classes of experiments, including the cloning of DNA from certain disease-causing organisms, the cloning of genes which code for toxins that harm vertebrate animals and the transference of antibiotic-resistance genes into organisms that cause diseases of humans, animals or plants.

The climate of opinion has now shifted dramatically towards an acceptance that genetic engineering—especially the kind that biotechnologists want to perform - poses no serious hazard. Indicative of this new attitude was the 1983 decision of the US Recombinant DNA Advisory Committee to allow scientists to carry out field tests involving genetically engineered plants and microbes. One such test involves growing tomato and tobacco plants which carry genes from bacteria and yeasts, and another will release genetically engineered microbes whose natural counterparts cause damage to the leaves and flowers of crops. The microbes found in the wild act as 'ice nucleation' centres—that is, they facilitate the formation of ice crystals on plants. The genetically altered microbes have been modified with the aim of removing this ability to promote ice crystallization. The decision to allow the release of these microbes was immediately challenged in the courts. After many months of acrimonious public and legal debates the cases were still unresolved by the beginning of 1985.

The certainty that other new proposals will be put forward by genetic engineers that require careful consideration, makes it unlikely that the present regulations will be swept away. Some scientists fiercely resent government-imposed guidelines, which they see as interfering with their 'traditional right' of free enquiry. Some have even compared their situation with the 'gagging' of Galileo, a claim which it is difficult to take seriously. The shock of finding themselves caught in the glare of public scrutiny has profoundly altered the way many biologists view their work and its place in society. Some bitterly regret that their action in drawing attention to the *conceivable* dangers of genetic engineering (as they were seen in the mid-seventies) were interpreted as a warning of imminent disaster. Others believe that it was to their credit that an awareness of their responsibilities enabled the whole matter to be discussed fully, thereby preventing a headlong plunge into unbridled experimentation.

Questions about genetic engineering will undoubtedly continue to be raised from time to time. However, current research, carried out under the regulations, carries very little risk, and its potential benefits amply justify the continuation of genetic engineering as a tool for biotechnologists. In this brief discussion it has only been possible to outline some of the main themes of a complex and often highly technical topic. The Bibliography lists some of the many publications that have examined the safety of genetic engineering from almost every perspective.

Biotechnology and the Military

War-either actual or prospective — has exerted a powerful influence on most areas of science and technology, and biotechnology is no exception. The development of fermentation processes for making chemicals used in the munitions industry during the First World War and the large scale production of penicillin during the Second World War are two examples mentioned in earlier chapters. It would be naive to assume that pressures from the 'defence' industry are not still operating today, especially since a large proportion of total research funding is channelled through the military in all its forms.

Because little of this research is ever discussed publicly one can only point to a few areas in which military considerations are playing a significant part in biotechnological research. Observers have paid a cosiderable amount of attention to the use of genetic engineering in biological warfare. Although biological warfare is banned by inter-national treaties, research aimed at protecting people from its effects is not prohibited. The kind of knowledge needed for truly defensive purposes is very similar to that required for offensive ones and hence it is virtually impossible to distinguish the two unless one has access to highly secret information and, indeed, to the minds of those directing the research. During 1984 a series of sensational stories appeared in the press concerning alleged preparations for biological warfare by the USSR. Much of the 'evidence' for this seems tendentious. In particular, revelations that Russian scientists are studying nerve toxins from snakes and other organisms loses much of its impact when one learns that dozens of scientists engaged in basic medical research in the West are studying these same chemicals. However, there is not much doubt that the military in both the East and West are taking a keen interest in genetic engineering and biotechnology in general.

Biological warfare has been employed on a small scale several times in the past, for example, by the Japanese in China. For a variety of reasons biological warfare has never been used widely. Among the most important reasons for this is the fact that biological warfare has been a relatively crude and ineffective way of waging war when compared with conventional explosive weapons and even more so in comparison with the hideously powerful forces of nuclear weaponry. A crucial question now faces us: how much will advances in genetic engineering increase the practical potential of biological warfare?

The most obvious drawback of biological warfare from a military point of view is its inherent uncontrollability. Epidemics are likely to strike friend and foe alike. There are two chief means of avoiding self-inflicted devastation on the aggressor's population, and neither appears to be very plausible. The first would be to develop a programme of mass immunization against the organism to be unleashed. It is unlikely that this could be done effectively in complete secrecy — and secrecy would be essential to avoid a pre-emptive attack from the opposition. It must be noted, however, that biotechnology is making it very much easier to produce vaccines against rare and fatal diseases. The second and even more horrifying plan would be to develop 'racially specific' biological weapons. Every racial group has its own characteristic genetic variations some of which may make them more susceptible than other groups to certain types of infection. Very little is known about these differences and their consequences and, of course, there is a great deal of genetic variation *within* each racial group. It is *conceivable* that relatively small and genetically homogenous populations could be attacked with specially developed strains of microbes, but for ethnically diverse populations this would appear to be a theoretical impossibility. In this context, it is noteworthy that both the American and Soviet populations are composed of many diverse groups. It must be emphasized that these are very speculative notions and that, as in all defence matters, hard facts are in short supply.

Two other aspects of biological weapons deserve special mention. One is their development and use by terrorist groups — either independent groups or those backed by governments whose moral stance differs little from terrorists'. The key factor here is the relative cheapness and simplicity of producing a biological weapon — which by no means has to involve genetic engineering or other aspects of high technology. A few thousand dollars is thought to be ample funding for a small-scale biological attack on a city.

The second aspect does relate more directly to the new technological capabilities: the use of biological weapons to undermine the health and economic stability of nations, especially in the less-developed world. It is almost certainly possible to develop microbes which are susceptible to the latest generation of antibiotics but are resistant to older types of antibiotics. Since much of the Third World War only has access to older antibiotics (if any) such microbes might devastate their populations, but any of the microbes which reached more advanced nations might be easily controlled by the newer drugs available there.

Similarly, the economies of certain countries could be ruined by an attack on crops or livestock especially important to these countries. Cuba has already accused the US of such actions following outbreaks of disease in their sugar canes and in pigs, and the US has claimed that the USSR has employed biological agents ('yellow rain') in South-East Asia. In neither case is there anything approaching proof of the accusations, but the mere fact that they were made indicates a great deal about the way governments are thinking about the issues.

The two other aspects of biotechnology that deserve special mention terms of warfare concern biosensors and biochips. Research into both of these is now being funded by the US Department of Defense to the extent of several million dollars a year, and doubtless the Department's counterparts in other developed countries are taking an equal interest.

Biosensors are specially suitable for detecting toxic gases, including some which might be employed as chemical weapons. Compared with existing methods of detecting such chemicals, biosensors are likely to offer advantages is terms of size, speed of response and high sensitivity.

The motive behind one fact of biochip research is even more chilling. Many or all of the metallic components of conventional silicon chips might be replaced by biochips built from biological materials. One possible consequence of this is that biochips could be immune to the destructive effects of the electromagnet pulses which are generated by nuclear explosions. The prospect of being able to construct control and guidance systems that would continue to operate during a nuclear holocaust would obviously appeal to military planners.

Reactions to these aspects of biotechnology will clearly be a matter of personal opinion. Those who find them disturbing and a perversion of what science and technology should be doing are faced with the problem of counteracting these developments. Efforts to alter the general political view of 'defence' are very important, as is any attempt to monitor the ways in which scientists are directed in their research. There are encouraging signs that at least some groups of scientists are conscious of these dangers. The US National Academy of Science, for example, refused to cooperate with the Army when asked to help in identifying 'agents of the future', a term which includes genetically engineered microbes. The reply referred to 'deep concern about the request to consider defenses against "agents of the future." ' While accepting the need for defensive measures it was considered 'a natural consequence of such a study to be the possible development of new offensive agents'. Parallels between biologists today and nuclear physicists fourty years ago are not difficult to draw. The fact that biological weapons are nowhere near as well developed as nuclear weapons in terms of sheer destructive power should perhaps be taken as a sign that it is not too late to prevent a further threat to mankind, rather that as an indication that there is no problem.

Genetic Data — Use and Abuse

The ability to discover a great deal about an individual's genetic constitution raises some profound ethical issues. There are two central questions: how should information be acquired and how should it be used?

The novel techniques of biotechnology, especially those involving DNA probes, make it feasible to analyse our genetic make-up in a way that was impossible with the older methods which depended on studying the characteristics of large numbers of related people. It is already becoming possible to tell if an individual carries certain defective genes. Previously geneticists were largely restricted to talking in terms of probabilities; now they are beginning to talk about certainties.

Some of the issues now coming to the fore are fairly straightforward in ethical terms, while others are quite novel and complex. The following few examples, ordered in increasing complexity, illustrate the kinds of questions we must face and answer.

The use of DNA probes to diagnose inherited diseases before starting an effective therapy is clearly little different in practical or moral terms than the hundreds of other diagnostic tests most people are quite prepared to accept. A little more problematic is the use of genetic analysis to tell young adults that they are carrying certain defective genes which do not have especially harmful consequences for the individual, but make it possible that any children they have may suffer from a genetic disease. 'Genetic counselling' is already available for people who are known to, or suspected to, carry certain 'recessive' genetic defects. If someone with such a defect has children by a partner who carries the same defective gene then each of their children will have a one in four chance of inheriting two sets of the defective gene and will thus suffer from the genetic disease itself. Counselling is designed to inform people about the risks of having affected children and the likely consequences; the decision is then left in the hands of the individuals concerned. The main impact of the new methods for collecting genetic information will be to widen the range of potential genetic diseases which can be identified and thus make genetic counselling a more common process. The obvious benefit is that fewer children might be born with genetic diseases, but on the other hand more couples would be faced with the dilemma of choosing whether to have children — or even whether they should stay together or look to another partner who does not carry the same risk factors.

More complex still are issues connected with diseases such as Hunting-ton's Chorea. This illness only becomes apparent in middle age and leads to a slow decline in mental functions and eventually to death. There is no known cure. The gene responsible for this disease has now been identified, but what is to be done with the information? It is quite likely that an individual will have had children before the disease strikes. Should the children then be checked to see if they carry the defective gene? Those that do not will be reassured and can carry on as normal and have children without qualms. But any children of the patient that do carry the gene can only be told that they will develop the disease and that there is virtually nothing that doctors can do to help them. The prospect of living under a threat which will surely strike in ten or twenty years is not one that many people could handle well. Thus genetic diagnosis for Huntingon's Chorea would have the advantage of relieving the fear of some of the patients' children while blighting the remaining years of others. Without such diagnostic tests all would have a mixture of hope and dread.

Genetic screening of workers in particular industries is already being given much attention. It seems that certain people are more likely than others to develop diseases caused by certain chemicals. At present there are no well-established examples, but if further research shows that susceptible individuals can be identified by genetic screening, what would follow? Would workers already in the industry be dismissed or have their insurance cover removed? Might certain people turn out to be susceptible to many different chemicals and hence find it difficult to get work? (The director of the US Oil, Chemical and Atomic Workers International Union has warned of 'a whole subculture of unemployable people if [genetic screening is] carried to the full extent.') Might employers relax their efforts to eliminate hazards if they think the workforce has been shown to be comparatively resistant to a particular occupational disease? We already accept that people who are especially at risk should not do certain jobs. Children are not allowed to work in factories, and most people would defend an employer's right to refuse a steeplejack's job to someone with vertigo. But how far down this path do we want to go?

Perhaps the most disturbing—and most futuristic-outcome of developments in genetic science might be the introduction of large-scale screening for many factors in the general population. At present only a few fairly specific genetic characteristics can be identified—mostly those which relate to diseases. In future it *may* be possible to extend this to cover a much wider range of features: likelihood of premature death from common diseases such as cancer, diabetes and heart attacks; suitability for certain occupations; and probability of developing mental lillnesses. No-one knows how much these things depend on an individual's genetic inheritance and how much on social and environmental factors. But if the genetic component is proved to be very significant one does not need much imagination to foresee that some people would call for large-scale testing of children to determine how each one should be treated. Many countries already place great emphasis on 'intelligence testing' at early ages for allocating children to certain types of education — usually more costly for those at the top of the group. How will society react to suggestions that children who are likely to die in middle age from cancer or heart attacks should not have as much 'invested' in their future as their more fortunate contemporaries? Would second-rate health care and education be deemed adequate for them?

It would be alarmist to imply that the most disturbing of these possibilities are just around the corner, but progress in these areas will need to be carefully monitored. The first requirement is undoubtedly some form of right to freedom of information—that no-one should collect genetic data about individuals without making clear the implications.

Where will the Bioriches Flow?

In the eighteenth century, the development of the steam engine launched Britain on its path towards domination of the world's economy, while Japan's recent meteoric rise in global trade has been boosted by the sophisticated use of microelectronics. Similarly, the innovations of mass production greately strengthened the US's power, and Germany reaped immense wealth from the growth of its chemical industry. The economic advantages of being among the first to capitalize on new technologies (not necessarily to invent them) are enormous — which nations will make the most of the bioindustrial revolution? The US and Japan — not necessarily in that order — are in the best position right now, with the US having a clear, but not yet commanding, lead in genetic engineering, and Japan having an edge in the kind of technology needed for large-scale fermentations, with about 80 per cent of the patents in this area. But it is certainly not too late for Britain, France, Switzerland, West Germany, Denmark and several other developed countries to seize a large slice of the cake. Furthermore, there is little reason why, given the necessary commitment, a number of developing countries should not become more involved in the lower-technology, less capital intensive areas of bio-technology.

The pattern of investment differs markedly between the US and Japan. In the former, especially in California and Massachusetts, dozens of small firms have been established in the last ten years. Many have considerable expertise, and are receiving massive financial backing from multinational corporations as diverse as Standard Oil, Dow Chemicals, International Nickel, General Foods and Bendix. Huge amounts of money are also being furnelled into basic research in universities and other public institutions, often in return for an option on any patent licences that may result from the investigations. For example, the German chemical giant Hoechst plans to plough $50 million into Massachusetts General Hospital's molecular biology laboratory.

Meanwhile in Japan, the government is instrumental in promoting biotechnology. The Ministry of International Trade and Industry is allocating $110 million over the next ten years to supplement the expertise of industrial firms in large and sophisticated fermentation processes. Their prowess in biotechnology is amply demonstrated by a virtual monopoly in world trade of some amino acids, enzymes and food additives, and Japan has more industrial experience in working with immobilized enzymes and cells than any other nation. Government funds of about $ 20 million in 1981 aided industry to increase its knowledge of monoclonal antibodies and genetic engineering, including the production of interferon. Japan already earns over $50 billion a year by exploiting microbes—about 5 per cent of its gross domestic product.

In Britain, the Spinks report, *Biotechnology*, published in 1980, called for a major commitment to biotechnology, and the Government's response was a skimpy White Paper which offered only bland platitudes and paid little more than lip-service to the idea that biotechnology could be the formation of vast and profitable industries within a few years. But more recently, the message seems to have struck home with a grant of £ 16 million from the Department of Energy to eight research institutions. Equally encouraging is the decision by the University Grants Committee, at a time when budgets are being slashed in virtually all areas, to earmark £ 2.4 million over 1983-5 for the development of biotechnology.

Recent estimates made by the EEC indicate that the UK is spending $46 million of public money on research and development centred on biotechnology. Comparable figures from other areas include West Germany ($36 million), France ($31 million), the EEC as a whole ($146 million), US ($200 million) and Japan ($50 million). However, when spending on wider research related to biotechnology is taken into account, the balance shifts. While the UK figure rises only to $59 million, West Germany's goes up to $132 million, France's to $84 million, the entire EEC's to $355 million and the US's to $550 million.

Of more than twenty firms set up in Britain to exploit new areas of biotechnology, the best-established is Celltech. This firm was created by the National Enterprise Board in 1980, and its financial backers include the Prudential Insurance Co. and the Midland Bank. Celltech has close links with the Medical Research Council which has some of the most prestigious laboratories in the world. This link has brought Celltech its first marketable product, an antibody which latches on to interferon molecules.

British industry is investing in biotechnology both through specialist research centres associated with universities and through the establishment of private research facilites. The breadth of interest in biotechnology is exemplified by the support being given to the new Biocentre at Leicester, which is receiving money from brewers (Whitbread), tobacco manufacturers (Gallaher), food processors (Dalgety-Spillers) and engineers (John Brown). Research and development is also expanding rapidly within industrial companies, ICI, Shell Glaxo, Burroughs Wellcome, G.D. Searle, and Tate and Lyle being just a few of the massive companies to take a stake in biotechnology.

Most meetings of biotechnologists over the last few years have included discussions about the desirability of government support for research and development (R & D). In general, US biotechnologists tend to believe that R & D investment is best left in the hands of private companies. They do, however, emphasize the need for a positive attitude from their governments — such as tax incentives and freedom from restrictive legislation. In addition, there is considerable concern about the lack of support for *basic* scientific research in the US which may undermine the country's pre-eminent position in many scientific disciplines.

By contrast, biotechnologists from Western Europe and Japan place more importance on direct government involvement in biotechnology R & D. It is worth noting that, despite a popular misconception to the contrary, the Japanese Government's proportion of support for R & D in general is lower than that of many other countries. Only 24 per cent of total R & D funding for all areas comes from the Japanese administration, whereas in the US the percentage is 46, in the UK it is 48 and in France it is 57. The crucial difference is found in the way government money is spent: in the UK and USA about a third of all government R & D expenditure goes into defence-related projects, while in Japan less than 2 per cent is spent in this way.

Britain's prosperity will certainly hinge on its ability to establish the industries that will underpin the world economy during the next century. The opportunities of the microchip were largely scorned and the chance to lead in that industrial revolution wasted, but there is no reason why the pattern should be repeated in biotechnology, so long as the political will and industrial imagination can be mustered.

Biotechnology and the Third World

The population of our planet is divided in many ways, but the most profound differences are to be found in health and access to effective medical treatment. Put bluntly, the inhabitants of the Third World die younger, in more pain and of different diseases than those of affluent countries. In much of Asia, Africa, and Central and South America, infectious and parasitic diseases are still rampant.

Millions are struck down each year by cholera, malaria, sleeping sickness and a hideous array of debilitating and fatal diseases. Biotechnology, as we have seen, *could* do much to lift the scourage of these diseases, but will the necessary commitment be forthcoming? If new vaccines and therapies are to be developed, a much higher priority must be assigned to these areas. The US Institute of Medicine calculates that cancer research in that country is funded at the rate of $209 a year for each case of the disease and cardiovascular disease receives $8 per case. The equivalent figures for schistosomiasis and malaria are 4.5 and 3 cents.

The development of a new drug can take years and cost millions of pounds. Once a substance has shown promise in laboratory experiments, a decision must be made on whether to initiate the work that consumes most time and money—testing its effects on animals and, eventually, human volunteers. Even the largest pharmaceutical companies can only follow up a few of the leads provided by their research scientists. Among the many criteria applied when taking the decision to continue development work on a potential new drug is one that weighs very heavily against the Third World — the profit potential of the product. There is little incentive to invest in a drug which will find its major markets in countries with hardly any money. Since most pharmaceutical firms are under a legal obligation to act in the interest of their shareholders, it is unlikely that this situation will change.

A disturbing example of the tensions between commercial profit and human health centred on research on the production of an anti-malaria vaccine. Work at New York University was funded in part by the World Health Organization which, as a matter of course, requires that such research must be publicly accessible. The California biotechnology firm Genentech had been greatly interested in helping to develop this vaccine, but when it became clear that the firm would not receive an exclusive licence for the vaccine Genentech's interest waned and they expressed doubts about the project's technical future. Attempts to find a similarly well equipped replacement firm with the expertise of Genentech have been unsuccessful so far.

Thus, if the Third World is to benefit fully from the opportunities that biotechnology presents for an attack on parasitic and infectious diseases, alternative funding must be found. The World Health Organization and other international institutions have research laboratories, but there is an urgent need for new non-profit-making development and testing facilities as well as for increased effort in basic research.

In 1984 the United Nations Industrial Development Organization (UNIDO) announced that it is to set up two international biotechnology research centres, in Trieste, Itlay and in Delhi, India. With a budget for the first five years of $40 million, these centres will focus on problems that are particularly important for developing countries. In the medical area, the primary need is to follow up more of the many discoveries that have been made in existing research centres and universities, and develop them into practical processes that could transorm the quality and length of life for millions. Although under the present system pharmaceutical companies could not justify funding development work themselves, they could be encouraged to hand over information on projects which they view as commercially unsound but of possible medical value. This is not as unrealistic a proposition as may at first appear. Pharmaceutical companies are acutely conscious of their public image, and a chance to be seen to aid the Third World, without any significant extra costs, could prove attractive.

Another factor which hinders the development of many types of drugs is the enormous investment in testing which is demanded by governments. Regulations take little account of the nature and extent of the illness concerned when stipulating the required safety and efficacy tests. It is clear that any drug which is intended to treat a relatively minor illness should be subject to the most intense scrutiny before it is licensed for general use, especially if other treatments are already available. When, however, we are considering common and fatal diseases — and not just those that primarily afflict the developing nations—the balance of risks versus benefits is very different. If governmental controls are such that no commercial company finds the research and development worthwhile, we shall clearly avoid the danger of unsafe drugs—there will not be any new drugs.

Any relaxation of the rules governing the commercial pharmaceutical industry is bound to be politically unacceptable, especially if it appeared that 'second-class' safety testing was considered adequate for some types of patient. Yet an international, non-profit-making research and production organization might be able to accelerate the introduction of some new drugs. It would be seen that medical benefits and risks were the main factors being taken into account and that considerations of sales income played no part in the decision.

It is important to note when discussing the 'safety' of drugs, that this is a relative term. No drug is absolutely safe in the sense that it will never produce some sort of undesirable side-effect in some patients. The public is certainly right to demand that any drugs put on the market should be tested as required by government regulations and any adverse effects which subsequently appear should be promptly and fully investigated. It is not realistic however to insist that all side-effects be eliminated—unless, that is, we are prepared to halt all work on new drugs and withdraw all the existing ones.

The famine which struck vast areas of Africa in 1984 and 1985 again brought into focus the immense suffering mankind still experiences due to food shortages. If crop losses due to predation by pests could be reduced — both in the Third World and elsewhere — much more food would be available. (Although there would still be no guarantee that food surpluses in some parts of the world would reach those who are short of food).

Many researchers in industry and universities are working on methods for decreasing pest losses, but are these approaches that will bring the most direct benefit to the world's hungry ? In this case of much industry-backed research the answer seems to be : no. There are two very different strategies in this area. One is to try to create crops which are resistant to various pests. The other is to develop crops which are resistant to *pesticides;* in this case powerful pesticides could be employed to kill weeds and other harmful organisms without damaging the crops themselves.

From a commercial point of view, it may be more profitable for agro-chemical companies to pursue the second approach since they could then gain income from the sale of both pesticide-resistant seeds *and* the pesticides themselves.

In the short-term, research into pesticide-resistant plants may be a better bet in terms of increased food production (as well as increased profit) than work on pest-resistant crops. In the longer term, however, the environmental and agricultural advantages of eliminating the need for pesticides ought to prevail.

The US Office of Technology Assessment estimates that by the end of the century biotechnology could be supplying 20 per cent of the nation's energy. If this target is achieved, fuel alcohol will play a major role, probably the dominant one. The main raw materials consumed during the manufacture of fuel alcohol — sugar cane and wheat in particular — are currently used as food. Does more fuel for the rich mean less food for the poor? At least until efficient methods can be designed which employ wood and other cellusose-rich plant materials to produce fuel alcohol, this will remain a very valid question. The Kenyan government withdrew its support for fuel alcohol programmes in 1982, when it discovered that food imports had soared to replace the agricultural produce diverted to supply the new industry. Countries should be wary of tailoring their agriculture towards crops which can yield fuel alcohol. Unfortunately, the temptation is especially great for developing countries which desperately need foreign currency.

Indeed, one of the major global economic problems that will be thrown up by the bioindustrial revolution concerns sugar cane — a major prop in the economies of many countries, especially in the Caribbean. Increased use of fuel alcohol should push up the demand for sugar cane, but the growth in the market for alternative sweeteners, including fructose and aspartame, will depress prices. The overall impact of these opposing forces can only be guessed at, but in the short term the already shaky economies of sugar-producing nations may well be further undermined.

The interaction of other areas of biotechnology are even more difficult to predict. Economic advisers in every country, especially in the Third World, will need to consider the likely effects on, for example, their energy needs. Methane generated from wastes should help cut oil imports and, in the longer term, the production of hydrogen from water will have the same effect. Yet a decreased death rate due to new medicines will increase the population and its energy demands. The extra people will also need to be fed. Diverting crops into the production of fuel alcohol will exacerbate food shortages, but the possibility of new crops grown on previously unusable land will have the opposite effect. Single-cell protein factories will consume plant starches, but yield more nutritious foods for domestic animals, or preferably, for humans.

The Scientist as Entrepreneur

No sooner were biologists recovering the shock of the genetic engineering debate than they engulfed in another controversy — the relationship between universities and industry. Little of the vast commercial potential of the newer biotechnologies would have been possible without the basic work undertaken by universities and other publicly funded research institutions. Until recently, virtually every significant development in genetic engineering and the production of monoclonal antibodies was supported by tax-payers or charities. Who is likely to reap the financial benefits that will accrue from the commercialization of this new knowledge? The answer in many cases is becoming increasingly clear — a rapidly growing band of top-rank scientists is leaving the universities to set up commercial enterprises whose main assets are the knowledge and skills of their researchers, knowledge and skills which were largely gained during years of publicly funded work. Even more controversy has been stirred up by the fact that some biologists have retained their university posts at the same time as sitting on the boards of such companies.

This trend has become particularly pronounced in the US, where most of the new biotechnology firms are staffed at the top scientific levels with people who only a year or so ago were heading major laboratories in the nation's finest universities. The desirability of various solutions to this problem, indeed the very perception that there is a problem at all, depends largely on the political stance of the observer.

Those in favour of the present system by which scientists move from academia to industry can put forward a convincing case. Biotechnology is the application of biology within an industrial context, and they argue that if there is no movement of scientists from basic research into applied development in commercial operations then the full potential of biotechnology will not be realized. Certainly there is little reason to believe that the universities are either willing or able to develop and market products which arise out of investigations within their laboratories. They also argue that scientists should have the same rights as everyone else to sell their labour (or knowledge) as best they can on an open market. Furthermore, if the interchange of scientists between academic and industrial life stimulates the growth of new industries based on biotechnology, everyone will gain through better medicine, more food, greater energy supplies and all the other benefits on offer. Finally, they note that, although these questions are new to biologists as a group, other types of scientists — chemists, for example — have for many years had close links with major industries and that there is no evidence that these arrangements have generally had adverse consequences.

The issues raised by those who believe that the present system is too much of a free-for-all are usually more complex, but are nonetheless very important. At one extreme—unlikely ever to be reached in practice—most of the best biotechnologists might be lured away to industry where commercial exigencies whould ensure that they devoted their efforts to high-profit programmes. These may often not be the most socially useful (however defined) applications of biotechnology. Certainly one could expect that, within the pharmaceutical area, more effort would be devoted to the quest for cold cures than therapies for leprosy. The profit motive can have more surprising implications. The US Office of Technology Assessment has noted, for example, that 'until recently, the US commercial seed companies, with one or two exceptions, have not been interested in wheat-breeding programmes as a profit making venture.' The reason for this is that a farmer need buy only one batch of seeds and then retain some of the harvested seeds for replanting next year; one-off sales are bad for business. Therefore, the development of higher-yielding wheat varieties had to come mainly through publicly funded research. Complete neglect of economically valuable, but unprofitable (in the narrow sense) research is a real danger; if the growth of biotechnoloy is almost entirely in the hands of commercial enterprises.

Within the universities the most active, and often acrimonious, debate concerns the distortion of traditional academic values of shared information and freedom of enquiry. These questions lie at the heart of scientific discovery—the very process that has powered the upsurge of biotechnology.

The tensions created by university scientists also being deeply involved in commercial enterprises was seen at its most acute in several US universities. The close links between Genentech, a genetic engineering firm, and staff at the Department of Biochemistry and Biophysics at the University of California in San Francisco caused much dissension. An investigating committee noted that some scientists they interviewed 'believed that the manner in which the particular contract [with Genentech] was carried out led to serious disruption within the department. A recurrent theme was that people were loath to ask questions or give suggestions in seminars or across the bench for there was a feeling that someone might take an idea and patent it; or that an individual's idea might be taken to make money for someone else.' They also foresaw greater problems when more than one firm had intimate links with researchers in a department. Subsequently, members of that department were serving as directors of three competing companies.

The souring of relationships as a result of such ties has been a pernicious influence in recent years. In the US, particularly, universities are searching almost frantically for solutions which will enable them to retain both their scientists and their reputations. Complex deals are constantly being proposed with the aim of gaining industrial funding, making sure that universities benefit from patents arising from their work, and relieving staff of the pressure to work on projects simply because they look profitable in the short-term. The conflict between the traditional values of university and industrial research is stark indeed. On the one hand there exist ideals of open communication of information and the freedom to pursue studies for their intrinsic interest; on the other we find the need to guard secrets and to devote precious time and money only to investigations which promise to be profitable.

Genetic Eldorado ?

In October 1980 shares in Genetech, a company with only a handful of employees and no products, were offered to the public. Wall Street was so smitten with the biotechnology bug that within twenty minutes of the start of dealing the share price for this apparently unprepossessing company had rocketed from $35 to $89. At the peak of the stock market's flirtation with biotechnology, the mere hint of a cloned gene was enough to send investors scurrying to their stockbrokers, even when neither of them knew the commercial viability of this latest wonder of science. Subsequent share offers were greeted with only slightly less exuberance as other biotechnology firms took the plunge into the stock market in the US. In Britain, the demand for shares in Amersham International, a company that supplies many of the chemicals used by genetic engineers, greatly exceeded the expectations of the Government's financial advisers when it was decided to invite private investment in this hitherto publicly owned firm.

The rush to take a stake in biotechnology made multimillionaries of several scientists who had set up small, high-technology companies. Predictably, there has been a backlash among investors as they began to realize that fortunes in biotechnological products were not to be made overnight. Share prices dropped back, but still remain at remarkably high levels when compared with major companies with established profit records over many years. Most people involved with biotechnology view this retreat as a desirable development; if the biobubble had become even further inflated its eventual, and inevitable, burst could have proved disastrous for the industry, destroying valuable and necessary confidence in its future.

Dozens of small biotechnology firms are in the race to produce genetically engineered microbes that will manufacture highly profitable proteins; some are bound to collapse over the next few years and a handful have already gone to the wall. Many of the successful ones are likely to be taken over by today's corporate giants, once they have solved the basic technical problems of producing new materials and require large infusions of capital to set up major manufacturing installations. The implications of this for employment are difficult to foresee.

The bioindustrial revolution is creating a huge demand for scientists and engineers who are trained in genetic engineering, fermentation technology, microbiology and numerous other specialities. In some areas the shortage of skilled personnel is already acute, and those with the right qualifications can command salaries far in excess of those thought possible only a decade ago. Governments and universities throughout the developed world are urgently seeking to expand high-level education in the subjects which underpin biotechnology. In Britain alone, nearly a dozen new courses at universities and colleges are being established and more are sure to follow. Thus, the employment prospects for those who are able to pursue these careers are bright.

The Office of Technology Assessment estimated in 1984 that there were about 5,000 people engaged in biotechnology research and development in the USA, while the Institute of Manpower Studies put the UK figure at 1,500-2,000. (The problem of defining 'a biotechnologist' makes it very difficult to compare figures from different sources). New jobs in the scientific and technological areas of biotechnology will undoubtedly be created, but the numbers are not large in comparison with the general employment — perhaps 500 or so a year will be required in the UK and one projection of total employment in the US 'synthetic genetic' industry is for 40,000 jobs by 1992.

The effects of the growth of biotechnology on employment as a whole are far less clear and no comprehensive estimates have been produced. The US Office of Technology Assessment suggests, for example, that 30,000 to 75,000 workers might be needed to produce $14.6 billion worth of chemicals from genetically engineered microbes. Since the products they considered are already being manufactured by the chemical factories, there would clearly be job losses in the more traditional industries. A rough estimate indicates that there would be little net gain or loss in employment caused by a shift towards biotechnology in this area.

The Organization for Economic Cooperation and Development (OECD) took a similar line in its 1982 report, stating that the growth of biotechnology will not substantially affect total employment in the short term. Looking further into the future, however, it is possible to foresee more jobs being created, as biotechnology yields entirely novel products rather than replacements, direct or indirect, for existing products and manufacturing methods.

Towards a more Stable World Economy ?

Since the industrial revolution, the global economy has been characterized by ever-increasing demands for energy and metals. Until now the demand for energy has been met chiefly by exploiting fossil fuels, while metals have been supplied by increasingly sophisticated mining techniques. We all know that there are limits on how long we can continue to pursue this path, even though we act most of the time as if no such limits existed. When people come to look back on the bioindustrial revolution of the late twentieth and early twenty-first centuries they will, perhaps, see its greatest contribution as being the transformation of the society from one dependent on non-renewable materials, and hence inherently unstable, into one based largely on a power source which will be with us for millions of year to come — the sun.

In the next century, plastics derived from living organisms may perform many of the jobs now done by metals. Since all organisms are, in effect, powered by the sun's energy, we can expect a virtually unlimited supply of these materials. Any remaining need for metals could be met by microbial mining and, more importantly, by using microbes to recycle the metals that have already been extracted from the Earth by conventional mining methods.

Our energy needs could similarly be met by sunlight with plants being used to manufacture fuel alcohol, and methane being generated from wastes. If reliable hydrogen generators can be made to work with an efficiency of 10 per cent — that is, converting 10 per cent of light energy into energy stored in hydrogen gas — an area of only 500,000 sq. km. (193,000 sq. miles) of such generators would be able to supply all of the world's current energy needs. This is only about 7 per cent of the area of Australia.

These are grandiose ideas, but no-one who has witnessed the almost incredible pace at which biotechnology has moved in the last decade would feel confident in dismissing them as mere pipe-dreams.